# Graph Symmetry

## Algebraic Methods and Applications

# NATO ASI Series

## Advanced Science Institutes Series

*A Series presenting the results of activities sponsored by the NATO Science Committee, which aims at the dissemination of advanced scientific and technological knowledge, with a view to strengthening links between scientific communities.*

The Series is published by an international board of publishers in conjunction with the NATO Scientific Affairs Division

| | |
|---|---|
| **A Life Sciences**<br>**B Physics** | Plenum Publishing Corporation<br>London and New York |
| **C Mathematical and Physical Sciences**<br>**D Behavioural and Social Sciences**<br>**E Applied Sciences** | Kluwer Academic Publishers<br>Dordrecht, Boston and London |
| **F Computer and Systems Sciences**<br>**G Ecological Sciences**<br>**H Cell Biology**<br>**I Global Environmental Change** | Springer-Verlag<br>Berlin, Heidelberg, New York, London,<br>Paris and Tokyo |

**PARTNERSHIP SUB-SERIES**

| | |
|---|---|
| **1. Disarmament Technologies** | Kluwer Academic Publishers |
| **2. Environment** | Springer-Verlag / Kluwer Academic Publishers |
| **3. High Technology** | Kluwer Academic Publishers |
| **4. Science and Technology Policy** | Kluwer Academic Publishers |
| **5. Computer Networking** | Kluwer Academic Publishers |

*The Partnership Sub-Series incorporates activities undertaken in collaboration with NATO's Cooperation Partners, the countries of the CIS and Central and Eastern Europe, in Priority Areas of concern to those countries.*

**NATO-PCO-DATA BASE**

The electronic index to the NATO ASI Series provides full bibliographical references (with keywords and/or abstracts) to more than 50000 contributions from international scientists published in all sections of the NATO ASI Series.
Access to the NATO-PCO-DATA BASE is possible in two ways:

– via online FILE 128 (NATO-PCO-DATA BASE) hosted by ESRIN, Via Galileo Galilei, I-00044 Frascati, Italy.

– via CD-ROM "NATO-PCO-DATA BASE" with user-friendly retrieval software in English, French and German (© WTV GmbH and DATAWARE Technologies Inc. 1989).

The CD-ROM can be ordered through any member of the Board of Publishers or through NATO-PCO, Overijse, Belgium.

**Series C: Mathematical and Physical Sciences – Vol. 497**

# Graph Symmetry

## Algebraic Methods and Applications

edited by

Geňa Hahn

Département d'Informatique et de Recherche opérationnelle,
Université de Montréal,
Montréal, Québec, Canada

and

Gert Sabidussi

Département de Mathématiques et de Statistique,
Université de Montréal,
Montréal, Québec, Canada

**Kluwer Academic Publishers**

Dordrecht / Boston / London

Published in cooperation with NATO Scientific Affairs Division

Proceedings of the NATO Advanced Study Institute and
Séminaire de mathématiques supérieures on
Graph Symmetry: Algebraic Methods and Applications
Montréal, Canada
July 1–12, 1996

A C.I.P. Catalogue record for this book is available from the Library of Congress

ISBN 0-7923-4668-8

---

Published by Kluwer Academic Publishers,
P.O. Box 17, 3300 AA Dordrecht, The Netherlands.

Sold and distributed in the U.S.A. and Canada
by Kluwer Academic Publishers,
101 Philip Drive, Norwell, MA 02061, U.S.A.

In all other countries, sold and distributed
by Kluwer Academic Publishers Group,
P.O. Box 322, 3300 AH Dordrecht, The Netherlands.

*Printed on acid-free paper*

---



Printed in the Netherlands

# Table of Contents

# Preface

The last decade has seen two parallel developments, one in computer science, the other in mathematics, both dealing with the same kind of combinatorial structures: networks with strong symmetry properties or, in graph-theoretical language, vertex-transitive graphs, in particular their prototypical examples, Cayley graphs.

In the design of large interconnection networks it was realised that many of the most frequently used models for such networks are Cayley graphs of various well-known groups. This has spawned a considerable amount of activity in the study of the combinatorial properties of such graphs. A number of symposia and congresses (such as the bi-annual IWIN, starting in 1991) bear witness to the interest of the computer science community in this subject.

On the mathematical side, and independently of any interest in applications, progress in group theory has made it possible to make a realistic attempt at a complete description of vertex-transitive graphs. The classification of the finite simple groups has played an important role in this respect.

In planning this ASI it has been our intention to bring these two lines of development together before an audience of mathematicians. On the one hand, we wanted graph theorists to be exposed to the combinatorial problems encountered in the design of large, highly symmetrical networks. On the other, we wanted to present an overview of the group theoretical methods that enter into the structural classification of such networks. The pervasiveness of group theory via Cayley graphs was the common thread that ran through all the lectures given at the ASI.

Twelve speakers presented mini-courses of four to five hours each. Two dealt with combinatorial problems of interconnection networks (Heydemann, Klasing), one with their group theoretic aspects (Lubotzky), three with the construction, classification and isomorphism problems of vertex-transitive graphs (Alspach, Praeger, Scapellato), two with eigenvalues of symmetric graphs (Godsil, Mohar), one with homomorphisms of graphs (Hahn), and two with infinite networks of high symmetry (Cameron, Watkins). Accounts of nine of these courses are included in the present volume.

The ASI was attended by 82 participants from 15 countries. We owe them our thanks for providing an appreciative, critical, and highly spirited audience and a lively forum for discussion. Our thanks also go to Aubert Daigneault and Ghislaine David who took care of the organisational side of the ASI with traditional smoothness.

Finally, we would like to express our gratitude to NATO for its support of this meeting through its Advanced Study Institute programme; also to the Université de Montréal and NSERC for additional funding. Our particular appreciation goes to Dr. Luis Veiga da Cunha of the Scientific and Environmental Affairs Division of NATO for his efforts on behalf of this ASI.

The Editors

1
2
3
4
5
6
7
8
9
10
11
12
13
14
15
16
17
18
19
20
21
22
23
24
25
26
27
28
29
30
31
32
33
34
35
36
37
38
39
40
41
42
43
44
45
46
47
48
49
50
51
52
53
54
55
56
57
58
59
60
61
62
63
64
65
66
67
68
69
70
71

# Key to group picture

1 X. Munoz
2 A. S. Lladó
3 J. Mykkeltveit
4 S. Thomassé
5 J. Fournier
6 M. Juvan
7 S. Gravier
8 D. De Caen
9 M. Chastand
10 C. Q. Zhang
11 R. Goldstone
12 J. Zaks
13 N. Marlin
14 F. Merola
15 F. Genest
16 X. G. Fang
17 Y. Quin
18 C. Tardif
19 A. Goodman
20 J. Janzen
21
22 P. Ille
23 I. Zaguia
24 W. Haemers
25
26 J. Morris
27 L. Nowitz
28 P. McKenzie
29 E. Dobson
30 H. Gavlas
31 R. Nedela
32 P. Neumann
33 R. Škoviera
34 O. Serra
35 R. Brewster
36 H.-J. Voss
37 E. van Dam
38 M. Weidner
39 P. Willenius
40 U. Baumann
41 M. Harborth
42 A. Niemeyer
43 J. Brunat
44 M. Mitjana
45 V. Gouaillier
46 C.-H. Li
47
48 E. Gabrieli
49 C. Rousseau
50 P. Boling
51 E. Ngimbi Ngembo
52 S. Chakrabarty
53 A. Chan
54 J. Puech
55 D. Sotteau
56 B. Nostrand
57 A. Daigneault
58 G. David
59 G. Sabidussi
60 M. E. Watkins
61 R. Scapellato
62 G. Hahn
63 A. D. Gardiner
64 C. D. Godsil
65 J. Howes
66 B. Alspach
67 P. Cameron
68 M.-C. Heydemann
69 C. E. Praeger
70 R. Klasing
71 B. Mohar

# Participants

Ulrike BAUMANN
Fb. Mathematik - Institut für Algebra
Technische Universität Dresden
Mommsenstr. 13
D-01062 Dresden
Germany
baumann@nalw01.math.tu-dresden.de

Olivier BODINI
CNRS - Équipe Combinatoire
Université Pierre et Marie Curie
4, Place Jussieu
Couloir 45-46, 3e
F-75005 Paris
France
bodini@ccr.jussieu.fr

Patricia BOLING
Dept. of Discrete & Statistical Sciences
Auburn University
Auburn, AL 36849
USA

Richard C. BREWSTER
Department of Mathematics
Capilano College
2055 Purcell Way
North Vancouver, BC, V7J 3H5
Canada
rbrewste@capcollege.bc.ca

Josep B. BRUNAT
Dept. de Matemàtica Aplicada II
Universitat Politècnica de Catalunya
c/. Pau Gargallo 5
E-08028 Barcelona
Spain
brunat@ma2.upc.es

Sanghamitra CHAKRABARTY
Department of Mathematics
Syracuse University
215 Carnegie Hall
Syracuse, NY 13244-1150
USA
schakrab@mailbox.syr.edu

Marc CHASTAND
Institut d'administration des entreprises
Université Jean Moulin (Lyon 3)
15, Quai Claude Bernard
B.P. 0638
F-69239 Lyon Cédex 02
France
chastand@jonas.univ-lyon1.fr

Dom DE CAEN
Department of Mathematics & Statistics
Queen's University
Kingston, Ont., K7L 3N6
Canada
decaend@qucdn.queensu.ca

Christian DELHOMMÉ
Lab. de Mathématiques discrètes - IMI
Université Claude Bernard (Lyon 1)
43 bd du 11 Novembre 1918
F-69626 Villeurbanne
France
delhomme@in2p3.fr

Maria Carla Di DOMENICO
Dipartimento di Matematica
Università di Catania
Viale A. Doria 6
I-95125 Catania
Italy
carla@dipmat.unict.it

Edward DOBSON
Department of Mathematics
Louisiana State University
Baton Rouge, LA 70803-4918
USA
dobson@math.lsu.edu

Bertrand DUCOURTHIAL
L.R.I. - Bâtiment 490
Université de Paris-Sud,
F-91405 Orsay Cédex
France
ducourth@lri.fr

Xin Gui FANG
Department of Mathematics
University of Western Australia
Nedlands, WA 6907
Australia
fang@maths.uwa.edu.au

Jérôme FOURNIER
Département de mathématiques
et de statistique
Université de Montréal
C.P. 6128, Succ. Centre-ville
Montréal, Qué., H3C 3J7
Canada
fournij@ere.umontreal.ca

Elisabetta GABRIELI
Dipartimento di Matematica
Università Cattolica del Sacro Cuore
Via Trieste 17
I-25121 Brescia
Italy
gabrieli@mathematik.tu-muenchen.de

Anthony D. GARDINER
Department of Mathematics
University of Birmingham
P.O. Box 363
Birmingham B15 2TT
United Kingdom
a.d.gardiner@bham.ac.uk

Heather GAVLAS
Dept. of Mathematics & Statistics
Western Michigan University
Kalamazoo, MI 49008-5152
USA
gavlas@ibis.math-stat.wmich.edu

François GENEST
Département de mathématiques
et de statistique
Université de Montréal
CP 6128, Succ. Centre-ville
Montréal, Qué., H3C 3J7
Canada
genestf@ere.umontreal.ca

Richard GOLDSTONE
Div. of Computer Science & Mathematics
Marist College
Poughkeepsie, NY 12601-1387
USA
richard.goldstone@marist.edu

Albert GOODMAN
Dept. of Mathematics & Statistics
University of Missouri-Rolla
Rolla, MO 65409-0020
USA
goodman@umr.edu

Valérie GOUAILLIER
Département de mathématiques
et de statistique
Université de Montréal
CP 6128, Succ. Centre-ville
Montréal, Qué., H3C 3J7
Canada
gouaillv@mathcn.umontreal.ca

Sylvain GRAVIER
Laboratoire Leibniz - IMAG
Université Joseph Fourier
BP 58
F-38041 Grenoble Cédex 9
France
sylvain.gravier@imag.fr

Willem HAEMERS
Departement of Econometrics
Tilburg University
P.O. Box 90153
NL-5000 LE Tilburg
The Netherlands
haemers@kub.nl

Martin HARBORTH
Institut für Algebra und Geometrie
Otto von Guericke Universität
Pf. 4120
D-39106 Magdeburg
Germany
martin.harborth@student.uni-magdeburg.de

Jennifer HOWES
Department of Mathematics
Syracuse University
215 Carnegie Hall
Syracuse, NY 13244-1150
USA
jahowes@mailbox.syr.edu

Pierre ILLE
Institut de Mathématiques de Luminy
CNRS - UPR 9016
163, Ave de Luminy
Case 930
F-13288 Marseille Cédex 9
France
ille@iml.univ-mrs.fr

Robert JAJCAY
Dept. of Mathematics & Computer Science
Indiana State University
Terre Haute, IN 47809
USA
jajcay@laurel.indstate.edu

Jens JANZEN
Inst. f. Informatik u. Praktische Mathematik
Universität Kiel
Olshausenstr. 40
D-24098 Kiel
Germany
jja@informatik.uni-kiel.de

Martin JUVAN
Department of Mathematics
University of Ljubljana
Jadranska 19
SI-1111 Ljubljana
Slovenia
martin.juvan@fmf.uni-lj.si

Hamamache KHEDDOUCI
L.R.I., Bâtiment 490
Université de Paris-Sud,
F-91405 Orsay Cédex
France
kheddouc@lri.lri.fr

Ralf KLASING
Inst. f. Informatik u. Praktische Mathematik
Universität Kiel
Olshausenstr. 40
D-24098 Kiel
Germany
rak@informatik.uni-kiel.de

Tom KRANTZ
École Normale Supérieure de Lyon - A23
46, Allée d'Italie
F-69364 Lyon Cédex 07
France
tom.krantz@ens-lyon.fr

Benoît LAROSE
LACIM - Département de mathématiques
Université du Québec à Montréal
CP 8888, Succ. Centre-ville
Montréal, Qué., H3C 3P8
Canada
larose@math.uqam.ca

Cai-Heng LI
Department of Mathematics
University of Western Australia
Nedlands, WA 6907
Australia
li@maths.uwa.edu.au

Alexander LUBOTZKY
Department of Mathematics
Hebrew University
91904 Jerusalem
Israel
alexlub@math.huji.ac.il

Maria Flavia MAMMANA
Dipartimento di Matematica
Università di Catania
Viale A. Doria 6
I-95125 Catania
Italy
flavia@dipmat.unict.it

Nausica-S. MARLIN
I3S - URA 1375 - CNRS
Université de Nice-Sophia Antipolis
950, route des Colles
B.P. 145
F-06903 Sophia Antipolis Cédex
France
nmarlin@essi.fr

Dragan MARUŠIČ
Department of Mathematics
University of Ljubljana
Jadranska 19
SI-1111 Ljubljana
Slovenia
dragan.marusic@uni-lj.si

Pierre McKENZIE
Département d'informatique
et de recherche opérationnelle
Université de Montréal
CP 6128, Succ. Centre-ville
Montréal, Qué., H3C 3J7
Canada
mckenzie@iro.umontreal.ca

Francesca MEROLA
Dipartimento di Matematica
Universitá di Palermo
Via Archirafi, 34
I-90123 Palermo
Italy
merola@ipamat.math.unipa.it

Lorenzo MILAZZO
Dipartimento di Matematica
Università di Catania
Viale A. Doria, 6
I-95125 Catania
Italy
milazzo@dipmat.unict.it

Margarida MITJANA
Dept. Matemàtica Aplicada I
Universitat Politècnica de Catalunya
Av. Gregorio Maranón 44-50
E-08028 Barcelona
Spain
margarida@ma1.upc.es

Joy MORRIS
Department of Mathematics & Statistics
Simon Fraser University
Burnaby, BC, V5A 1S6
Canada
morris@cs.sfu.ca

Xavier MUNOZ
Dept. Matemàtica Aplicada i Telemàtica
Universitat Politècnica de Catalunya
c/. Gran Capitá s/n
Mòd. C3 Campus Nord
E-08034 Barcelona
Spain
xml@mat.upc.es

Johannes MYKKELTVEIT
Rogaland Research
P.O. Box 2503 Ullandhaug
N-4004 Stavanger
Norway
johannes.mykkeltveit@rf.no

Roman NEDELA
Department of Mathematics
Matej Bel University
Tajovského 40
SK-97549 Banská Bystrica
Slovakia
nedela@bb.sanet.sk

Peter M. NEUMANN
Institute of Mathematics
Queen's College
Oxford OX1 4AW
United Kingdom
neumann@maths.ox.ac.uk

Emmanuel NGIMBI NGEMBO
Institut de mathématiques
Université de Liège
15, Ave des Tilleuls
B-4000 Liège
Belgique
ngimbi@montefiore.ulg.ac.be

Alice NIEMEYER
Department of Mathematics
University of Western Australia
Nedlands, WA 6907
Australia
alice@maths.uwa.edu.au

Igor NIKOLAEV
Mathematical Institute
Academy of Sciences
5 Akademicheskaya St.
277028 Kishinev
Moldavia
15nikol@mathem.moldova.su

Barbara NOSTRAND
de Moivre Institute
84 Lowther Ave., apt. 203
Toronto, Ont., M5R 1E1
Canada
bnostran@lynx.neu.edu

Lewis A. NOWITZ
2345 Broadway, apt.526
New York, NY 10024-3263
USA
lanowitz@pipeline.com

Mohamed OULD BEDDI
Dép. de mathématiques et informatique
Université de Nouakchott
B.Pl. 5026
Nouakchot
Mauritanie

Walter PACCO
Dipartimento di Matematica
Politecnico di Milano
Piazza Leonardo da Vinci 32
I-20133 Milano
Italy

Stéphane PERENNES
I3S - URA 1375 - CNRS
Université de Nice-Sophia Antipolis
950, route des Colles
B.P. 145
F-06903 Sophia Antipolis Cédex
France
sp@unice.fr

Joel PUECH
Département de Mathématiques
Bâtiment 425
Université de Paris-Sud,
F-91405 Orsay Cédex
France
puech@lri.lri.fr

Yusheng QIN
Department of Mathematics & Statistics
Simon Fraser University
Burnaby, BC, V5A 1S6
Canada
yqin@cs.sfu.ca

Gert SABIDUSSI
Département de mathématiques
et de statistique
Université de Montréal
CP 6128, Succ. Centre-ville
Montréal, Qué., H3C 3J7
Canada
sab@mathcn.umontreal.ca

Anna SANCHEZ-LLADÓ
Dept. Matemàtica Aplicada i Telemàtica
Universitat Politècnica de Catalunya
c/. Gran Capitá s/n
Mòd. C3 Campus Nord
E-08034 Barcelona
Spain
ana@mat.upc.es

Oriol SERRA
Dept. Matemàtica Aplicada i Telemàtica
Universitat Politècnica de Catalunya
c/. Gran Capitá s/n
Mòd. C3 Campus Nord
E-08034 Barcelona
Spain
oriol@mat.upc.es

Martin SKOVIERA
Department of Computer Science
Comenius University
Mlynská dolina
SK-84215 Bratislava
Slovakia
skoviera@fmph.uniba.sk

Dominique SOTTEAU
L.R.I. - Bâtiment 490
Université de Paris-Sud,
F-91405 Orsay Cédex
France
sotteau@lri.fr

Stéphan THOMASSÉ
Lab. de Mathématiques discrètes - Bât. 1
Université Claude Bernard (Lyon 1)
43 bd du 11 Novembre 1918
F-69626 Villeurbanne
France
thomasse@jonas.univ-lyon1.fr

Edwin VAN DAM
Departement of Econometrics
Tilburg University
P.O. Box 90153
NL-5000 LE Tilburg
The Netherlands
edwin.vandam@kub.nl

Heinz-Jürgen VOSS
Fb. Mathematik - Institut für Algebra
Technische Universität Dresden
Mommsenstr. 13
D-01062 Dresden
Germany
voss@math.tu-dresden.de

Michael WEIDNER
Mathematisches Institut
Albert Ludwigs Universität
Albertstr. 23b
D-79104 Freiburg
Germany
weidner@ibm2.mathematik.uni-freiburg.de

Per WILLENIUS
Institut für Mathematik
Universität Kaiserslautern
Pf. 3049
D-67653 Kaiserslautern
Germany
willeniu@mathematik.uni-kl.de

Dave WITTE
Department of Mathematics
Oklahoma State University
Stillwater, OK 74078
USA
dwitte@math.okstate.edu

Imed ZAGUIA
Lab. de Mathématiques discrètes - Bât. 1
Université Claude Bernard (Lyon 1)
43 bd du 11 Novembre 1918
F-69626 Villeurbanne
France
zaguia@in2p3.fr

Joseph ZAKS
Deartment of Mathematics
University of Haifa
31905 Haifa
Israel
jzaks@mathcs2.haifa.ac.il

Cun-Quan ZHANG
Department of Mathematics
West Virginia University
P.O.Box 6310
Morgantown, WV 26506-7529
USA
cq.zhang@math.wvu.edu

# Contributors

Brian ALSPACH
Department of Mathematics & Statistics
Simon Fraser University
Burnaby, BC, V5A 1S6
Canada
alspach@cs.sfu.ca

Peter J. CAMERON
School of Mathematical Sciences
Queen Mary and Westfield College
University of London
Mile End Rd.
London E1 4NS
United Kingdom
p.j.cameron@qmw.ac.uk

Ada CHAN
Department of Combinatorics
and Optimization
University of Waterloo
Waterloo, Ont., N2L 3G1
Canada
ssachan@barrow.uwaterloo.ca

Chris D. GODSIL
Department of Combinatorics
and Optimization
University of Waterloo
Waterloo, Ont., N2L 3G1
Canada
cgodsil@math.uwaterloo.ca

Geña HAHN
Département d'informatique
et de recherche opérationnelle
Université de Montréal
C.P. 6128, Succ. Centre-ville
Montréal, Qué., H3C 3J7
Canada
hahn@iro.umontreal.ca

Marie-Claude HEYDEMANN
L.R.I. Bâtiment 490
Université de Paris-Sud,
F-91405 Orsay Cédex
France
mch@lri.fr

Bojan MOHAR
Department of Mathematics
University of Ljubljana
Jadranska 19
SI-1111 Ljubljana
Slovenia
bojan.mohar@uni-lj.si

Cheryl E. PRAEGER
Department of Mathematics
University of Western Australia
Nedlands, WA 6907
Australia
praeger@maths.uwa.edu.au

Raffaele SCAPELLATO
Dipartimento di Matematica
Politecnico di Milano
Piazza Leonardo da Vinci 32
I-20133 Milano
Italy
rafsca@mate.polomi.it

Claude TARDIF
Département de mathématiques et de statistique
Université de Montréal
C.P. 6128, Succ. Centre-ville
Montréal, Québec, H3C 3J7
Canada
sfb9@Mathematik.Uni-Bielefeld.DE

Mark E. WATKINS
Department of Mathematics
Syracuse University
215 Carnegie Hall
Syracuse, NY 13244-1150
USA
mwatkins@summon.syr.edu

# Isomorphism and Cayley graphs on abelian groups

Brian ALSPACH*

*Department of Mathematics and Statistics*
*Burnaby, BC, V5A 1S6*
*Canada*

**Abstract**

We study the problem of determining when two Cayley graphs on a given abelian group are isomorphic. The emphasis is on abelian $p$-groups.

## 1 Circulant graphs

Let us begin with a few basic definitions.

**1.1 Definition** Let $S$ be a subset of a finite group $G$ satisfying

(1) $1 \notin S$, where 1 denotes the identity of $G$, and

(2) $S = S^{-1}$, that is, $s \in S$ implies that $s^{-1} \in S$.

A subset $S$ satisfying the above conditions is called a *Cayley subset.* The *Cayley graph* $X(G; S)$ is defined to be the graph whose vertices correspond to the elements of $G$ with an edge between $g$ and $h$ if and only if $h = gs$ for some $s \in S$. We call $S$ the *connection set* and say that $X(G; S)$ is a *Cayley graph on the group G.*

**1.2 Definition** A permutation of the vertex set of a graph $X$ preserving adjacency is called an *automorphism* of the graph. The group of automorphisms of $X$ is denoted $\mathrm{Aut}(X)$. A graph $X$ is said to be *vertex-transitive* if $\mathrm{Aut}(X)$ acts transitively on the vertex set of $X$. Likewise, $X$ is said to be *edge-transitive* if $\mathrm{Aut}(X)$ acts transitively on the edge set of $X$.

**1.3 Proposition** *Cayley graphs are vertex-transitive.*

**Proof** It is easy to see that left multiplication by a group element on which the Cayley graph is defined is an automorphism of the graph. The result follows easily from this. □

**1.4 Definition** A Cayley graph $X(G; S)$ is called a *circulant* graph when $G$ is a cyclic group.

*This work was partially supported by the Natural Sciences and Engineering Research Council of Canada under Grant A-4792. The author wishes to thank Aubert Daigneault, Geňa Hahn, Gert Sabidussi and the Department of Mathematics of the Université de Montréal for their support and organizational skills.



*G. Hahn and G. Sabidussi (eds.), Graph Symmetry,* 1–22.

We shall use additive notation when dealing with circulant graphs. Thus, given that $|G| = n$, the connection set $S$ is a subset of $\{1, 2, \ldots, n-1\}$ satisfying $s \in S$ implies that $n - s \in S$, and $i$ is adjacent to $i + s$ for all $s \in S$ and $i = 0, 1, \ldots, n-1$. In this special case, we use the notation $X(n; S)$.

Circulant graphs are Cayley graphs on the most straight-forward of all groups. This leads one to believe that circulant graphs are a good testing ground for questions about Cayley graphs. We shall put them to the test regarding isomorphism. The basic question is what can one say about the isomorphism of $X(n; S)$ and $X(n; S')$. The first case we consider is when $n$ is a prime, but before doing so we introduce some notation.

The ring of integers modulo $n$ will be denoted by $Z_n$, and $Z_n^*$ will denote the multiplicative subgroup of units in $Z_n$. The mapping $T_{a,b}$ acting on $Z_n$ is defined by $T_{a,b}(x) = ax + b$. If $a \in Z_n^*$, then $T_{a,b}$ is a permutation of $Z_n$.

**1.5 Definition** Let $G_1$ and $G_2$ be permutation groups acting on the object sets $\Omega_1$ and $\Omega_2$, respectively. We say that $G_1$ and $G_2$ are *isomorphic as permutation groups,* and write $G_1 \equiv G_2$, provided that there is a bijection $t : \Omega_1 \to \Omega_2$ and an isomorphism $f : G_1 \to G_2$ so that

$$f(\sigma)(t(x)) = t(\sigma(x))$$

for every $\sigma \in G_1$ and every $x \in \Omega_1$.

In simple terms, the previous definition is telling us that if we rename the elements of $\Omega_1$ in all the permutations of $G_1$ according to some bijection, then we obtain exactly the permutations of $G_2$. A basic and powerful tool is the following theorem of Burnside [25].

**1.6 Theorem** *If $G$ is a transitive permutation group acting on a prime number $p$ of points, then either $G$ is doubly transitive or*

$$G \equiv \{T_{a,b} : a \in H < Z_p^* \text{ and } b \in Z_p\}.$$

We now shall consider some of the many applications of Theorem 1.6. The following corollary is proved easily by noting that the only mappings of the form $T_{a,b}$ without a fixed point are those for which $a = 1$.

**1.7 Corollary** *If $G$ is a transitive permutation group acting on a prime number $p$ of points and $G$ is not doubly transitive, then $G$ contains a unique subgroup of order $p$.*

**1.8 Theorem** *A graph $X$ of order $p$ is vertex-transitive if and only if $X$ is a circulant graph.*

**Proof** If $X$ is a circulant graph, it is vertex-transitive because of Proposition 1.3. On the other hand, let $X$ be any vertex-transitive graph of order $p$. If $\mathrm{Aut}(X)$ is doubly transitive, $X$ is either $K_p$, the complete graph of order $p$, or the complement $\overline{K}_p$. Both of these graphs certainly are circulants. Otherwise, $\mathrm{Aut}(X)$ is not doubly transitive and contains a subgroup of order $p$ by Corollary 1.7. The subgroup of order $p$ must be generated by a permutation of the form $\rho = (0\ 1\ \cdots\ p-1)$. If $i$ is adjacent to $j$ in $X$, then $i + s$ is adjacent to $j + s$ for all $s$ modulo $p$. Hence, $S = \{j : 0j \in E(X)\}$ completely determines all the edges of $X$, and $X$ is isomorphic to the circulant graph of order $p$ with connection set $S$. □

The preceding result characterizes vertex-transitive graphs of prime order. This characterization will be even more useful if we can find a reasonable condition for two circulants of prime order $p$ to be isomorphic. The next result, first proved by J. Turner [28], does this. The proof given here is quite different from that in [28].

**1.9 Theorem** *Two circulant graphs $X = X(n; S)$ and $X' = X(n; S')$ of order $p$, $p$ a prime, are isomorphic if and only if there is an $a \in Z_p^*$ such that $S' = aS$.*

**Proof** We assume without loss of generality that both $X$ and $X'$ are circulant graphs on the same set $\{0, 1, \ldots, p-1\}$ with the permutation $\rho = (0\ 1\ \cdots\ p-1)$ corresponding to $S$ and $S'$. First suppose that there is an $a \in Z_p^*$ such that $S' = aS$. It is then easy to verify that the mapping $f : i \mapsto ai$ is an isomorphism of $X$ to $X'$.

If $X = K_p$ or $X = \overline{K}_p$, then $S = \{1, 2, \ldots, p-1\}$ or $S = \emptyset$, respectively. Clearly, $S' = S$ in both cases. Thus, we suppose that $X$ is neither $K_p$ nor $\overline{K}_p$, and that $f$ is an isomorphism from $X$ onto $X'$. This implies that $(f(0)\ f(1)\ \cdots\ f(p-1)) \in \mathrm{Aut}(X')$. By Corollary 1.7, the group $\langle\rho\rangle$ generated by $\rho$ is the unique subgroup of $\mathrm{Aut}(X')$ of order $p$. Thus, we have

$$(f(0)\ f(1)\ \cdots\ f(p-1)) = \rho^a = (0\ a\ \cdots\ a(p-1))$$

for some $a \neq 0$. Therefore, once we know that $f(0) = ak$, for some $k \in Z_p$, we obtain that $f(i) = a(k+i)$. Hence,

$$\begin{aligned} s \in S \quad &\text{iff} \quad i \text{ is adjacent to } i+s \\ &\text{iff} \quad f(i) \text{ is adjacent to } f(i+s) \\ &\text{iff} \quad a(k+i) \text{ is adjacent to } a(k+i+s) \\ &\text{iff} \quad as \in S'. \end{aligned}$$

Therefore, $S' = aS$. □

Turner's Theorem allows us to carry out some interesting enumeration results. We are going to employ Pólya's Enumeration Theorem, a good account of which can be found in [7]. The simplest setting for this theorem is as follows. One is given two finite sets $D$ and $R$ and the set $R^D$ of all functions from $D$ to $R$. Furthermore, there is a permutation group $G$ acting on the domain $D$. This induces a permutation group acting on $R^D$. Of course, the latter group in general is acting on a much larger set than $G$ is acting on and, thus, will have many more orbits than $G$ has. The miraculous fact about Pólya's Theorem is that it allows us to count the latter orbits by considering only $G$.

**1.10 Definition** Let $G$ be a permutation group acting on a set $\Omega$. With each $\sigma \in G$ we associate a monomial $x(\sigma)$. Consider the disjoint cycle decomposition of $\sigma$. Suppose there are $b_k$ cycles of length $k$. Then $x(\sigma) = x_1^{b_1} x_2^{b_2} x_3^{b_3} \cdots x_m^{b_m}$, where $m = |\Omega|$. Note that $b_1 + 2b_2 + \cdots + mb_m = |\Omega|$. The *cycle index* $\mathcal{Z}(G)$ of the permutation group $G$ is defined by

$$\mathcal{Z}(G) = \frac{1}{|G|} \sum_{\sigma \in G} x(\sigma).$$

**1.11 Example** Let $G$ be the permutation group generated by the 6-cycle $(0\ 1\ 2\ 3\ 4\ 5)$. The monomial corresponding to the identity permutation is $x_1^6$. The monomial corresponding to the two powers which are 6-cycles is $x_6^1$. The monomial corresponding to the two powers which are products of two 3-cycles is $x_3^2$. Finally, the remaining power yields the monomial $x_2^3$. The cycle index is given by

$$\mathcal{Z}(G) = \frac{1}{6}(x_1^6 + x_2^3 + 2x_3^2 + 2x_6^1).$$

**1.12 Theorem (Pólya)** *Let $D$ and $R$ be finite sets, $R^D$ denote the set of all functions from $D$ to $R$, and let $G$ be a permutation group acting on $D$. For each $\sigma \in G$, define a permutation $\tilde{\sigma}$ acting on $R^D$ by*

$$(\tilde{\sigma}(f))(x) = f(\sigma(x))$$

*for all $f \in R^D$ and all $x \in D$. The mapping taking $\sigma$ to $\tilde{\sigma}$ is a homomorphism of $G$ to a permutation group $\tilde{G}$. The number of orbits of the permutation group $\tilde{G}$ is given by evaluating the cycle index of $G$ with each variable set equal to $|R|$.*

Let us now apply Pólya's Theorem to counting the number of circulant graphs of order $p$. The domain $D$ is the set of unordered pairs

$$\{1, p-1\}, \{2, p-2\}, \ldots, \{(p-1)/2, (p+1)/2\}$$

and the range $R$ is the set $\{0, 1\}$. A function $f$ in $R^D$ corresponds to a circulant graph $X(n; S)$ by putting $i$ and $p - i$ in $S$ if and only if $f$ maps the pair $\{i, p-i\}$ to 1. Multiplication by $a \in Z_p^*$ generates a permutation of the elements of $D$. Since multiplication by $a$ has the same effect as multiplication by $-a$, the permutation group acting on $D$ is just the cyclic group of order $(p-1)/2$.

**1.13 Theorem (Turner [28])** *Let $p$ be a prime. The number of circulant graphs (thus, vertex-transitive graphs) of order $p$, to within isomorphism, is*

$$\frac{2}{p-1} \sum_{d|(p-1)/2} \phi(d) 2^{(p-1)/2d},$$

*where $\phi$ denotes the Euler $\phi$-function and the sum is taken over all divisors $d$ of $(p-1)/2$.*

**Proof** Interpreting the circulant graphs of order $p$ as functions in $R^D$ as described above, two circulant graphs of order $p$ are isomorphic if and only if they are in the same orbit of the permutation group on $R^D$ induced by the cyclic group of order $(p-1)/2$ acting on $D$. The cycle index of the latter group is

$$\frac{2}{p-1} \sum_{d|(p-1)/2} \phi(d) x_d^{(p-1)/2}.$$

We now substitute $2 = |R|$ for each variable, according to Pólya's Theorem, and obtain the desired result. □

**1.14 Example** Consider $p = 37$. The cycle index of the cyclic group of order 18 is

$$\frac{1}{18}(x_1^{18} + x_2^9 + 2x_3^6 + 2x_6^3 + 6x_9^2 + 6x_{18}^1).$$

Thus, the number of non-isomorphic circulant graphs of order 37 is

$$\frac{1}{18}(2^{18} + 2^9 + 2^8 + 3 \cdot 2^3 + 3 \cdot 2^2) = 14,602.$$

Pólya's Theorem can be refined further in the following way. Assign a weight to each element of the range. A good choice is to use distinct commuting variables $w_1, w_2, \ldots, w_{|R|}$. Then substitute the sum $w_1^i + w_2^i + \cdots + w_{|R|}^i$ for the variable $x_i$ in the cycle index. The resulting coefficients of the terms in the expanded cycle index is the number of orbits of functions with a given weight of the form $w_1^{e_1} w_2^{e_2} \cdots w_{|R|}^{e_{|R|}}$. The following example should make this clear.

**1.15 Example** Return to the example above when $p = 37$. The cycle index is given above. Associate the weights $w_0$ and $w_1$ with 0 and 1 in $R$, respectively. Making the substitutions described above yields

$$\frac{1}{18}((w_0 + w_1)^{18} + (w_0^2 + w_1^2)^9 + 2(w_0^3 + w_1^3)^6 + 2(w_0^6 + w_1^6)^3 5g + 6(w_0^9 + w_1^9)^2 + 6(w_0^{18} + w_1^{18}))$$

$$= w_0^{18} + w_0^{17} w_1 + 9 w_0^{16} w_1^2 + 46 w_0^{15} w_1^3 + \cdots + w_1^{18}.$$

The coefficient of $w_0^{15} w_1^3$ is 46 which means there are forty-six non-isomorphic circulant graphs of order 37 with weight $w_0^{15} w_1^3$. This weight means that the function used $1 \in R$ three times and 0 fifteen times. Hence, the corresponding graphs have degree 6 and we see that there are forty-six non-isomorphic circulant graphs of order 37 and degree 6.

Burnside's Theorem also gives us an easy proof of the following result first proved by J. L. Berggren [5].

**1.16 Theorem** *A circulant graph $X = X(p; S)$ of order $p$ is edge-transitive if and only if $S = \emptyset$ or $S$ is a coset of an even order subgroup $H \leq Z_p^*$. Thus, to within isomorphism, the number of circulant graphs of order $p$ which are also edge-transitive is* 1 *plus the number of even divisors of $p - 1$.*

**Proof** Both $K_p$ and $\overline{K}_p$ are edge-transitive. Their connection sets are $\emptyset$ and $Z_p^*$, respectively. Since $Z_p^*$ is cyclic, it has one subgroup of order $d$ for each divisor $d$ of $p - 1$. Let $S$ be the subgroup of $Z_p^*$ of order $d$, where $d$ is a proper even divisor of $p-1$. We claim that $X = X(p; S)$ is edge-transitive. The mapping $T_{s,0} \in \mathrm{Aut}(X)$, for each $s \in S$, because $i$ is adjacent to $j$ if and only if $j - i \in S$ if and only if $s(j - i) \in S$ if and only if $T_{s,0}(i)$ is adjacent to $T_{s,0}(j)$. Now $T_{s,0}$ fixes 0 and maps 1 to $s$, that is, it maps the edge 01 to the edge $0s$. This implies that all edges incident with 0 are in the same orbit. Therefore, $X$ is edge-transitive. By Theorem 1.9, any circulant graph of order $p$ whose connection set is a coset of $S$ is isomorphic to $X$. Thus, for each even divisor of $p - 1$, we have found one edge-transitive circulant of order $p$. The result will now follow if we can show that the connection set of any edge-transitive circulant of order $p$ different from $\overline{K}_p$ is a coset of an even order subgroup of $Z_p^*$.

Let $X \neq \overline{K}_p$ be an edge-transitive circulant of order $p$. If $X = K_p$, then the connection set is $Z_p^*$ which has even order. We may assume $X = X(p;S)$ is neither $\overline{K}_p$ nor $K_p$. By Burnside's Theorem, $\mathrm{Aut}(X) = \{T_{a,b} : a \in H < Z_p^*, b \in Z_p\}$. Since $T_{-1,0} \in \mathrm{Aut}(X)$, $-1 \in H$ and $H$ has even order. If $s \in S$, then $0s$ is an edge of $X$. The mapping $T(h,0)$ fixes 0 and maps $s$ to $hs$. Thus, the entire coset $sH$ is contained in $S$. On the other hand, the only mappings in $\mathrm{Aut}(X)$ which map edges incident with 0 to edges incident with 0 are those of the form $T_{h,0}$. Thus, the only edges incident with 0 must be those whose other end vertex is an element of the coset $sH$. This implies that $S = sH$ completing the proof. □

D. König [17] asked which groups are automorphism groups of some graph. It was shown quickly [12] that for every finite group $G$ there are infinitely many graphs whose automorphism group is isomorphic to $G$ as abstract groups. It was shown that this still holds when one places restrictions on the graphs [13, 15, 26]. However, if one considers the same question for permutation groups, as first asked by J. Petersen [19], the situation is different. That is, if $G$ is a permutation group of degree $n$, trying to decide whether or not there is a graph $X$ of order $n$ such that $G \equiv \mathrm{Aut}(X)$ is difficult.

For example, consider the permutation group of order $n$ generated by the $n$-cycle $\rho = (0\ 1\ \cdots\ n-1)$. Determine the orbits of this group acting on the unordered pairs from $\{0,1,\ldots,n-1\}$. If one is going to find a graph of order $n$ with $\langle\rho\rangle$ as its automorphism group, then if an edge $ij$ appears in the graph, the entire orbit containing the unordered pair $\{i,j\}$ must appear as edges in the graph. So the orbits of the group acting on the unordered pairs play a major role. In this case, we observe that the dihedral group $D_n$ has the same orbits acting on the unordered pairs implying that there is no graph of order $n$ with $\langle\rho\rangle$ as its automorphism group. One will always have at least $D_n$ in the automorphism group.

In the case of transitive permutation groups of degree $p$, something reasonable can be done [2]. The first step is to use Pólya's Theorem to count the number of circulant graphs whose automorphism groups contain the given permutation group. Then we use inclusion-exclusion to eliminate those whose automorphism groups properly contain the given group. We need the definition of a simple manifestation of the Möbius function.

**1.17 Definition** The Möbius function $\mu$ is defined on the positive integers according to

$$\mu(n) = \begin{cases} 0 & \text{if } n \text{ is not square-free,} \\ (-1)^r & \text{if } n \text{ is a product of } r \text{ distinct primes,} \\ 1 & \text{if } n = 1. \end{cases}$$

**1.18 Definition** Let $F_m$ denote the cycle index of the cyclic permutation group generated by an $m$-cycle. Thus,

$$F_m = \frac{1}{m}\sum_{d|m} \phi(m/d) x_{m/d}^d,$$

where the sum is taken over the divisors of $m$. Let $F_m(\overline{2})$ denote the value of $F_m$ with each $x_i = 2$.

**1.19 Lemma** *Let $G \equiv \{T_{a,b} : a \in H \leq Z_p^*, b \in Z_p\}$, where $|H| = h$ is even. Then the number of circulant graphs $X$ of order $p$ such that $G \leq \mathrm{Aut}(X)$ is given by $F_{(p-1)/h}(\overline{2})$.*

**Proof** To obtain circulant graphs $X(p;S)$ of order $p$ whose automorphism groups contain $G$, we must ensure that $S$ is a union of cosets of $H$. There are $(p-1)/h$ cosets and each coset can be either chosen or not chosen to go into $S$. Isomorphism is still characterized by Theorem 1.9 so that we can apply Pólya's Theorem to count the number of non-isomorphic circulant graphs of order $p$ whose automorphism groups contain $G$. The group acting on the cosets is the permutation group generated by the cycle of length $(p-1)/h$. The cycle index of this group is $F_{(p-1)/h}$ from which the result follows immediately. □

**1.20 Theorem** *Let $G$ be a transitive permutation group of degree $p$, $p$ a prime, and let $n(G)$ denote the number of non-isomorphic circulant graphs $X$ of order $p$ such that* $\mathrm{Aut}(X) \equiv G$. *Then $n(S_p) = 2$, $n(G) = 0$ if $G \neq S_p$ is doubly-transitive or $|G|$ is odd, and*

$$n(G) = \sum_{d|(p-1)/h} \mu(d) F_{(p-1)/dh}(\overline{2}),$$

*where $|G| = hp$, $h$ is even, and $1 < h < p-1$.*

**Proof** If $G = S_p$, then the complete graph $K_p$ and its complement $\overline{K}_p$, and no other graphs of order $p$, have their automorphism groups isomorphic as permutation groups to $G$. Thus, $n(S_p) = 2$.

If $G \neq S_p$ and $G$ is doubly transitive, then no graph $X$ of order $p$ can satisfy $\mathrm{Aut}(X) \equiv G$ because if $\mathrm{Aut}(X)$ contains a doubly transitive group, then $X$ is either complete or has no edges. Hence, $n(G) = 0$ in this case.

Since $T_{-1,0}$ is an automorphism of any circulant graph $X$ of order $p$, $|\mathrm{Aut}(X)|$ is always even. Thus, $n(G) = 0$ if $|G|$ is odd.

Now let $G \equiv \{T_{a,b} : a \in H \leq Z_p^*, b \in Z_p\}$, where $|H| = h$ is even. The term $F_{(p-1)/h}(\overline{2})$ in the above formula counts all non-isomorphic circulant graphs whose automorphism groups contain $G$. Suppose $X$ is a circulant graph such that $G < \mathrm{Aut}(X)$. Then the connection set $S$ of $X$ is a union of cosets of a supergroup $K$ of $H$. Let $|K| = hp_1^{e_1}p_2^{e_2}\cdots p_i^{e_i}$. It is easy to verify that $X$ is counted zero times in the formula. □

**1.21 Example** Let $G$ be the dihedral group of order 26 and degree 13. Then $h = 2$ and $(p-1)/2 = 6$. The divisors of 6 are 1, 2, 3 and 6. The respective cycle indices are

$$F_6 = \frac{1}{6}(x_1^6 + x_2^3 + 2x_3^2 + 2x_6^1),\ F_3 = \frac{1}{3}(x_1^3 + 2x_3^1),\ F_2 = \frac{1}{2}(X_1^2 + x_2^1),\ F_1 = x_1^1.$$

Using the formula of Theorem 1.20, we obtain that there are

$$\mu(1)F_6(\overline{2}) + \mu(2)F_3(\overline{2}) + \mu(3)F_2(\overline{2}) + \mu(6)F_1(\overline{2}) = 9$$

circulant graphs of order 13 whose automorphism groups are all isomorphic to the dihedral group of order 26.

A. Ádám [1] asked if Theorem 1.9 holds for all $n$. It is easy to construct counter-examples. Consider the two circulant graphs $X$ and $X'$ of order 25 whose respective connection sets are $S = \{1,4,5,6,9,11,14,16,19,20,21,24\}$ and $S' = \{1,4,6,9,10,11,14,15,16,19,21,24\}$.

They are isomorphic since both are wreath products of a 5-cycle with a 5-cycle. However, any multiplier $a$ relating the two symbols would have to be congruent to 2 or 3 modulo 5 in order to map 5 and 20 to 10 and 15, whereas, it would have to be congruent to 1 or 4 modulo 5 to map the other numbers to where they have to go. It is easy to see that this can be extended to any other $p^2$, where $p$ is a prime larger than 5. Note that all these counterexamples are connected. Not only does it not work for either 4 or 9, but Theorem 1.9 holds for $n \in \{4, 8, 9\}$. It is easy to verify this directly.

Now suppose one has a counterexample for $n$. Let us see how to find a counterexample for any multiple of $n$. Let $X = X(n; S)$ and $X' = X(n; S')$ be counterexamples of order $n$. Let $Y$ be $m$ vertex-disjoint copies of $X$ and $Y'$ be $m$ vertex-disjoint copies of $X'$, that is, $Y = X(mn; mS)$ and $Y' = X(mn; mS')$. It is easy to see that $Y$ and $Y'$ are isomorphic. Now there can be no multiplier $a \in Z^*_{mn}$ such that $a(mS) = mS'$, for then $a$ reduced modulo $n$ would be a multiplier $\overline{a}$ satisfying $\overline{a}S = S'$. Hence, we have a counterexample for $mn$. Taking the complement of the preceding graph yields a connected counterexample and by the above remark, if $n$ is divisible by the square of any prime $p > 3$, there is a connected counterexample of order $n$.

Now we construct a counterexample of order $8m$, for all $m > 1$, based on an example of order 16 given in [10].

**1.22 Example** Let $m > 1$. Consider

$$X = X(8m; \{1, 2, 4m - 1, 4m + 1, 8m - 2, 8m - 1\})$$

and

$$X' = X(8m; \{2, 2m - 1, 2m + 1, 6m - 1, 6m + 1, 8m - 2\}).$$

It is easy to see that $X$ and $X'$ are isomorphic by observing that the mapping $f : V(X) \to V(X')$ defined by $f(2i) = 2i$ and $f(2i + 1) = 2i + 2m + 1$ is an isomorphism, that is, both consist of two vertex-disjoint cycles of length $4m$, arising from the element 2 in the respective symbols, joined together in the "same" way. The only possible multipliers are $\pm(2m - 1)$ and $\pm(2m + 1)$. It is easy to verify that none of these work. Note that these examples are all connected.

Now we want to construct a counterexample of order $9m$, for all $m > 2$, for Theorem 1.9.

**1.23 Example** This example appears in [24]. Let

$$\begin{aligned} S_0 &= \{1, 3, 3m - 1, 3m + 1, 6m - 1, 6m + 1, 9m - 3, 9m - 1\}, \\ S_1 &= \{1, 3m - 1, 3m + 1, 3m + 3, 6m - 3, 6m - 1, 6m + 1, 9m - 1\}, \\ S_2 &= \{1, 3m - 3, 3m - 1, 3m + 1, 6m - 1, 6m + 1, 6m + 3, 9m - 1\}. \end{aligned}$$

We claim that $X(9m; S_0)$ is isomorphic to $X(9m; S_1)$ when 3 does not divide $m + 1$, or is isomorphic to $X(9m; S_2)$ when 3 does not divide $2m + 1$.

The circulant graph $X(9m; S_0)$ is made up of three vertex-disjoint $3m$-cycles (from $3 \in S_0$), with each vertex of each cycle adjacent to three equally spaced vertices from each of the other two cycles. If 3 does not divide $m + 1$, the element $3m + 3 \in S_1$ produces three vertex-disjoint $3m$-cycles in $X(9m; S_1)$ which are joined amongst each other as they are in

$X(9m; S_0)$. Hence, the mapping that sends $3k + j$, where $0 \le j \le 2$, to $3k(m+1)+j$ is an isomorphism from $X(9m; S_0)$ to $X(9m; S_1)$.

When 3 does not divide $2m + 1$, the element $3m - 3 \in S_2$ produces the three vertex-disjoint $3m$-cycles. In this case, the isomorphism from $X(9m; S_0)$ to $X(9m; S_2)$ is given by $3k + j \mapsto 3k(m-1)+j$, where $0 \le j \le 2$.

Thus, the isomorphisms hold as claimed. We now show that the connection sets are not multiplicatively related. If 3 does not divide $m + 1$, any multiplier between $S_0$ and $S_1$ must take 3 to either $3m + 3$ or $6m - 3$ as the latter two are the only elements of $S_1$ divisible by 3. The only possible multipliers are the members of $\{m+1, 4m+1, 7m+1, 2m-1, 5m-1, 8m-1\}$. None of them map 1 to an element of $S_1$. Thus, $S_0$ and $S_1$ are not multiplicatively related.

When 3 does not divide $2m + 1$ a similar analysis leads to the only possible multipliers being an element of $\{2m+1, 5m+1, 8m+1, m-1, 4m-1, 7m-1\}$. None of them work so that $S_0$ and $S_1$ are not multiplicatively related.

From the preceding results, we see that Ádám's Conjecture can hold only for numbers of the form $2^e m$, where $e \in \{0, 1, 2\}$ and $m$ is odd and square-free, other than the exceptional values of 8, 9 and 18 for which the conjecture also holds. That it holds for these values is the subject of the next theorem.

**1.24 Theorem (Muzychuk [21, 22])** *Given two isomorphic circulant graphs $X(n; S)$ and $X(n; S')$ of order $n$, there is an $a \in Z_n^*$ such that $S' = aS$ if and only if $n \in \{8, 9, 18\}$ or $n = 2^e m$, where $e \in \{0, 1, 2\}$ and $m$ is odd and square-free.*

The proof of the preceding result is long and we shall not go into it here. There is another result we wish to mention in this section. It again points out the role played by Sylow $p$-subgroups.

**1.25 Definition** Let $G$ be a group of order $ab$ where $\gcd(a, b) = 1$. Any subgroup of order $a$ is called a *Hall* subgroup.

Hall subgroups are generalizations of Sylow subgroups. They have similar properties in certain situations. The following proposition is useful in the proof of the subsequent result.

**1.26 Proposition** *Let $G$ be a finite group. If $H_1$ and $H_2$ are two Hall subgroups of the same order and both are nilpotent, then $H_1$ and $H_2$ are conjugate in $G$.*

**1.27 Theorem (Morris [20])** *Let $n = p_1^{e_1} p_2^{e_2} \cdots p_r^{e_r}$, where $p_1, p_2, \ldots, p_r$ are distinct primes. Let $X = X(n; S)$ be a circulant graph of order $n$. If the Sylow $p_i$-subgroups of $\mathrm{Aut}(X)$ have order $p_i^{e_i}$ for $i = 1, 2, \ldots, r$, then $X' = X(n; S')$ is isomorphic to $X$ if and only if there exists an $a \in Z_n^*$ satisfying $S' = aS$.*

**Proof** It is easy to see that the existence of an $a \in Z_n^*$ satisfying $S' = aS$ yields an isomorphism between $X$ and $X'$. For the other direction, let $f$ be an isomorphism from $X' = X(n; S')$ to $X$. Then $H_1 = \langle (f(0)\ f(1)\ \cdots\ f(n-1)) \rangle \le \mathrm{Aut}(X)$. Also, $H_2 = \langle (0\ 1\ \cdots\ n-1) \rangle \le \mathrm{Aut}(X)$. Then $H_1$ and $H_2$ are Hall subgroups of the same order. Since both are nilpotent, they are conjugate by Proposition 1.26. Thus, there exists $g \in \mathrm{Aut}(X)$ such that $gH_1g^{-1} = H_2$. This implies that

$$(gf(0)\ gf(1)\ \cdots\ gf(n-1)) = (0\ 1\ \cdots\ n-1)^a = (0\ a\ \cdots\ (n-1)a)$$

for some $a$ satisfying $\gcd(a,n)=1$. Assume that $gf(0)=ka$. Then

$$\begin{aligned} j-i\in S' \quad &\text{iff} \quad ij\in E(X') \\ &\text{iff} \quad f(i)f(j)\in E(X) \\ &\text{iff} \quad gf(i)gf(j)\in E(X) \\ &\text{iff} \quad (k+i)a(k+j)a\in E(X) \\ &\text{iff} \quad a(j-i)\in S. \end{aligned}$$

Thus, $S'=a^{-1}S$. □

Note that when $n$ is even, any circulant graph $X$ of order $n$ has the automorphism $\tau=(0)(n/2)(1\ n-1)(2\ n-2)\cdots(n/2-1\ n/2+1)$. Thus, the Sylow 2-subgroups of $\mathrm{Aut}(X)$ have order bigger than the power of 2 in the prime factorization of $n$. Thus, Theorem 1.27 does not work for even $n$. However, the conclusion of Theorem 1.27 still holds as long as the order of the Sylow 2-subgroups of $\mathrm{Aut}(X)$ does not get too large.

**1.28 Theorem** *Let $n=2^e p_1^{e_1}\cdots p_r^{e_r}$, where $p_1,p_2,\ldots,p_r$ are distinct odd primes. Let $X=X(n;S)$ be a circulant graph of order $n$. If the Sylow $p_i$-subgroups of $\mathrm{Aut}(X)$ have order $p_i^{e_i}$ for $p_1,p_2,\ldots,p_r$, and the Sylow 2-subgroups have order $2^{e+1}$, then $X'=X(n;S')$ is isomorphic to $X$ if and only if there exists an $a\in Z_n^*$ satisfying $S'=aS$.*

**Proof** Let $f: V(X')\to V(X)$ be an isomorphism. Considering the proof of Theorem 1.27, it is sufficient to show that $H_1=\langle (f(0)\ f(1)\ \cdots\ f(n-1))\rangle$ and $H_2=\langle (0\ 1\ \cdots\ n-1)\rangle$ are conjugate in $\mathrm{Aut}(X)$. Let $\tau_1$ and $\tau_2$ be the respective involutions described above. Then let $D_1=\langle H_1,\tau_1\rangle$ and $D_2=\langle H_2,\tau_2\rangle$. $D_1$ and $D_2$ are both dihedral groups of order $2n$, and both are subgroups of $\mathrm{Aut}(X)$. It is sufficient to prove that they are conjugate in $\mathrm{Aut}(X)$ since $H_1$ and $H_2$ are respective unique cyclic subgroups.

Both $D_1$ and $D_2$ contain cyclic subgroups of order $m=p_1^{e_1}p_2^{e_2}\cdots p_r^{e_r}$. Since these cyclic subgroups of order $m$ are Hall subgroups, they are conjugate in $\mathrm{Aut}(X)$. Hence, there is an $f\in \mathrm{Aut}(X)$ so that $fD_1f^{-1}$ and $D_2$ contain the same cyclic subgroup $C$ of order $m$. Thus, both $fD_1f^{-1}$ and $D_2$ are contained in the normalizer $N_G(C)$ of $C$. This implies that the Sylow 2-subgroup $P_2$ of $fD_1f^{-1}$ is conjugate to a Sylow 2-subgroup of $D_2$ inside $N_G(C)$, that is, $gP_2g^{-1}\le D_2$ for some $g\in N_G(C)$. This gives us that

$$g(fD_1f^{-1})g^{-1}=g(P_2C)g^{-1}=(gP_2g^{-1})(gCg^{-1})=gP_2g^{-1}C\le D_2.$$

Since, $|D_2|=|fD_1f^{-1}|$, we have that $g(fD_1f^{-1})g^{-1}=D_2$. □

We conclude this section with an old conjecture of S. Toida [27].

**1.29 Conjecture** If $S\subset Z_n^*$, then whenever $X(n;S')$ is isomorphic to $X(n;S)$, there exists an $a\in Z_n^*$ satisfying $S'=aS$.

## 2 Abelian groups

The previous section dealt with circulant graphs, that is, Cayley graphs on cyclic groups. It is natural to attempt to duplicate the previous isomorphism results for a more general class of groups. We first consider abelian groups.

Also, we want to look more carefully at what we were doing in the previous section. Note that a multiplier $a \in Z_n^*$ determines an automorphism of the additive group $Z_n$. This is the direction we want to take.

**2.1 Theorem** *Let $X = X(G;S)$ be a Cayley graph on the group $G$. If $a \in \mathrm{Aut}(G)$, then $a$ is an isomorphism from the graph $X$ to the graph $X' = X(G;a(S))$.*

**Proof** We have that $uv \in E(X)$ if and only if $v = us$ for some $s \in S$. Since $a$ is a group automorphism, $v = us$ if and only if $a(v) = a(u)a(s)$ if and only if $a(u)a(v) \in E(X')$. Thus, $a$ is an isomorphism from $X$ to $X'$ as required. □

In the preceding context, when two Cayley graphs on a given group $G$ are isomorphic, we would like to find a group automorphism that yields an isomorphism between them. The previous section deals with the cyclic group.

**2.2 Definition** Let $G$ be a finite group, and let $X = X(G;S)$ be a Cayley graph on $G$. If $X$ isomorphic to $X'$ implies that there exists a group automorphism $a$ of $G$ such that $a : V(X) \to V(X')$ is an isomorphism, then we say that $S$ satisfies the *Cayley isomorphism property*. This will be denoted by saying $S$ is a *CI-subset*.

**2.3 Definition** If every Cayley subset of $G$ is a CI-subset, then we say that $G$ satisfies the *Cayley isomorphism property*. We denote this by saying that $G$ is a *CI-group*.

In order to determine whether or not $G$ is a CI-group, we must find a way to see when isomorphism between two Cayley graphs on $G$ can be realized by a group automorphism. The next theorem translates this problem into examining subgroups of the group $G$. It provides a very useful tool.

**2.4 Definition** A transitive permutation group $G$ of degree $n$ is said to be *regular* if $|G| = n$.

**2.5 Proposition** *The following conditions for a transitive group $G$ to be regular are equivalent:*

(1) $|G| = n$,

(2) $|G_x| = 1$ *for every $x \in \Omega$, and*

(3) *For any two $x, y \in \Omega$, there exists a unique permutation $\sigma \in G$ such that $\sigma(x) = y$.*

**Proof** This is left as an exercise for the reader. □

It was mentioned in the proof of Proposition 1.3 that left multiplication by an element of $G$ is an automorphism of any Cayley graph on $G$. We shall use the notation $g_L$ to denote left multiplication by $g \in G$. We then let $G_L = \{g_L : g \in G\}$.

**2.6 Proposition** *If $X$ is a Cayley graph on the group $G$, then $G_L$ is a regular subgroup of* $\mathrm{Aut}(X)$.

**Proof** This is left to the reader. □

The next result is very useful in determining whether or not a Cayley subset is a CI-subset. It transforms the problem from having to compare two Cayley graphs to a problem of conjugacy in the automorphism group of a Cayley graph.

**2.7 Theorem (Babai [3])** *Let $X = X(G;S)$ be a Cayley graph on the finite group $G$. Then $S$ is a CI-subset of $G$ if and only if given any $\varphi \in S_G$ such that $\varphi G_L \varphi^{-1} \leq \mathrm{Aut(X)}$, $G_L$ and $\varphi G_L \varphi^{-1}$ are conjugate in* $\mathrm{Aut}(X)$.

**Proof** Assume that $S$ is a CI-subset of $G$. Suppose there exists $\varphi \in S_G$ such that $\varphi G_L \varphi^{-1} \leq \mathrm{Aut}(X)$. Let $Y = \varphi^{-1}(X)$, that is, relabel the vertices of $X$ according to $\varphi^{-1}$. Then $\mathrm{Aut}(Y) = \varphi^{-1}\mathrm{Aut}(X)\varphi \geq G_L$ so that $Y$ is a Cayley graph on $G$. It certainly is isomorphic to $X$. There is a group automorphism $\alpha$ such that $\alpha(X) = Y$ because $S$ is a CI-subset. Then $\alpha(X) = \varphi^{-1}(X)$ implying that $\beta = \varphi\alpha \in \mathrm{Aut}(X)$. Thus,

$$\beta G_L \beta^{-1} = \beta\alpha^{-1} G_L \alpha\beta^{-1} = \varphi G_L \varphi^{-1}$$

so that $G_L$ and $\varphi G_L \varphi^{-1}$ are conjugate in $\mathrm{Aut}(X)$.

Now prove the other direction. Let $\varphi : Y \to X$ be an isomorphism from $Y = X(G;S')$ to $X$. We have $\varphi G_L \varphi^{-1} \leq \mathrm{Aut}(X)$ because $G_L \leq \mathrm{Aut}(Y)$. By assumption, there exists $\beta \in \mathrm{Aut}(X)$ such that

$$\varphi G_L \varphi^{-1} = \beta G_L \beta^{-1}.$$

Let $\gamma = \beta^{-1}\varphi$ so that

$$\gamma G_L \gamma^{-1} = G_L.$$

We want to find a group automorphism that will map $S$ to $S'$. We start with $\gamma$ which is an element of $\mathrm{Aut}(X)$. Let $\gamma(1) = g$ and define $\delta$ by $\delta = g_L^{-1}\gamma$. Then $\delta(1) = 1$ and $\delta G_L \delta^{-1} = G_L$. So $\delta$ is a candidate to yield an automorphism of $G$. It induces a mapping $\alpha : G \to G$ defined as follows:

$$\delta g_L \delta^{-1} = \alpha(g)_L.$$

We claim that $\alpha \in \mathrm{Aut}(G)$. First, $\delta 1_L \delta^{-1} = 1_L$, implying that $\alpha(1) = 1$. Second,

$$\begin{aligned} \alpha(gh)_L &= \delta(gh)_L\delta^{-1} &&= \delta(g_L h_L)\delta^{-1} \\ &= \delta g_L \delta^{-1}\delta h_L \delta^{-1} &&= \alpha(g)_L \alpha(h)_L. \end{aligned}$$

Thus, $\alpha$ is a homomorphism from $G$ to itself. If $\delta g_L \delta^{-1} = 1_L$, then $g = 1$ by considering the disjoint cycle decomposition of $1_L$. We conclude that $\alpha \in \mathrm{Aut}(G)$ as claimed.

Next we wish to show that $\alpha(g)_L = \alpha g_L \alpha^{-1}$. Now

$$\begin{aligned} \alpha(g)_L \alpha(x) &= \alpha(g)\alpha(x) \\ &= \alpha(gx) && \text{since } \alpha \in \mathrm{Aut}(G) \\ &= (\alpha g_L)x && \text{definition of } g_L \\ &= (\alpha g_L \alpha^{-1})\alpha(x). \end{aligned}$$

Since the above holds for all $x$, $\alpha(g)_L = \alpha g_L \alpha^{-1}$.

We then have that

$$\delta g_L \delta^{-1} = \alpha(g)_L = \alpha g_L \alpha^{-1}.$$

From this it follows that $\alpha^{-1}\delta g_L = g_L \alpha^{-1}\delta$ which in turn implies that $\alpha^{-1}\delta$ is in the centralizer of $G_L$. But the centralizer of $G_L$ is $G_R$. However, $\alpha^{-1}\delta$ fixes $1 \in G$, and since $G_R$ is regular, $\alpha^{-1}\delta = 1 \in \mathrm{Aut}(G)$.

We are now ready to prove that $\alpha \in \mathrm{Aut}(G)$ is an isomorphism from $Y$ to $X$. Let $u$ and $v$ be adjacent in $Y$. Then

$$\alpha(u) = (\alpha(\alpha^{-1}\delta)u) = \delta(u) = (g_L^{-1}\gamma)u = g^{-1}\gamma(u).$$

Similary, $\alpha(v) = g^{-1}\gamma(v)$. We know that $g^{-1}\gamma(u)$ and $g^{-1}\gamma(v)$ are adjacent in $X$. Hence, $\alpha$ is an isomorphism from $Y$ to $X$ and $\alpha(S') = S$. Therefore, $S$ is a CI-subset. □

The first groups we consider in this section are $Z_p^2 = Z_p \times Z_p$ and $Z_p^3 = Z_p \times Z_p \times Z_p$. The group $Z_p^2$ was shown to be a CI-group by C. Godsil [14]. His proof used a deep theorem about primitive groups due to M. O'Nan. The group $Z_p^3$ was shown to be a CI-group by both M.-Y. Xu [30] and E. Dobson [9]. Xu uses the classification of finite simple groups in his proof. Dobson gives a more combinatorial proof and avoids using the classification of finite simple groups; still his proof is somewhat complex. In a recent letter from L. Nowitz an outline of the following proof for the $Z_p^3$ case was given. It is indeed more elementary than either of the two published proofs. We apply the idea to the $Z_p^2$ case first.

**2.8 Definition** Consider permutation groups $G$ and $H$ acting on $\Omega_1 = \{x_1, x_2, \ldots, x_m\}$ and $\Omega_2 = \{y_1, y_2, \ldots, y_n\}$, respectively. The *wreath product* of $G$ and $H$, denoted $G \wr H$, is a permutation group acting on $\Omega_1 \times \Omega_2$ as follows. For $g \in G$ and $h_1, h_2, \ldots, h_m \in H$ define the permutation $(g; h_1, h_2, \ldots, h_m)$ by

$$(g; h_1, h_2, \ldots, h_m)(x_i, y_j) = (g(x_i), h_i(y_j)),$$

for $i = 1, 2, \ldots, m$. Then

$$G \wr H = \{(g; h_1, h_2, \ldots, h_m) : g \in G \text{ and } h_i \in H \text{ for } i = 1, 2, \ldots, m\}.$$

The way to picture the wreath product defined above is to think of replacing each element of $\Omega_1$ with a copy of $\Omega_2$, and then taking all permutations that permute the copies of $\Omega_2$ according to a permutation from $G$ and independently permute the elements within each copy of $\Omega_2$ according to permutations from $H$. Note that $|G \wr H| = |G||H|^m$.

**2.9 Theorem** *The group $Z_p^2$, $p$ a prime, is a CI-group.*

**Proof** Let $X = X(Z_p^2; S)$ be a Cayley graph on the group $Z_p^2$. Suppose that $\sigma(Z_p^2)_L\sigma^{-1} \leq \mathrm{Aut}(X)$ for some $\sigma \in S_{Z_p^2}$. We want to show that $(Z_p^2)_L$ and $\sigma(Z_p^2)_L\sigma^{-1}$ are conjugate in $\mathrm{Aut}(X)$.

The symmetric group $S_{p^2}$ has order $p^{p+1}m$, where $\gcd(p, m) = 1$, that is, the Sylow $p$-subgroups of $S_{p^2}$ have order $p^{p+1}$. Since $Z_p \wr Z_p$ has order $p^{p+1}$ and it is a subgroup of $S_{p^2}$, the Sylow subgroups of $S_{p^2}$ are the conjugates of $Z_p \wr Z_p$. Note the important fact that $Z_p \wr Z_p$ is imprimitive with blocks of length $p$ and that the block system is uniquely determined.

Let $P$ and $P'$ be Sylow $p$-subgroups of Aut$(X)$ containing $(Z_p^2)_L$ and $\sigma(Z_p^2)_L\sigma^{-1}$, respectively. Then there exists $f \in$ Aut$(X)$ such that $fP'f^{-1} = P$. Thus, $(Z_p^2)_L$ and $f\sigma(Z_p^2)_L\sigma^{-1}f^{-1}$ are in the same Sylow $p$-subgroup $P$ of Aut$(X)$. Now $P$ itself is contained in some Sylow $p$-subgroup of $S_{p^2}$ which has unique imprimitive blocks of length $p$. Let the blocks be

$$\begin{aligned} B_0 &= \{x_{0,0}, x_{0,1}, \ldots, x_{0,p-1}\}, \\ B_1 &= \{x_{1,0}, x_{1,1}, \ldots, x_{1,p-1}\}, \\ &\vdots \\ B_{p-1} &= \{x_{p-1,0}, x_{p-1,1}, \ldots, x_{p-1,p-1}\}. \end{aligned}$$

Let the permutation $\theta_i = (x_{i,0}\ x_{i,1}\ \cdots\ x_{i,p-1})$, $i = 0, 1, \ldots, p-1$. Without loss of generality, we may assume that the blocks have been coordinatized so that $\tau_1 = \theta_0\theta_1\cdots\theta_{p-1}$ and the permutation defined by $\tau_2 : x_{i,j} \mapsto x_{i+1,j}$ for all $i, j \in \{0, 1, \ldots, p-1\}$ are elements of $(Z_p^2)_L$.

We shall say that two blocks of imprimitivity are *wreathed* if either there is no edge with an end vertex in each of the two blocks or every possible edge with an end vertex in each block is in the graph.

Consider the automorphism $\tau_1' \in f\sigma(Z_p^2)_L\sigma^{-1}f^{-1}$ which maps $x_{0,0}$ to $x_{0,1}$. We must have $\tau_1' = \theta_0\theta_1^{e_1}\cdots\theta_{p-1}^{e_{p-1}}$ since it has order $p$ and is from a regular group. If $e_1 = e_2 = \cdots = e_{p-1} = 1$, then $\tau_1' = \tau_1$. If some $e_i = r \neq 1$, then it is easy to see that any two blocks $B_k$ and $B_j$ are wreathed. Let $\tau_2'$ be the element of $f\sigma(Z_p^2)_L\sigma^{-1}f^{-1}$ mapping $x_{0,0}$ to $x_{1,0}$. Since $P$ is a subgroup of $Z_p \wr Z_p$, $\tau_2'(B_i) = B_{i+1}$ for all $i$. Furthermore, the mapping $\tau_2^i\tau_2'^{-i}$ is an automorphism of $X$ mapping $x_{i,ar}$ to $x_{i,a+j}$ implying that the connection set $S_i$ for the subcirculant graph $X_i$ induced on $B_i$ satisfies $rS_i = S_i$. Thus, the mapping $g_i'$ given by $g_i'(x_{i,k}) = x_{i,r^{-1}k}$ is an automorphism of $X_i$. It can be extended to an automorphism $g_i$ of $X$ by letting it fix all other vertices. Thus, conjugation of $\tau_1'$ by $f_1 = g_1g_2\cdots g_{p-1}$ yields $\tau_1$. Therefore, $f_1f\sigma(Z_p^2)_L\sigma^{-1}f^{-1}f_1^{-1}$ contains $\tau_1$.

If $\tau_2'(x_{i,0}) = x_{i+1,0}$ for all $i$, then $\tau_2' = \tau_2$ and we are finished. Hence, we consider the cycle $\zeta = (x_{0,0}\ x_{1,0}\ \cdots)$ in the disjoint cycle decomposition of $\tau_2'$. Let $x_{i,d_i} \in \zeta$, $0 \leq i \leq p-1$. Since some $d_i \neq 0$, it is easy to see that any two blocks $B_j$ and $B_k$ are wreathed. We want to cyclically relabel each of the blocks $B_i$ by $-d_i$. Let $g_i$ denote $\theta_i^{-d_i}$ extended to all of $X$ by letting it fix any vertex not in $B_i$. Then $f_2 = g_2g_3\cdots g_{p-1}$ conjugates $\tau_2'$ to $\tau_2$ and leaves $\tau_1$ unaltered. Therefore, $(Z_p^2)_L = f_2f_1f\sigma(Z_p^2)_L\sigma^{-1}f^{-1}f_1^{-1}f_2^{-1}$ and we are finished. □

The proof of the analogous result for $p^3$ is essentially the same. There are a few technicalities to worry about. As a useful aside for the forthcoming proof, we note that the preceding theorem also is true for Cayley color digraphs.

**2.10 Theorem** *The group $Z_p^3$, $p$ a prime, is a CI-group.*

**Proof** Similar to the comments in the proof of the preceding theorem, the Sylow $p$-subgroups of $S_{p^3}$ are conjugate to $Z_p \wr Z_p \wr Z_p$. This implies that every Sylow $p$-subgroup of $S_{p^3}$ has a unique block system of $p^2$ blocks each of length $p$, and a unique block system of $p$ blocks each of length $p^2$.

Let $X = X(Z_p^3; S)$ be a Cayley graph on the group $Z_p^3$. Suppose that $\sigma(Z_p^3)_L\sigma^{-1} \leq \mathrm{Aut}(X)$ for some $\sigma \in S_{Z_p^3}$. We want to show that $(Z_p^3)_L$ and $\sigma(Z_p^3)_L\sigma^{-1}$ are conjugate in $\mathrm{Aut}(X)$.

By taking a conjugate of $\sigma(Z_p^3)_L\sigma^{-1}$ if necessary, as mentioned in the proof of Theorem 2.9, we may assume that $(Z_p^3)_L$ and $\sigma(Z_p^3)_L\sigma^{-1}$ are in the same Sylow $p$-subgroup $P$ of $\mathrm{Aut}(X)$. The group $P$ is contained in some Sylow $p$-subgroup $P^*$ of $S_{Z_p^3}$ which has unique imprimitive blocks of length $p$. Let the blocks be

$$B_{i,j} = \{x_{i,j,k} : 0 \leq k \leq p-1\}, 0 \leq i,j \leq p-1.$$

Furthermore, it also has unique imprimitive blocks of length $p^2$. Let

$$B_i = B_{i,0} \cup B_{i,1} \cup \cdots \cup B_{i,p-1},$$

for $0 \leq i \leq p-1$, be the blocks of length $p^2$.

Let the permutation $\theta_{i,j} = (x_{i,j,0}\ x_{i,j,1}\ \cdots\ x_{i,j,p-1})$, for $0 \leq i,j \leq p-1$. Since both $(Z_p^3)_L$ and $\sigma(Z_p^3)_L\sigma^{-1}$ are abelian and act transitively on the set of blocks $\{B_{i,j} : 0 \leq i,j \leq p-1\}$, they must act regularly on this set of blocks. Thus, without loss of generality, we may assume that the blocks have been coordinatized so that the permutations $\tau_1 : x_{i,j,k} \mapsto x_{i,j,k+1}$, $\tau_2 : x_{i,j,k} \mapsto x_{i,j+1,k}$ and $\tau_3 : x_{i,j,k} \mapsto x_{i+1,j,k}$, for $0 \leq i,j,k \leq p-1$, are elements of $(Z_p^3)_L$. Note that

$$\tau_1 = \prod_{0 \leq i,j \leq p-1} \theta_{i,j}$$

and $(Z_p^3)_L = \langle \tau_1, \tau_2, \tau_3 \rangle$.

Suppose we have found a conjugate $H$ of $\sigma(Z_p^3)_L\sigma^{-1}$ such that $\tau_1, \tau_2'', \tau_3'' \in H$, where $\tau_2''$ maps $B_{0,j}$ to $B_{0,j+1}$, $\tau_3'' : B_i \to B_{i+1}$, $0 \leq i \leq p-1$, and both $\tau_2''$ and $\tau_3''$ fix the third subscript of every $x_{i,j,k}$. This means that $\langle \tau_2, \tau_3, \tau_2'', \tau_3'' \rangle$ maps blocks of length $p$ to blocks of length $p$, keeping all third subscripts fixed. We then can complete the proof as follows.

Define a Cayley color digraph $\vec{X}$ on $Z_p^2$ whose vertices are the blocks $B_{i,j}$, $0 \leq i,j \leq p-1$. Define the $p$-tuple $u_{i,j,k,\ell} = (u_0, u_1, \ldots, u_{p-1})$ by $u_t = 1$ if $x_{i,j,0}$ is adjacent to $x_{k,\ell,t}$ in $X$, and $u_t = 0$ otherwise. Assign the color $u_{i,j,k,\ell}$ to the arc from $B_{i,j}$ to $B_{k,\ell}$. If $\vec{f} \in \mathrm{Aut}(\vec{X})$, then $\vec{f}$ determines an $f \in \mathrm{Aut}(X)$ defined by $f(x_{i,j,r}) = x_{k,\ell,r}$ if $\vec{f} : B_{ij} \to B_{k,\ell}$. Since Theorem 2.9 holds for Cayley color digraphs on $Z_p^2$, there is an $\vec{f} \in \mathrm{Aut}(\vec{X})$ such that $\vec{f}\langle \vec{\tau_2''}, \vec{\tau_3''} \rangle \vec{f}^{-1} = \langle \vec{\tau_2}, \vec{\tau_3} \rangle$. Thus, $f\langle \tau_2'', \tau_3'' \rangle f^{-1} = \langle \tau_2, \tau_3 \rangle$. Since $f\tau_1 f^{-1} = \tau_1$, $fHf^{-1} = (Z_p^3)_L$ completing the proof.

Let $\tau_1' \in \sigma(Z_p^3)_L\sigma^{-1}$ be the element mapping $x_{0,0,0}$ to $x_{0,0,1}$. Then

$$\tau_1' = \prod_{0 \leq i,j \leq p-1} \theta_{i,j}^{e_{i,j}},$$

where $e_{0,0} = 1$. If two blocks $B_{i,j}$ and $B_{k,\ell}$ are not wreathed, then $e_{i,j} = e_{k,\ell}$ must hold. Form an auxiliary graph $Y$ whose vertices are the blocks of length $p$ and with an edge joining two blocks if and only if the two blocks are not wreathed.

If $Y$ is connected, then all $e_{i,j}$ are equal implying that $\tau_1' = \tau_1$ and we are finished. If $Y$ is totally disconnected, then we can independently conjugate each block of length $p$ to obtain $\theta_{i,j}$ as in the proof of Theorem 2.9. This yields a conjugate of $\sigma(Z_p^3)_L\sigma^{-1}$ containing $\tau_1$ in this case.

Finally, let $Y$ have non-trivial components. It is easy to see that each component intersects each $B_i$, $0 \le i \le p-1$, in exactly one block. Let $C$ be a component and $C \cap B_i = B_{i,j_i}$, $0 \le i \le p-1$. Let $e_{i,j_i} = r$ and $g = \theta^r_{0,j_0}\theta^r_{1,j_1} \cdots \theta^r_{p-1,j_{p-1}}$. So $g$ is part of the disjoint cycle decomposition of $\tau_1'$. By mapping $C$ onto the component containing $B_{0,0}$, using the mapping $\tau_2' \in \sigma(Z_p^3)_L\sigma^{-1}$ that maps $x_{0,0,0}$ to $x_{0,1,0}$, and mapping it back using $\tau_2$, we obtain an automorphism of $C$ which conjugates $g$ to $\theta_{0,j_0}\theta_{1,j_1} \cdots \theta_{p-1,j_{p-1}}$. If we do this for each component, we find a conjugate $f\sigma(Z_p^3)_L\sigma^{-1}f^{-1}$ of $\sigma(Z_p^3)_L\sigma^{-1}$ containing $\tau_1$.

Let $\tau_2'$ be the element of $f\sigma(Z_p^3)_L\sigma^{-1}f^{-1}$ mapping $x_{0,0,0}$ to $x_{0,1,0}$. Then $\tau_2'(B_{0,j}) = B_{0,j+1}$ for $0 \le j \le p-1$ because $P^*$ is unaltered. Consider the cycles $\rho_i$ beginning $(x_{i,0,0} \;\cdots)$, $0 \le i \le p-1$, in the disjoint cycle decomposition of $\tau_2'$. If $x_{i,j,k}$, $k \ne 0$, appears in $\rho_i$, then conjugation of $\rho_i$ by $\theta_{i,j}^{-k}$ cyclically relabels the cycle $\theta_{i,j}$ so that $x_{i,j,0}$ appears in $\rho_i$ and $\tau_1$ is unaltered. If all pairs of blocks of length $p$ are wreathed, then we can do the preceding to all $\theta_{i,j}$ independently to obtain a $\tau_2''$ of the desired form.

If some $x_{0,j,k}$, $k \ne 0$, appears in $\rho_0$, then any two blocks of length $p$ inside the same block of length $p^2$ are wreathed by considering the actions of $\tau_2$, $\tau_3$ and $\tau_2'\tau_2^{-1}$. If $\tau_2'(B_{i,j}) \ne B_{i,j+1}$ for some $i$, then any two blocks of length $p$ are wreathed. So we may assume $\tau_2'(B_{i,j}) = B_{i,j+1}$ for all $1 \le i \le p-1$.

Let $\rho_0 = (x_{0,0,0}\; x_{0,1,0}\; x_{0,2,a_2} \;\cdots\; x_{0,p-1,a_{p-1}})$. Define the *profile* of $\rho_0$ to be the sequence $(0, a_2, a_3 - a_2, \ldots, -a_{p-1})$. The profile of $\rho_i$ is defined similarly. If $B_{0,0}$ and $B_{i,j}$ are not wreathed, $i \ne 0$, then $B_{0,k}$ and $B_{i,j+k}$ also are not wreathed for $1 \le j \le p-1$. Since $\tau_2'\tau_2^{-1}$ must be the same power of $\theta_{0,k}$ and $\theta_{i,j+k}$ for each $k$, the profile of $\rho_i$ must be a cyclic shift of the profile of $\rho_0$. Since the length of a profile is prime, no two cyclic shifts of the profile are the same so that each block $B_{i,j}$ of length $p$ can have at most one block $B_{k,\ell}$, $k \ne i$, of $B_k$ with which it is not wreathed. If we define the same auxiliary graph $Y$ as before, each component intersects each $B_i$ in exactly one block of length $p$ because of the uniqueness of the profile. So if $x_{0,i,a_i}$ is the vertex of $\rho_0$ in a given component, conjugate each block of the component by $\theta_{i,j}^{-a_i}$. Conjugating $\tau_2'$ by the product over the components yields a $\tau_2''$ in which all the third subscripts are fixed and $\tau_2''$ has the same action on $\{B_{i,j} : 0 \le i, j \le p-1\}$ as $\tau_2'$.

This leaves us with the case that the profile of $\rho_0 = (0, 0, \ldots, 0)$. If any other $\rho_i$, $i \ne 0$, does not fix the third subscript, then any two blocks of length $p$ are wreathed and we can conjugate $\tau_2'$ to an appropriate $\tau_2''$. Otherwise, $\tau_2'$ already has the proper form. Therefore, we have a conjugate $gf\sigma(Z_p^3)_L\sigma^{-1}f^{-1}g^{-1}$ containing $\tau_1$ and $\tau_2'' : B_{0,j} \to B_{0,j+1}$ so that $\tau_2''$ fixes all third subscripts.

Let $\tau_3' \in gf\sigma(Z_p^3)_L\sigma^{-1}f^{-1}g^{-1}$ be the element mapping $x_{0,0,0}$ to $x_{1,0,0}$. If $\tau_3'(x_{i,0,0}) = x_{i+1,b_i,a_i}$, then $\tau_3'(x_{i,j,k}) = x_{i+1,j+b_i,k+a_i}$ for all vertices of $B_i$ because $\tau_1, \tau_2''$ both belong to $gf\sigma(Z_p^3)_L\sigma^{-1}f^{-1}g^{-1}$. Define the profile of $\tau_3'$ to be the sequence $(0, a_1, a_2, \ldots, a_{p-1})$.

If any two blocks of length $p$ are wreathed, we can conjugate each $\theta_{i,j}$, $0 \le j \le p-1$, by $\theta_{i,j}^{-\ell_i}$, where $\ell_i = a_1 + a_2 + \cdots + a_{i-1}$ for $2 \le i \le p-1$. This leaves $\tau_1$ and $\tau_2''$ unaltered while changing $\tau_3'$ to a $\tau_3''$ of the appropriate form.

If $a_1 = a_2 = \cdots = a_{p-1} = 0$, then $\tau_3'$ already fixes every third subscript as required. So we may assume some $a_i \ne 0$. The element $\tau_2^{-1}\tau_2'$ fixes each vertex of $B_{0,0}$ and maps $x_{i,j,k}$ to $x_{i,j+b_i,k}$ so that the connection set between $B_{0,0}$ and $B_{i,j}$ is the same as that between $B_{0,0}$ and $B_{i,j+b_i}$. Hence, the connection set between any $B_{0,\ell}$ and any $B_{i,r}$ is the same.

The element $\tau_3^{-1}\tau_3'$ fixes each vertex of $B_{0,0}$ and maps $x_{i,j,k}$ to $x_{i,j+b_i,k+a_i}$. This implies that every block of length $p$ in $B_0$ is wreathed with every block of length $p$ in $B_i$. Therefore, any two blocks of length $p$ in different blocks of length $p^2$ are wreathed, and we can find a conjugate $hgf\sigma(Z_p^3)_L\sigma^{-1}f^{-1}g^{-1}h^{-1}$ of $\sigma(Z_p^3)_L\sigma^{-1}$ containing $\tau_1$, and $\tau_2''$ and $\tau_3''$ of the desired form. □

L. Babai and P. Frankl [4] conjectured that all of the abelian groups $Z_p^n$, $p$ a prime and $n \geq 1$, are CI-groups. The following result of L. Nowitz [23] provides a counterexample to the conjecture, but suggests many open questions.

**2.11 Theorem** *The group $Z_2^6$ is not a CI-group.*

**Proof** We are going to define a Cayley graph $X$ on $Z_2^6$ and a subgroup of $\mathrm{Aut}(X)$ which is conjugate to $(Z_2^6)_L$ in $S_{Z_2^6}$, but is not conjugate to $(Z_2^6)_L$ is $\mathrm{Aut}(X)$. This proves the theorem because of Theorem 2.7.

Let

$$\begin{aligned} S_1 &= \{(x_1, x_2, \ldots, x_6) \in Z_2^6 : x_1 = x_2 = 0 \text{ and } x_3 \neq x_4\}, \\ S_2 &= \{(x_1, \ldots, x_6) \in Z_2^6 : (x_1, x_2) \neq (0,0), (x_3, x_4) \neq (1,1) \\ &\qquad \text{and } (x_5, x_6) \in \{(0,0), (x_1, x_2)\}\}, \text{ and} \\ S_3 &= \{(x_1, x_2, 1, 1, x_5, x_6) \in Z_2^6 : (x_1, x_2) \neq (0,0) \\ &\qquad \text{and } (x_5, x_6) \notin \{(0,0), (x_1, x_2)\}\}. \end{aligned}$$

Note that $|S_1| = 8$, $|S_2| = 18$ and $|S_3| = 6$. Now let $S = S_1 \cup S_2 \cup S_3$. Since $(0,0,0,0,0,0) \notin S$ and each element of $Z_2^6$ is its own inverse, $S$ is a connection set for a Cayley graph on $Z_2^6$. Let $X = X(Z_2^6; S)$. Then $X$ has order 64 and degree 32. Let $(Z_2^6)_L$ denote the group of left translations of $Z_2^6$.

We write $\tau(x_1, \ldots, x_6)$ rather than $\tau((x_1, \ldots, x_6))$ in the remainder of the proof, when $\tau$ is a function on $Z_2^6$. Let $s$ be a permutation acting on $Z_2^2$. We shall write $(s(x_1, x_2), x_3, x_4, x_5, x_6)$ to indicate that $s(x_1, x_2)$ is to be taken as the first two coordinates of the 6-tuple.

Let $s$ be a permutation of $Z_2^2$. Define the mapping $\tau_{1,s}$ by

$$\tau_{1,s}(x_1, x_2, x_3, x_4, x_5, x_6) = (s(x_1, x_2), x_3, x_4, s(x_5, x_6)).$$

Define $\tau_2$ by

$$\tau_2(x_1, x_2, x_3, x_4, x_5, x_6) = (x_1, x_2, x_4, x_3, x_5, x_6).$$

Finally, let $x = (y, z) \neq (0,0)$. Define $\tau_{3,x}$ as follows, where $(x_1, x_2) \neq (y, z)$ and $(x_1, x_2) \neq (0,0)$ hold:

$$\begin{aligned} \tau_{3,x}(0,0,0,x_4,x_5,x_6) &= (0,0,0,x_4,x_5,x_6) \\ \tau_{3,x}(0,0,1,x_4,x_5,x_6) &= (0,0,1,x_4,x_5+y,x_6+z) \\ \tau_{3,x}(x_1,x_2,0,x_4,x_5,x_6) &= (x_1,x_2,0,x_4+1,x_5,x_6) \\ \tau_{3,x}(y,z,0,x_4,x_5,x_6) &= (y,z,0,x_4,x_5,x_6) \\ \tau_{3,x}(x_1,x_2,1,x_4,x_5,x_6) &= (x_1,x_2,1,x_4+1,x_5+y,x_6+z) \\ \tau_{3,x}(y,z,1,x_4,x_5,x_6) &= (y,z,1,x_4,x_5+y,x_6+z) \end{aligned}$$

Though tedious, it can be shown that $\tau_{1,s}$, $\tau_2$ and $\tau_{3,x}$ are automorphisms of $X$. Let $A_1 = \{\tau_{1,s} : s \in S_{Z_2^2}\}$, $A_2 = \{\tau_2\}$ and $A_3 = \{\tau_{3,x} : x \in Z_2^2 \text{ and } x \neq (0,0)\}$. Note that $\tau_{1,s}$

fixes $(0,0,0,0,0,0)$ if and only if $s$ fixes $(0,0)$, $\tau_2$ fixes $(0,0,0,0,0,0)$, and every element of $A_3$ fixes $(0,0,0,0,0,0)$.

We now consider the orbits of $\mathrm{Aut}(X)_{(0,0,0,0,0,0)}$, that is, the stabilizer of $(0,0,0,0,0,0)$. Let

$$\Delta_1 = \{(0,0,0,0,x_5,x_6) : (x_5,x_6) \neq (0,0)\}.$$

The three vertices of $\Delta_1$ are in the same orbit under the action of automorphisms in $A_1$.

Let

$$\Delta_2 = \{(0,0,1,1,x_5,x_6) : (x_5,x_6) \in Z_2^2\}.$$

These four vertices are in the same orbit. Automorphisms from $A_1$ work when $(x_5,x_6) \neq (0,0)$, and $\tau_{3,x}$ maps $(0,0,1,1,0,0)$ to $(0,0,1,1,1,0)$ for $x = (1,0)$.

Let

$$\Delta_3 = \{(0,0,x_3,x_4,x_5,x_6) : x_3 \neq x_4\}.$$

These eight vertices are in the same orbit using automorphisms from $A_2$ and $A_3$.

Let $\Delta_4$ be the twenty-four vertices adjacent to $(0,0,0,0,0,0)$ arising from $S_2$ and $S_3$. They are in the same orbit using automorphisms of $A_1$, $A_2$ and $A_3$.

Let $\Delta_5$ be the remaining twenty-four vertices. They are in the same orbit using automorphisms of $A_1$, $A_2$ and $A_3$.

There are twenty paths of length 2 from $(0,0,0,0,0,0)$ to any vertex of $A_2$, while there are sixteen paths of length 2 from $(0,0,0,0,0,0)$ to any vertex of $\Delta_1 \cup \Delta_5$. Vertices of $\Delta_1$ have no neighbors in $\Delta_2$, whereas, vertices of $\Delta_5$ have two neighbors in $\Delta_2$. So $\Delta_1$, $\Delta_2$ and $\Delta_5$ are the three orbits of $\mathrm{Aut}(X)_{(0,0,0,0,0,0)}$ comprising the non-neighbors of $(0,0,0,0,0,0)$.

Vertices of $\Delta_3$ are adjacent to everything in $\Delta_2$, and vertices of $\Delta_4$ are adjacent to two vertices of $\Delta_2$. Hence, $\Delta_0 = \{(0,0,0,0,0,0)\}, \Delta_1, \Delta_2, \Delta_3, \Delta_4$ and $\Delta_5$ are the orbits of $\mathrm{Aut}(X)_{(0,0,0,0,0,0)}$.

The $\delta$ appearing in the next function definition is the Kronecker delta. For $\mathbf{y} = (y_1, y_2, y_3, y_4, y_5, y_6) \in Z_2^6$, define $f_\mathbf{y}$ by

$$f_\mathbf{y}(x_1,x_2,x_3,x_4,x_5,x_6) = (u_1,u_2,u_3,u_4,u_5,u_6),$$

where

$$\begin{aligned}
u_1 &= x_1 + y_1, \\
u_2 &= x_2 + y_2, \\
u_3 &= x_3 + y_3, \\
u_4 &= x_4 + y_4 + \delta((x_1,x_2),(1,1)) + \delta((x_1+y_1,x_2+y_2),(1,1)), \\
u_5 &= x_5 + y_5 + \delta(y_3,1)(x_1+y_1) + \delta(x_3,1)y_1, \\
u_6 &= x_6 + y_6 + \delta(y_3,1)(x_2+y_2) + \delta(x_3,1)y_2).
\end{aligned}$$

It can be verified, tedious though it be, that $f_\mathbf{y} \in \mathrm{Aut}(X)$ and that the mapping $y \mapsto f_\mathbf{y}$ is a homomorphism from $Z_2^6$ to $\tilde{G} = \{f_\mathbf{y} : \mathbf{y} \in Z_2^6\}$.

Suppose that $f_\mathbf{y}(x_1,x_2,x_3,x_4,x_5,x_6) = (x_1,x_2,x_3,x_4,x_5,x_6)$ for some $(x_1,\ldots,x_6) \in Z_2^6$. This immediately forces $y_1 = y_2 = y_3 = 0$. Also, $y_4 = 0$ must hold because $\delta((x_1,x_2),(1,1)) = \delta((x_1+y_1,x_2+y_2),(1,1))$. It is then easy to see that $y_5 = y_6 = 0$ is true as well. This implies that $f_{(0,0,0,0,0,0)}$ is the only automorphism in $\tilde{G}$ having fixed points. Therefore, we conclude that $\tilde{G}$ is a regular group which is isomorphic to $Z_2^6$.

It is clear that $(Z_2^6)_L$ and $\tilde{G}$ are conjugate in $S_{Z_2^6}$. Thus, it suffices to prove that $(Z_2^6)_L$ and $\tilde{G}$ are not conjugate in $\mathrm{Aut}(X)$.

If $(Z_2^6)_L$ and $\tilde{G}$ are conjugate in $\mathrm{Aut}(X)$, then they are conjugate via an element in $\mathrm{Aut}(X)_{(0,0,0,0,0,0)}$. We complete the proof by showing there is no element of $(Z_2^6)_L$ conjugate to the particular element $f_{(0,0,1,0,0,0)}$.

Note that

$$f_{(0,0,1,0,0,0)}(x_1,x_2,x_3,x_4,x_5,x_6) = (x_1,x_2,x_3+1,x_4,x_5+x_1,x_6+x_2).$$

This implies that $f_{(0,0,1,0,0,0)}$ has the transposition

$$(0,0,0,0,0,0) \leftrightarrow (0,0,1,0,0,0)$$

in its disjoint cycle decomposition. Thus, if $\pi \in \mathrm{Aut}(X)_{(0,0,0,0,0,0)}$ conjugates $(z_1,z_2,z_3,z_4,z_5,z_6)_L$ to $f_{(0,0,0,0,0,0)}$, $\pi$ must map $(z_1,z_2,\ldots,z_6)$ to $(0,0,1,0,0,0)$ as $(Z_1,z_2,\ldots,z_6)_L$ has the transposition

$$(0,0,0,0,0,0) \leftrightarrow (z_1,z_2,z_3,z_4,z_5,z_6).$$

Since $(0,0,1,0,0,0) \in \Delta_3$, $(z_1,z_2,\ldots,z_6)$ must be in $\Delta_3$ too. In particular, $z_1 = z_2 = 0$.

We know there are eight possibilities for $(0,0,z_3,z_4,z_5,z_6)_L$. We can cut this in half by observing that

$$\tau_2(x_1,x_2,x_3,x_4,x_5,x_6)_L\tau^{-1} = (x_1,x_2,x_4,x_3,x_5,x_6)_L.$$

This implies that $\tau_2(Z_2^6)_L\tau^{-1} = (Z_2^6)_L$ and so we may assume $(z_1,z_2,\ldots,z_6)_L$ has the form $(0,0,1,0,z_5,z_6)_L$.

Let

$$V_{x_1,x_2,x_4} = \{(x_1,x_2,\ldots,x_6) : x_3,x_5,x_6 \in \{0,1\}\}.$$

It is easy to see that these sets form blocks of imprimitivity for both $(Z_2^6)_L$ and $\tilde{G}$. Consider the cycle decomposition of $(0,0,1,0,z_5,z_6)_L$. There are four transpositions in eacg block of imprimitivity. For example,

$$\begin{aligned}
(0,0,0,0,0,0) &\leftrightarrow (0,0,1,0,z_5,z_6)\\
(0,0,1,0,0,0) &\leftrightarrow (0,0,0,0,z_5,z_6)\\
(0,0,0,0,1,1) &\leftrightarrow (0,0,1,0,z_5+1,z_6+1)\\
(0,0,1,0,1,1) &\leftrightarrow (0,0,0,0,z_5+1,z_6+1)
\end{aligned}$$

are the transpositions for $V_{0,0,0}$.

There are two types of adjacencies between vertices of the blocks of imprimitivity. Either each vertex in one block is adjacent to one vertex in each of the four transpositions of another block, or each vertex in one block is adjacent to both vertices in two of the transpositions of another block. Form a graph whose vertices are the eight blocks of imprimitivity. Make two blocks adjacent if the second possibility occurs. For each $(0,0,1,0,z_5,z_6)_L$, the graph obtained consists of two vertex-disjoint complete graphs of order 4. If we do the same for the transpositions determined by $f_{(0,0,1,0,0,0)}$, we obtain a connected graph. Therefore, $\tilde{G}$ and $(Z_2^6)_L$ cannot be conjugate in $\mathrm{Aut}(X)$. □

When $p > 3$ is a prime, we saw that $Z_{p^2}$ is not CI, whereas, $Z_p^2$ is CI. This means there are graphs which are Cayley graphs on $Z_{p^2}$ and are not CI with respect to $Z_{p^2}$, but simultaneously are Cayley graphs on $Z_p^2$ and are CI with respect to $Z_p^2$. This adds interest to the problem of determining which graphs are Cayley on both $Z_{p^2}$ and $Z_p^2$.

**2.12 Definition** A finite group $G$ is said to be a *Burnside group* provided that any permutation group $H$ containing a regular subgroup isomorphic to $G$ is imprimitive whenever it is not doubly transitive.

**2.13 Definition** Let $X$ and $Y$ be graphs of orders $m$ and $n$, respectively. Define the *wreath product* $X \wr Y$ of $X$ and $Y$ as follows: $V(X \wr Y) = V(X) \times V(Y)$ and $(x, y)$ is adjacent to $(u, v)$ if and only if either $x$ is adjacent to $u$ in $X$, or $x = u$ and $y$ is adjacent to $v$ in $Y$.

**2.14 Theorem (Joseph [16])** *A graph $X$ of order $p^2$, $p$ a prime, is a Cayley graph on both $Z_{p^2}$ and $Z_p \times Z_p$ if and only if $X$ is a wreath product of two circulants of order $p$.*

**Proof** Assume that $X \cong X(p; S_1) \wr X(p; S_2)$. Then $X \cong (p^2; (S_1 + \overline{P} \cup pS_2)$, where $\overline{P} = \{0, p, 2p, \ldots, (p-1)p\}$, $A + B = \{a + b : a \in A, b \in B\}$, and $pS_2 = \{pi : i \in S_2\}$. Also,

$$X \cong X(Z_p^2; \{(0, j) : j \in S_2\} \cup \{(i, j) : i \in S_1 \text{ and } 0 \leq j \leq p-1\}).$$

So $X$ is a Cayley graph on both $Z_{p^2}$ and $Z_p^2$.

Assume that $X$ is a Cayley graph on both $Z_p^2$ and $Z_{p^2}$. If $X = K_{p^2}$ or $X = \overline{K}_{p^2}$, then it is certainly a wreath product of two circulants of order $p$. Therefore, we may assume $\mathrm{Aut}(X)$ is not doubly transitive. Since $Z_{p^2}$ is a Burnside group [29], $\mathrm{Aut}(X)$ is imprimitive. The blocks of imprimitivity must have length $p$ and are uniquely determined because $\rho = (0\ 1\ \cdots p^2-1) \in \mathrm{Aut}(X)$. Let $P$ be a Sylow $p$-subgroup of $\mathrm{Aut}(X)$. Since there are isomorphs of both $Z_p^2$ and $Z_{p^2}$ in $P$, $|P| > p^2$. Thus, the stabilizer $P_0$ of 0 contains a non-trivial automorphism whose order is $p$. Using the same technique employed in the proof of Theorem 2.9, we can see that either there are no edges between two blocks or all possible edges between two blocks. The subgraphs induced on the blocks are circulants because of the action of $\rho^p$ and isomorphic because of the action of $\rho$. Therefore, $X$ is isomorphic to a wreath product of two circulant graphs of order $p$. □

Let $\mathrm{Cay}(G)$ denote the set of Cayley graphs on the group $G$.

**2.15 Corollary** *The graph $X \in \mathrm{Cay}(Z_{p^2}) \cap \mathrm{Cay}(Z_p^2)$ if and only if the Sylow $p$-subgroups of $\mathrm{Aut}(X)$ have order $p^{p+1}$. Furthermore, the Sylow $p$-subgroups of $\mathrm{Aut}(X)$ have order $p^2$ when $X \in \mathrm{Cay}(Z_{p^2}) \Delta \mathrm{Cay}(Z_p^2)$.*

**Proof** This follows easily from the fact that the Sylow $p$-subgroups of the automorphism group of the wreath product of two circulant graphs of order $p$ have order $p^{p+1}$, and from the proof of the preceding result. □

J. Morris [20] has generalized Theorem 2.14. We shall not give the proof here as it is similar to that of Theorem 2.14 with additional technical details needing to be addressed.

**2.16 Theorem** *Let $p$ be an odd prime, and $k_1, k_2, \ldots, k_m$ a partition of $n$. The following two statements are equivalent.*

(1) *The graph $X$ is a Cayley graph on $Z_{p^{k_1}} \times \cdots \times Z_{p^{k_m}}$ and on $Z_{p^n}$.*

(2) *There is a refinement $i_1, \ldots, i_r$ of $k_1, \ldots, k_m$ and circulant graphs $X_j$ of order $p^{i_j}$, $j = 1, \ldots, r$, such that $X \cong X_1 \wr X_2 \wr \cdots \wr X_r$.*

Note that the preceding theorem tells us that $X$ is a Cayley graph on every abelian group of order $p^n$ of the form $Z_{p^{\ell_1}} \times \cdots \times Z_{p^{\ell_t}}$, where $i_1, \ldots, i_r$ is a refinement of $\ell_1, \ldots, \ell_t$.

## 3 Degree restrictions

There are many groups that are not CI-groups. It is natural to ask if there is some way of measuring how badly a group is not CI. For example, there may be groups such that almost all Cayley subsets are CI. Another possibility is that some restriction on the degree may render all Cayley subsets satisfying the restriction CI. The latter situation is what we investigate at this time.

**3.1 Definition** A group $G$ is called an *$m$-CI-group* if, for every Cayley subset of $G$ cardinality at most $m$, $X(G;S) \cong X(G;T)$ implies $\sigma(S) = T$ for some $\sigma \in \mathrm{Aut}(G)$.

Note that a group is CI if it is $m$-CI for $m = |G| - 1$.

**3.2 Definition** A finite $p$-group $G$ is said to be *homocyclic* if it can be expressed as a direct product of cyclic $p$-groups of the same order.

X.-G. Fan and M.-Y. Xu [11] have proved the following result.

**3.3 Theorem** *A finite abelian group $G$ is 1-CI if and only if the Sylow 2-subgroup is homocyclic. The group $G$ is $m$-CI, $2 \leq m \leq 4$, if and only if every Sylow subgroup is homocyclic and, in addition, the Sylow 2-subgroup is cyclic or elementary abelian.*

An immediate corollary of the preceding result settles a conjecture of F. Boesch and R. Tindell [6]. It was first proved in [8].

**3.4 Corollary** *The cyclic group $Z_n$ is 4-CI.*

## References

[1] A. Ádám, Research Problem 2-10, *J. Combin. Theory* **2** (1967), 393.

[2] B. Alspach, Point-symmetric graphs and digraphs of prime order and transitive permutation groups of degree prime degree, *J. Combin. Theory* **15** (1973), 12–17.

[3] L. Babai, Isomorphism problem for a class of point-symmetric structures, *Acta Math. Acad. Sci. Hungar.* **29** (1977), 329–336.

[4] L. Babai and P. Frankl, Isomorphisms of Cayley graphs I, in: *Combinatorics (Proc. Fifth Hungarian Colloq., Keszthely 1976)*, Colloq. Math. Soc. János Bolyai **18**, North-Holland, Amsterdam, 1978, 35–52.

[5] J. L. Berggren, An algebraic characterization of symmetric graphs with $p$ points, $p$ and odd prime, *Bull. Austral. Math. Soc.* **7** (1972), 131–134.

[6] F. Boesch and R. Tindell, Circulants and their connectivities, *J. Graph Theory* **8** (1984), 487–499.

[7] N. G. de Bruijn, Pólya's theory of counting, in: *Applied Combinatorial Mathematics*, Wiley, New York, 1964.

[8] C. Delorme, O. Favaron and M. Maheo, Isomorphisms of Cayley multigraphs of degree 4 on finite abelian groups, *European J. Combin.* **13** (1992), 59–61.

[9] E. Dobson, Isomorphism problem for Cayley graphs of $Z_p^3$, *Discrete Math.* **147** (1995), 87–94.

[10] B. Elspas and J. Turner, Graphs with circulant adjacency matrices, *J. Combin. Theory* **9** (1970), 297–307.

[11] X.-G. Fan and M.-Y. Xu, On isomorphisms of Cayley graphs of small valency, *Algebra Colloq.* **1** (1994), 67–76.

[12] R. Frucht, Herstellung von Graphen mit vorgegebener abstrakter Gruppe, *Compositio Math.* **6** (1938), 239–250.

[13] R. Frucht, Graphs of degree three with a given abstract group, *Canad. J. Math.* **1** (1949), 365–378.

[14] C. D. Godsil, On Cayley graph isomorphisms, *Ars Combin.* **15** (1983), 231–246.

[15] H. Izbicki, Unendliche Graphen endlichen Grades mit vorgegebenen Eigenschaften, *Monatsh. Math.* **63** (1959), 298–301.

[16] A. Joseph, The isomorphism problem for Cayley digraphs on groups of prime-squared order, *Discrete Math.* **141** (1995), 173–183.

[17] D. König, *Theorie der endlichen und unendlichen Graphen*, Akad. Verlagsgesellschaft, Leipzig 1936; reprinted by Chelsea, New York, 1950.

[18] C.-H. Li, Isomorphisms and classification of Cayley graphs of small valencies on finite abelian groups, *Austral. J. Combin.* **12** (1995), 3-14.

[19] J. Lützen, G. Sabidussi and B. Toft, Julius Petersen 1839-1910: A biography, *Discrete Math.* **100** (1992), 9-82.

[20] J. Morris, Isomorphic Cayley graphs on nonisomorphic groups, preprint, Simon Fraser University 1996.

[21] M. Muzychuk, Ádám's conjecture is true in the square-free case, *J. Combin. Theory Ser. A* **72** (1995), 118–134.

[22] M. Muzychuk, On Ádám's conjecture for circulant graphs, *Discrete Math.*, to appear.

[23] L. Nowitz, A non-Cayley-invariant Cayley graph of the elementary Abelian group of order 64, *Discrete Math.* **110** (1992), 223–228.

[24] P. P. Pálfy, Isomorphism problem for relational structures with a cyclic automorphism, *European J. Combin.* **8** (1987), 35–43.

[25] D. S. Passman, *Permutation Groups*, Benjamin, Menlo Park, CA, 1968.

[26] G. Sabidussi, Graphs with given group and given graph-theoretical properties, *Canad. J. Math.* **9** (1957), 515–525.

[27] S. Toida, A note on Ádám's conjecture, *J. Combin. Theory Ser. B* **23** (1977), 239–246.

[28] J. Turner, Point-symmetric graphs with a prime number of points, *J. Combin. Theory* **3** (1967), 136–145.

[29] H. Wielandt, *Finite Permutation Groups*, Academic Press, New York, 1964.

[30] M.-Y. Xu, On isomorphisms of Cayley digraphs and graphs of groups of order $p^3$, *Adv. in Math. (China)* **17** (1988), 427–428.

# Oligomorphic groups and homogeneous graphs

Peter J. CAMERON

*School of Mathematical Sciences*
*Queen Mary and Westfield College*
*Mile End Road*
*London E1 4NS*
*U.K.*

**Abstract**

These notes are, in a sense, an elaboration on the comments on pages 232–234 of the book [10] on *Distance-regular Graphs* by Brouwer, Cohen and Neumaier. Apart from these brief comments, the book is exclusively concerned with finite graphs. Many of these finite graphs are geometrically defined, and involve a finite field and sometimes a finite dimension. By allowing one or both of these parameters to be infinite, we often obtain infinite distance-transitive graphs. However, entirely new phenomena occur in the infinite case, which have no finite analogues.

In these notes, I discuss infinite graphs (and other structures) with a large amount of symmetry. The discussion concentrates on the purely infinite phenomena, though we glance briefly at the infinite versions of finite graphs. The celebrated 'random graph' is typical of the new situation; the first section summarises some of the remarkable properties of this graph. Following this, we survey some basic material from permutation groups and model theory. Then we discuss various constructions and characterisations of infinite highly symmetric graphs, and connections with several topics in finite combinatorics, including random graphs, enumeration, and graph reconstruction.

## 1 The random graph

The point of this section is to lead you to believe that there is only one interesting countably infinite graph!

### 1.1 Introduction

An infinite set $S$ has the property that, if it is partitioned into two parts, then one of the parts has the same cardinality as $S$. (This assertion requires the Axiom of Choice in its proof; but it holds for countably infinite sets without any extra assumption.) A graph theorist might ask: are there infinite graphs with this property? (We should replace 'has the same cardinality as' by 'is isomorphic to' to provide the natural analogue.) Of course, any infinite complete or null graph satisfies the condition; but this is just the set-theoretic assertion in light disguise. Are there any non-trivial graphs? We restrict our attention to the countably infinite case.

**Theorem 1.1** *There is a unique countable graph $X$ (up to isomorphism), other than the complete and null graphs, with the property that, if the vertex set of $X$ is partitioned into two parts, then the induced subgraph on one of the parts is isomorphic to $X$.*



*G. Hahn and G. Sabidussi (eds.), Graph Symmetry*, 23–74.

This unique graph and its properties form the topic of the first chapter of these notes.

**Proof** We consider the following graph-theoretic property.

($*$) If $U$ and $V$ are finite disjoint sets of vertices, then there exists a vertex $z$ joined to every vertex in $U$ and to no vertex in $V$.

The strategy of proof of Theorem 1.1 is as follows. First we show that any two countable graphs satisfying ($*$) are isomorphic. Then we show that the condition of the Theorem is equivalent to ($*$). Finally, we give the dramatic non-constructive existence proof of Erdős and Rényi [32] for such a graph.

*Step 1* We begin with a translation of ($*$) into a form which will later be easily generalised to structures other than graphs.

($\dagger$) Let $A$ and $B$ be finite graphs with $A$ the induced subgraph of $B$ obtained by deleting one vertex. Then any embedding of $A$ as induced subgraph of $X$ extends to an embedding of $B$.

As a warmup, we show:

**Proposition 1.2** *Let $X$ be a graph satisfying ($\dagger$). Then any finite or countable graph is embeddable as an induced subgraph of $X$.*

We say that $X$ is *universal* (for finite and countable graphs) if it satisfies the conclusion of this result.

**Proof** Let $Y$ be a finite or countable graph. Enumerate its vertices as $y_1, y_2, \ldots$, and let $Y_n$ be the induced subgraph on $\{1, \ldots, n\}$. Now we inductively embed $Y_n$ into $X$, using ($\dagger$) to extend the embedding from $Y_n$ to $Y_{n+1}$. The end result is an embedding of $Y$.

To complete Step 1, we use the model-theoretic trick known as *back-and-forth*. Let $X$ and $Y$ be two countable graphs satisfying ($\dagger$), and enumerate their vertices as $x_1, x_2, \ldots$ and $y_1, y_2, \ldots$. We construct an isomorphism $\theta$ from $X$ to $Y$ in stages as follows. We begin with $\theta_0$ the empty map.

Let $n$ be odd, and assume we have a partial isomorphism $\theta_{n-1}$ from $X$ to $Y$ (that is, an isomorphism of induced subgraphs). Let $A$ be the domain of $\theta_{n-1}$. Choose the first vertex in the enumeration of $X$ which is not in $A$, say $x_r$. By ($\dagger$), we can extend $\theta_{n-1} : A \to Y$ to $\theta_n : A \cup \{x_r\} \to Y$.

If $n$ is even, we go back from $Y$ to $X$. Choose the first vertex of $Y$ which is not in the range of $\theta_{n-1}$, say $y_s$, and extend $\theta_{n-1}^{-1}$ so that its domain contains $y_s$.

Let $\theta$ be the union of the partial maps $\theta_n$. The construction ensures that every vertex of $X$ is eventually put into the domain of $\theta$, and every vertex of $Y$ into the range. So $\theta$ is an isomorphism from $X$ to $Y$.

*Step 2* Suppose that the vertex set of $X$ can be partitioned into two parts $X_1$ and $X_2$, neither of which has property ($*$). Thus, for $i = 1, 2$, there are disjoint finite subsets $U_i$ and $V_i$ of $X_i$ for which no vertex $z \in X_i$ satisfies the conclusion of ($*$). Now set $U = U_1 \cup U_2$ and $V = V_1 \cup V_2$. No vertex in either part now satisfies the conclusion of ($*$) relative to these two sets.

Conversely, suppose that $X$ is not complete or null, and that for any partition of the vertices into two parts, one of the parts induces a subgraph isomorphic to $X$. Since any graph can be partitioned into a null graph and a graph with no isolated vertices, we see that $X$ has no isolated vertices; and the same is true of its complement.

Assume that $X$ does not satisfy Condition $(*)$, and let $U$ and $V$ be finite disjoint sets of vertices for which no choice $z$ witnesses $(*)$, and with $m = |U| + |V|$ chosen minimal subject to this. Clearly $m > 1$: for $m = 1$, $(*)$ just says that no vertex is isolated in $X$ or its complement. Replacing the graph by its complement if necessary (noting that both hypothesis and conclusion are unaffected), we may assume that $U \neq \emptyset$.

Choose $u \in U$. Let $A$ consist of $u$ together with its non-neighbours other than those in $U \cup V$, and let $B$ consist of the remaining vertices. Then $u$ is isolated in the induced subgraph on $A$, so this subgraph is not isomorphic to $B$. So the induced subgraph on $B$ is isomorphic to $X$. Hence, by minimality of $m$, $B$ contains a vertex $z$ joined to every vertex of $U \setminus \{u\}$ and no vertex of $V$. But $z$ is also joined to $u$, by construction. So $(U, V)$ was not after all a counterexample to $(*)$.

*Step 3* Following Erdős and Rényi, we take a countable set $S$ of vertices and choose a random graph on $S$ as follows: list all the pairs of vertices, and choose for each pair, independently with probability $1/2$, whether to join the two vertices with an edge. The claim is that condition $(*)$ holds with probability 1 (that is, fails with probability 0) in this random graph.

There are only countably many choices of disjoint finite sets $U$ and $V$ in $S$. Since the union of countably many null sets is a null set, it is enough to prove that, for fixed $U$ and $V$, the probability that no suitable $z$ exists is zero. Enumerate the points outside $U \cup V$ as $z_1, z_2, \ldots$. If $|U \cup V| = m$, the probability that each $z_n$ is not correctly joined is $(1 - 1/2^m)$. Since these events (for different $n$) are independent, the probability that no $z_n$ is suitable is

$$\lim_{n \to \infty} (1 - 1/2^m)^n = 0,$$

as required.

If something happens with probability 1, then it certainly happens; so there exists a countable graph which satisfies condition $(*)$. Moreover, by Step 1, this graph is unique up to isomorphism. We call it the *random graph*, and denote it by $R$.

If a non-constructive existence proof of such an important graph is regarded as unsatisfactory, here is an explicit construction of it, due to Rado [60]. The vertices of Rado's graph are the natural numbers (including zero). Suppose that $x < y$. To decide whether to join $x$ to $y$, we write $y$ to the base 2, that is, we express $y$ as the sum of distinct powers of 2. Then join $x$ to $y$ if and only if $2^x$ is one of these powers (that is, if $y$ in base 2 has a 1 in position $x$). The joining rule is asymmetric, but we regard the graph as undirected.

Let $U$ and $V$ be finite disjoint sets of natural numbers. By enlarging $U$ if necessary, we can assume that $\max(U) > \max(V)$. Let $z = \sum_{u \in U} 2^u$. By definition, $z$ is joined to every vertex in $U$. Since $z > v$ for all $v \in V$, we see that $z$ is joined to no vertex in $V$.

Rado's construction is independent of the existence proof of Erdős and Rényi, so it is reasonable that $R$ stands for 'Rado' as well as for 'random'. (Rado was interested in the universal property of the graph.)

The following property will be crucial for us. A graph $X$ is said to be *homogeneous* if any isomorphism between finite subgraphs of $X$ can be extended to an automorphism of $X$.

**Theorem 1.3** *R is homogeneous.*

**Proof** The proof is by back-and-forth. Instead of starting with the empty isomorphism between two different graphs satisfying $(*)$, as we did before, start with the given finite partial automorphism from $R$ to itself. Then the back-and-forth procedure extends this to an isomorphism from $R$ to $R$, that is, an automorphism of $R$.

In particular, $R$ is vertex-transitive, edge-transitive, ...

I conclude this section with two problems.

**Problem 1** Which countable graphs $X$ have the property that, whenever the vertices are partitioned into two subsets $A$ and $B$, at least one of $A$ and $B$ contains an induced subgraph isomorphic to $X$?

**Problem 2** Which countable graphs $X$ have the property that $X - v \cong X$ for all vertices $v$? (This is the special case where one part of the partition is a singleton. The case where one part is finite follows.)

## 1.2 Further constructions

As we noted above, any countable graph with property $(*)$ is isomorphic to the random graph. Using this, we give several constructions of $R$ taken from various fields.

*Set theory* From a logical point of view, set theory is just the theory of the membership relation $\in$, a binary relation which satisfies the Zermelo–Fraenkel axioms. Now the downward Löwenheim–Skolem theorem of first-order logic (see Section 2.1) asserts that, provided that set theory is consistent, then there is a countable model of it. This assertion is known as the *Skolem paradox*, since there are theorems of set theory asserting the existence of uncountable sets. The resolution is that a model $M$ of set theory may be countable 'in the real world', but no set within the model is a bijective function between the entire model and the set of natural numbers (in $M$).

The membership relation is asymmetric: if we had $x \in y$ and $y \in x$, then $\ldots \in x \in y \in x$ would be an infinite descending chain, contradicting the Axiom of Foundation. So we can regard membership as an oriented graph on the vertex set $M$. Now, if $M$ is countable, then the underlying undirected graph is necessarily isomorphic to $R$, no matter which countable model of set theory is chosen. So first-order set theory is just the theory of certain special orientations of $R$.

To prove that $(*)$ holds, let $U$ and $V$ be finite disjoint sets of sets, and take $z = U \cup \{V\}$. Certainly all members of $U$ are in $z$. Suppose some $v \in V$ is joined to $z$. Then either $v \in z$, whence $v \in U$ or $v = V$; or $z \in v$. The first contradicts our disjointness assumption, while the second and third lead to violations of the Axiom of Foundation (by $v \in v$ or $z \in v \in V \in z$, respectively).

This example is related to Rado's construction from the last section. We have not used all of the Zermelo–Fraenkel axioms in the above discussion, merely the Empty Set, Pairing and Union axioms (to build finite sets) and the Axiom of Foundation. Now a standard model for the theory of finite sets (set theory with the negation of the Axiom of Infinity) takes the elements to be the natural numbers, with $x \in y$ whenever $2^x$ occurs in the base 2 expansion of $y$. The undirected version of this gives Rado's construction.

*Number theory* One of the neatest constructions is as follows. Let $\mathbb{P}_1$ be the set of prime numbers congruent to 1 mod 4. According to the law of Quadratic Reciprocity, if $p, q \in \mathbb{P}_1$, then $p$ is a quadratic residue mod $q$ if and only if $q$ is a quadratic residue mod $p$. We join $p$ to $q$ whenever this occurs. Now the resulting graph satisfies $(*)$, and so is isomorphic to $R$. For let $U$ and $V$ be finite disjoint subsets of $\mathbb{P}_1$. Choose a quadratic residue $a_i$ mod $u_i$ for each $u_i \in U$ (for example, $a_i = 1$), and a quadratic non-residue $b_j$ mod $v_j$ for each $v_j \in V$. Consider the congruences

$$\begin{aligned} z &\equiv a_i \pmod{u_i} \text{ for all } u_i \in U, \\ z &\equiv b_j \pmod{v_j} \text{ for all } v_j \in V, \\ z &\equiv 1 \pmod{4}. \end{aligned}$$

There is a unique solution $z$ mod $N$, where $N = 4\prod u_i \prod v_j$, by the Chinese Remainder Theorem. Since $(z, N) = 1$, Dirichlet's Theorem guarantees that there is a prime $z$ in this residue class. Now $z$ is the vertex required by $(*)$.

If instead we use $\mathbb{P}_{-1}$, the set of primes congruent to $-1$ mod 4, then $p$ is a quadratic residue mod $q$ if and only if $q$ is a quadratic non-residue mod $p$. So a similar construction gives a tournament, which turns out to be the unique countable random tournament (see Section 3.2). Moreover, the bipartite graph defined on the bipartite blocks $\mathbb{P}_1$ and $\mathbb{P}_{-1}$ in the analogous way is the unique countable random bipartite graph (see Section 3.5).

*Probability* We already saw the connection between $R$ and probability theory: for a natural model of random countable graphs, almost all graphs are isomorphic to $R$. There are many similar constructions. The one given now has the side-effect of giving information about automorphisms of $R$.

There is a natural bijection between subsets $S$ of the set $\mathbb{N}$ of positive integers and countable zero-one sequences $s = s_1 s_2 \ldots$. (To the set $S$ corresponds its *characteristic function*, the sequence $s$ with $s_n = 1$ if and only if $n \in S$.)

Now, to each set $S$, we construct a graph $X(S)$ as follows. The vertices of $X(S)$ are the integers, and $x \sim y$ in $X(S)$ if and only if $|x - y| \in S$. Note that the map $x \mapsto x + 1$ (the *cyclic shift*) is an automorphism of $X(S)$, and that every graph on $\mathbb{Z}$ which admits the cyclic shift as an automorphism is of the form $X(S)$ for some set $S$ (in fact, $S = \{x > 0 : x \sim 0\}$).

Suppose that $X$ is a countable graph which has two cyclic automorphisms $\sigma_1$ and $\sigma_2$, so can be represented in two different ways as $X(S)$, say $X(S_1)$ and $X(S_2)$. Then $S_1 = S_2$ if and only if $\sigma_1$ and $\sigma_2$ are conjugate in the automorphism group of $X$. Hence, if $S_1 \neq S_2$ but $X(S_1) \cong X(S_2)$, then the corresponding cyclic shifts are non-conjugate automorphisms of this graph.

An infinite zero-one sequence is *universal* if every finite zero-one sequence occurs as a consecutive subsequence. Now $X(S)$ is isomorphic to the random graph if and only if the characteristic function of $S$ is universal. Moreover, if the entries in the sequence record the result of independent coin tosses, then the sequence is almost surely universal. So almost every sequence produces $R$ by this construction. Also, an event with probability 1 is uncountable (indeed, of cardinality $2^{\aleph_0}$). Combining all this, we have the result:

**Theorem 1.4** *$R$ admits $2^{\aleph_0}$ pairwise non-conjugate cyclic automorphisms.*

It is easy to give an explicit construction of $R$ based on these ideas: a universal sequence can be found, for example, by enumerating all finite zero-one sequences (for example, by length, and lexicographically for fixed length), and concatenating them.

## 1.3 Properties

The random graph has many striking properties. This section makes no pretence at completeness. I have concentrated on two aspects: 'indestructibility', and representation of $R$ as a Cayley graph. For much more, see [18].

Many of these properties depend on the observation that, given finite disjoint sets $U$ and $V$ of vertices, the set of all vertices $z$ as in $(*)$ is infinite. (Given any finite number of such vertices, say $z_1, \ldots, z_m$, replace $U$ by $U \cup \{z_1, \ldots, z_m\}$, and use $(*)$ to find yet another $z$.) In fact, it is easy to show that the set of all such $z$ induces a subgraph isomorphic to $R$. (If $Z$ is the set of such vertices, and $U'$ and $V'$ are finite disjoint subsets of $Z$, apply $(*)$ to the sets $U \cup U'$ and $V \cup V'$ to find $z \in Z$ joined to all vertices of $U'$ and to none in $V'$. So $Z$ satisfies $(*)$.)

**Theorem 1.5** *Each of the following operations, applied to $R$, yields a graph isomorphic to $R$:*

(a) *removing finitely many vertices;*

(b) *adding or removing a finite number of edges;*

(c) *interchanging adjacency and non-adjacency between a finite set of vertices and the rest of the graph;*

(d) *complementing the graph.*

The property of being unchanged by the addition or removal of finitely many edges does not characterise $R$. The disjoint union of countably many copies of all finite connected graphs also has this property.

It would be interesting to study the extent to which graphs with these properties can be characterised.

We now turn to the representation of $R$ as a Cayley graph. Let $G$ be a group, $S$ a subset of $G$ which contains the inverse of each of its elements but not the identity. The *Cayley graph* $\mathrm{Cay}(G, S)$ has vertex set $G$, and edge set $\{\{g, sg\} : g \in G, s \in S\}$. (The conditions on $S$ ensure that the graph is undirected and has no loops.) It is invariant under right multiplication by elements of $G$, and hence vertex-transitive. It is connected if and only if the subset $S$ generates $G$. Moreover, any graph on the vertex set $G$ which is invariant under right multiplication is a Cayley graph of $G$; and hence any graph which admits an action of $G$ regular (sharply transitive) on the vertices is isomorphic to a Cayley graph.

Now reconsider the graph $X(S)$ defined in Section 1.2. Clearly, this is a Cayley graph for the additive group of integers; indeed, $X(S) = \mathrm{Cay}(\mathbb{Z}, S \cup (-S))$.

We can choose a random Cayley graph for $G$ as follows. Divide the non-identity elements into inverse 'pairs'. (Such a 'pair' has a single element $s$ if $s^2 = 1$.) Then include these pairs in $S$ independently with probability $1/2$. It is known that, for most finite groups $G$, the automorphism group of a random Cayley graph is precisely $G$ with high probability [4]. By contrast, we have the following result for countable groups.

**Theorem 1.6** *The following three conditions on the countable group $G$ are equivalent:*

(a) *some Cayley graph of $G$ is isomorphic to the random graph;*

(b) *with probability 1, a random Cayley graph of $G$ is isomorphic to the random graph;*

(c) *for any two finite disjoint subsets $U$ and $V$ of $G$, there exists $z \in G$ such that, for all $u \in U$, $v \in V$, we have $(zu^{-1})^2 \neq vu^{-1}$.*

**Proof** Since (b) clearly implies (a), we prove that (a) implies (c) and (c) implies (b).

Suppose that $\mathrm{Cay}(G, S) \cong R$, and let $U$ and $V$ be finite disjoint subsets of $G$. Choose $z$ according to property $(*)$. For any $u \in U$ and $v \in V$, if $(zu^{-1})^2 = vu^{-1}$, then $zu^{-1} = vz^{-1}$; so $\{z, u\}$ and $\{v, z\}$ are both edges or both non-edges (depending on whether $zu^{-1} \in S$), contrary to the choice of $z$. So (c) holds.

Conversely, suppose that (c) holds. There are infinitely many elements $z$ satisfying the condition of (c), since we can add finitely many such elements to $U$ and find another one. For any such $z$, it never occurs that $zu^{-1} = (zv^{-1})^{-1}$. So the probability that $z$ is not correctly joined in a random Cayley graph is at most $1-1/2^m$, where $m = |U \cup V|$. (The probability may be smaller, since it may happen that, say, $zu_1^{-1} = (zu_2^{-1})^{-1}$, in which case $\{z, u_1\}$ and $\{z, u_2\}$ are both edges or both non-edges.) Now, given finitely many such $z$, say $z_1, \ldots, z_n$, there are only finitely many further elements $z$ for which the joins of $z$ to $U \cup V$ are not independent of those of $z_1, \ldots, z_n$. For such dependence would require that $zx^{-1} = (z_i y^{-1})^{\pm 1}$ for some $i$ and some $x, y \in U \cup V$, and any such equation has a unique solution $z$. So we can construct an infinite sequence of $z$ for which these events are independent, and conclude that with probability 1 some $z$ is correctly joined. Since there are only countably many choices of $U$ and $V$, we see that $\mathrm{Cay}(G, S) \cong R$ with probability 1.

We say that the group $G$ is *R-genic* if the equivalent conditions of this theorem hold. Condition (c) is stated in purely group-theoretic terms.

¿From this theorem, we can deduce a sufficient condition for a group to be $R$-genic which is easier to check. The *square root set* $\sqrt{g}$ of the element $g \in G$ is the set $\{x \in G : x^2 = g\}$. It is *non-principal* if $g \neq 1$. A *translate* of a square root set is obtained by multiplying it on the right by a fixed group element. We note that the equation $(zu^{-1})^2 = vu^{-1}$ of Theorem 1.6 says simply that $z \in (\sqrt{vu^{-1}})u$. So we have the following corollary:

**Theorem 1.7** *Suppose that the countable group $G$ cannot be expressed as the union of finitely many translates of non-principal square root sets together with a finite set. Then $G$ is $R$-genic.*

If $G$ is abelian, then the square root set of the identity is a subgroup $G_2$ of $G$ (consisting of all elements with order 1 or 2), and any other square root set is empty or a coset of $G_2$. Hence, if $G_2$ has infinite index in $G$ (in particular, if $H$ has unbounded exponent), then $G$ is $R$-genic.

It is not known whether or not the sufficient condition of Theorem 1.7 is actually necessary for $R$-genicity of a countable group.

Some simple properties of $R$-genic groups are given in the next result.

**Proposition 1.8** *If $G$ is $R$-genic, then so is any subgroup of finite index in $G$, or any quotient of $G$ by a finite normal subgroup.*

**Proof** Let $G$ be a regular subgroup of $\mathrm{Aut}(R)$. If $H$ is a subgroup of finite index in $G$, then the induced subgroups of $R$ on the orbits of $R$ are all Cayley graphs for $H$, and at least one of them is isomorphic to $R$, by Theorem 1.1.

Suppose that $K$ is a normal subgroup of finite index in $G$. Let $\overline{R}$ be the quotient graph on the orbits of $K$: its vertices are the orbits (corresponding to the cosets of $K$ in $G$), and an edge joins two orbits if and only if they have representatives adjacent in $R$. Now let $\overline{U}$ and $\overline{V}$ be finite disjoint sets of vertices of $\overline{R}$, and let $U$, $V$ be the unions of the corresponding orbits. Choose $z$ adjacent to all vertices in $U$ and none in $V$. Then the orbit of $z$ is adjacent to all vertices of $\overline{U}$ and none of $\overline{V}$. So $\overline{R} \cong R$.

Theorem 1.6 shows that, if $G$ is any regular permutation group, either no $G$-invariant graph is isomorphic to $R$, or almost all are. Does the same statement hold for other permutation groups? Given any permutation group $G$ on a countable set, we choose a random $G$-invariant graph as follows: list the orbits of $G$ on unordered pairs of vertices, and choose independently with probability $1/2$ whether to join the pairs in an orbit or not.

We cannot expect an exact analogue. For example, let $G$ be the full automorphism group of $R$. Then $G$ has just two orbits on pairs, the edges and non-edges of $R$. So there are four equally likely outcomes: the null graph, the complete graph, $R$, or the complement of $R$ (which is of course isomorphic to $R$). Thus, the probability of obtaining a graph isomorphic to $R$ is $1/2$. So the appropriate question should be:

**Problem** Is it true that, for any permutation group $G$ of countable degree, either no $G$-invariant graph is isomorphic to $R$, or the random $G$-isomorphic graph is isomorphic to $R$ with positive probability?

An argument like the one above shows that this is true if $G$ has only finitely many orbits on 2-sets.

Another case where the solution is affirmative is when $G$ is a cyclic group. All the possible cycle structures of automorphisms of $R$ were determined by Truss [77]. In each case, it is possible to show that a random graph invariant under a permutation with this cycle structure is isomorphic to $R$ with positive probability. To take a specific example, suppose that a permutation $g$ has one fixed point and two infinite cycles. Then the fixed point is joined to all or no points in each cycle. We certainly don't obtain $R$ if it is joined to both, or to neither. On the other hand, it can be shown that if the fixed point is joined to one but not the other, then we almost surely obtain $R$. So the probability of getting $R$ is $1/2$.

# 2 Preliminaries

In this section, we take a quick tour through the background topics of permutation groups and model theory, give the basic results about homogeneous and countably categorical structures, and examine the age of a relational structure from the viewpoint of probability and Baire category.

## 2.1 Model theory

Model theory deals with the relations between the logical properties of a set of formulae and the mathematical properties of structures satisfying these formulae. In these notes, the only kind of logic used is 'first-order logic'.

The language in which formulas are expressed consists of symbols of the following types:

*Relation symbols, function symbols, constant symbols*: A relation symbol $R$ or function symbol $f$ comes equipped with an 'arity', the number of arguments it expects, which is a positive integer. A constant symbol of course requires no arguments. We can regard a constant symbol as a function symbol of arity zero. However, for technical reasons, we do not allow a relation symbol to have arity zero. The precise collection of symbols of these types varies according to the type of mathematics we are discussing. For graph theory, we might have a binary relation for adjacency. For group theory, we might have a binary function for the group operation, a unary function for inversion, and a constant for the identity. However, we always assume that the language contains a binary relation symbol for equality.

*Variables*: We simply require an infinite supply of these.

*Connectives*: We use some or all of the standard logical connectives: $\vee$ ('or'), $\wedge$ ('and'), $\neg$ ('not'), $\Rightarrow$ ('implies'), $\Leftrightarrow$ ('if and only if').

*Quantifiers*: $\forall$ ('for all'), $\exists$ ('there exists').

*Parentheses*: (, ) (to make formulae unambiguous).

Formulae (or 'well-formed formulae') are defined inductively, in two stages. First, a *term* is a variable, a constant symbol, or of the form $F(t_1, \ldots, t_n)$, where $F$ is an $n$-ary function symbol and $t_1, \ldots, t_n$ are terms. Then an *atomic formula* is of the form $R(t_1, \ldots, t_n)$, where $R$ is an $n$-ary relation symbol and $t_1, \ldots, t_n$ are terms; and a *formula* is either an atomic formula or of the form $(\phi \vee \psi)$, $(\phi \wedge \psi)$, $(\neg\phi)$, $(\phi \Rightarrow \psi)$, $(\phi \Leftrightarrow \psi)$, $(\forall x)\phi$, or $(\exists x)\phi$, where $\phi$ and $\psi$ are formulae and $x$ is a variable.

The occurrence of the variable $x$ in the quantifier and all its occurrences in the formula $\phi$ within a subformula $(\forall x)\phi$ or $(\exists x)\phi$ are said to be *bound*, and any other occurrences of a variable are *free*. A formula is a *sentence* if it has no free occurrences of any variables.

Given a first-order language $L$, a *structure* $M$ over $L$ consists of a set $X$ on which are defined relations, functions and constants corresponding to the symbols in the language, the relations and functions having the appropriate arity (so that an $n$-ary function maps $X^n$ to $X$, while an $n$-ary relation is just a subset of $X^n$), and the constants are just elements of $X$. By convention, the relation symbol $=$ is always interpreted by the relation of equality. A *valuation* is an assignment of values in $X$ to the variables of the language; it inductively assigns values in $X$ to the terms, and truth values ('true' or 'false') to the formulae. I will not spell out in detail the rules for this; it does exactly what you would expect. The truth value assigned to a formula is independent of the values assigned to variables not occurring free in the formula. In particular, a sentence is either true or false in $M$, independent of the valuation. We say $M \models \sigma$ (read '$M$ models $\sigma$') if the sentence $\sigma$ is true in $M$. If $\Sigma$ is a set of sentences, we say that $M$ *is a model for* $\Sigma$, and write $M \models \Sigma$, if $M \models \sigma$ for all $\sigma \in \Sigma$.

For example, in the language of graphs, the sentence

$$(\forall x)(\forall y)(((x = y) \vee (x \sim y)) \vee (\exists z)((x \sim z) \wedge (z \sim y)))$$

is true in a graph $M$ if and only if $M$ has diameter at most 2. Incidentally, we commonly abbreviate formulae without losing clarity: for example, the above sentence could be written as

$$(\forall x, y)((x = y) \vee (x \sim y) \vee (\exists z)(x \sim z \sim y)).$$

Thus, first-order sentences say something about structures. For a survey of what can and can't be said, see the chapter on graphs and logic in [6].

The third aspect of logic, after the syntax and semantics, is the process of deduction or inference. The simplest, Henkin-style deductive systems specify that certain formulae are *axioms*, and there are *rules of inference* by which formulae can be deduced from others. (The classic example is *modus ponens*, where $\psi$ is inferred from $\phi$ and $\phi \Rightarrow \psi$.) In a deductive system, a formula $\phi$ is a *theorem* if it is either an axiom, or a consequence of theorems using a rule of inference of the system. Alternatively, a *proof* is a finite sequence of formulae, each of which is either an axiom, or a consequence of earlier formulae using a rule of inference; and a *theorem* is the last line of a proof. More generally, we can consider deductions from a set $\Sigma$ of formulae, where formulae in $\Sigma$ can be quoted in proofs as if they were axioms. The set $\Sigma$ is called *consistent* if no contradiction can be deduced from it.

A formula is called *logically valid* if it is true in all valuations in all structures over the appropriate language. A deductive system is called *sound* if every theorem is logically valid, and *complete* if every logically valid formula is a theorem. The existence of a deductive system with the most desirable properties is given by the following result.

**Theorem 2.1 (Gödel–Henkin Completeness Theorem)** *There is a deductive system for first-order logic which is sound and complete.*

*Moreover, this system has the following properties:*

(a) *A sentence $\phi$ can be deduced from a set $\Sigma$ of sentences if and only if every structure satisfying $\Sigma$ also satisfies $\phi$.*

(b) *A set $\Sigma$ has a model if and only if it is consistent.*

From this theorem and its proof come two central results of model theory.

**Theorem 2.2 (Compactness Theorem)** *A set $\Sigma$ of sentences has a model if and only if every finite subset has a model.*

For we can replace 'has a model' by 'is consistent'. Now, if a contradiction can be deduced from $\Sigma$, then (since proofs are finite), only finitely many sentences in $\Sigma$ are used in the proof, and so a finite subset of $\Sigma$ is inconsistent.

**Theorem 2.3 (Downward Löwenheim–Skolem Theorem)** *If a set of sentences in a countable language has a model, then it has a finite or countable model.*

This is a consequence of the proof of Theorem 2.1, since the models constructed there are no larger than the cardinality of the language.

Actual mathematical structures can be represented in various ways as first-order structures. For example, we could choose to represent a group as described earlier, using a binary function for the group operation, a unary operation for inverse, and a constant for the identity. Alternatively, we could use just a binary function for the group operation, and describe the identity and inverses by axioms (as is usual in the axioms for a group). Yet again, we could describe the group operation by a ternary relation $R$, such that $(x, y, z) \in R$ if and only if $x \circ y = z$.

What difference does it make? One significant difference is in the class of substructures. A *substructure* of a first-order structure is a subset which contains all the constants and is closed under all the functions: it carries a structure over the same language. With the first description, the substructures of a group are the subgroups; with the second, they are the sub-semigroups (closed under the group operation); with the third, every subset carries a substructure, which is a 'partial group'.

However, the automorphisms of a group are exactly the same, however we choose to regard the group. (An *automorphism* of a first-order structure $M$ on $X$ is a permutation $g$ of $X$ which preserves every relation (regarded as a subset of $X^n$), fix every constant, and commute with every function (in the sense that $F(x_1g, \ldots, x_ng) = F(x_1, \ldots, x_n)g$ for all $x_1, \ldots, x_n \in X$).)

A language or structure is called *relational* if there are no functions or constants. Relational structures have the property that every subset carries a substructure (which, for graphs, would be called the 'induced subgraph'). This property will be very important later.

## 2.2 Permutation groups

We briefly recount some of the basic concepts of permutation group theory, and some possibly less familiar material on the topology of the infinite symmetric group.

A *permutation group* on a set $\Omega$ is a subgroup of the symmetric group $\mathrm{Sym}(\Omega)$, the group of all permutations of $\Omega$ (with the operation of composition). We write permutations 'on the right', so that the image of $\alpha$ under the permutation $g$ is $\alpha g$; this ensures that composition of permutations works from left to right (that is, $gh$ means 'first apply $g$, then $h$').

Sometimes we have to consider the more general concept of an *action*, or *permutation representation*, of an abstract group $G$ on a set $\Omega$. This is a homomorphism from $G$ to $\mathrm{Sym}(\Omega)$. The image of an action is a permutation group. Usually we suppress the name of the action and write $\alpha g$ for the image of $\alpha$ under the permutation obtained by applying the action to $g$.

Conversely, given a permutation group $G$ on $\Omega$, the identity map is an action of $G$ on $\Omega$.

Most of the concepts that follow are defined for permutation groups, but the definitions can be easily transferred to actions (so that an action is transitive, or whatever, if its image is).

An *orbit* of a permutation group $G$ on $\Omega$ is an equivalence class of the equivalence relation $\equiv$ defined as follows: $\alpha \equiv \beta$ if there exists $g \in G$ such that $\alpha g = \beta$. If $\Delta$ is an orbit, then there is an action of $G$ on $\Delta$, defined by restricting the domain of each permutation to $\Delta$. We denote by $G^\Delta$ the image of this action; this group is called a *transitive constituent* of $G$. (It is a transitive permutation group on $\Delta$.) Then we have

(a) $\Omega$ is a disjoint union of the orbits of $G$;

(b) $G$ is a *subcartesian product* of its transitive constituents (a subgroup of the cartesian product which projects onto each factor).

The *stabiliser* of a point $\alpha \in \Omega$ is the subgroup

$$\{g \in G : \alpha g = \alpha\}$$

of $G$. Its right cosets correspond naturally to the points of the orbit containing $\alpha$, by the rule

$$\beta \leftrightarrow \{g \in G : \alpha g = \beta\}.$$

An *isomorphism* between two actions of $G$ on sets $\Omega_1$ and $\Omega_2$ is a bijection $\theta : \Omega_1 \to \Omega_2$ such that $(\alpha g)\theta = (\alpha\theta)g$ for all $\alpha \in \Omega_1$ and $g \in G$. (In other words, if $\Omega_1$ and $\Omega_2$ are identified according to the bijection, then each element of $G$ induces the same permutation on each set.)

Now let $H$ be a subgroup of $G$. The *coset space* $G : H$ is the set of all right cosets of $H$ in $G$. There is an action of $G$ on it, defined by $(Hx)g = Hxg$. The coset spaces are representatives for all the isomorphism types of transitive actions:

(a) Any transitive action of a group $G$ is isomorphic to the action on the coset space $G : H$, where $H$ is the stabiliser of a point.

(b) The actions of $G$ on coset spaces $G : H$ and $G : K$ are isomorphic if and only if $H$ and $K$ are conjugate subgroups.

The action of $G$ on $\Omega$ is *regular* if it is transitive and the stabiliser of a point is the identity. By the above, any regular action of $G$ is isomorphic to the action of $G$ on itself by right multiplication. (We used this implicitly in our discussion of Cayley graphs in the last section.)

Let $G$ be a permutation group on $\Omega$. Then there are actions of $G$ on each of the sets $\Omega^n$ (all $n$-tuples of elements of $\Omega$), $\Omega^{(n)}$ (all $n$-tuples of distinct elements of $\Omega$), and $\Omega^{\{n\}}$ (all $n$-element subsets of $\Omega$). For example, we define

$$(\alpha_1, \ldots, \alpha_n)g = (\alpha_1 g, \ldots, \alpha_n g).$$

The group $G$ is said to be *oligomorphic* if the number of orbits of $G$ on each of these sets is finite for all $n$.

Note that the finiteness of the numbers of orbits on $\Omega^n$, $\Omega^{(n)}$, and $\Omega^{\{n\}}$ are equivalent, for any given value of $n$. In fact, if $F_n^*$, $F_n$ and $f_n$ denote these numbers of orbits, then we have $f_n \le F_n \le n! f_n$, since each orbit on $n$-sets corresponds to at least one and at most $n!$ orbits on $n$-tuples. Also,

$$F_n^* = \sum_{k=1}^{n} S(n,k) F_k,$$

where $S(n, k)$ is the Stirling number of the second kind, the number of partitions of an $n$-set into $k$ parts. For an orbit $(\alpha_1, \ldots, \alpha_n)^G$ on $n$-tuples determines, and is determined by, a partition of $\{1, \ldots, n\}$ into (say) $k$ parts (where $i$ and $j$ lie in the same part if $\alpha_i = \alpha_j$) and an orbit on $k$-tuples of distinct elements.

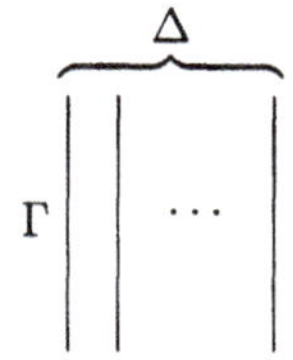


Figure 1: $\Gamma \times \Delta$ as a covering of $\Delta$

The group $G$ is said to be *n-transitive* if it is transitive on $\Omega^{(n)}$, and *n-set transitive* (or *n-homogeneous*) if it is transitive on $\Omega^{\{n\}}$. (I will avoid the term '$n$-homogeneous', since homogeneity in a different sense will be very important later.) Also, $G$ is *highly transitive* if it is $n$-transitive for all $n$, and *highly set transitive* if it is $n$-set transitive for all $n$. Note that transitivity implies set transitivity but not conversely.

Suppose that $G$ is a permutation group on $\Omega$. A *G-congruence* is an equivalence relation on $\Omega$ which is invariant under $G$ (so that $G$ permutes among themselves the equivalence classes of the relation). The *trivial* congruences (which are preserved by any permutation group) are equality $E$ and the 'universal' relation $U$, defined by

(a) $(\alpha, \beta) \in E$ if and only if $\alpha = \beta$;

(b) $(\alpha, \beta) \in U$ for all $\alpha, \beta \in \Omega$.

We say that $G$ is *primitive* if the only $G$-congruences are the trivial ones, or *imprimitive* otherwise. A congruence class of a $G$-congruence is called a *block* for $G$. Note that a primitive group on a set of more than two points is transitive, since the orbit partition is always a congruence.

Let $H$ be a permutation group on $\Gamma$, $K$ a permutation group on $\Delta$. We define the *wreath product* $H$ Wr $K$ as a permutation group on $\Gamma \times \Delta$ as follows. Regard $\Gamma \times \Delta$ as a covering of $\Delta$, with fibres $\Gamma_\delta = \Gamma \times \{\delta\}$ indexed by $\Delta$ and each fibre naturally bijective with $\Gamma$ (see Figure 1). Then define the *base group* $B$ to be the Cartesian product of a family of copies of $H$ indexed by $\Delta$, where the copy $H_\delta$ acts on $\Gamma_\delta$ in the same way that $H$ acts on $\Gamma$. Also define the *top group* $T$ to be the permutation group on $\Gamma \times \Delta$ induced by $K$, acting on the labels of the fibres. The wreath product is now the group of permutations generated by $B$ and $T$. (In fact $T$ normalises $B$ and $B \cap T = \{1\}$, so it is the semidirect product of these groups.)

The wreath product is imprimitive, having a congruence whose blocks are the fibres. The stabiliser of a fibre acts on it as $H$ acts on $\Gamma$, while the whole wreath product acts on the set of fibres as $K$ acts on $\Delta$.

Conversely, let $G$ be transitive but imprimitive. Let $\Gamma$ be a block, and $H$ the permutation group induced on $\Gamma$ by its setwise stabiliser in $G$. Also, let $\Delta$ index the set of blocks, and let $K$ be the permutation group induced on $\Delta$ by $G$. Then there is a bijection between $\Omega$ and $\Gamma \times \Delta$ which takes $G$ to a subgroup of $H$ Wr $K$.

There is a topology on the symmetric group on an infinite set $\Omega$, whose basic open sets are of the form

$$\{g \in G : (\alpha_1, \ldots, \alpha_n)g = (\beta_1, \ldots, \beta_n)\},$$

where $(\alpha_1, \ldots, \alpha_n), (\beta_1, \ldots, \beta_n) \in \Omega^{(n)}$. Informally, two permutations are close if they agree on a long tuple of elements.

With this topology, $\mathrm{Sym}(\Omega)$ becomes a topological group. So the topology can be defined more concisely by saying that a basis of neighbourhoods of the identity consists of the stabilisers of finite tuples.

If $\Omega$ is countable (the case of most interest), the topology is derived from a metric, which can be chosen so that $\mathrm{Sym}(\Omega)$ is a complete metric space. In this case, a sequence $(g_n)$ of permutations converges to a limit $g$ if and only if it converges pointwise: that is, for any $\alpha$, there exists $n_0$ such that $\alpha g_n = \alpha g$ for all $n \geq n_0$.

The most significant fact for us is the following result.

**Theorem 2.4** *The following three conditions on a permutation group $G$ on $\Omega$ are equivalent:*

(a) *$G$ is a closed subgroup of* $\mathrm{Sym}(\Omega)$*;*

(b) *$G = \mathrm{Aut}(M)$ for some first-order structure $M$ on $\Omega$;*

(c) *$G = \mathrm{Aut}(M)$ for some relational structure $M$ on $\Omega$.*

**Proof** As explained in the last section, (b) and (c) are equivalent, since an $n$-ary function can be described by an $(n+1)$-ary relation.

For simplicity, we assume that $\Omega$ is countable, and take $\Omega = \mathbb{N}$. Suppose first that $G = \mathrm{Aut}(M)$, where $M$ is a relational structure on $\mathbb{N}$. Let $(g_n)$ be a sequence of automorphisms converging to a limit $g$. For any $n$-ary relation $R$ and $n$-tuple $(\alpha_1, \ldots, \alpha_n)$, choose $m_0$ such that $\alpha_i g_m = \alpha_i g$ for all $m \geq m_0$ and $i = 1, \ldots, n$. Since the $g_m$ are automorphisms, it follows that $(\alpha_1, \ldots, \alpha_n)g \in R$ if and only if $(\alpha_1, \ldots, \alpha_n) \in R$. So $g \in \mathrm{Aut}(M)$.

Conversely, suppose that $G$ is any permutation group on $\Omega$. We define a relational structure $M$ on $\Omega$, called the *canonical relational structure* for $G$, and show that $\mathrm{Aut}(M)$ is the closure of $G$ in $\mathrm{Sym}(\Omega)$. So, if $G$ is closed, then $\mathrm{Aut}(M) = G$. To construct $M$, we take a relational language containing, for each $n$, a set of $n$-ary relation symbols in bijective correspondence with the orbits of $G$ on $\Omega^n$: we decree that the relation symbol corresponding to the orbit $O_i$ on $\Omega^n$ is satisfied by just those $n$-tuples in the orbit $O_i$. It is clear that $G \leq \mathrm{Aut}(M)$. Hence, since $\mathrm{Aut}(M)$ is closed, it contains the closure of $G$.

Suppose that $g \in \mathrm{Aut}(M)$, and take any $n$ points $\alpha_1, \ldots, \alpha_n$. Let $R$ be the unique $n$-ary relation of $M$ satisfied by $(\alpha_1, \ldots, \alpha_n)$. Then, since $g \in \mathrm{Aut}(M)$, the tuple $(\alpha_1 g, \ldots, \alpha_n g)$ also satisfies $R$. Since the tuples satisfying $R$ form an orbit of $g$, there exists $g_n \in G$ such that $\alpha_i g_n = \alpha_i g$ for $i = 1, \ldots, n$. Hence $g$ lies in the closure of $G$, as required.

Closed subgroups of the symmetric group are much better behaved than arbitrary subgroups. For example, a closed subgroup is finite or countable or else has order $2^{\aleph_0}$. In some circumstances, we can obtain classifications. As an example, we consider a theorem of Thomas [76].

If $N$ is an expansion of $M$ (that is, $N$ is obtained from $M$ by adding new relations, functions or constants), then $\mathrm{Aut}(N)$ is a closed subgroup of $\mathrm{Aut}(M)$. Conversely, if $G$ is a closed supergroup of the closed group $\mathrm{Aut}(N)$, then we can regard $N$ as an expansion of a structure $M$ with $\mathrm{Aut}(M) = G$; we call $M$ a *reduct* of $N$. Thomas has determined the reducts of various structures. We will examine his result on the random graph. First, we give some examples. Let $R$ be the random graph. Instead of structures, we may talk about closed supergroups of $\mathrm{Aut}(R)$.

One such group consists of all the automorphisms and *anti-automorphisms* of $R$ (the latter being isomorphisms from $R$ to its complement). It contains $\mathrm{Aut}(R)$ as a subgroup of index 2, and is 2-transitive (but not 3-transitive).

The operation of *switching* a graph $X$ with respect to a set $A$ of vertices was introduced by Seidel [68] for studying strongly regular graphs. It consists of changing all edges between $A$ and its complement into non-edges, and all non-edges into edges, while leaving adjacencies within or outside $A$ unaltered. Being related by switching is an equivalence relation on the set of all graphs on a given vertex set; the equivalence classes are called *switching classes*. Higman(see [69]) defined a *two-graph* on a set $V$ to be a set $T$ of 3-subsets of $V$ with the property that any 4-set contains an even number of members of $T$. For any graph $X$, the set of triples inducing subgraphs with an odd number of edges in $X$ is a two-graph; every two-graph arises in this way. Furthermore, graphs $X_1$ and $X_2$ give rise to the same two-graph under this construction if and only if they belong to the same switching class. So there is a natural bijection between two-graphs and switching classes. If the random graph $R$ is switched with respect to a random set of vertices, the result is almost surely isomorphic to $R$. So the two-graph $T$ derived from $R$ is a proper reduct of $R$. Its automorphism group is 2-transitive but not 3-transitive.

If we allow both complementation and switching, we obtain a reduct whose automorphism group is 3-transitive but not 4-transitive.

**Theorem 2.5 (Thomas' Theorem)** *There are exactly five closed supergroups of* $\mathrm{Aut}(R)$, *namely* $\mathrm{Aut}(R)$, *the group of automorphisms and anti-automorphisms of $R$, the automorphism group of the two-graph $T$ derived from $R$, the group of automorphisms and anti-automorphisms of $T$, and the symmetric group on $R$.*

## 2.3 Homogeneity

One of the most important properties of the random graph $R$ is that it is *homogeneous*: every isomorphism between finite induced subgraphs extends to an automorphism of $R$. In this section, we describe a construction of Fraïssé, which produces a large number of homogeneous structures, and proves their uniqueness, in an analogous way. Because Fraïssé's work predated the construction of $R$, he used a different prototype: the rational numbers $\mathbb{Q}$ (as ordered set). Given two $n$-tuples of rationals arranged in increasing order, say $a_1 < \ldots < a_n$ and $b_1 < \ldots < b_n$, we can extend the map $a_i \mapsto b_i$ to a piecewise linear function on the whole of $\mathbb{Q}$ by interpolating a linear function in each gap and shifting the two ends appropriately.

At the heart of Fraïssé's method is the concept of the *age* of a relational structure $M$, the class of all finite structures (over the same language) which are embeddable in $M$ as induced substructures. (Fraïssé uses the term 'relation' for what I have called 'relational structure', and in addition to 'the age of a relation' he speaks of a 'rich relation', and so on.) We write the age of $M$ as $\mathrm{Age}(M)$.

The age $\mathcal{A}$ of a countable relational structure has three obvious properties:

(A1) $\mathcal{A}$ is closed under isomorphism.

(A2) $\mathcal{A}$ is closed under taking induced substructures.

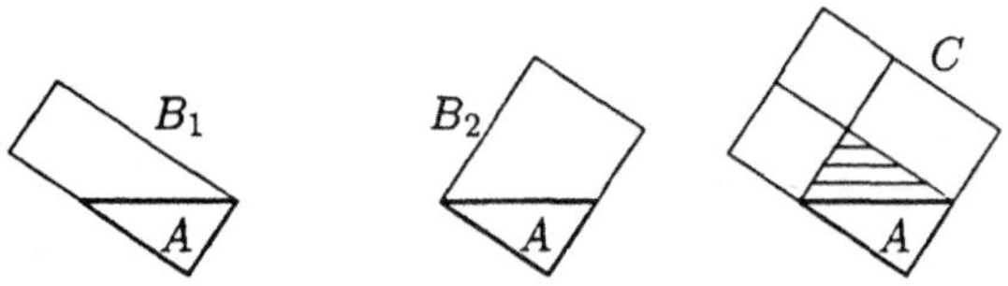


Figure 2: The Amalgamation Property

(A3) $\mathcal{A}$ contains at most countably many non-isomorphic structures.

The next property is a little less obvious:

(A4′) $\mathcal{A}$ has the *joint embedding property*: given $A, B \in \mathcal{A}$, there exists $C \in \mathcal{A}$ such that $A$ and $B$ are embeddable in $C$.

To see this, embed $A$ and $B$ into $M$ and let $C$ be the union of their images.

These properties characterise ages:

**Theorem 2.6** *A class $\mathcal{A}$ of finite relational structures is the age of a finite or countable relational structure if and only if it satisfies* (A1)–(A3) *and* (A4′).

**Proof** Enumerate representatives of the isomorphism types: $A_1, A_2, \ldots$. Define $B_0 = \emptyset$ and, inductively, let $B_n \in \mathcal{A}$ embed $B_{n-1}$ and $A_n$. Let $M = \bigcup B_n$. Every finite substructure of $M$ is contained in $B_n$ for some $n$, and hence is in $\mathcal{A}$, by (A1) and (A2). By construction, every member of $\mathcal{A}$ is embeddable in $M$.

To construct a homogeneous structure, we need what appears to be only a slight strengthening of (A4′), namely

(A4″) $\mathcal{A}$ has the *amalgamation property*, that is, given $A, B_1, B_2 \in \mathcal{A}$ and embeddings $f_i : A \to B_i$ for $i = 1, 2$, there exists $C \in \mathcal{A}$ and embeddings $g_i : B_i \to C$ for $i = 1, 2$ such that $f_1 g_1 = f_2 g_2$.

In other words, given two structures in $\mathcal{A}$ with a common substructure, we can glue them together along the substructure (and possibly some more). To see that this holds for the age of a homogeneous structure $M$, let $h_i$ be an embedding of $B_i$ into $M$ for $i = 1, 2$. Then the images of $A$ under $f_1 h_1$ and $f_2 h_2$ are isomorphic substructures of $M$. By homogeneity, there is an automorphism $k$ such that $f_1 h_1 k$ and $f_2 h_2$ agree on $A$. Replacing the embedding $h_1$ by $h_1 k$, we can assume that $k$ is the identity. Then take $C = B_1 h_1 \cup B_2 h_2$, and let $g_1$ and $g_2$ be $h_1$ and $h_2$, regarded as embeddings into $C$.

Note that (A4′) is a special case of (A4″), obtained by taking $A$ to be the empty structure. (Note that it is usual in model theory to disallow the empty structure. If we did that, we would have to keep both conditions.)

Fraïssé showed that these conditions are also sufficient:

**Theorem 2.7 (Fraïssé's Theorem)** *A class $\mathcal{A}$ is the age of a countable homogeneous relational structure $M$ if and only if it satisfies* (A1)–(A3) *and* (A4″). *If this holds, then the structure $M$ is unique up to isomorphism.*

I will outline the proof of this theorem, because of its importance in what follows. We have proved the necessity of the conditions, and have to show their sufficiency and the uniqueness of $M$. The tool for this is a property which is the exact analogue of condition (*) for the random graph. This is called the *I-property* because of its similarity to injectivity in homological algebra.

Let $M$ be a countable (or finite) structure with age $\mathcal{A}$. We say that $M$ has the *I-property* if, whenever $A, B \in \mathcal{A}$ with $A \subset B$ and $|B| = |A| + 1$, any embedding of $A$ into $M$ can be extended to an embedding of $B$ into $M$.

*Step 1* Any two countable (or finite) structures with the same age, both having the I-property, are isomorphic. Any structure with the I-property is homogeneous.

This is proved by back-and-forth, just as for the random graph.

*Step 2* A homogeneous relational structure has the I-property.

For let $M$ be homogeneous, and let $A, B \in \text{Age}(M)$ satisfy the hypotheses of the I-property. Let $f$ be an embedding of $A$ into $M$. There is an embedding $g$ of $B$ into $M$. The images of $A$ under $f$ and the restriction of $g$ are isomorphic finite substructures of $M$. By homogeneity, there is an automorphism $h$ of $M$ such that $f$ and $gh$ agree on $A$. Now take $gh$ as the required embedding of $B$.

*Step 3* If $\mathcal{A}$ is a class satisfying (A1)–(A3) and (A4″), then there is a structure $M$ with age $\mathcal{A}$ having the I-property.

To prove this, we just 'shovel everything in'. Enumerate all triples $(A, B, f)$ such that $A, B \in \mathcal{A}$, $|B| = |A| + 1$, and $f : A \to B$ an embedding (up to isomorphism). If we have at any stage built a structure $X \in \mathcal{A}$ and have an embedding $g : A \to X$, the amalgamation property guarantees a structure $Y$ containing $X$ and an embedding $h : B \to Y$ such that $fh = g$. Now we start with $X_0 = \emptyset$. If $X_n$ has been constructed, we list all embeddings $A \to X_n$, where $A$ is the first component of one of the first $n$ triples in the list. Applying the above amalgamation to all of them, we end up with a structure $X_{n+1}$ containing $X_n$ such that all of these embeddings have been lifted to the second component of the triples. Then take $M$ to be the union of all the $X_n$.

One type of relational structure we already saw is necessarily homogeneous. This type consists of the canonical structures for permutation groups, discussed in Section 2.2. Recall that the $n$-ary relations in the canonical structure for $G$ are the orbits of $G$ on $n$-tuples. Now suppose that the map $f : \alpha_i \mapsto \beta_i$ (for $i = 1, \ldots, n$) is an isomorphism of induced substructures. In particular, the unique relations satisfied by the $n$-tuples $(\alpha_1, \ldots, \alpha_n)$ and $(\beta_1, \ldots, \beta_n)$ are equal. But this means that the two $n$-tuples lie in the same orbit of $G$; so $f$ is induced by an element of $G$.

Note that the language for the canonical structure has infinitely many relation symbols.

**Problem** Which closed permutation groups are the automorphism groups of homogeneous structures over relational languages with only finitely many relation symbols?

There is an important strengthening of the amalgamation property, as follows:

(A4‴) $\mathcal{A}$ has the *strong amalgamation property*, that is, if $A, B_1, B_2 \in \mathcal{A}$ and $f_i : A \to B_i$ are embeddings for $i = 1, 2$, then there exists $C \in \mathcal{A}$ and embeddings $gi : B_i \to C$ for $i = 1, 2$ such that $f_1g_1 = f_2g_2$ and, if $b_1 \in B_1$, $b_2 \in B_2$ and $b_1g_1 = b_2g_2$, then there exists $a \in A$ such that $b_i = af_i$ for $i = 1, 2$.

In other words, the shaded region in Figure 2 can be chosen to be empty. The definition of the amalgamation property allows the possibility that, when $B_1$ and $B_2$ are glued together on their common substructure $A$, some additional points also become identified. The strong amalgamation property guarantees that the amalgamation can be performed in such a way that no extraneous identifications are made.

**Theorem 2.8** *Let $M$ be a countable homogeneous structure. Then the following conditions are equivalent:*

(a) $\mathrm{Age}(M)$ *has the strong amalgamation property;*

(b) *the stabiliser in* $\mathrm{Aut}(M)$ *of any finite number of points has no additional fixed points;*

(c) $M\backslash\{x\} \cong M$ *for any point $x$ of $M$.*

For the proof, see [14] for (a) and (b), [30] for (c).

The strong amalgamation property is very useful for constructions. Suppose that $M_1, \ldots,$ $M_k$ are homogeneous structures over languages $L_1, \ldots, L_k$ with pairwise disjoint sets of relation symbols, all of whose ages have the strong amalgamation property. Then there is a homogeneous structure $M$ over the union $L$ of the languages, whose reduct to $L_i$ is $M_i$ for each $i$. We simply take the class of finite structures over $L$ whose reduct to $L_i$ belongs to $\mathrm{Age}(M_i)$ for each $i$. Conditions (A1)–(A3) are trivial, and we can form (strong) amalgams by taking a strong amalgam of the reducts to each $L_i$.

So, for example, there is a dense total order (without endpoints) of the vertices of $R$, such that any two finite sets which are related by an order-preserving isomorphism of induced subgraphs are equivalent under the group of order-preserving automorphisms of $R$.

Recently, Hrushovski [41] has developed an important weakening of the amalgamation property, where only certain amalgamations are required to exist. This theory enables the construction of many new types of first-order structures and permutation groups. It also turns out, unexpectedly, to be connected with the theory of sparse random graphs. See Wagner's paper in [45] or the chapter on graphs and logic in [6]. We give a brief account in Section 3.8.

## 2.4 Categoricity

An obvious question about the relation between logic and mathematics is the extent to which mathematical structures can be axiomatised by logical formulae. If we are using first-order logic, we quickly observe that we can't say anything about the cardinality of a structure. We saw in Section 2.1 the downward Löwenheim–Skolem Theorem, according to which if a set of first-order sentences in a countable language is consistent, then it has a finite or countable model. There is also an *upward Löwenheim–Skolem Theorem*, asserting that if a set of sentences has an infinite model, then it has models of arbitrarily large cardinality. (If a structure is finite, then we can describe it completely by a single sentence.)

Accordingly, the best we can hope to do with first-order axioms is to specify that any structure with the same cardinality as $M$ is isomorphic to $M$. If $|M| = \lambda$, we say that $M$ is *$\lambda$-categorical* if this holds. Logicians more commonly apply this concept to a *theory* (a set of sentences) than to a structure. According to a theorem of Vaught [79], if a theory in a countable language is $\lambda$-categorical for any uncountable cardinal $\lambda$, then it is $\mu$-categorical for any uncountable cardinal $\mu$. Hence there are only two different types of categoricity: countable and uncountable.

These two conditions have strong but very different implications for the structure of models of the theory. According to results of Zil'ber [81], uncountable categoricity implies a structural description for models, resembling that for vector spaces over a fixed field $F$, where a single cardinal number parameter, the dimension, determines the model up to isomorphism. (If the field is countable, then a countable vector space has finite or countable dimension, but an uncountable vector space has dimension equal to its cardinality. So the theory is uncountably but not countably categorical.) In general, a collection of parameters is required.

Countable categoricity, however, is equivalent to a very strong symmetry condition, as shown by the following theorem proved independently by Engeler, Ryll-Nardzewski and Svenonius [31], [65], [74]:

**Theorem 2.9** *The countable first-order structure $M$ is countably categorical if and only if* $\mathrm{Aut}(M)$ *is oligomorphic. Moreover, if these conditions hold, then two tuples lie in the same orbit of* $\mathrm{Aut}(M)$ *if and only if they satisfy the same $n$-variable first-order formulae.*

I will not give the proof here, but remark only on one aspect which resembles what we have seen before. Suppose that $M$ is a first-order structure. Call two $n$-tuples in $M$ *equivalent* if they satisfy the same first-order formulae. Suppose that, for each $n$, there are only finitely many equivalence classes of $n$-tuples. Now it is easy to show that, if $(\alpha_1, \ldots, \alpha_n)$ and $(\beta_1, \ldots, \beta_n)$ are equivalent then, for any $\alpha$, there exists $\beta$ such that $(\alpha_1, \ldots, \alpha_n, \alpha)$ and $(\beta_1, \ldots \beta_n, \beta)$ are equivalent. Then back-and-forth can be used to show that the equivalence classes are orbits of $\mathrm{Aut}(M)$. We will use this fact later.

**Corollary 2.10** *Let $M$ be a countable homogeneous relational structure having only a finite number of isomorphism types of $n$-element substructures for all $n$. Then $M$ is $\aleph_0$-categorical. In particular, this holds if there are only finitely many relation symbols in the language of $M$.*

This follows immediately from Theorem 2.9, since each isomorphism type of $n$-element substructure contributes at most $n!$ orbits on $M^{(n)}$.

The prototypes for this result are the homogeneous structures $\mathbb{Q}$ (as ordered set) and $R$ (the random graph). By a famous theorem of Cantor, $\mathbb{Q}$ is characterised by countability and the axioms for a total order together with the three sentences

$$(\forall x, y)((x < y) \Rightarrow (\exists z)(x < z < y)),$$
$$(\forall x)(\exists y)(y < x),$$
$$(\forall x)(\exists y)(x < y).$$

These say, respectively, that the order is dense and has no least or greatest element.

The random graph is characterised by countability and the axioms for a graph (an irreflexive, symmetric binary relation) together with sentences $\sigma_{m,n}$ for $m, n \geq 0$, asserting that for any vertices $x_1, \ldots, x_m, y_1, \ldots, y_n$ there exists $z$ joined to all the $x$'s and none of the $y$'s.

It is possible to describe a set of sentences which characterise any structure $M$ with the properties of the Corollary as countable structure. These are of three types:

(a) For each $n$-element structure in Age($M$), an existential sentence asserting that it occurs as a substructure.

(b) For each $n$, a (universal) sentence asserting that, for each choice of $n$ points, the induced substructure is one of the (finitely many) $n$-element structure in Age($M$).

(c) A sentence asserting that each instance of the I-property holds (one for each choice of $A$ and $B$).

Conditions (a) and (c) can be enforced by first-order sentences for any $M$; but (b) depends on the finiteness of the number of $n$-element structures in the age.

## 2.5 Ages and ubiquity

The age Age($M$) of a relational structure $M$ can be regarded as a tree. The nodes of the tree occur on 'levels' indexed by the non-negative integers. On level $n$, we put all the structures in the age which have point set $\{1, 2, \ldots, n\}$. A node on level $n+1$ (corresponding to a structure $S$ on the set $\{1, \ldots, n+1\}$ is adjacent to the unique node on level $n$ which is obtained by deleting the point $n+1$, that is, the induced substructure on $\{1, \ldots, n\}$.

The root of the tree is the empty set, the unique structure on the empty set. Any infinite path in the tree corresponds to a structure $S$ on $\mathbb{N}$ whose induced substructure on $\{1, \ldots, n\}$ belongs to the age Age($M$) for all $n$. Since the age is closed under taking induced substructures, every finite substructure of $S$ is contained in Age($M$); that is, Age($S$) $\subseteq$ Age($M$). We say that $S$ is *younger than* $M$ if this condition holds.

Conversely, let $S$ be any countable structure which is younger than $M$. If the point set of $S$ is indexed by the natural numbers, then the induced substructure of $S$ on $\{1, \ldots, n\}$ is a node of the tree for each $n$, and nodes on consecutive levels are adjacent. Hence $S$ is represented by a path in the tree. (Of course, a different enumeration of the points of $S$ could lead to a different path.)

So we conclude that the infinite paths from the root represent the (labelled) structures younger than $M$.

Now let $\mathcal{Y}(M)$ denote the set of such paths. We define a metric on $\mathcal{Y}(M)$ by the rule that two paths which agree up to level $n$ and diverge thereafter should have distance $1/2^n$. (The precise decreasing function of $n$ that we use is not crucial; this is chosen so that the diameter of the space is 1.) Now $\mathcal{Y}(M)$, with this metric, is a complete metric space.

In a metric space, a set $X$ is said to be *residual*, or *comeagre*, if it contains an intersection of countably many open dense sets.

**Theorem 2.11 (Baire Category Theorem)** *In a complete metric space, a residual set is non-empty.*

This theorem implies the apparently stronger statement that residual sets are in some sense 'large'; for example, a residual set has non-empty intersection with every open dense set. So residual sets behave like the complements of null sets in measure spaces, and are used in similar situations (for example, in non-constructive existence proofs).

In the case of the tree defined by Age($M$), it is possible to translate the concepts of openness and denseness in a more combinatorial way. A set $S$ of paths (that is, a set of structures younger than $M$) is open if and only if it is *finitely determined*, in the following sense:

For any $S \in \mathcal{S}$, there exists $n$ such that any structure in $\mathcal{Y}(M)$ which induces the same structure on $\{1, \ldots, n\}$ as $S$ belongs to $\mathcal{S}$.

(For the ball of radius $1/2^n$ containing $S$ consists of all structures agreeing with $S$ on $\{1, \ldots, n\}$.) A set $\mathcal{S}$ is dense if and only if it is *always reachable*, in the sense:

Any finite structure, say on $\{1, \ldots, n\}$, is induced by some member of $\mathcal{S}$.

(This simply says that $\mathcal{S}$ meets all open balls.)

We now illustrate these concepts by showing the following:

**Proposition 2.12** *Let $R$ be the random graph. Then the set of graphs (on the vertex set $\mathbb{N}$) which are isomorphic to $R$ is residual in $\mathcal{Y}(R)$.*

Note that $\mathcal{Y}(R)$ is the set of all graphs on $\mathbb{N}$. This result says that almost all countable graphs, in the sense of Baire category, are isomorphic to $R$; it is the analogue of the Erdős–Rényi result asserting the same in the sense of measure.

**Proof** We show that, for fixed finite disjoint sets $U$ and $V$, the set $\mathcal{S}(U, V)$ of graphs for which there exists $z$ joined to every vertex in $U$ and to none in $V$, is open and dense in $\mathcal{Y}(R)$. Since a graph is isomorphic to $R$ if and only if it lies in the intersection of all these sets, the result follows. (Note that this gives another non-constructive existence proof.)

*$\mathcal{S}(U, V)$ is open:* Let $S$ be a graph in this set, and $z$ the witnessing vertex. Let $m = \max(U \cup V \cup \{z\})$. Then any graph agreeing with $S$ on $\{1, \ldots, m\}$ is in $\mathcal{S}(U, V)$.

*$\mathcal{S}(U, V)$ is dense:* Suppose we have a graph on $\{1, \ldots, n\}$, where without loss of generality $\max(U \cup V) \leq n$. Then we can join the next vertex $n + 1$ correctly to $U$ and not $V$, and guarantee membership in $\mathcal{S}(U, V)$.

The principle involved in this proof can be widely generalised. A sentence of first-order logic is called *inductive* if it takes the form $(\forall x_1, \ldots, x_m)(\exists y_1, \ldots, y_n)\phi$, where $\phi$ is quantifier-free.

**Theorem 2.13** *Let $M$ be a countable relational structure, and $\Sigma$ a countable set of inductive sentences in the language of $M$. Then the set of models of $\Sigma$ is residual in $\mathcal{Y}(M)$.*

**Proof** Since a countable intersection of residual sets is residual, it suffices to prove this for a single sentence. Now follow the argument used above for the inductive sentence $\sigma_{m,n}$ in $R$.

**Corollary 2.14** *Let $M$ be a countable relational structure. Then the set of structures $S$ satisfying* $\mathrm{Age}(S) = \mathrm{Age}(M)$ *is residual in $\mathcal{Y}(M)$.*

This is proved by taking, for each finite structure $X$ of $M$, the (existential) sentence saying that $X$ occurs as a substructure. The corollary says, informally, that almost all structures younger than $M$ actually have the same age.

**Corollary 2.15** *Let $M$ be a countable homogeneous relational structure. Then the set of structures isomorphic to $M$ is residual in $\mathcal{Y}(M)$.*

A structure $M$ is said to be *ubiquitous in category* if the conclusions of this corollary hold, that is, almost all structures younger than $M$ are isomorphic to $M$. In this language, the corollary says that countable homogeneous structures are ubiquitous in category. Since we know that almost all structures in $\mathcal{Y}(M)$ have the same age as $M$, we have only to observe that $M$ is characterised by the I-property, which can be expressed as a countable number of inductive sentences.

The random graph has a similar property expressed in terms of measure. We know that $\mathcal{Y}(R)$ is the class of countable graphs, and that almost all countable graphs (in the sense of either measure or category) are isomorphic to $R$. This suggests that there is a measure-theoretic analogue of the above considerations. However, the development of this quickly leads to very difficult questions which have not been resolved.

For a discussion of the relationship between measure and Baire category, see Oxtoby [58].

We can use these ideas to generalise the universal property of the random graph. Let $M$ be a relational structure. We say that $M$ is *universal* if every structure which is younger than $M$ is embeddable in $M$. (Fraïssé's term is 'rich for its age'.)

Of course, if $S$ is embeddable in $M$, then clearly $S$ is younger than $M$. But the converse is false. If $M$ is the two-way infinite path, then Age($M$) consists of all disjoint unions of finite paths, and $\mathcal{Y}(M)$ consists of all disjoint unions of finite or countable paths. The union of two two-way infinite paths is not embeddable in $M$. However, the following holds.

**Theorem 2.16** *Let $M$ be a countable relational structure. Assume that either $M$ is homogeneous, or $M$ is $\aleph_0$-categorical. Then $M$ is universal.*

This theorem is proved in [15]. The case where $M$ is homogeneous is straightforward, following the argument for $R$ in the first section. The $\aleph_0$-categorical case is a bit more substantial, requiring the use of König's Infinity Lemma. For completeness, here is a sketch of the proof. Let $M$ be $\aleph_0$-categorical, and let $S$ be a structure in $\mathcal{Y}(M)$, on the set $\mathbb{N}$. We construct a tree in which the nodes on level $n$ are the embeddings of the induced structure of $S$ on $\{1, \ldots, n\}$ into $M$, modulo automorphisms of $M$ (essentially orbits of Aut($M$) on $n$-tuples carrying the same structure as $(1, \ldots, n)$ carries in $S$). Since Aut($M$) is oligomorphic, there are only finitely many nodes at level $n$. Adjacency is defined by the adjacency of orbit representatives in the age of $M$. So every node lies on a path from the root. König's Lemma guarantees an infinite path in the tree. Now we can choose the orbit representatives for the nodes on this path consistently, in such a way that we have a path in Age($M$), defining an embedding of $S$ into $M$.

## 2.6 Random graphs

Any graph property defined by a first-order sentence satisfies a zero-one law: either almost all finite graphs satisfy it, or almost none do. The explanation is provided by the countable random graph $R$: almost all finite graphs satisfy $\phi$ if and only if $R$ satisfies $\phi$. In this section, we discuss the first-order theory of random graphs, and give a recent application which also uses Thomas' Theorem 2.5 on reducts of $R$.

**Theorem 2.17** *Let $\phi$ be a sentence in the first-order language of graphs. Then the proportion of $n$-vertex graphs satisfying $\phi$ tends to 1 (as $n \to \infty$) if $R$ satisfies $\phi$, and to 0 otherwise.*

**Proof** $R$ is the unique countable graph having property (*). We express this property by first-order sentences. Let $\sigma_{m,n}$ be the sentence asserting that, for any choice of distinct vertices $u_1, \ldots, u_m, v_1, \ldots, v_n$, there is a vertex $z$ joined to all of the $u$'s and none of the $v$'s. These sentences characterise $R$ as a countable graph. By Gödel's Completeness Theorem, if $\phi$ is a sentence true in $R$, then $\phi$ can be deduced from the set $\{\sigma_{m,n} : m, n \in \mathbb{N}\}$. Since deductions are finite, $\phi$ can be deduced from a finite set of these sentences. If we can show that each $\sigma_{m,n}$ holds in almost all finite graphs, then the same will be true for any finite set of these sentences, and hence for $\phi$.

The probability that a random $N$-vertex graph does not satisfy $\sigma_{m,n}$ is at most

$$\binom{N}{m+n}(1 - 1/2^{m+n})^{N-(m+n)},$$

which does tend to zero as $N \to \infty$.

Finally, if $\phi$ does not hold in $R$, then $\neg\phi$ does hold in $R$; hence $\neg\phi$ holds in almost all finite graphs, and $\phi$ holds in almost none.

We use this to derive a result about finite graphs, which uses $R$ in its proof.

Let $\mathcal{F}$ be a finite set of finite graphs. For any graph $X$, let $\mathcal{F}(X)$ denote the hypergraph on the vertex set of $X$ whose hyperedges are all those subsets for which the induced subgraph belongs to $\mathcal{F}$. We also define five equivalence relations on the class of finite graphs: $E_1$, equality; $E_2$, equality or complementation; $E_3$, switching (see Section 2.2); $E_4$, switching possibly combined with complementation; $E_5$, same vertex set.

**Theorem 2.18** *Let $\mathcal{F}$ be any finite set of finite graphs. Then there is an index $i \in \{1, \ldots, 5\}$ such that, for almost all finite graphs $X$, we have $\mathcal{F}(X) = \mathcal{F}(Y)$ if and only if $(X, Y)$ are related by $E_i$.*

For example, if $\mathcal{F} = \{K_3\}$, and $X$ is any graph in which any two vertices are the ends of a path of length 2, then knowledge of the triangles of $X$ determines the edges; almost all finite graphs satisfy the condition (which is just the sentence $\sigma_{2,0}$). Again, if $\mathcal{F}$ consists of $K_3$ and the 3-vertex graph with one edge, then $\mathcal{F}(X)$ is the two-graph of $X$, which (as we saw) determined $X$ up to switching for all graphs $X$. Finally, if $\mathcal{F}$ consists of all four 3-vertex graphs, then $\mathcal{F}(X)$ is the complete 3-uniform hypergraph, and tells us nothing about $X$ beyond its vertex set.

For another example, the 'semi-strong perfect graph theorem' of Reed [62] states that, if $\mathcal{F} = \{P_4\}$, where $P_4$ is the path on four vertices, then the following holds: if $X$ is a perfect graph and $\mathcal{F}(X) = \mathcal{F}(Y)$, then $Y$ is perfect. According to Theorem 2.18, for almost all graphs, this is equivalent to the 'weak perfect graph theorem' of Lovász [51], asserting that the complement of a perfect graph is perfect. (In fact, Reed's proof gives a more structural description.)

We outline the proof. First, consider $\mathrm{Aut}(\mathcal{F}(R))$. This is a closed permutation group containing $\mathrm{Aut}(R)$. Now Thomas' Theorem gives us five possibilities for this group, corresponding to the five cases in the theorem. We consider the first case, when $\mathrm{Aut}(\mathcal{F}(R)) = \mathrm{Aut}(R)$; the others are similar.

Now $\mathrm{Aut}(\mathcal{F}(R))$ is oligomorphic. As we saw in the discussion of the theorem of Engeler, Ryll-Nardzewski and Svenonius, each orbit on $n$-tuples is defined by a first-order formula. The set of edges of $R$ is an orbit; so there is a formula $\phi(x,y)$, in the language of hypergraphs appropriate for $\mathcal{F}(R)$, such that $x$ and $y$ are joined if and only if $\phi(x,y)$ holds in $\mathcal{F}(R)$.

The hypergraph language contains a relational $R_k$ symbol for each cardinality $k$ of hyperedges. So any formula can be 'translated' into graph language by replacing $R_k(x_1,\ldots,x_k)$ by the assertion that the induced subgraph on $\{x_1,\ldots,x_k\}$ is one of the $k$-vertex graphs in $\mathcal{F}$. Let $\phi'(x,y)$ be the 'translation' of $\phi(x,y)$. Then the sentence

$$(\forall x,y)((x \sim y) \Leftrightarrow \phi'(x,y))$$

holds in $R$. Hence it holds in almost all finite graphs. But this asserts that, in almost all finite graphs $X$, the edges are precisely the pairs $x,y$ for which $\phi(x,y)$ holds in $\mathcal{F}(X)$. So $\mathcal{F}(X)$ determines $X$ for almost all $X$.

The basic results on first-order theory of random graphs are due to Glebskii *et al.* [37], Blass and Harary [8], and Fagin [34]. The application is due to Cameron and Martins [19].

# 3 Constructions

## 3.1 Analogues of finite structures

There are many families of finite structures with a lot of symmetry, whose definition involves some parameters (perhaps positive integers or finite fields). Often, these structures have infinite analogues, obtained by allowing some of the finite parameters to become infinite. Sometimes a little care is required. I do not intend to make a complete survey here, but will describe two topics: poset block structures (which arise in statistical design, see Speed and Bailey [73]), and smoothly approximable structures (arising in model theory, see Kantor, Liebeck and Macpherson [44], Hrushovski [42]).

Let $P$ be a finite partially ordered set. A subset $Q$ of $P$ is called *ancestral* if $q \in Q$ and $r \geq q$ imply $r \in Q$. The *boundary* $\partial Q$ of an ancestral set $Q$ is the set of maximal elements of $P \setminus Q$. Now a *poset block structure* based on $P$ is defined as follows. Take a set $X(p)$ with $|X(p)| > 1$ for each $p \in P$, and let $X$ be the Cartesian product $\prod_{p\in P} X(p)$. Now define a relation $R_Q$ on $X$ for each ancestral set $Q$ as follows: for $f,g \in X$,

$$(x,y) \in R_Q \qquad \Longleftrightarrow \qquad x(p) = y(p) \text{ for } p \in Q.$$

The relations $R_Q$ are equivalence relations; and $X$, equipped with these relations, is homogeneous.

If all the sets $X(p)$ are finite, these relations define an association scheme. The relations of the association scheme are the ones obtained from the $R_Q$ by removing all finer relations: specifically,

$$(x,y) \in S_Q \qquad \Longleftrightarrow \qquad \begin{cases} x(p) = y(p) & \text{for } p \in Q, \\ x(p) \neq y(p) & \text{for } p \in \partial Q. \end{cases}$$

The dimensions of the eigenspaces of this scheme, the orthogonal projections, and the $P$ and $Q$ matrices in the sense of Delsarte [26], can all be computed using Möbius inversion on the poset of ancestral subsets of $P$. See [73], [5].

Here are two simple examples. First, let $P$ consist of two incomparable elements $p$ and $q$. Then $X = X(p) \times X(q)$ can be regarded as a rectangular array with rows and columns indexed by $X(p)$ and $X(q)$ respectively. There are four ancestral sets, namely $P$, $\{p\}$, $\{q\}$, and $\emptyset$. It is easily checked that $S_P$ is the relation of equality; $S_{\{p\}}$ is the relation 'different elements of the same row'; $S_{\{q\}}$ is 'different elements of the same column', and $S_\emptyset$ is 'different row and different column'.

Now let $P = \{p,q\}$ with $p < q$. There are just three ancestral sets, $P$, $\{q\}$, and $\emptyset$; as before, $S_P$ is equality, but now $S_{\{q\}}$ is 'different elements of the same column', and $S_\emptyset$ is 'different columns'. So the rows have no significance.

Statisticians refer to these two operations as *crossing* and *nesting* respectively.

Let $M$ be any homogeneous structure. If $N$ is a finite substructure of $M$, then all automorphisms of $N$ extend to automorphisms of $M$ fixing $N$; so two tuples of $N$ lying in the same orbit of $\mathrm{Aut}(N)$ also lie in the same orbit of $\mathrm{Aut}(M)$. We say that $N$ is a *finite homogeneous substructure* of $M$ if the converse holds. (So a finite homogeneous substructure of $M$ is certainly a homogeneous structure in its own right.) Now $M$ is *smoothly approximable* if it is the union of a chain of finite homogeneous substructures of itself. The prototype is a projective space of countable dimension over a finite field, which is the union of a chain of finite-dimensional subspaces.

The poset block structures defined above, with some or all of the sets $X(p)$ countably infinite, are easily seen to be $\aleph_0$-categorical and smoothly approximable. Their automorphism groups are 'generalised wreath products', and are highly imprimitive; but all the primitive 'building blocks' are symmetric groups. At the other extreme is the following result of Kantor, Liebeck and Macpherson [44], which determines the smoothly approximable structures whose automorphism groups are primitive (and oligomorphic). This result is a generalisation of the famous Cherlin–Harrington–Lachlan theorem [23] on $\aleph_0$-categorical and $\aleph_0$-stable structures. The statement given below is rather less detailed than that given in [44], which runs to several pages.

**Theorem 3.1** *Let $M$ be an $\aleph_0$-categorical, smoothly approximable structure. Suppose that $G = \mathrm{Aut}(M)$ is primitive. Then one of the following holds:*

(a) *$G$ is the symmetric group of countable degree, acting on $k$-subsets;*

(b) *$G$ is a classical group of countable degree over a finite field, acting on an orbit of subspaces in its natural module (or possibly pairs of subspaces of complementary dimension);*

(c) $G \leq G_1 \,\mathrm{Wr}\, S_m$, *where $G_1$ is as in* (a) *or* (b)*;*

(d) *$G$ is the semidirect product of a classical group of countable degree over a finite field by its natural module;*

(e) *$G$ is the semidirect product of $V_1 \otimes V_2$ by $G_1 * G_2$ (central product), where $(G_1, V_1)$ is as in* (d) *and $G_2$ is an irreducible linear group on a finite vector space over the same field;*

(f) $G \leq G_1 \,\mathrm{Wr}\, S_m$, *where $G_1$ is as in* (d) *or* (e).

The assumption of primitivity has been removed by Hrushovski [42]. He obtains no explicit classification, but is able to show that, for any given $b$, the finite permutation groups having at most $b$ orbits on 4-tuples fall in to a finite number of families, each family being parametrised by a finite number of 'dimensions'. Each family is characterised by a unique countable smoothly approximable structure, obtained when all the dimension parameters are infinite. Poset block structures based on a fixed poset $P$ form one of these families; the 'dimensions' are the cardinalities of the sets $X(p)$ for $p \in P$.

## 3.2 Hypergraphs and hypertournaments

Let $k$ be a positive integer. A *k-uniform hypergraph* consists of a set of vertices and a set of *hyperedges*, each a $k$-subset of the vertex set. Alternatively, it is a relational structure with a $k$-ary relation which holds only if its arguments are all different and which is completely symmetric.

Clearly the class of finite $k$-uniform hypergraphs satisfies the hypotheses of Fraïssé's Theorem. So there is a unique countable universal homogeneous $k$-uniform hypergraph $R_k$. For $k = 2$, this is just the random graph. In general, it shares many properties of the random graph. Its automorphism group is $(k-1)$-transitive (since the induced substructure on a $(k-1)$-set is trivial) but not $k$-transitive (since no automorphism can map a hyperedge to a non-edge). Indeed, this is the simplest possible proof that infinite permutation groups with all possible degrees of transitivity exist. Moreover, every countable group acts regularly as a group of automorphisms of $R_k$ for $k > 2$.

The closed supergroups of $\mathrm{Aut}(R_k)$ (that is, the reducts of $R_k$) have been determined by Thomas; all (apart from the symmetric group) are defined by suitably generalised switching. In particular, $R_k$ is isomorphic to its complement (interchanging edges and non-edges), and so the group of automorphisms and anti-automorphisms is $k$-transitive. This shows that the theorem that a non-trivial normal subgroup of a $k$-transitive group is $(k-1)$-transitive for $k > 4$ is best possible. (This theorem for finite groups is due to Jordan, and essentially the same proof works in general.)

A $k$-uniform *hypertournament* is a $k$-ary relation $R$ which holds only if all its arguments are distinct, and which is *alternating* (that is, the truth value of $R(a_1, \ldots, a_k)$ is reversed if two of the arguments are interchanged). For $k = 2$, this is just a *tournament*, which assigns an orientation to every 2-set. Fraïssé's Theorem implies the existence of a universal homogeneous $k$-uniform hypertournament for every $k > 1$. Its automorphism group is $k$-set transitive but not $k$-transitive, but is a normal subgroup of index 2 in a $k$-transitive group (consisting of permutations which either preserve or reverse the orientation of all $k$-tuples).

A remarkable theorem of Lachlan [49] (see also Cherlin [22]) determines the countable homogeneous tournaments. There is an example, variously called a *local order*, *locally transitive tournament*, or *vortex-free tournament*, defined as follows. Let $S^*$ be the set of all complex roots of unity. Let $S$ be a subset of $S^*$ which has the property that it contains exactly one of each antipodal pair of points of $S^*$ and is dense in the unit circle. (If we choose at random one element of each antipodal pair, the resulting set is dense with probability 1). Now, for $x, y \in S$, let $R(x, y)$ hold if and only if the argument of $yx^{-1}$ is between 0 and $\pi$. By choice of $S$, we obtain a tournament. The age of this countable tournament consists of all finite tournaments which do not contain a directed 3-cycle dominating or dominated by a vertex.

It follows that the isomorphism type of the tournament is independent of the choice of the set $S$.

**Theorem 3.2** *There are exactly three countable homogeneous tournaments, up to isomorphism. These are the ordered set* $\mathbb{Q}$*; the homogeneous local order; and the homogeneous universal tournament.*

See Knuth [47] for an unexpected occurrence of local orders in computational geometry.

## 3.3 Linear and circular structures

The ordered set of rational numbers is the unique countable dense unbounded linear order, by a famous theorem of Cantor. Later, Skolem proved that there is a unique way (up to order-preserving permutations) to colour the rationals red and blue in such a way that the colour classes are mutually dense and unbounded (that is, between any two red points, or above or below any red point, is a blue point, and similarly with the colours reversed).

There is a curious fact about this which is worth mentioning. It is now quite standard to use the 'back-and-forth' technique to prove results of this kind. However, Cantor did not do this, but went 'forth' only. In other words, he showed that, if two countable dense unbounded linear orders $A$ and $B$ are enumerated as $(a_1, a_2, \ldots)$ and $(b_1, b_2, \ldots)$, and if a map $\theta : A \to B$ is defined inductively by the rule that $\theta(a_{n+1})$ is the first element (in the $B$ enumeration) in the same relation to $\theta(a_1), \ldots, \theta(a_n)$ as $a_{n+1}$ has to $a_1, \ldots, a_n$, then $\theta$ is an order-isomorphism. However, this style of argument will not succeed in proving Skolem's result; back-and-forth is required.

Skolem's result is subsumed by Fraïssé's theorem, since it is easily shown that the class of finite 2-coloured linear orders satisfies the amalgamation property, and that the I-property of the countable homogeneous object is equivalent to the colour classes being mutually dense. More generally, this works with any finite number $n$ of colours. Let $H_n$ be the resulting structure.

It turns out that $\mathrm{Aut}(H_n)$ has a transitive extension, which can be constructed as follows.

Let $\Omega$ be the set of all complex roots of unity. Given $n > 1$, let $Y_n$ denote the subgroup consisting of the $n$-th roots of unity. Then any coset of $Y_n$ consists of $n$ points equally spaced around the unit circle. Let $L_n$ be a subset of $\Omega$ with the properties

(a) $L_n$ is a transversal for $Y_n$ in $\Omega$, that is, every coset of $L_n$ contains exactly one element of $Y_n$;

(b) $L_n$ is dense in $\Omega$.

Such a set $L_n$ exists, and is unique up to a permutation of $\Omega$ preserving the circular order. Indeed, if we choose independently at random one element of each coset of $L_n$, then with probability 1 the set of chosen elements will be dense.

Now define binary relations $R_j$, for $0 \leq j \leq n-1$, on $L_n$ by the rule that $(x, y) \in R_j$ if the argument of $x^{-1}y$ (the angular distance from $x$ to $y$) lies between $2\pi j/n$ and $2\pi(j+1)/n$. The relational structure consisting of $L_n$ with the circular order and the relations $R_0, \ldots, R_{n-1}$ is homogeneous, and the stabiliser of a point in its automorphism group is isomorphic to $\mathrm{Aut}(H_n)$. This can be seen as follows. Take any point $x \in L_n$; without loss, we can suppose that $x = 1$. Now project the remaining points of $Y_n$ onto the circular arc from 1 to $e^{2\pi i/n}$ by reducing the argument of each point mod $2\pi/n$. Assign to the projection of $y$ the colour $c_i$ if $(x, y) \in R_i$. Then the projected points form a countable dense ordered set coloured with $n$ colours so that the colour classes are mutually dense; and it can be shown that any automorphism of this structure can be 'lifted' to an automorphism of $L_n$ fixing $x$.

For $k = 2$, the relations $R_1$ and $R_2$ form a pair of converse (and isomorphic) tournaments; each is the countable homogeneous *local order* (see Section 3.2).

There is a reduct of $\mathrm{Aut}(L_n)$ which consists of permutations which reverse the circular order and exchange the converse relations $R_j$ and $R_{n-1-j}$ for $j = 0, \ldots, n-1$. Using this we see that various distance-transitive graphs can be obtained. The graph $R_0 \cup R_{n-1}$ is always distance-transitive; and, if $n$ is odd, the 'Borsuk-type' graph $R_{(n-1)/2}$ is also distance-transitive. This recalls how, in the finite case, if $n = 2m + 1$ is odd, the Johnson graph $J(n, m)$ and the 'odd graph' $O_{m+1}$ give two different distance-transitive structures on the same set. (In each case, the vertices are the $m$-subsets of an $n$-set; two vertices are adjacent in $J(n, m)$ if their intersection has maximum cardinality $m - 1$, and adjacent in $O_{m+1}$ if they are disjoint.)

For $n = 3$, the digraph $R_0$ is homogeneous, and arises as a 'sporadic' example in Cherlin's classification of homogeneous digraphs [21] (see Section 3.6). The undirected version ($R_0 \cup R_2$) has a curious property. A graph is *n-set homogeneous* if, whenever $X$ and $Y$ are isomorphic induced subgraphs with at most $n$ vertices, there is an automorphism of the graph mapping $X$ to $Y$. (We do not require that all isomorphisms from $X$ to $Y$ extend to automorphisms.) A graph is *set homogeneous* if it is $n$-set homogeneous for all $n$. Droste *et al.* [28] proved:

**Theorem 3.3**

(a) *The graphs $X$, $\overline{X}$ obtained from the above construction are set-homogeneous but not 3-homogeneous.*

(b) *Any countable graph which is 8-set homogeneous but not 3-homogeneous is isomorphic to one of these.*

## 3.4 Treelike objects

The structures described in this section have arisen in several different contexts, in differing disguises: the $\Lambda$-trees of Alperin and Bass [3], the direction sets or D-relations of Adeleke and Neumann [2], the projective valuations of Tits [75]. There are some differences in detail. This section gives an outline of these ideas. I begin with an approach based on ages, as in [13].

Let $T$ be a finite tree. We let $V(T)$, $E(T)$ and $L(T)$ be the sets of vertices, edges, and leaves (vertices of valency 1), respectively. Each of these sets carries a relational structure.

On $L(T)$ there is a quaternary relation, defined as follows. Let $w, x, y, z$ be four distinct leaves. There are two possibilities for the tree formed by the paths joining them, shown in Figure 3. We write $(wx|yz)$ to indicate the first possibility: the paths $wx$ and $yz$ are disjoint (but neither of the other pairs are).

On $V(T)$ there is a *betweenness relation*: we write $(x; yz)$ to indicate that $x, y, z$ are all distinct and $x$ lies on the unique path joining $y$ to $z$ in the tree. Also, given four vertices such that no three of them satisfy the betweenness relation, we have the situation described above for the leaves, and can define a quaternary relation in the same way.

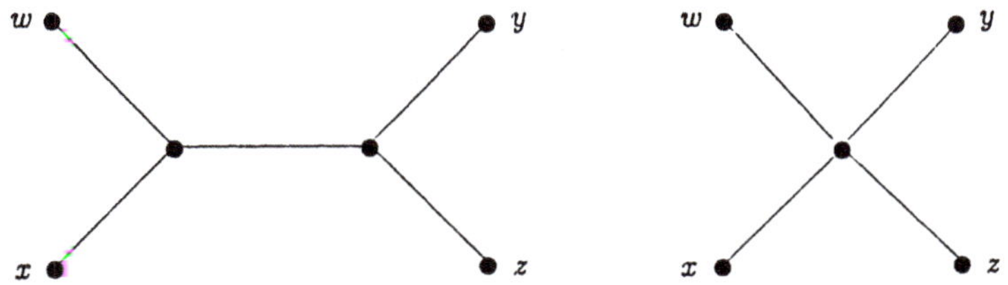


Figure 3

The betweenness and the quaternary relation on $E(T)$ are defined analogously.

**Theorem 3.4** *Each of the three classes of relational structures just described, based on either the class of all finite trees, or the class of finite trees of valency at most $n$ (where $n > 2$), satisfies the hypotheses of Fraïssé's Theorem.*

Let $G$ be a permutation group on a set $\Omega$. A *Jordan set* $\Delta$ is a subset of $\Omega$ such that the pointwise stabiliser of the complement of $\Delta$ is transitive on $\Delta$. We say that $G$ is a *Jordan group* if it has a Jordan set $\Delta$ but is not $(|\Omega\setminus\Delta| + 1)$-transitive.

An example is the group of order-preserving permutations of $\mathbb{Q}$: the subgroup fixing pointwise all the non-positive rationals acts transitively on the positive ones.

Treelike objects of the type considered above have the property that their automorphism groups are Jordan groups. For example, consider the homogeneous structure whose age is the class of leaf structures $L(T)$ for finite trees $T$. The subgroup corresponding to the stabiliser of an internal vertex $v$ is the cartesian product of groups corresponding to ends in each possible direction at $v$.

Conversely, the conclusion of a major research effort by Adeleke, Macpherson and Neumann [2], [1] has been a determination of the primitive Jordan groups of countable degree.

A *D-relation* (or *direction relation*) on a set $\Omega$ is a quaternary relation $(ab|cd)$ on $\Omega$ like the one considered above: in more detail, we require:

if $(ab|cd)$ holds, then $(ba|cd)$ and $(cd|ab)$ hold;

at most one of $(ab|cd)$, $(ac|db)$ and $(ad|bc)$ hold;

if $(ab|cd)$ holds and $e$ is another point, then either $(ab|ce)$ or $(cd|ae)$ holds.

A *C-relation* (or *chain relation*) is a ternary relation obtained by distinguishing one point in a D-relation: that is, $(a|bc)$ holds if and only if $(\infty a|bc)$ holds in the D-relation.

A *semilinear order* is a partial order in which any two elements have an upper bound, and for any $x$ the set $\{y : y \geq x\}$ is totally ordered. In a semilinear order, we can define a *B-relation* or *generalised betweenness*) by the rule that $(x; yz)$ holds if either

$$y < x < z \text{ or } z < x < y;$$

or

$$\text{either } x > y \text{ or } x > z, \text{ and } x \leq b \text{ for any upper bound } b \text{ of } y \text{ and } z.$$

Now the set of maximal chains (totally ordered subsets) of a semilinear order carries a natural C-relation, and every C-relation can be embedded in one derived from a semilinear order in this way. Similarly, the set of 'ends' of a B-relation carries a natural D-relation, and every D-relation can be embedded in one of this form.

All these relations are axiomatised, and the assertions above proved, in Adeleke and Neumann [2].

Now the theorem of Adeleke and Macpherson [1] is as follows.

**Theorem 3.5** *Let $G$ be an infinite Jordan group on $\Omega$ which is primitive but not highly homogeneous. Then one of the following occurs:*

(a) *$G$ preserves or reverses a linear or circular order;*

(b) *$G$ preserves a semilinear order, B-, C- or D-relation;*

(c) *$G$ preserves a Steiner system;*

(d) *$G$ preserves a 'limit' of B-relations, D-relations, or Steiner systems.*

The concept of 'limit' required here is rather technical, and I will not attempt to describe it. The method of Hrushovski (see Section 3.8) shows that there are groups with arbitrary degrees of transitivity which preserve Steiner systems. This result is discussed further in Macpherson's article in [45].

A recent development in combinatorial group theory has been the introduction of $\Lambda$-trees. See Alperin and Bass [3] or the forthcoming monograph by Chiswell [24] for details of this. We give a brief account. Let $\Lambda$ be an ordered abelian group: that is, an abelian group with a total order invariant under translation (that is, $a < b$ implies $a + c < b + c$). A *$\Lambda$-metric space* is a metric space $(M, d)$, where the metric $d$ takes values in the non-negative elements of $\Lambda$ instead of the non-negative reals, satisfying the standard axioms:

- $d(x, y) \geq 0$, with equality if and only if $x = y$;
- $d(x, y) = d(y, x)$;
- $d(x, z) + d(z, y) \geq d(x, y)$.

A *$\Lambda$-tree* is a $\Lambda$-metric space $T$ satisfying the further conditions:

- for any $x, y$, the set
$$[x, y] = \{z : d(x, z) + d(z, y) = d(x, y)\}$$
is isometric to $\{a \in \Lambda : 0 \leq a \leq d(x, y)\}$ (such a set is called an *interval*);
- for any $x, y, z$, there exists $w$ with $[x, y] \cap [x, z] = [x, w]$;
- if $[x, y] \cap [x, z] = \{x\}$, then $[x, y] \cup [x, z] = [y, z]$.

Any $\Lambda$-tree gives rise to a B-relation, where $x$ is between $y$ and $z$ if and only if $d(x,y) + d(y,z) = d(x,z)$. Conversely, any of our homogeneous B-relations can be made into a $\mathbb{Q}$-tree in many different ways, since its intervals and paths are all embeddable in $\mathbb{Q}$; and any B-relation is embeddable into a $\Lambda$-tree for some $\Lambda$.

Also, the set of ends of a $\Lambda$-tree carries a D-relation. A *ray* is a subset isometric to $\{a \in \Lambda : a \geq 0\}$, and as usual an *end* is an equivalence class of rays, where two rays are equivalent if their intersection is a ray. Given three ends $a, b, c$, there is a unique point $v = v(a,b,c)$ such that the ends contain rays meeting pairwise only in $v$. Four ends $a, b, c, d$ satisfy the relation $(ab|cd)$ if $v(a,b,c) = v(a,b,d) \neq v(a,c,d) = v(b,c,d)$. Conversely, any set with a D-relation can be realised as a subset of the set of ends of a $\Lambda$-tree. In our homogeneous examples, we can take $\Lambda = \mathbb{Q}$.

The difference in outlook between the theory of $\Lambda$-trees and what is described here is that the $\Lambda$-tree theorists consider groups of isometries; these are subgroups of the groups preserving the betweenness relations in the corresponding B-structures.

The projective line over a field $F$ carries a quaternary relation given by cross-ratio. The *cross-ratio* of four distinct points $a, b, c, d \in F \cup \{\infty\}$ is given by

$$\mathrm{cr}(a,b,c,d) = \frac{(a-c)(d-b)}{(a-d)(c-b)}.$$

(This version of the cross-ratio has the property that $\mathrm{cr}(\infty, 0, 1, x) = x$.) The cross-ratio takes (in general) six values under the twenty-four permutations of its arguments, the stabiliser of any one value being the Klein group of order 4. If $x$ is one of these values, then the others are $1-x$, $1/x$, $1/(1-x)$, $(x-1)/x$, and $x/(x-1)$.

A *valuation* on the field $F$ is a map $v$ from $F \backslash \{0\}$ to an ordered abelian group $\Lambda$ satisfying

(a) $v(xy) = v(x) + v(y)$;

(b) $v(x+y) \geq \min\{v(x), v(y)\}$ if $x + y \neq 0$.

We exclude the case when $v(x) = 0$ for all $x$, and assume without loss of generality that $v$ is onto $\Lambda$. The valuation is *discrete* or *indiscrete* according as the ordered abelian group $\Lambda$ is discrete or indiscrete. By convention, we sometimes put $v(0) = +\infty$, where $+\infty$ is greater than every element of $\Lambda$. Note that, if $v(x) \neq v(y)$, then $v(x+y) = \min\{v(x), v(y)\}$. For suppose that $v(x) > v(y)$. It can be shown that $v(-x) = v(x)$; so $v(y) \geq \min\{v(x+y), v(-x)\}$, whence $v(y) \geq v(x+y) \geq v(y)$, and equality holds.

The set $R = \{x \in F : v(x) \geq 0\}$ is a subring of $F$ called the *valuation ring*. The set $I = \{x \in F : v(x) > 0\}$ is an ideal of $R$. Also, if $v(x) = 0$ then $v(x^{-1}) = 0$; so every element of $R \setminus I$ is a unit of $R$, and $I$ is the unique maximal ideal of $R$.

Note that, of the six values of the cross-ratio of four points, either all six lie in $R \setminus I$, or else two (say $x$ and $x/(x-1)$ lie in $I$, two ($1-x$ and $1/(1-x)$) lie in $R \setminus I$, and two ($1/x$ and $(x-1)/x$) lie in $F \setminus R$. Now define a relation on the projective line over $F$ by the rule that $(ab|cd)$ holds if $v(\mathrm{cr}(a,b,c,d)) = 0$ but not all permutations of $a, b, c, d$ have this property. Some calculation shows that the relation satisfies the axioms of Adeleke and Neumann for a D-relation.

The $\Lambda$-tree for which this is the 'structure at infinity' can also be defined algebraically. The points are equivalence classes of *R-lattices* (that is, $R$-submodules spanned by an $F$-basis of $F^2$); two $R$-lattices are equivalent if one is obtained from the other by multiplying by a non-zero element of $F$. Given any two points $a$ and $b$, we can find representative $R$-lattices $A$ and $B$ such that $A \subseteq B$ and $B/A$ is a rank 1 module. Now given $a, b, c$, we say that $a$ is between $b$ and $c$ if we can choose representative lattices $A, B, C$ pairwise satisfying this condition such that $B \subseteq A \subseteq C$. If the valuation is discrete, we obtain an ordinary tree. This generalises the familiar trees on which $\mathrm{SL}(2, \mathbb{Q})$ acts. See Serre [70], Ronan and Tits [63].

More generally, Tits [75] has defined a *projective valuation* on a set $\Omega$ to be a function on 4-tuples of elements of $\Omega$ with values in $\Lambda$, satisfying conditions like those satisfied by $v(\mathrm{cr}(a, b, c, d))$ in the above example. Then the set of ends of a $\Lambda$-tree carries a projective valuation; any projective valuation gives rise to a D-relation; and any D-relation is representable by a subset of the set of ends of a $\Lambda$-tree.

For more details, see Tits [75], Koudela [48].

## 3.5 Graphs

First, we survey examples of homogeneous graphs of at most countable cardinality. The best-known is the random graph. There are some trivial examples: the disjoint union of $m$ complete graphs of size $n$, where $m$ and $n$ are finite or countable cardinals. (This graph is infinite provided that at least one of $m$ and $n$ is infinite.) Henson [38] noted that, for any $k \geq 3$, the class of graphs containing no complete subgraph of size $k$ satisfies Fraïssé's hypotheses. (To amalgamate two $K_k$-free graphs $B$ and $C$ over a common subgraph $A$, take their union with no further identifications of vertices and no additional edges; then any complete subgraph will lie in either $B$ or $C$.) Hence there is a unique coountable homogeneous graph $H_k$ which contains no $K_k$ but contains every $K_k$-free finite graph.

The I-property for Henson's graph $H_k$ reads as follows:

($*_k$) *Let $U$ and $V$ be disjoint sets of vertices, such that $U$ contains no $K_{k-1}$. Then there is a vertex $z$ joined to every vertex in $U$ and to no vertex in $V$.*

In addition, the complement of a homogeneous graph is homogeneous. Of the list so far, only the random graph is isomorphic to its complement. The complement of a disjoint union of complete graphs is a complete multipartite graph.

There are two 'sporadic' finite homogenous graphs, both isomorphic to their complements: the 5-cycle, and the $3 \times 3$ grid (the line graph of the complete bipartite graph $K_{3,3}$). The homogeneity of these graphs is proved by inspection.

Now we have the following characterisations:

**Theorem 3.6 (Sheehan, Gardiner)** *A finite homogeneous graph is a disjoint union of complete graphs, a complete multipartite graph, a 5-cycle, or the line graph of $K_{3,3}$.*

**Theorem 3.7 (Lachlan and Woodrow)** *A coountable homogeneous graph is a disjoint union of complete graphs, a complete multipartite graph, one of Henson's graphs $H_k$ or its complement (for $k \geq 3$), or the random graph.*

The first theorem can be proved by induction since, if $X$ is homogeneous, then the neighbours and non-neighbours of any vertex both induce homogeneous graphs, so we have to solve a number of jigsaw puzzles. In fact, the theorem has been considerably strengthened.

It is only necessary to assume that isomorphisms between sets of at most five vertices extend to automorphisms. Even this can be replaced by a purely combinatorial hypothesis. We say that the graph $X$ is *k-tuple regular* if, whenever $(x_1, \ldots, x_k)$ and $(y_1, \ldots, y_k)$ are two $k$-tuples such that the map $x_i \mapsto y_i$ $(i = 1, \ldots, k)$ is an isomorphism of induced subgraphs, then the number of common neighbours of $x_1, \ldots, x_k$ is equal to the number of common neighbours of $y_1, \ldots, y_k$. Thus, 1-tuple regularity is regularity in the usual sense, while 2-tuple regularity is strong regularity as defined by Bose [9]. Using a combination of linear algebra and the 'jigsaw puzzle' technique, Cameron [12] showed:

**Theorem 3.8** *A 5-tuple regular finite graph is a disjoint union of complete graphs, or complete multipartite, or isomorphic to a 5-cycle or the line graph of $K_{3,3}$.*

In the infinite case, induction is not available, and the proof of Lachlan and Woodrow is much more complicated. Rather than deal directly with the graphs, these authors determine which classes of finite graphs satisfy Fraïssé's conditions.

**Problem** Does there exist a number $N$ such that, if a countable graph is $n$-homogeneous for all $n \leq N$, then it is homogeneous?

We will see in the next section that the analogous statement is false for posets: any degree of homogeneity is possible.

A bipartite graph cannot be homogeneous unless it is complete bipartite or null: for two non-adjacent vertices in different blocks of the bipartition cannot be mapped by an automorphism to two vertices in the same block. For bipartite graphs, it is reasonable to specify the bipartition, so that pairs as above are not considered equivalent. More generally, we say that the graph $X$ is *almost homogeneous* if there are a finite number of relations $R_1, \ldots, R_k$ such that

(a) each $R_i$ is first-order definable (without parameters) in $X$;

(b) the structure $(X, R_1, \ldots, R_k)$ is homogeneous.

For example, suppose that a bipartite graph $X$ has diameter at most 3. Then the bipartition $B$ can be defined by

$$B = \{(x, y) : (\exists z)(x \sim z \sim y)\}.$$

The *random bipartite graph* is the unique countable graph $X$ such that, if we take two disjoint countable sets of vertices and join pairs of vertices in different sets independently with probability 1/2, the resulting graph is almost surely isomorphic to $X$. It is universal for finite bipartite graphs, and almost homogeneous; it is made homogeneous by adding a bipartition relation defined as above. It is characterised by property I, which reads as follows:

($*_B$) *If $U$ and $V$ are finite disjoint sets of vertices in the same bipartite block, then there exists a vertex $z$ in the opposite block joined to every vertex in $U$ and no vertex in $V$.*

Now we have the following characterisation:

**Theorem 3.9** *Let $X$ be a bipartite graph with the properties that the bipartite blocks are countably infinite, and the subgroup of* Aut$(X)$ *fixing them is highly transitive on each of them. Then one of the following holds:*

(a) *X is null or complete bipartite;*

(b) *X is a matching, or the complement of one in the complete bipartite graph;*

(c) *X is the random bipartite graph.*

**Proof** If condition $(*_B)$ holds, then $X$ is the random bipartite graph. So suppose that it fails, and let $n$ be the smallest value of $|U| + |V|$ for which no such $z$ exists. We have $n > 0$ trivially. By complementing if necessary, we may assume that $U \neq \emptyset$. Also note that, by hypothesis, for any pair $U', V'$ of disjoint sets in the same bipartite block as $U$ and $V$ and having the same cardinalities, there is no vertex $z$ as in $(*_B)$.

Thus there is a vertex $z$ joined to every vertex in $U \setminus \{u\}$ and to no vertex in $V$. Moreover, for any other vertex $u'$ in the same bipartite block $B$ as $u$, $z$ is not joined to $u'$. So $z$ has finite valency $|U| - 1 = k$, say. Since $B$ is fixed by a $k$-transitive group, every $k$-subset of $B$ is the neighbour set of some vertex in the same bipartite block $A$ as $z$.

Now we have $k \leq 1$. For, if $k \geq 2$, then there are two vertices of $A$ whose neighbour sets are intersecting $k$-sets, and two whose neighbour sets are disjoint $k$-sets; no automorphism can map the first pair to the second. A similar argument shows that, if $k = 1$, then distinct vertices in $A$ have distinct neighbours in $B$. So the graph is either null or a matching, and the proof is complete.

Another example of an almost homogeneous graph was discovered by Covington [25]. A graph is N-*free* if it contains no induced path of length 3. Note that, in the terminology of Theorem 2.18, if $\mathcal{F}$ consists of the path of length 3, then $X$ is N-free if and only if $\mathcal{F}(X) = \mathcal{F}(N)$, where $N$ is the null graph on the same vertex set as $X$. So N-free graphs are among those ignored in Theorem 2.18. Also, by the 'semi-strong perfect graph theorem' of Reed [62], they are all perfect.

There is no homogeneous N-free graph, by the theorem of Lachlan and Woodrow. The reason is that, in an N-free graph, any set of three vertices has (at least potentially) one distinguished vertex. This is clear (and the vertex is distinguished internally) if the induced subgraph has one or two edges (the distinguished vertex being the isolated vertex in the first case and the vertex of valency 2 in the second). Suppose that $\{x, y, z\}$ carries a complete graph. There cannot exist two vertices $u$ and $v$ such that the neighbours of $u$ and $v$ in $\{x, y, z\}$ are different 1-sets. For suppose that $u$ is joined to $x$, and $v$ to $y$. If $u \not\sim v$, then $\{u, x, y, v\}$ is a path of length 3; if $u \sim v$, then $\{v, u, x, z\}$ is such a path. Hence two subgraphs like $\{x, y, z, u\}$ and $\{x, y, z, v\}$ cannot be amalgamated. The fix is to distinguish one of the three vertices as the only possible single neighbour of an additional vertex (actual or potential). The case where $\{x, y, z\}$ is a null graph is similar.

Covington axiomatised this ternary relation and showed that there is a unique countable universal N-free graph which is made homogeneous by the addition of such a relation. It turns out that this is a structure we have met before.

Recall from Section 3.4 that there is a quaternary relation (a D-relation) on an infinite set whose age consists of the D-relations on the leaves of 'boron trees' (finite trees whose internal vertices are trivalent). We now refine this structure. Colour the internal vertices of the boron tree black and white (that is, imagine that there are two isotopes of boron). Consider the ternary relation consisting of all triples $(x, y, z)$ of leaves such that the paths joining $x, y, z$ meet at a black vertex. The resulting structure is a two-graph (see Section 2.2), that is, any

4-set contains an even number of triples which satisfy the relation. The two-graphs which arise in this way are precisely those which contain no *pentagon*. (The pentagon two-graph is the two-graph associated with the switching class containing a 5-cycle; in other words, up to isomorphism, it has points $\{a, b, c, d, e\}$ and triples $\{acd, bde, cea, dab, ebc\}$.) The class of finite structures obtained in this way (two-graph plus D-relation) satisfies Fraïssé's conditions, so there is a countable homogeneous object $M$.

If we now distinguish one vertex $x$ of $M$, the triples of the two-graph containing $x$ define a graph on $M \setminus \{x\}$, which is Covington's universal N-free graph; and the instances of the D-relation containing $x$ define a B-relation on this set, which is Covington's ternary relation.

See Cameron [13], [16] for further discussion.

## 3.6 Digraphs and posets

We saw in Section 3.2 that there are only three countable homogeneous tournaments: these are $\mathbb{Q}$ (a transitive tournament), a locally transitive tournament, and the universal tournament.

Another class of digraphs which has been extensively investigated is the class of posets. (We regard a poset as a digraph with an arc from $x$ to $y$ whenever $x < y$. The homogeneous posets were determined by Schmerl [67]. All but one are either trivial (an antichain), or minor modifications of $\mathbb{Q}$ (the disjoint union of a number of copies of $\mathbb{Q}$, or the result of blowing up each point into an antichain of fixed size). The one interesting example is the countable universal homogeneous poset (the Fraïssé limit of the class of finite posets).

The symmetry assumptions in Schmerl's theorem were substantially weakened by Droste [27], who classified the countable posets with the property that any two $k$-sets which induce isomorphic sub-posets lie in the same orbit of the automorphism group. (This condition is weaker than $k$-homogeneity, since we do not require that all isomorphisms between 3-sets extend to automorphisms. Somewhat misleadingly, Droste calls such posets *k-transitive*.)

The following statement is weaker than Droste's result, but avoids the somewhat complicated statement Droste gives.

**Theorem 3.10** *A countable* 2*-homogeneous poset* $P$ *satisfies one of the following:*

(a) $P$ *is an antichain;*

(b) $P$ *is a disjoint union of copies of* $\mathbb{Q}$*;*

(c) $P$ *is obtained from* $\mathbb{Q}$ *by replacing each point by an antichain of fixed size;*

(d) $P$ *or its dual is a semlinear order;*

(e) $P$ *contains the product of two chains of size* 2.

Schmerl's theorem follows from this and the facts

(1) a semilinear order cannot be 3-homogeneous;

(2) a homogeneous poset containing the product of two chains contains all finite posets.

Semilinear orders were defined in Section 3.4: they are posets in which any two elements have an upper bound, and the elements lying above any fixed element are linearly ordered. The semilinear orders which arise here are determined by two parameters: the first is the *ramification number*, the number of cones below each branchpoint, which is an integer greater than 1 or else $\aleph_0$; the second is the boolean parameter which is true if the semilinear order contains pairwise least upper bounds.

The last case of these theorems includes many interesting examples. Droste and Macpherson [29] showed that, for every $n$, there is a poset which is $n$-homogeneous but not $(n+1)$-homogeneous. (Recall that a structure is *$n$-homogeneous* if every isomorphism between sets of cardinality at most $n$ extends to an automorphism.) These and related matters are described in the survey article by Mekler [57].

In contrast to these classifications, Henson [39] showed that there are $2^{\aleph_0}$ non-isomorphic homogeneous digraphs, even if we exclude the possibility that two vertices are joined in both directions by an edge. The construction is similar to his construction of homogeneous graphs. If we amalgamate digraphs $B_1$ and $B_2$ in such a way that no directed edges are added between $B_1 \setminus B_2$ and $B_2 \setminus B_1$ or *vice versa*, we never create a tournament which is not present in $B_1$ or $B_2$. So, if $\mathcal{T}$ is a class of tournaments such that no member of $\mathcal{T}$ contains another, then the class of digraphs containing no member of $\mathcal{T}$ has the amalgamation property. It is not difficult to find a countable set $\mathcal{T}$ with this property. Then all of its $2^{\aleph_0}$ subsets define distinct ages, and hence distinct countable homogeneous digraphs.

The existence of uncountably many countable structures is traditionally supposed to mean that the structures are unclassifiable. However, Cherlin [20] suggested that it is possible to give a classification for the homogeneous digraphs. Subsequently he has carried out this program. An outline appears in [21]; a more detailed account is forthcoming.

## 3.7 Metric spaces

The metric spaces which mainly concern us in this section are those which arise from distance-transitive graphs. First, we must recognise these 'internally'.

We consider *integral* metric spaces, that is, metric spaces where all distances are non-negative integers. In such a metric space $(M, d)$, we form a graph (the *distance*-1 *graph* of $M$) by joining two points whose distance is 1. If the graph is connected, then the graph distance $d_g$ defines another non-negative integral metric on $M$, and we have $d \leq d_g$, as is clear by applying the triangle inequality to a shortest path in the graph between any two vertices.

The metric space $(M, d)$ is 2-homogeneous, or *distance-transitive*, if whenever $d(x, y) = d(u, v)$, there is an isometry of $M$ mapping $(x, y)$ to $(u, v)$. If $M$ is distance-transitive and the distance-1 graph is connected, then $d = d_g$. (This is proved by induction: by definition, if $d(x, y) = 1$, then also $d_g(x, y) = 1$. Suppose the equality holds for $d(x, y) \leq m$. Let $X_n = \{y : d(x, y) \leq n\}$. By connectedness, there is an edge $\{y, z\}$ with $y \in X_m$, $z \notin X_m$. Then $d(x, y) = m$ and $d(x, z) = m + 1$. Since $X_{m+1} \setminus X_m$ is an orbit of the stabiliser of $x$, every point $u$ of this set is joined to a point of $X_m$, and so satisfies $d(x, u) = m + 1$. So the equality holds for distance $m + 1$, and the inductive step is proved.)

We say that the integral metric space $M$ is *graphic* if the distance in $M$ is equal to the distance in its distance-1 graph $X$; that is, if $X$ is connected and $d = d_g$.

**Theorem 3.11** *The class of finite integral metric spaces has the amalgamation property; so there exists a unique countable homogeneous integral metric space $M$. Moreover, $M$ is graphic.*

**Proof** Let $A, B_1, B_2$ be finite integral metric spaces with $A$ a subspace of both $B_1$ and $B_2$. It suffices to consider the case where $B_1 = A \cup \{x\}$, $B_2 = A \cup \{y\}$: the general case can be done by a number of steps of this kind. We have to define an integral distance $d(x, y)$ so that the triangle inequality is satisfied. This means that, for any $a \in A$, we must have

$$|d(a,x) - d(a,y)| \leq d(x,y) \leq d(a,x) + d(a,y).$$

This could only fail if there exist $a, b \in A$ such that

$$|d(a,x) - d(a,y)| > d(b,x) + d(b,y).$$

Assume that this holds and, without loss, that $d(a,x) > d(a,y)$. Then

$$d(a,x) - d(b,x) > d(a,y) + d(b,y),$$

contradicting the fact that

$$d(a,x) - d(b,x) \leq d(a,b) \leq d(a,y) + d(b,y).$$

So the choice of $d(x, y)$ is always possible.

To show that the resulting homogeneous metric space $M$ is graphic, take any two points $x, y$, and let $d(x, y) = m$. By applying the I-property $m - 1$ times, we find $z_1, \ldots, z_{m-1}$ such that $d(z_{i-1}, z_i) = 1$ for $i = 1, \ldots, m$ (where $x = z_0$ and $y = z_m$). Hence $d_g(x, y) \leq m$, and since $d \leq d_g$ we have $d = d_g$ and $M$ is graphic.

The construction can be varied in a number of ways.

**Example 1** Consider the class of finite integral metric spaces with diameter at most $n$, for any fixed $n$. For, in the above argument, the only possible failure of amalgamation would occur if $|d(a, x) - d(a, y)| > n$, which is impossible if $d(a, x)$ and $d(a, y)$ are both at most $n$.

If we choose $n = 2$, the distance-1 graph is the random graph.

**Example 2** Consider the class of finite integral metric spaces in which each triangle has even perimeter. In this case,

$$d(a,x) - d(b,x) \equiv d(a,b) \equiv d(a,y) + d(b,y) \pmod{2},$$

and so the upper and lower bounds for $d(x, y)$ have the same parity, and we can choose $d(x, y)$ to have the same parity as these bounds.

Note that the distance-1 graph of this metric space is bipartite. Hence the distance-2 graph on a bipartite block is also distance-transitive, and in fact is isomorphic to the distance-1 graph of the universal homogeneous metric space.

We can combine this example with the previous one to produce a homogeneous metric space of diameter $n$ in which all triangles have even perimeter. For $n = 3$, the distance-1 graph is the universal almost-homogeneous bipartite graph (see Section 3.5).

**Example 3** The upper bound for $d(x,y)$ is always at least 2, so we are never forced to prescribe $d(x,y) = 1$. Hence, for fixed $m$, the class of integral metric spaces containing no $m$ points mutually at distance 1 has the amalgamation property, and we obtain a distance-transitive $K_m$-free graph. We can combine this with a diameter bound as in Example 1. For $n = 2$, we obtain Henson's homogeneous $K_m$-free graph.

We constructed bipartite distance-transitive graphs of arbitrary diameter above. What about antipodal graphs, the other type of imprimitive distance-transitive graphs? (A graph of diameter $d$ is *antipodal* if the relation of being equal or at diameter $d$ is an equivalence relation on the vertex set.) Unfortunately, simple examples exist to show that the class of finite integral metric spaces of diameter at most $d$ which are antipodal, even for $d = 3$, fail to have the amalgamation property. So further cleverness is required. Below, I give a sketch of some constructions; further details will appear elsewhere. Readers familiar with the theory of voltage graphs will recognise these concepts in another language.

Let $\Gamma$ be a graph, and $\Lambda$ be an abelian group. A 0-*cochain* on $\Gamma$ (with coefficients in $\Lambda$) is a function from vertices of $\Gamma$ to $\Lambda$, while a 1-*cochain* is a function from ordered edges of $\Gamma$ to $\Lambda$ which changes sign when the order of the edge is reversed. The 1-*coboundary* $\partial e$ of a 0-chain $e$ is given by $\partial e(x,y) = e(y) - e(x)$, while the 2-*coboundary* of a 1-chain $f$ is the function on oriented induced cycles whose value on $C$ is obtained by summing the values of $f$ on the ordered edges of $C$. Now two 1-cochains have the same 2-coboundary if and only if they differ by the 1-coboundary of a 0-chain.

**Remark** Suppose that $\Lambda$ is the cyclic group of order 2, with elements 0 and 1, and that $\Gamma$ is a complete graph. Then a 0-cochain is the characteristic function of a set of vertices, and a 1-cochain the characteristic function of the edge set of a graph. If $e$ is a 1-cochain, then $\partial e$ is a complete bipartite graph, and adding it to $f$ corresponds to switching the graph represented by $f$ with respect to the subset represented by $e$. The 2-coboundary of $f$ can be identified with the set of triples containing an odd number of edges of the graph, which is the associated two-graph. So we have generalised the relationship between switching classes of graphs and two-graphs.

For any 1-chain $f$ on $\Gamma$, we construct a graph $\Gamma_f$ as follows. The vertex set of $\Gamma_f$ is $V \times \Lambda$, where $V$ is the vertex set of $\Gamma$; vertices $(x,\alpha)$ and $(y,\beta)$ are adjacent if $f(x,y) = \beta - \alpha$. This graph is a cover of $\Gamma$, with covering projection $p$ given by $p(x,\alpha) = x$. If $f' = f + \partial e$ for some 0-cochain $e$, then the map $(x,\alpha) \mapsto (x, \alpha + e(x))$ is an isomorphism from $\Gamma_f$ to $\Gamma_{f'}$. So the covering graph depends only on the coboundary of $f$.

Now the following results hold.

**Theorem 3.12** *If $\Gamma$ is a countable graph whose age has the strong amalgamation property* (A4$'''$), *and $\Lambda$ a finite or countable abelian group, then there exists a universal homogeneous 2-coboundary on $\Gamma$* (*with coefficients in $\Lambda$*).

**Theorem 3.13** *Let $\Gamma$ denote the countable complete graph, or the countable complete bipartite graph* (*with both bipartite blocks infinite*). *Let $\Lambda$ be a finite or countable direct power of either a cyclic group of prime order or the additive group of $\mathbb{Q}$. Let $f$ be the 2-coboundary of the preceding theorem. Then the graph $\Gamma_f$ is distance-transitive and* $(d-2)$*-arc transitive of diameter $d$, where $d = 3$ or 4 is the girth of $\Gamma$.*

**Remark** In the case where $\Gamma$ is complete, the induced subgraph on the neighbours of any vertex in $\Gamma_f$ is isomorphic to the random graph $R$.

**Problem** Can other distance-transitive graphs be constructed by similar methods?

**Remark** Henson's triangle-free graph $H_3$ has a distance-transitive antipodal double cover of diameter 5, namely $K_2 \times H_3$, corresponding to the $\mathbb{Z}_2$-cochain which takes the value 1 on every edge.

An important theorem about infinite distance-transitive graphs is due to Macpherson [52]. Let $r$ and $s$ be integers greater than 1, and let $T_{r,s}$ be the unique tree with the property that every vertex in one bipartite block $B_r$ has valency $r$, and every vertex in the other block $B_s$ has valency $s$. Let $M_{r,s}$ be the graph with vertex set $B_r$, in which two vertices are adjacent whenever they lie at distance 2 in $T_{r,s}$. Then $M_{r,s}$ is not 2-connected: its blocks (maximal 2-connected subgraphs) are complete graphs of size $s$ (the neighbour sets in $T_{r,s}$ of vertices in $B_s$), and any vertex lies in $r$ of these blocks.

**Theorem 3.14** *Any infinite distance-transitive graph of finite valency is isomorphic to $M_{r,s}$ for some integers $r, s > 1$.*

## 3.8 Hrushovski's method

Hrushovski [41] has proved an extension of Fraïssé's Theorem which allows us to construct a range of objects which cannot be obtained by the techniques discussed so far. I will discuss the method only as it applies to graphs here (see also the chapter on graphs and logic in [6]), though the method is much more general (see the articles by Macpherson and Wagner in [45]). The essence of the method is that we only require some instances of amalgamation to be possible.

Let $\alpha$ be a positive real number. Define the *predimension* $\delta(G)$ of a graph $G$ by $\delta(G) = |V(G)| - \alpha|E(G)|$. For any non-negative function $f$, let $\mathcal{C}_f$ be the class of graphs $G$ with $\delta(G') \geq f(|V(G')|)$ for all induced subgraphs $G'$ of $G$. This class is clearly closed under isomorphism and under induced subgraphs. We require $f(x) \leq x$ for all $x$, to ensure that $\mathcal{C}_f$ contains arbitrarily large graphs.

We have $\delta(G \cup H) + \delta(G \cap H) \leq \delta(G) + \delta(H)$, with equality if and only if there are no edges between $G \setminus H$ and $H \setminus G$; if equality holds, we say that $G \cup H$ is the *free amalgamation* of $G$ and $H$ over $G \cap H$.

Suppose that $G, H \in \mathcal{C}_0$. We write $G \subseteq H$ to mean that $G$ is an induced subgraph of $H$. We say that $G$ is *self-sufficient* in $H$ (written $G \leq H$) if $\delta(G) \leq \delta(G')$ for all $G'$ with $G \subseteq G' \subseteq H$. Now we have:

(a) $\emptyset \leq H$ for any $H$;

(b) $G \subseteq H' \subseteq H$ and $G \leq H$ imply that $G \leq H'$;

(c) $G_1, G_2 \subseteq H$, $G_2 \leq H$ imply that $G_1 \cap G_2 \leq G_1$.

The class $\mathcal{C}$ is *good* if there is no infinite chain $G_0 \subseteq G_1 \subseteq G_2 \subseteq \ldots$ of graphs in $\mathcal{C}$ such that $\delta(G_i)$ is strictly decreasing.

Let $M$ be an infinite graph whose age is contained in $\mathcal{C}$. For finite $G \subseteq M$, we write $G \leq M$ if $G \leq H$ for all finite $H$ with $G \subseteq H \subseteq M$. If $\mathcal{C}$ is good then, for any $G \subseteq M$, there is a unique smallest $G' \leq M$ containing $G$; we call $G'$ the *closure* of $G$ in $M$, written $G' = \mathrm{cl}_M(G)$. Now Hrushovski proved the following analogue of Fraïssé's Theorem:

**Theorem 3.15** *Suppose that $\mathcal{C}$ is a good subclass of $\mathcal{C}_0$ which is closed under free amalgamation of two graphs over a self-sufficient subgraph. Then there is a countable graph $M$ with age $\mathcal{C}$, having the property that if $G \leq H$ are finite graphs, then any self-sufficient embedding of $G$ in $M$ extends to a self-sufficient embedding of $H$. Moreover, $M$ is unique up to isomorphism; and any isomorphism between finite closed subgraphs of $M$ extends to an automorphism of $M$.*

In practice, there are two situations in which $\mathcal{C}$ is good. If $\alpha$ is rational, then the values of $\delta(G)$ form a discrete set, and $\mathcal{C}$ is automatically good. For example, suppose that $\alpha = 1$ and $\mathcal{C} = \mathcal{C}_1 \cup \{\emptyset\}$. Then $\mathcal{C}$ is the class of finite forests. In this case, $\delta(G)$ is the number of connected components of $G$; we have $G \leq H$ if and only if distinct components of $G$ are contained in distinct components of $H$; and $\mathrm{cl}_M(G)$ is the union of the paths in $M$ joining vertices of $G$. The graph $M$ given by Theorem 3.15 is the disjoint union of countably many trees of countably infinite degree.

A more subtle situation occurs when $\alpha$ is irrational. In this case, the class $\mathcal{C}_f$ is good if and only if $f$ is unbounded. However, the requirement that $\mathcal{C}_f$ is closed under free amalgamation puts a constraint on $f$. This is most easily seen geometrically. We represent a graph $G$ by the point $P(G) = (|V(G)|, \delta(G))$ in the plane. Each graph in $\mathcal{C}_f$ is represented by a point between the curve $y = f(x)$ and the line $y = x$. The point representing the free amalgam of $G_1$ and $G_2$ is the fourth point of the parallelogram through $P(G_1 \cap G_2)$, $P(G_1)$ and $P(G_2)$. Now closure under free amalgamation requires that $f$ is convex and that, if $G \leq H$, then $f'(|H|)$ does not exceed the slope of the segment from $P(G)$ to $P(H)$.

Hrushovski showed, using partial fractions and the Baire category theorem, that, for continuum many irrational numbers $\alpha$ in any interval, there is a function $f = f_\alpha$ for which $f$ is unbounded and $\mathcal{C}_f$ is closed under free amalgamation.

Note that, if these conditions hold, then the graph $M$ given by Theorem 3.15 is $\aleph_0$-categorical. For, given $G \subseteq M$, there is a bound for $|V(\mathrm{cl}_M(G))|$ in terms of $|V(G)|$, since

$$f(|V(\mathrm{cl}_M(G))|) \leq \delta(\mathrm{cl}_M(G)) \leq \delta(G) \leq |V(G)|,$$

and $f$ is unbounded. Now $\mathrm{Aut}(M)$ has only finitely many orbits on self-sufficient subgraphs of bounded size, and any such subgraph contains only finitely many subgraphs of order $|V(G)|$; so $\mathrm{Aut}(M)$ is oligomorphic. This shows the existence of $2^{\aleph_0}$ countable $\aleph_0$-categorical graphs. Some of these graphs are distance-transitive.

The construction is much more flexible than indicated; it can be extended by using other classes $\mathcal{C}$, by using relational structures other than graphs, or by further intricacies in the construction itself. See Wagner's article for an account of these.

# 4 Enumeration

Let $M$ be a countable homogeneous structure with age $\mathcal{A}$ and automorphism group $G$. Then two finite subsets of $M$ lie in the same orbit of $G$ if and only if the induced substructures are isomorphic (by the definition of homogeneity). So the sequence enumerating *unlabelled $n$-element* members of $\mathcal{A}$ (that is, up to isomorphism) is identical with the sequence enumerating the orbits of $G$ on unordered $n$-element subsets of $M$. In the same spirit, the number of *labelled* $n$-element members of $\mathcal{A}$ (that is, members of $\mathcal{A}$ on the set $\{1, 2, \ldots, n\}$) is equal to the number of orbits of $G$ on ordered $n$-tuples of distinct elements of $M$.

This means that counting orbits of an oligomorphic permutation group on ordered or unordered $n$-sets is the same problem as enumerating labelled or unlabelled structures in a class which satisfies Fraïssé's hypotheses. This includes many classical combinatorial enumeration problems. The present approach is complementary to the traditional methods of combinatorial enumeration; each enriches the other. Some examples will be given here. For a more detailed study, see [17].

We let $F_n$ denote the number of orbits on $n$-tuples of distinct points of the oligomorphic group $G$, or the number of labelled $n$-element structures in the age of a homogeneous structure. (If the particular group or structure is significant, we embellish the notation to include its name.) Similarly, $f_n$ denotes the number of orbits on $n$-sets, or of unlabelled $n$-element structures.

Thus, a permutation group is *$n$-transitive* if $F_n = 1$, and *$n$-homogeneous* (or *$n$-set transitive*) if $f_n = 1$. (The term 'homogeneous' is used here in a different sense from that previously considered, so I will use the term 'set transitive' instead.) Trivially, we have

$$f_n \leq F_n \leq n! f_n,$$

since an $n$-element structure can be labelled in $n!$ different ways, though not all of these may be distinct. In particular, an $n$-transitive group is $n$-set transitive. We recall that $G$ is *highly transitive* if it is $n$-transitive for all $n$, and *highly set transitive* if it is $n$-set transitive for all $n$.

## 4.1 Growth rates

Examples of classes of structures satisfying Fraïssé's conditions may seem to be so many and varied that no general statements can be made about the growth of the sequences counting them. The fact that there are general theorems on growth rates is one of the major contributions of this point of view to enumeration theory. Of course, we cannot expect the results to be very precise!

Many of these results require that the permutation group $G = \mathrm{Aut}(M)$ is primitive in its action on $M$. As we saw earlier, primitivity is not a first-order notion. We can, however, describe it completely if $M$ is oligomorphic. In this case, there are only finitely many isomorphism types of 2-element structures, say $X_1, \ldots, X_r$, where $r = f_2$. We require that each of them defines a connected orbital graph. In other words, given $i, j$ with $1 \leq i, j \leq r$, there should exist a structure $X_{ij} = \{x_1, \ldots, x_k\} \in \mathrm{Age}(M)$ such that $\{x_s, x_{s+1}\} \cong X_i$ for $i = 1, 2, \ldots, k-1$, while $\{x_1, x_k\} \cong X_j$.

For example, the automorphism group of the random graph is primitive. For there are just two 2-vertex graphs, an edge and a non-edge; and we can take $X_{12}$ to be a path of length 2, and $X_{21}$ its complement.

It is known that $f_{n+1} \geq f_n$ for all $n$, that is, the sequence $(f_n)$ is non-decreasing. Two completely different proofs are known: an algebraic proof, relying on properties of binomial coefficients; and a proof using Ramsey's Theorem. See [14] for discussion. The conceptual outline of the algebraic proof is given in Section 4.3.

The first result on growth rate concerns sequences which don't grow at all (Cameron [11]):

**Theorem 4.1** *Let $G$ be a permutation group of countable degree which is highly set transitive but not highly transitive. Then $G$ preserves or reverses either the linear order of the rational numbers, or the circular order of the complex roots of unity.*

Here the circular order is the ternary relation $\gamma$ for which $\gamma(x,y,z)$ holds if and only if $x,y,z$ occur in that order proceeding in the anticlockwise sense round the unit circle starting at $x$. In particular, $G$ is not 4-transitive. The theorem has been quantified by Macpherson [55]: The same conclusion holds if $G$ is $(n+3)$-set transitive but not $n$-transitive. For every $n$, there is a group which is $n$-set transitive but not $n$-transitive; and there is a group which is 5-set transitive but not 4-transitive.

Macpherson [53] showed that there is a surprising gap in growth rates, between constant and exponential:

**Theorem 4.2** *There is a constant $c > 1$ with the following property. Let $G$ be an oligomorphic permutation group which is primitive but not highly set transitive. Then $f_n > c^n$ for all sufficiently large $n$.*

Macpherson's proof is a substantial argument which uses Ramsey's theorem to create large substructures with some uniformity, and then encode finite trees into these substructures. He shows that the theorem holds with $c = 2^{1/5} - \epsilon$. The smallest known growth rates for primitive groups have $c = 2 - \epsilon$.

There is a version of this theorem which does not require primitivity. To motivate it, here are two examples of imprimitive groups.

(a) For $G = S \text{ Wr } S_k$, the homogeneous structure is a countable set partitioned into $k$ countable subsets. The age consists of finite sets carrying a partition into at most $k$ parts. It is known that the number of these is asymptotic to $n^{k-1}/k!$. Exactly the same numbers, with the same asymptotics, are obtained for the group $S_k \text{ Wr } S$, where we count partitions whose parts have size at most $k$.

(b) For $G = S \text{ Wr } S$, the honogeneous structure is a set partitioned into countably many countable parts. The age consists of arbitrary partitioned finite sets. So $f_n$ is the partition number $p(n)$, which is asymptotic to $c\exp(\pi(2n/3)^{1/2})/n$. Similarly, $F_n$ is the Bell number $B(n)$.

The following result is due to Pouzet [59] and Macpherson [54]:

**Theorem 4.3** *Let $G$ be an oligomorphic permutation group. Then either*

(a) $c_1 n^d \leq f_n \leq c_2 n^d$ *for all $n$, where $d$ is a non-negative integer and $c_1, c_2$ positive constants; or*

(b) $f_n \geq \exp(n^{1/2-\epsilon})$ *for all sufficiently large $n$.*

Known examples suggest that there are further gaps below exponential in the growth rate spectrum: all lie between $\exp(n^{p/(p+1)-\epsilon})$ and $\exp(n^{p/(p+1)+\epsilon})$ for some non-negative integer $p$. Perhaps these results hold for ages of arbitrary (not necessarily homogeneous) graphs. For example:

**Problem** Is it true that if the number $f_n$ of $n$-vertex induced subgraphs of a countable graph satisfies $f_n \geq \exp(n^{1/2+\epsilon})$, then $f_n \geq \exp(n^{2/3-\epsilon})$?

There are also growth rates faster than any fractional exponential $\exp(n^c)$ for any $c < 1$ but slower than straight exponential $c^n$ for $c > 1$. Indeed, $H \operatorname{Wr} S$ has this property, for any group $H$ with growth rate faster than polynomial but slower than exponential (for example, $H = S \operatorname{Wr} S$).

Macpherson [56] has considered growth rates faster than exponential, and established connections with model-theoretic properties related to stability.

## 4.2 Cycle index

In combinatorial enumeration, it is customary to represent counting sequences by generating functions. For unlabelled structures, we take the *ordinary generating function* $f(t) = \sum_{n=0}^{\infty} f_n t^n$, and for labeled structures the *exponential generating function* $F(t) = \sum_{n=0}^{\infty} F_n t^n/n!$. In the case where $f_n$ and $F_n$ count structures in an age, these are both specialisations of a formal power series in infinitely many variables, the *cycle index*. We now define the cycle index and give some of its important properties.

Let $g$ be a permutation of an $n$-element set. The *cycle index* of $G$ is a monomial (in indeterminates $s_1, s_2, \ldots$) recording its cycle lengths:

$$z(g) = s_1^{c_1(g)} s_2^{c_2(g)} \cdots ,$$

where $c_i(g)$ is the number of cycles of $g$ of length $i$. The *cycle index* of the finite permutation group $G$ is the average of the cycle indices of its elements:

$$Z(G) = (1/|G|) \sum_{g \in G} z(g).$$

We cannot make directly any analogous definition for an infinite permutation group. However, if $G$ is oligomorphic, we can proceed as follows. For any finite set $\Delta$, let $G_\Delta^\Delta$ be the (finite) permutation group induced on $\Delta$ by its setwise stabiliser in $G$. Now define

$$\mathcal{Z}(G) = \sum_{\Delta} Z(G_\Delta^\Delta),$$

where $\Delta$ runs through a set of representatives for the $G$-orbits on finite sets. Note that this sum is well-defined if and only if $G$ is oligomorphic. For any individual monomial $s_1^{m_1} s_2^{m_2} \cdots$ arises from terms in the sum with $|\Delta| = m_1 + 2m_2 + \cdots$; and $G$ is oligomorphic if and only if there are only finitely many such terms involving any given monomial.

If $G$ is finite, we obtain nothing new: it can be shown that

$$\mathcal{Z}(G) = Z(G; s_i \leftarrow s_i + 1),$$

where the notation means 'substitute $s_i + 1$ for $s_i$ for all $i$'.

If $G$ is the automorphism group of a homogeneous structure $M$, then $\mathcal{Z}(G)$ is the sum, over all unlabelled structures $X$ in $\operatorname{Age}(M)$, of the cycle index $Z(\operatorname{Aut}(X))$ of the automorphism group of $X$. For example, $\mathcal{Z}(\operatorname{Aut}(R))$ is the sum of the cycle indices of all finite graphs.

The generating functions of the numbers $f_n$ and $F_n$ of orbits on unordered and ordered $n$-tupes are both specialisations of $\mathcal{Z}$:

$$\sum_{n\geq 0} f_n t^n = \mathcal{Z}(G; s_i \leftarrow t^i);$$

$$\sum_{n\geq 0} F_n t^n/n! = \mathcal{Z}(G; s_1 \leftarrow t, s_i \leftarrow 0 \text{ for } i > 1).$$

Also, the cycle index of a direct product (in its intransitive action) or a wreath product (in its imprimitive action) can be calculated from the cycle indices of the factors. For a direct product, we simply multiply the cycle indices of the factors.

The cycle indices of a number of particular groups have been calculated. For further details, see [17].

## 4.3 Algebra and reconstruction

In this section and the next, we explore some aspects of the sequences $(f_n)$ and $(F_n)$ attached to oligomorphic permutation groups. As well as counting things, sequences of positive integers often record the dimensions of the homogeneous components of graded algebras. We construct such an algebra for the sequence $(f_n)$.

Let $\Omega$ be an infinite set. Let $\binom{\Omega}{n}$ denote the set of $n$-element subsets of $\Omega$, $V_n$ the vector space of functions from $\Omega$ to $\mathbb{Q}$. Set $\mathcal{A} = \bigoplus_{n\geq 0} V_n$, with multiplication defined as follows: for $f \in V_n$, $g \in V_m$, and $X \in \binom{\Omega}{n+m}$,

$$(fg)(X) = \sum_{Y\in\binom{X}{n}} f(Y)g(X\backslash Y).$$

This is the *reduced incidence algebra* of the poset of finite subsets of $\Omega$ (Rota [64]). It is a commutative and associative algebra with identity, but is far from an integral domain: any function with finite support is nilpotent. However, there is a particular element $e$ which is known not to be a zero divisor: the constant function in $V_1$ with value 1.

Now, if $G$ is any permutation group on $\Omega$, let $\mathcal{A}^G = \bigoplus_{n\geq 0} V_n^G$, where $V_n^G$ consists of the functions in $V_n$ which are $G$-invariant (where $G$ acts on $V_n$ in the natural way: $f^g(X) = f(Xg^{-1})$). Now a function in $V_n$ is fixed by $G$ if and only if it is constant on the $G$-orbits. So, if $G$ is oligomorphic, then $\dim(V_n^G) = f_n(G)$ is the number of orbits of $G$ on $\binom{\Omega}{n}$. Since $e \in V_1^G$ is not a zero-divisor, multiplication by $e$ is a monomorphism from $V_n^G$ to $V_{n+1}^G$ for all $n$. This implies that the numbers $f_n$ are non-decreasing: $f_{n+1} \geq f_n$ for all $n$, as noted in Section 4.1.

In some special cases, the structure of the algebra can be determined. Here are a few examples. For the direct product (in its intransitive action), we have

$$\mathcal{A}^{G\times H} = \mathcal{A}^G \otimes_{\mathbb{Q}} \mathcal{A}^H.$$

If $G = \mathrm{Aut}(R)$, then $\mathcal{A}^G$ is a polynomial algebra, whose free generators correspond to the connected finite graphs. This is a special case of a general result, where a 'good theory of connectedness' for $\mathrm{Age}(M)$ guarantees that $\mathcal{A}^{\mathrm{Aut}(M)}$ is a polynomial algebra. If $G$ is the wreath product of an infinite symmetric group with a finite permutation group $H$, then $\mathcal{A}^G$

is isomorphic to the classical ring of invariants of $H$. If $G$ is the automorphism group of one of the circular structures of Section 3.3, then $\mathcal{A}^G$ is the so-called 'shuffle algebra' arising in the theory of free Lie algebras. See my paper in [33] for further details.

A twenty-year-old conjecture asserts that, if $G$ is any permutation group with no finite orbits, then $\mathcal{A}$ is an integral domain.

The celebrated Vertex Reconstruction Conjecture for graphs, due to Kelly [46], can also be formulated in algebraic terms. Take $G = \mathrm{Aut}(R)$, so that a basis for $V_n^G$ consists of the characteristic functions $\chi_X$ of all unlabelled $n$-vertex graphs $X$. Consider the map $\theta = \theta_n : V_n^G \to V_{n-1}^G$ which is dual to multiplication by $e$:

$$\theta(\chi_X) = \sum_{x \in X} \chi_{X \setminus \{x\}}.$$

Thus, $\theta$ maps the graph $X$ to the sum of the graphs in its deck of vertex-deleted subgraphs. Since multiplication by $e$ is one-to-one, $\theta$ is onto. A similar definition can be made for any oligomorphic permutation group $G$: in place of the characteristic functions of $n$-vertex graphs, we use the orbits of $G$ on $n$-sets.

It is easy to show that a vector of weight 2 in the kernel of $\theta$ (that is, a linear combination $a\chi_X + b\chi_Y$) must satisfy $b = -a$, and $X$ and $Y$ must have the same decks of vertex-deleted subgraphs. So the Vertex Reconstruction Conjecture asserts that, for $n > 2$, the kernel of $\theta$ has minimum weight greater than 2. This immediately suggests the problem: what is the minimum weight of $\ker(\theta)$? Even if we cannot prove the Conjecture, can we find a good upper bound for it?

A number of variants of the Conjecture can be posed, for graphs and other classes of structures. See [17] for details.

The algebraic content of this section has been formalised in various ways. On the one hand, Joyal [43] and his school study species, which are formal power series where the coefficients are combinatorial objects rather than numbers: specialisation from an object to a numberical property includes virtually everything we have done above. Also, in the case where the sets of interest are edge sets of graphs, an approach going back to Whitney [80] and Tutte [78] for counting subgraphs (see also Biggs [7]) has been put into a Hopf algebra framework by Ray [61]. But it would take us too far afield to survey these developments.

## 4.4 Derangements and differential equations

The exponential generating function $F(t) = \sum F_n t^n/n!$ for the numbers $F_n$ is, in many respects, better-behaved than the ordinary generating function for the numbers $f_n$. In this section we consider some of its properties.

For finite groups, it has a close connection with the distribution of the numbers of fixed points of elements. Let $G$ be a finite permutation group of degree $n$. Let $x_m$ be the proportion of elements of $G$ having exactly $m$ fixed points, and let $X(t)$ be their generating function: $X(t) = \sum x_m t^m$. Thus $X(0) = x_0$ is the probability that a random element of $G$ is a *derangement* or fixed-point-free element.

**Theorem 4.4** $X(t) = F(t-1)$.

**Proof** According to the Orbit-Counting Lemma, the number $F_m$ of orbits of $G$ on $m$-tuples of distinct elements is the average number of $m$-tuples fixed by an element of $G$. An element with $l$ fixed points has $l(l-1)\cdots(l-m+1)$ fixed $m$-tuples, and there are $x_l|G|$ such elements. So we have

$$F_m = \sum_{l=m}^{n} x_l l(l-1)\cdots(l-m+1).$$

Multiplying by $t^m/m!$ and some manipulation gives

$$F(t) = \sum_{m=0}^{n}\sum_{l=m}^{n}\binom{l}{m}x_l t^m = X(t+1),$$

from which the result follows.

In particular, the probability that an element of $G$ is a derangement is given by $F(-1)$. This gives, for example, a simple proof for the well-known probability that an element of $S_n$ is a derangement. For $G = S_n$, we have $F_m = 1$ for $m \leq n$ and $F_m = 0$ otherwise; so

$$F(t) = \sum_{m=0}^{n} t^m/m!$$

is the truncated exponential series. Putting $t=-1$ shows that the probability is close to $1/\mathrm{e}$.

For an infinite permutation group $G$, the numbers $x_n$ make no sense (except in the very special case where there is a natural probability measure on $G$, and every element of $G$ has only finitely many fixed points). However, if $G$ is oligomorphic, then $F(t)$ is defined as a formal power series, and we may be able to calculate $F(-1)$, either as a value of the function $F(t)$, or by some summability method (such as analytic continuation). It is never very clear what the significance of this number is. Four examples illustrate the point.

**Example 1** Let $G = S$, an infinite symmetric group. Then $F(t) = \mathrm{e}^t$, and $F(-1) = 1/\mathrm{e}$. This is the limiting probability of derangements in finite symmetric groups.

**Example 2** Let $G = A$, the group of order-preserving permutations of $\mathbb{Q}$. There is just one orbit on $n$-sets, giving rise to $n!$ orbits on $n$-tuples (corresponding the the $n!$ orderings of its elements). So

$$F(t) = \sum t^n = \frac{1}{1-t},$$

and $F(-1) = 1/2$. Note that $F(t)$ has radius of convergence 1, with a pole at $t=1$, and can be analytically continued to the rest of the complex plane (giving Euler's 'formula' $1-1+1-1+\cdots = \frac{1}{2}$). I do not know of any 'interpretation' of this number connected to derangements.

**Example 3** Let $G^{\{n\}}$ and $G$ denote the permutation groups induced on 2-element subsets by the symmetric groups $S_n$ and $S$ respectively, and $F^{\{n\}}(t)$ and $F(t)$ the corresponding generating functions. Now $F_m^{\{n\}}$ is the number of graphs with $n$ unlabelled vertices and $m$ labelled edges. This number increases with $n$ until $n = 2m$ and then remains constant; this constant limiting value is $F_m$. (We ignore isolated vertices here.) So the power series $F^{\{n\}}(t)$ converge termwise to $F(t)$, in a strong sense. Moreover, known results on the distribution of cycles in random permutations show that $F^{\{n\}}(-1)$ tends to the limit $2\mathrm{e}^{-3/2}$ as $n \to \infty$.

**Problem:** Is it true that $F(-1) = 2\mathrm{e}^{-3/2}$ (in some sense)?

**Example 4** For $G = \mathrm{Aut}(R)$, we have $F_n = 2^{n(n-1)/2}$ (the number of graphs on $n$ labelled vertices). This grows so fast that $F(t)$ does not converge in any neighbourhood of the origin. Is there any summability method which gives a value to $F(-1)$?

The series $F(t)$ for a direct product (in the intransitive action) or a wreath product (in the imprimitive action) can be calculated from those of the factors:

$$\begin{aligned} F_{G\times H}(t) &= F_G(t)F_H(t), \\ F_{G\mathrm{Wr}H}(t) &= F_H(F_G(t)-1). \end{aligned}$$

Also, if $G$ is transitive, then the series for the point stabiliser $G_\alpha$, acting on the remaining points, can be calculated:

$$F_{G_\alpha}(t) = \frac{\mathrm{d}}{\mathrm{d}t}F_G(t).$$

If $G$ is intransitive, then the derivative of $F_G(t)$ gives the sum of $F_{G_\alpha}(t)$, over a set of representatives $\alpha$ of the orbits of $G$.

These properties sometimes allow us to express a group-theoretic property of $G$ as a differential equation for $F_G(t)$. Here are two examples.

**Example 1** What can be said about a group $G$ for which $G_\alpha \cong G \times G$? For such a group, we have $F'(t) = F(t)^2$. This equation, with the initial condition $F(0) = 1$, has the solution $F(t) = 1/(1-t)$, so $F_n = n!$ for all $n$. We recognise that this holds for the group $G = A = \mathrm{Aut}(\mathbb{Q}, <)$. Indeed, this group satisfies the condition, since, for any rational $\alpha$, both $\{q \in \mathbb{Q} : q < \alpha\}$ and $\{q \in \mathbb{Q} : q > \alpha\}$ are order-isomorphic to $\mathbb{Q}$, and the stabiliser of $\alpha$ is the direct product of groups acting on these two sets.

There is another permutation group $G$ with $F_n = n!$, namely $G = C \mathrm{\ Wr\ } S$, where $C$ preserves the cyclic order of the roots of unity and $S$ is the infinite symmetric group. (An orbit of $G$ on $n$-tuples, that is, a labelled $n$-element structure in the age of the homogeneous structure preserved by $G$, consists of a partition of $\{1,\ldots,n\}$ into non-empty subsets and a cyclic order on each part. Such structures correspond to the cycle decomposition of permutations in $S_n$, and so there are $n!$ of them.) This group does not satisfy the condition $G_\alpha \cong G \times G$. Nevertheless, both of these groups satisfy the related equation $G_\alpha \cong G \times A$.

**Example 2** The equation $G_\alpha \cong G \mathrm{\ Wr\ } A$ is satisfied by a permutation group, the automorphism group of one of the treelike structures in Section 3.4. We take the structure $M$ derived from the leaves of trivalent trees (the Fraïssé limit of the structures $L(T)$, where $T$ is a finite tree all of whose non-leaves are trivalent). Now distinguish one point $a$ in $M$, and let $G$ be its stabiliser, acting on the remaining points. (Thus $G$ is the automorphism group of a C-relation.) Given any other point $x$ in $M$, there is a 'path' from $a$ to $x$ (the union of two rays meeting at a point) in the treelike object; this path is dense, and so is order-isomorphic to $\mathbb{Q}$. At each point $q$ of the path, one treelike object or C-relation isomorphic to $M_a$ (not containing the rays $[q,a)$ or $[q,x)$) grows, with $q$ as its distinguished point. Any further leaf $y$ of $M$ lies in a unique such treelike object, growing at the point $v(a,x,y)$, in our earlier notation. So the leaves different from $a$ and $x$ carry a congruence, where each congruence class carries a structure isomorphic to $M_a$, and the set of classes carries a dense linear order. This gives the stated equation for $G = \mathrm{Aut}(M_a)$.

**Exercise:** Use the equation $G_\alpha \cong G \mathrm{\ Wr\ } A$ to calculate $F_G(t)$, and hence show that $F_n = (2n-3)!! = 1\cdot 3\cdot 5\cdots(2n-3)$ for $n \geq 2$.

## References


[1] Adeleke, S. A. and Macpherson, H. D., Classification of infinite primitive Jordan permutation groups, *Proc. London Math. Soc. (3)* **72** (1996), 63–123.

[2] Adeleke, S. A. and Neumann, P. M., Relations related to betweenness: their structure and automorphisms, *Memoirs Amer. Math. Soc.*, to appear.

[3] Alperin, R. C. and Bass, H., Length functions of group actions on Λ-trees, in: *Combinatorial Group Theory and Topology* (S. M. Gersten and J. R. Stallings, eds.), Princeton University Press, Princeton, 1987, 265-378.

[4] Babai, L. and Goodman, A. J., On the abstract group of automorphisms, in: *Coding Theory, Design Theory, Group Theory* (D. Jungnickel and S. A. Vanstone, eds.), Wiley, New York, 1993, 121–143.

[5] Bailey, R. A., Orthogonal partitions in designed experiments, *Des. Codes Cryptogr.* **8** (1996), 45–77.

[6] Beineke, L. W. and Wilson, R. J., *Graph Connections*, Oxford University Press, Oxford, 1996.

[7] Biggs, N. L., *Interaction Models*, Cambridge University Press, Cambridge, 1977.

[8] Blass, A. and Harary, F., Properties of almost all graphs and complexes, *J. Graph Theory* **3** (1979), 225–240.

[9] Bose, R. C., Strongly regular graphs, partial geometries, and partially balanced designs, *Pacific J. Math.* **13** (1963), 389–419.

[10] Brouwer, A. E., Cohen, A. M. and Neumaier, A., *Distance-Regular Graphs*, Springer, Berlin, 1989.

[11] Cameron, P. J., Transitivity of permutation groups on unordered sets, *Math. Z.* **48** (1976), 127–139.

[12] Cameron, P. J., 6-transitive graphs, *J. Combin. Theory Ser.B* **28** (1980), 168–179.

[13] Cameron, P. J., Some treelike objects, *Quart. J. Math. (Oxford) (2)* **38** (1987), 155–183.

[14] Cameron, P. J., *Oligomorphic Permutation Groups*, London Math. Soc. Lecture Note Ser. **152**, Cambridge University Press, Cambridge, 1990.

[15] Cameron, P. J., The age of a relational structure, *Discrete Math.* **95** (1991), 49–67.

[16] Cameron, P. J., Two-graphs and trees, *Discrete Math.* **127** (1994), 63–74.

[17] Cameron, P. J., Stories about groups and sequences, *Des. Codes Cryptogr.* **8** (1996), 109–134.

[18] Cameron, P. J., The random graph, in: *The Mathematics of Paul Erdős*, Springer, Berlin, 1996.

[19] Cameron, P. J. and Martins, C., A theorem on reconstructing random graphs, *Combin. Probab. Comput.* **2** (1993), 1–9.

[20] Cherlin, G., Homogeneous directed graphs, *J. Symbolic Logic* **52** (1987), 296.

[21] Cherlin, G., Homogeneous directed graphs, in: *Finite and Infinite Combinatorics in Sets and Logic* (N. Sauer, R. E. Woodrow and B. Sands, eds.), Kluwer, Dordrecht, 1993, 81–95.

[22] Cherlin, G., Homogeneous tournaments revisited, *Geom. Dedicata* **26** (1988), 231–239.

[23] Cherlin, G., Harrington L. and Lachlan, A. H., $\aleph_0$-categorical, $\aleph_0$-stable structures, *Ann. Pure Appl. Logic* **28** (1985), 103–135.

[24] Chiswell, I. M., Λ-trees, *QMW Maths Notes*, Queen Mary and Westfield College, to appear.

[25] Covington, J., A universal structure for N-free graphs, *Proc. London Math. Soc. (3)* **58** (1989), 1–16.

[26] Delsarte, P., The association schemes of coding theory, *Philips Res. Rept. Suppl.* **10** (1973).

[27] Droste, M., *Structure of Partially Ordered Sets With Transitive Automorphism Groups, Memoirs Amer. Math. Soc.* **334**, American Mathematical Society, Providence, RI, 1985.

[28] Droste, M., Giraudet, M, Macpherson, H. D., and Sauer, N. W., Set-homogeneous graphs, *J. Combin. Theory Ser.B* **62** (1994), 63–95.

[29] Droste, M. and Macpherson, H. D., On $k$-homogeneous posets and graphs, *J. Combin. Theory Ser.A* **56** (1991), 1–15.

[30] El-Zahar, M. and Sauer, N. W., Ramsey-type properties of relational structures, to appear in *Discrete Math.*

[31] Engeler, E., Äquivalenzklassen von $n$-Tupeln, *Z. Math. Logik Grundlag. Math.* **5** (1959), 340–345.

[32] Erdős, P. and Rényi, A., Asymmetric graphs, *Acta Math. Acad. Sci. Hungar.* **14** (1963), 295–315.

[33] Evans, D. M. (ed.), *Proc. RESMOD Conference on Models and Groups (Blaubeuren 1995)*, to appear.

[34] Fagin, R., Probabilities on finite models, *J. Symbolic Logic* **41** (1976), 50–58.

[35] Fraïssé, R., Sur certains relations qui généralisent l'ordre des nombres rationnels, *C. R. Acad. Sci. Paris* **237** (1953), 540–542.

[36] Gardiner, A. D., Homogeneous graphs, *J. Combin. Theory Ser.B* **20** (1976), 94–102.

[37] Glebskii, Y. V., Kogan, D. I., Liogon'kii, M. I. and Talanov, V. A., Range and degree of realizability of formulas in the restricted predicate calculus, *Kibernetika* **2** (1969), 17–28.

[38] Henson, C. W., A family of countable homogeneous graphs, *Pacific J. Math.* **38** (1971), 69–83.

[39] Henson, C. W., Countable homogeneous relational structures and $\aleph_0$-categorical theories, *J. Symbolic Logic* **37** (1972), 494–500.

[40] Hodges, W. A., *Model Theory*, Cambridge University Press, Cambridge, 1993.

[41] Hrushovski, E., A new strongly minimal set, *Ann. Pure Appl. Logic* **62** (1993), 147–166.

[42] Hrushovski, E., Finite structures with few types, in: *Finite and Infinite Combinatorics in Sets and Logic* (N. W. Sauer, R. E. Woodrow, B. Sands, eds.), Kluwer, Dordrecht, 1996.

[43] Joyal, A., Une theorie combinatoire des séries formelles, *Adv. Math.* **42** (1981), 1–82.

[44] Kantor, W. M., Liebeck, M. W. and Macpherson, H. D., $\aleph_0$-categorical structures smoothly approximable by their finite substructures, *Proc. London Math. Soc. (3)* **59** (1989), 439–463.

[45] Kaye R. W. and Macpherson, H. D. (eds.), *Automorphisms of First-Order Structures*, Oxford University Press, Oxford, 1994.

[46] Kelly, P. J., On some mappings related to graphs, *Pacific J. Math.* **14** (1964), 191–194.

[47] Knuth, D. E., *Axioms and Hulls*, Lecture Notes in Comput. Sci. **606**, Springer, Berlin, 1992.

[48] Koudela, E., Ph.D. thesis, University of London, 1996.

[49] Lachlan, A. H., Countable homogeneous tournaments, *Trans. Amer. Math. Soc.* **284** (1984), 431–461.

[50] Lachlan, A. H. and Woodrow, R. E., Countable ultrahomogeneous undirected graphs, *Trans. Amer. Math. Soc.* **262** (1980), 51–94.

[51] Lovász, L., Normal hypergraphs and the perfect graph conjecture, *Discrete Math.* **2** (1972), 253–267.

[52] Macpherson, H. D., Infinite distance-transitive graphs of finite valency, *Combinatorica* **2** (1982), 63–69.

[53] Macpherson, H. D., The action of an infinite permutation group on the unordered subsets of a set, *Proc. London Math. Soc. (3)* **46** (1983), 471–486.

[54] Macpherson, H. D., Growth rates in infinite graphs and permutation groups, *Proc. London Math. Soc. (3)* **51** (1985), 285–294.

[55] Macpherson, H. D., Homogeneity in infinite permutation groups, *Period. Math. Hungar.* **17** (1986), 211–233.

[56] Macpherson, H. D., Permutation groups of rapid growth, *J. London Math. Soc. (2)* **35** (1987), 276–286.

[57] Mekler, A. H., Homogeneous partially ordered sets, in: *Finite and Infinite Combinatorics in Sets and Logic* (N. Sauer, R. E. Woordow and B. Sands, eds.), Kluwer, Dordrecht, 1993, 279–288.

[58] Oxtoby, J. C., *Measure and Category*, Springer, Berlin, 1980.

[59] Pouzet, M., Application de la notion de relation presque-enchaînable au dénombrement des restrictions finies d'une relation, *Z. Math. Logik Grundlag. Math.* **27** (1981), 289–332.

[60] Rado, R., Universal graphs and universal functions, *Acta Arith.* **9** (1964), 331–340.

[61] Ray, N, Tutte algebras of graphs and formal group theory, *Proc. London Math. Soc. (3)* **65** (1992), 23–45.

[62] Reed, B., A semi-strong perfect graph theorem, *J. Combin. Theory Ser.B* **43** (1987), 223–240.

[63] Ronan, M. A. and Tits, J., Twin trees I, *Invent. Math.* **116** (1994), 463–479.

[64] Rota, G.-C , On the foundations of combinatorial theory, I: Theory of Möbius functions, *Z. Wahrscheinlichkeitstheorie und Verw. Gebiete* **2** (1964), 340–368.

[65] Ryll-Nardzewski, C., On category in power $< \aleph_0$, *Bull. Acad. Polon. Sci. Sér. Math. Astronom. Phys.* **7** (1959), 545–548.

[66] Sabidussi, G., Vertex-transitive graphs, *Monatsh. Math.* **68** (1964), 426–438.

[67] Schmerl, J H., Countable homogeneous partially ordered sets, *Algebra Universalis* **9** (1979), 317–321.

[68] Seidel, J. J., Strongly regular graphs of $L_2$-type and of triangular type, *Proc. Kon. Nederl. Akad. Wetensch. Ser. A* **70** (= *Indag. Math.* **29**) (1967), 188–196.

[69] Seidel, J. J., A survey of two-graphs, in: *Colloquio Internazionale sulle Teorie Combinatorie (Roma 1973)*, Accad. Naz. Lincei, Rome, 1976, 481–511.

[70] Serre, J.-P., *Trees*, Springer, Berlin, 1980.

[71] Sheehan, J., Smoothly embeddable subgraphs, *J. London Math. Soc. (2)* **9** (1974), 212–218.

[72] Skolem, S., Logisch-kombinatorische Untersuchungen über die Erfüllbarkeit oder Beweisbarkeit mathematischer Sätze, nebst einem Theoreme über dichte Mengen, *Krist. Vid. Selsk. Skr.* **1** (1920), Nr.4.

[73] Speed, T. P. and Bailey, R. A., On a class of association schemes derived from lattices of equivalence relations, in: *Algebraic Structures and Applications* (P. Schultz, C. E. Praeger, R. P., Sullivan, eds.), Marcel Dekker, New York, 1982, 55–74.

[74] Svenonius, L., $\aleph_0$-categoricity in first-order predicate calculus, *Theoria* **25** (1959), 82–94.

[75] Tits, J., Immeubles de type affine, in: *Buildings and the Geometry of Diagrams* (L. Rosati, ed.), Lecture Notes in Math. **1181**, Springer, Berlin, 1986, 159–190.

[76] Thomas, S. R., Reducts of the random graph, *J. Symbolic Logic* **56** (1991), 176–181.

[77] Truss, J. K., The group of the countable universal graph, *Math. Proc. Cambridge Philos. Soc.* **98** (1985), 213–245.

[78] Tutte, W. T., A ring in graph theory, *Proc. Cambridge Philos. Soc.* **43** (1947), 26–40.

[79] Vaught, R. L., Models of complete theories, *Bull. Amer. Math. Soc.* **69** (1963), 299–313.

[80] Whitney, H., The colouring of graphs, *Ann. Math.* **33** (1932), 688–718.

[81] Zil'ber, B. I., The structure of models of uncountably categorical theories, *Proc. Internat. Congr. Math.* (Warsaw 1983), 359–368.

# Symmetry and eigenvectors

Ada CHAN and Chris D. GODSIL *

*Department of Combinatorics and Optimization*
*University of Waterloo*
*Waterloo, Ontario, N2L 3G1*
*Canada*

**Abstract**

We survey some of what can be deduced about automorphisms of a graph from information on its eigenvalues and eigenvectors. Two of the main tools are convex polytopes and a number of matrix algebras that can be associated to the adjacency matrix of a graph.

## 1 Eigenvectors and eigenpolytopes

We study eigenvectors of the adjacency matrix of a graph, and how these interact with the graph's automorphisms. The treatment in the first four sections is based loosely on [10, 16].

### 1.1 Eigenvalues and automorphisms

Our aim in this section is to introduce the idea that the eigenvectors of the adjacency matrix of a graph are best viewed as a class of functions on the vertices of the graph. So we present some definitions and then use them to derive some classical results concerning simple eigenvalues of vertex-transitive graphs.

Let $X$ be a graph with vertex set $V$. If $i$ and $j$ are vertices in $X$, we write $i \sim j$ to denote that $i$ is adjacent to $j$. Let $F(V)$ denote the space of real functions on $V$. We denote the characteristic vector of the vertex $i$ of $X$ by $e_i$; the vectors $e_i$ form the *standard basis* of $F(V)$. The *adjacency operator* $\mathcal{A}$ is the linear operator on $F(V)$ defined by

$$(\mathcal{A}f)(i) := \sum_{j \sim i} f(j).$$

The matrix that represents $\mathcal{A}$ relative to the standard basis is the usual adjacency matrix $A$ of $X$.

The inner product of two functions $f$ and $g$ from $F(V)$ is

$$\sum_{i \in V} f(i)g(i);$$

*Support from a Natural Sciences and Engineering Research Council of Canada operating grant is gratefully acknowledged.



*G. Hahn and G. Sabidussi (eds.), Graph Symmetry*, 75–106.

the adjacency operator is self-adjoint relative to this inner product (equivalently, the adjacency matrix is symmetric). We denote the function on $V$ that is constant and equal to 1 by $\mathbf{1}$. A function $f$ is an eigenvector for $\mathcal{A}$ if it is not zero and there is a constant $\theta$ such that, for each vertex $i$ in $V$,

$$\theta f(i) = \sum_{j \sim i} f(j),$$

or $\mathcal{A}f = \theta f$. The constant $\theta$ is, of course, the eigenvalue of $f$. The graph $X$ is regular if and only if $\mathbf{1}$ is an eigenvector for $\mathcal{A}$, with the corresponding eigenvalue equal to the valency of $X$.

Let $f$ be an eigenvector for $\mathcal{A}$ and let $i$ be the vertex of $X$ such that $|f(i)| \geq |f(j)|$, for all $j \in V$. Then

$$|\theta||f(i)| = |\sum_{j \sim i} f(j)| \leq \sum_{j \sim i} |f(j)|,$$

and therefore,

$$|\theta| \leq \sum_{j \sim i} \frac{|f(j)|}{|f(i)|}.$$

If $X$ is connected then equality holds if and only if $|f|$ is a constant function. If $f$ itself is constant, then $X$ is regular, otherwise $X$ is regular and bipartite. In both cases, the function $|f|$ is an eigenvector. Moreover, by our choice of $i$, we can conclude that $|\theta|$ is less than or equal to the valency of $i$ and, thus, the absolute value of all eigenvalues for $\mathcal{A}$ is bounded above by the maximum valency of $X$.

If $X$ is a complete graph on the vertex set $V$ of size $n$, then since it is regular, it has eigenvector $\mathbf{1}$ with eigenvalue $(n-1)$. Let $f \in F(V)$ such that $\sum_{i \in V} f(i) = 0$. Then for all $i \in V$,

$$\sum_{j \sim i} f(j) = -f(i),$$

and therefore $-1$ is an eigenvalue for $X$ with $n-1$ linearly independent eigenvectors. If $X$ is a circuit on $n$ vertices and $\tau$ is the primitive $n^{th}$ root of unity, then the function $f(i) = \tau^i$, for $i = 0 \ldots n-1$, is an eigenvector for $X$ with eigenvalue $(\tau + \tau^{-1})$.

We denote the automorphism group of $X$ by $\mathrm{Aut}(X)$; if $i \in V$ and $a \in \mathrm{Aut}(X)$ then $ia$ is the image of $i$ under $a$. If $f \in F(V)$ and $a \in \mathrm{Aut}(X)$ then the composition $f \circ a$ is again a real function on $X$, with

$$(f \circ a)(i) = f(ia).$$

The map

$$f \mapsto f \circ a$$

is an invertible linear mapping of $F(V)$ onto itself. If we need to distinguish the above mapping from the automorphism $a$, we will denote it by $\hat{a}$. Our conventions imply that

$$\widehat{ab} f = \hat{a}\hat{b} f.$$

If $a \in \mathrm{Sym}(V)$ then

$$\hat{a} e_i = e_{ia^{-1}},$$

which is probably not what we would expect.

Now we show that if $a \in \mathrm{Sym}(V)$, then $a \in \mathrm{Aut}(X)$ if and only if $\hat{a}$ and $\mathcal{A}$ commute. Equivalently, if $a \in \mathrm{Aut}(X)$ then $\hat{a}$ lies in the *centralizer* $C(\mathcal{A})$ of $\mathcal{A}$. Consider for any $f \in F(V)$ and $i \in V$,

$$(\mathcal{A}\hat{a})f(i) = \sum_{j \sim i} \hat{a}f(j) = \sum_{j \sim i} f(ja)$$

and

$$(\hat{a}\mathcal{A})f(i) = \mathcal{A}f(ia) = \sum_{l \sim ia} f(l).$$

Then $\mathcal{A}$ and $\hat{a}$ commute if and only if the right hand side of the above equations are equal. This occurs exactly when $i \sim j$ if and only if $ia \sim ja$ for all $i \in V$; that is, $a \in \mathrm{Aut}(X)$.

If $f$ is an eigenvector with eigenvalue $\theta$ and $a \in \mathrm{Aut}(X)$, then $f \circ a$ is again an eigenvector with eigenvalue $\theta$; simply note that

$$\sum_{j \sim i}(f \circ a)(j) = \sum_{j \sim i} f(ja) = \sum_{\ell \sim ia} f(\ell) = \theta(f \circ a)(i).$$

Although we have not developed much theory yet, we can use what we have to derive some results concerning the eigenvalus of vertex-transitive graphs.

Assume $X$ is a vertex-transitive graph with automorphism group $G$. Let $\theta$ be a simple eigenvalue of $\mathcal{A}$, and let $f$ be a corresponding eigenvector. If $a \in G$ then $f \circ a$ is again an eigenvector with eigenvalue $\theta$; as $\theta$ is simple it follows that

$$f \circ a = cf \tag{1.1}$$

for some constant $c$. Now there is a least positive integer $r$ such that $a^r = 1$; from (1.1) we then find that $c^r = 1$. Hence $c$ is an $r$-th root of unity. As both $f$ and $f \circ a$ are real functions this implies that $c = \pm 1$.

**1.1 Theorem (Petersdorf and Sachs [26])** *Let $X$ be a vertex-transitive graph on $n$ vertices with valency $k$, and let $\theta$ be a simple eigenvalue of $\mathcal{A}$. Then $k - \theta$ is an even integer. If $\theta \neq k$, then $|V|$ is even.*

**Proof** Let $f$ be an eigenvector for $\theta$. If $a \in \mathrm{Aut}(X)$ and $f \circ a = f$, then $f$ is constant on the orbits of $a$. If $f \circ a = -f$ and $S$ is an orbit of $a$, then either $f$ is zero on $S$, or $f(ia) = -f(i)$ for each $i$ in $S$.

If $f$ is constant then $\mathcal{A}f = kf$, i.e., $\theta = k$. So we may assume that $f$ is not constant. As $\mathrm{Aut}(X)$ is vertex-transitive, it follows that there is a constant $c$ such that $f(i) = \pm c$, for all vertices $i$ in $X$. We may assume without loss of generality that $c = 1$. Let $i$ be a vertex of $X$ and let $\ell_i$ be the number of neighbours $j$ of $i$ such that $f(j) = -f(i)$. Then

$$\theta f(i) = \sum_{j \sim i} f(j) = (k - 2\ell_i)f(i). \tag{1.2}$$

This proves our claim concerning $\theta$.

Finally, as $X$ is regular, the all ones vector $\mathbf{1}$ is an eigenvector for $\mathcal{A}$. As $\mathcal{A}$ is self-adjoint, eigenvectors with distinct eigenvalues are orthogonal. Hence

$$0 = \langle f, \mathbf{1} \rangle = \sum_{i \in V} f(i)$$

and thus we deduce that $|V|$ must be even. □

It is an interesting exercise to show that, if $X$ is vertex-transitive and has two simple eigenvalues, neither equal to the valency $k$, then $|V(X)|$ must be divisible by 4. It can also be shown that if $X$ has at least three vertices, then it has at most $|V(X)|/2$ simple eigenvalues. This is another reasonable exercise. (Proofs of both these results are given in [16].)

A natural question is how many simple eigenvalues a vertex-transitive graph can have.

Note that (1.2) implies that $\ell_i$ is independent of the vertex $i$. Let $\pi$ be the partition of $V(X)$, where vertices $i$ and $j$ are in the same cell of $\pi$ if $f(i) = f(j)$. Then $\pi$ has exactly two cells, and the number of neighbours in a given cell of a vertex $i$ of $X$ is determined by the cell in which $i$ lies. This means, in the terminology of Section 2.1, that $\pi$ is an equitable partition of $X$.

## 1.2 Eigenspaces

If real-valued functions on $V(X)$ are interesting, then it is not unreasonable to consider functions on $V(X)$ with values in the vector space $\mathbb{R}^m$, in particular those functions $f$ from $V$ to $\mathbb{R}^m$ such that, for some constant $\theta$,

$$\theta f(i) = \sum_{j \sim i} f(j). \tag{1.3}$$

By way of example, consider a regular dodecahedron in 3-space, with centre of mass at the origin. Let $X$ be the 1-skeleton of this solid and let $p$ be the map that associates each vertex of $X$ with its image in $\mathbb{R}^3$. From the symmetry of the solid, we see that for each vertex $i$ there is a constant $c_i$ such that

$$c_i p(i) = \sum_{j \sim i} p(j).$$

Since all vertices of the solid are equivalent under its symmetry group, $c_i$ is independent of $i$ and so (1.3) holds.

Our aim is to generalize this example. We use $A$ to denote the adjacency matrix of $X$. Let $\theta$ be an eigenvalue of $A$ with multiplicity $m$ and let $U_\theta$ be a $|V| \times m$ matrix whose columns form an orthonormal basis for the eigenspace belonging to $\theta$. As each column of $U_\theta$ is an eigenvector for $A$ we have $\theta U_\theta = AU_\theta$. Hence, if $u_\theta(i)$ denotes the $i$-th row of $U_\theta$, then (1.3) holds (with $u_\theta$ in place of $f$). We call the function $u_\theta$ from $V$ to $\mathbb{R}^m$ the *representation* belonging to $\theta$. As the columns of $U_\theta$ form an orthonormal set of vectors,

$$U_\theta^T U_\theta = I$$

and therefore, if we define $E_\theta := U_\theta U_\theta^T$, then $E_\theta = E_\theta^T$ and

$$E_\theta E_\tau = \begin{cases} E_\theta, & \text{if } \tau = \theta; \\ 0, & \text{otherwise.} \end{cases}$$

In fact, $E_\theta$ is the matrix representing orthogonal projection onto the column space of $U_\theta$. This implies that $E_\theta$ is an invariant of the eigenspace, and does not depend on the orthonormal basis we chose to form $U_\theta$.

Next, if $x$ and $y$ are vectors in $\mathbb{R}^m$, for some $m$, then $\langle x, y\rangle$ denotes their inner product. If $i$ and $j$ are two vertices of $X$ then

$$(E_\theta)_{i,j} = \langle u_\theta(i), u_\theta(j)\rangle.$$

This shows that the Euclidean distance between the points $u_\theta(i)$ and $u_\theta(j)$ is determined by $i$, $j$ and $\theta$.

A *convex polytope* is the convex hull of a finite set of points in $\mathbb{R}^m$. A *vertex* of a polytope is a point in it which is an endpoint of any closed line segment that contains it and is contained in the polytope. (Equivalently, a vertex is an extreme point.)

If $S$ is a finite subset of $\mathbb{R}^m$, then the vertices of the convex hull $C$ of $S$ are a subset of $S$. Any linear mapping of $\mathbb{R}^m$ that fixes $S$ as a set must fix the polytope generated by $S$, and a linear mapping fixes all points in $C$ if and only if it fixes each vertex of $C$. The simplest example occurs when $m = 1$. If $|S| \geq 2$ then $C$ has exactly two vertices and certainly any linear mapping that fixes these two vertices fixes all points in $C$.

The convex hull of the set of points $u_\theta(i)$, for all $i$ in $V(X)$, is the *eigenpolytope* belonging to $\theta$. Figure 1 illustrates the four eigenpolytopes of the cube. The corresponding matrices $U_\theta$ are listed in the following table. If $h$ is an eigenvector for $\mathcal{A}$ with eigenvalue $\theta$, and $S$ is the set of vertices on which $h$ takes its maximum value, then the convex hull of the points $u_\theta(i)$, for $i \in S$, is a face of the eigenpolytope. Conversely, each face can be obtained from an eigenvector, since we can define the faces of a polytope as the set of points on which a linear function takes its maximum value. Note that the eigenvector $-h$ determines a face disjoint from that gotten from $h$. Eigenpolytopes of distance-regular graphs are studied at some length in [14].

## 1.3 Walks

A *walk* of length $r$ in a graph $X$ is a sequence $x_0, \dots, x_r$ of vertices from $X$, such that consecutive vertices are adjacent. A walk is *closed* if its first and last vertices are equal. In this section we show that the geometry of each eigenpolytope of $X$ is determined by the number of walks of certain types in $X$. The number of walks $W_{i,j}(X,r)$ of length $r$ from vertex $i$ to vertex $j$ in $X$ is equal to

$$\langle e_i, \mathcal{A}^r e_j\rangle.$$

We will derive an alternative expression for this number. The *walk generating function* $W_{i,j}(X,t)$ is defined to be the formal sum

$$\sum_{r\geq 0} W_{i,j}(X,r)t^r.$$

The geometry of the eigenpolytopes of $X$ is completely determined by these walk generating functions. Our main goal in this section is to demonstrate this.

If $\theta$ is an eigenvalue of $\mathcal{A}$, let $E_\theta$ denote orthogonal projection onto the associated eigenspace. For each eigenvalue $\theta$ of $\mathcal{A}$, we have

$$\mathcal{A}E_\theta = E_\theta\mathcal{A} = \theta E_\theta. \tag{1.4}$$

The *spectral decomposition* of $\mathcal{A}$ is equivalent to the identity

$$\mathcal{A} = \sum_\theta \theta E_\theta,$$

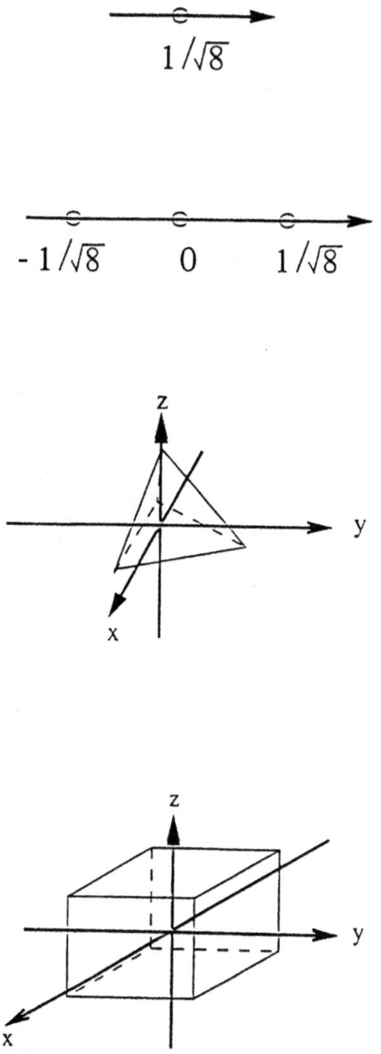


Figure 1: Eigenpolytopes of the cube

| $\theta$ | $m$ | $U_\theta$ |
|---|---|---|
| 3 | 1 | $\frac{1}{\sqrt{8}}\mathbf{1}$ |
| $-3$ | 1 | $\begin{pmatrix} \frac{-1}{\sqrt{8}} \\ \frac{1}{\sqrt{8}} \\ \frac{1}{\sqrt{8}} \\ \frac{-1}{\sqrt{8}} \\ \frac{1}{\sqrt{8}} \\ \frac{-1}{\sqrt{8}} \\ \frac{-1}{\sqrt{8}} \\ \frac{1}{\sqrt{8}} \end{pmatrix}$ |
| 1 | 3 | $\begin{pmatrix} \frac{-1}{\sqrt{8}} & \frac{-1}{\sqrt{8}} & \frac{-1}{\sqrt{8}} \\ \frac{-1}{\sqrt{8}} & \frac{-1}{\sqrt{8}} & \frac{1}{\sqrt{8}} \\ \frac{-1}{\sqrt{8}} & \frac{1}{\sqrt{8}} & \frac{-1}{\sqrt{8}} \\ \frac{-1}{\sqrt{8}} & \frac{1}{\sqrt{8}} & \frac{1}{\sqrt{8}} \\ \frac{1}{\sqrt{8}} & \frac{-1}{\sqrt{8}} & \frac{-1}{\sqrt{8}} \\ \frac{1}{\sqrt{8}} & \frac{-1}{\sqrt{8}} & \frac{1}{\sqrt{8}} \\ \frac{1}{\sqrt{8}} & \frac{1}{\sqrt{8}} & \frac{-1}{\sqrt{8}} \\ \frac{1}{\sqrt{8}} & \frac{1}{\sqrt{8}} & \frac{1}{\sqrt{8}} \end{pmatrix}$ |
| $-1$ | 3 | $\begin{pmatrix} \frac{1}{2} & \frac{-1}{\sqrt{12}} & \frac{-1}{\sqrt{24}} \\ 0 & \frac{1}{\sqrt{3}} & \frac{-1}{\sqrt{24}} \\ 0 & 0 & \sqrt{\frac{3}{8}} \\ \frac{-1}{2} & \frac{-1}{\sqrt{12}} & \frac{-1}{\sqrt{24}} \\ \frac{-1}{2} & \frac{-1}{\sqrt{12}} & \frac{-1}{\sqrt{24}} \\ 0 & 0 & \sqrt{\frac{3}{8}} \\ 0 & \frac{1}{\sqrt{3}} & \frac{-1}{\sqrt{24}} \\ \frac{1}{2} & \frac{-1}{\sqrt{12}} & \frac{-1}{\sqrt{24}} \end{pmatrix}$ |

where we sum over all eigenvalues $\theta$ of $\mathcal{A}$. This is a standard result, of course, which can be found in many texts. (See [21], for example.) From (1.4) and the spectral decomposition we obtain, for any non-negative integer $r$:

$$\mathcal{A}^r = \sum_\theta \theta^r E_\theta.$$

Consequently

$$\langle e_i, \mathcal{A}^r e_j \rangle = \sum_\theta \theta^r \langle e_i, E_\theta e_j \rangle. \tag{1.5}$$

As $E_\theta = U_\theta U_\theta^T$, we obtain our next result.

**1.2 Theorem** *For any two vertices $i$ and $j$ in the graph $X$ and any non-negative integer $r$, we have*

$$W_{i,j}(X,r) = \sum_\theta \theta^r \langle u_\theta(i), u_\theta(j) \rangle,$$

*where we sum over all eigenvalues $\theta$ of $\mathcal{A}$.* □

**1.3 Corollary** *The inner product $\langle u_\theta(i), u_\theta(j)\rangle$ is determined by $\theta$ and the sequence $W_{i,j}(X,r)$ for $r = 0, 1, \dots$.*

**Proof** From (1.5), we have $\langle e_i, p(\mathcal{A}) e_j \rangle = \sum_\theta p(\theta) \langle u_\theta(i), u_\theta(j) \rangle$, for any polynomial $p$. Let

$$p_\theta(x) := \prod_{\lambda \neq \theta} \frac{(x-\lambda)}{(\theta - \lambda)} = \begin{cases} 1, & \text{if } x = \theta; \\ 0, & \text{if } x \neq \theta,\ x \text{ an eigenvalue of } \mathcal{A}, \end{cases}$$

be a product over all eigenvalues of $\mathcal{A}$. Then $\langle u_\theta(i), u_\theta(j)\rangle = \langle e_i, p_\theta(\mathcal{A}) e_j \rangle$ and the result follows. □

The significance of Corollary 1.3 is that it shows that the geometry of the set of points

$$\{u_\theta(i) : i \in V\}$$

is determined by $\theta$ and the walk generating functions $W_{i,j}(X,t)$ for all vertices $i$ and $j$ of $X$.

As one useful consequence we note that, if $X$ is a vertex-transitive graph, the walk generating function $W_{i,i}(X,t)$ is independent of the vertex $i$, and therefore the length of the vector $u_\theta(i)$ does not depend on $i$. Thus in this case all points $u_\theta(i)$ for $i$ in $V$ lie on a sphere with centre at the origin. The same conclusion will hold more generally for graphs $X$ with the property that the walk generating functions $W_{i,i}(X,t)$ are independent of $i$. We will call such graphs *walk-regular*. They need not be vertex-transitive, but nonetheless have a number of properties in common with vertex-transitive graphs. For example, the Petersdorf-Sachs theorem 1.1 holds for walk-regular graphs. For if $X$ is walk-regular and $\theta$ is a simple eigenvalue for $X$, then $|u_\theta(i)|$ is independent of $i$, and the rest of the proof follows. (For further information, see [16, 11].)

### 1.4 Automorphisms

It follows from the spectral decomposition of $\mathcal{A}$ that each projection $E_\theta$ is a polynomial in $\mathcal{A}$. Therefore each automorphism of $X$, viewed as a linear mapping, commutes with each projection $E_\theta$. As $E_\theta = U_\theta U_\theta^T$, we thus have

$$\hat{g}U_\theta = \hat{g}U_\theta U_\theta^T U_\theta = U_\theta U_\theta^T \hat{g}U_\theta,$$

which implies that

$$u_\theta(ig) = u_\theta(i)U_\theta^T \hat{g}U_\theta.$$

Here $U_\theta^T \hat{g}U_\theta$ is an $m \times m$ matrix that is easily shown to be orthogonal. Let us denote it by $g_\theta$. The mapping $g \mapsto g_\theta$ is thus a representation of the automorphism group into the group of $m \times m$ real orthogonal matrices; we obtain one such representation from each eigenspace of $X$.

An automorphism $g$ lies in the kernel of the representation if and only if $\hat{g}U_\theta = U_\theta$, i.e., if and only if $\hat{g}$ fixes each column of $U_\theta$. It follows that if $g_\theta$ is the identity for each eigenvalue $\theta$, then $\hat{g}$ fixes every eigenvector of $\mathcal{A}$, which implies that it is the identity mapping on $F(V)$.

Now consider the eigenpolytope of $X$ associated to an eigenvalue $\theta$ of $X$, with multiplicity $m$.

If $m = 1$ then this is either a single point or a closed line segment in $\mathbb{R}^1$. In the first case the representation of $\mathrm{Aut}(X)$ afforded by the eigenspace is trivial (each automorphism is mapped to the identity). In the second case, the image of an automorphism either fixes each endpoint of the line segment, and hence induces the identity mapping, or it swaps the two endpoints and its square is the identity. Thus the representation of $\mathrm{Aut}(X)$ determines a homomorphism of $\mathrm{Aut}(X)$ onto the group of order 2, and therefore $\mathrm{Aut}(X)$ contains a subgroup of index 2. If each eigenvalue of $X$ has multiplicity 1, it follows that $\mathrm{Aut}(X)$ must be an elementary abelian 2-group. (It is an interesting problem just how large $\mathrm{Aut}(X)$ can be in this case. Note that we saw in Section 1.1 that the eigenvalues of a vertex-transitive graph cannot all be simple, hence $|V| - 1$ is an upper bound.)

If $m = 2$ then the eigenpolytope is either a closed line segment or a convex polygon. In the first case our previous analysis suffices, in the second we obtain a representation of $\mathrm{Aut}(X)$ as a subgroup of a dihedral group. If each eigenvalue of $X$ has multiplicity at most 2, then $\mathrm{Aut}(X)$ is therefore a subgroup of a direct product of dihedral groups.

Babai, Grigoryev and Mount [1] have proved that if the multiplicities of the eigenvalues are bounded by some constant, then $\mathrm{Aut}(X)$ can be computed in polynomial time.

## 2 Equitable partitions

Let $C_1, \ldots, C_m$ be the orbits of a group of automorphisms of $X$. Any two vertices in $C_i$ must have the same number of neighbours in $C_j$, for all $1 \le i, j \le m$. Equitable partitions are partitions of $V(X)$ with this regularity property, that are not necessarily orbit partitions. Our treatment in this and the next section follows [15] and [11, Chapter 5].

### 2.1 Basics

If $\pi$ is a partition of a set, let $|\pi|$ denote the number of cells of $\pi$. A partition is *discrete* if its cells are all singletons.

Let $X$ be a graph with vertex set $V$ and let $\pi$ be a partition of $V$ with cells $C_1, \ldots, C_m$. We say that $\pi$ is *equitable* if, for each $i$ and $j$, the number of neighbours in $C_j$ of a given vertex $u$ in $C_i$ is determined by $i$ and $j$, and is independent of the choice of $u$ in $C_i$. If $G$ is a subgroup of $\mathrm{Aut}(X)$, then the orbits of $G$ form a partition $\pi$ of $V$ with the properties that the subgraph induced by a cell of $\pi$ is regular, and the subgraph formed by the edges joining any two cells of $\pi$ is bipartite and semiregular. Hence, the orbits of an automorphism group of $X$ form an equitable partition. However, the converse is not true; Figure 2 shows an example of an equitable partition that is not an orbit partition.

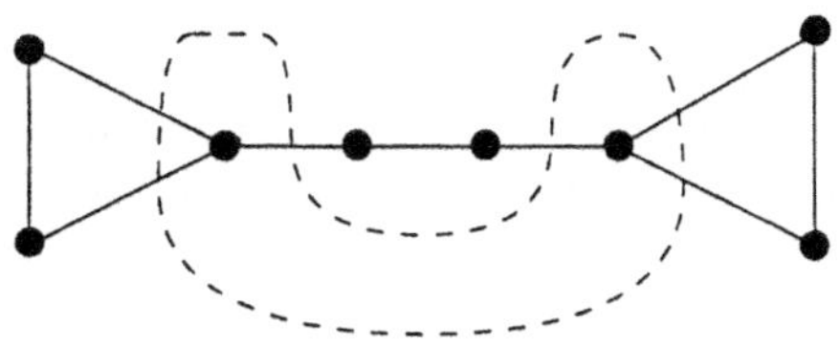

Figure 2: An equitable partition that is not an orbit partition

The *quotient* $X/\pi$ of $X$ relative to the equitable partition $\pi$ is the directed graph with the cells $C_1, \ldots, C_m$ of $\pi$ as its vertices and $c_{i,j}$ arcs from cell $i$ to cell $j$, where $c_{i,j}$ is the number of vertices in $C_j$ adjacent to a given vertex in $C_i$. The adjacency matrix of $X/\pi$ is the $m \times m$ matrix with $ij$-entry equal to $c_{i,j}$.

Figure 3 lists four equitable partitions of the Petersen graph with their quotients.

If $\pi$ is a partition of $V$, let $F(V, \pi)$ denote the space of real functions on $V(X)$ that are constant on the cells of $\pi$.

**2.1 Lemma** *A subspace $U$ of $F(V)$ equals to $F(V, \pi)$ for some partition $\pi$ of $V$ if and only if $U$ is a subalgebra of $F(V)$ which contains the constant functions.*

**Proof** If $\pi$ a partition of $V$ then $F(V, \pi)$ is a subspace of $F(V)$ that contains all the constant functions; it is also closed under multiplication and is thus a subalgebra of $F(V)$.

Conversely, suppose $U$ is a subalgebra of $F(V)$ which contains the constant functions and with basis $\{x_1, \ldots, x_m\}$. For each vector $x_i$, let $\pi_i$ be the partition of $V$ such that vertices $u, v$ belong to the same cell if and only if $x_i(u) = x_i(v)$. Let $\pi$ be the partition of $V$ whose cells are the non-empty pairwise intersections of the cells of $\pi_1, \ldots, \pi_m$. Let

$$c = bx_1 + b^2 x_2 + \ldots + b^m x_m,$$

where the real number $b$ is chosen large enough that $c$ takes distinct values on vertices in distinct cells of $\pi$. Suppose $v_1, \ldots, v_r$ are vertices in the cells $C_1, \ldots, C_r$ of $\pi$ respectively, and for $k = 1, \ldots, r$ let

$$y_k(v) = \prod_{j \neq k} \frac{(c(v) - c(v_j))}{(c(v_k) - c(v_j))} = \begin{cases} 1, & \text{if } v \in C_k; \\ 0, & \text{otherwise.} \end{cases}$$

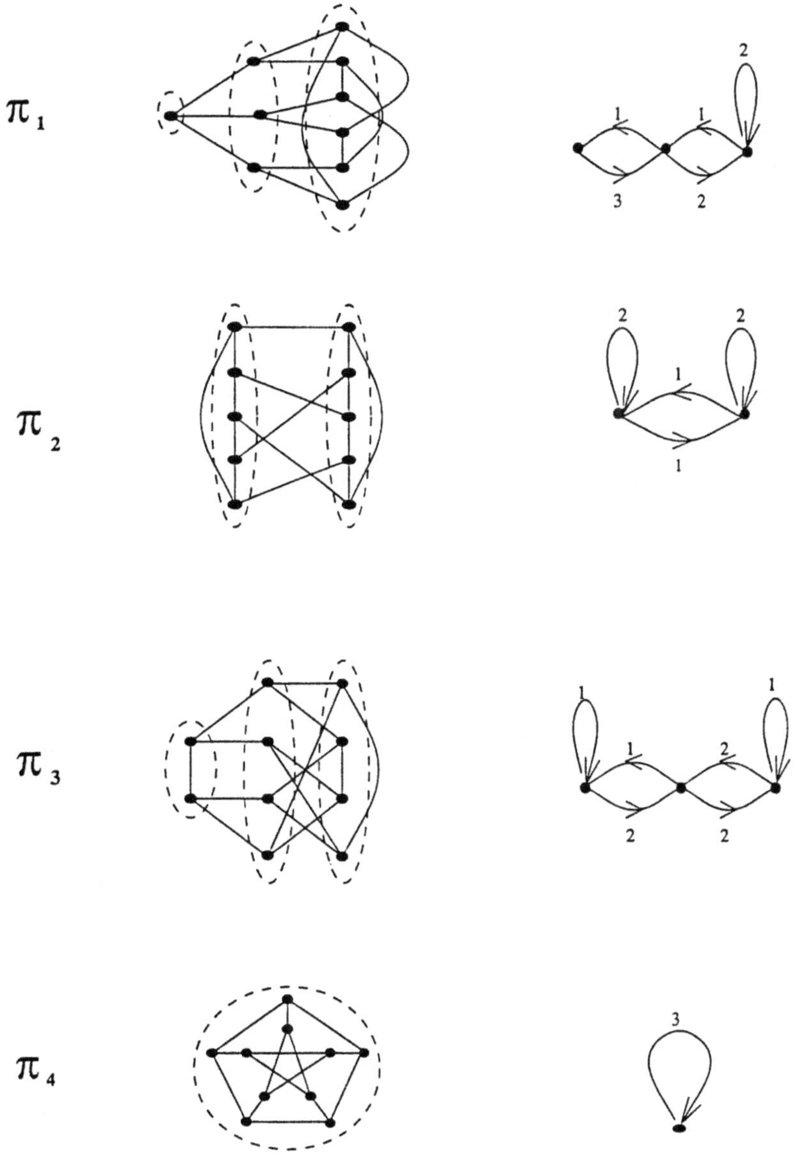


Figure 3: Equitable partitions of Petersen graph

The functions $y_1, \ldots, y_r$ are the characteristic functions of $C_1, \ldots, C_r$, respectively. They all belong to $U$, because $U$ contains the constant functions $c(v_j)$. Moreover, they are linearly independent in $U$. Since each cell of $\pi$ is contained in a cell of $\pi_i$, for $i = 1, \ldots, m$, we can express the $x_i$'s as a linear combination of $y_1, \ldots, y_r$. As a result, the $y_j$'s form a basis of $U$, and $U$ is equal to $F(V, \pi)$. □

**2.2 Lemma** *Let $X$ be a graph and let $\pi$ be a partition of $V(X)$. Then $\pi$ is equitable if and only if $F(V, \pi)$ is $\mathcal{A}$-invariant.*

**Proof** Suppose that $x$ is the characteristic function of the subset $S$, a cell of $\pi$ of $V(X)$. If $i \in V$ then $(\mathcal{A}x)(i)$ equals the number of vertices in $S$ adjacent to $i$. So $\mathcal{A}x$ is constant on the cells of $\pi$ if and only if the number of neighbours in $S$ of a vertex $i$ is determined by the cell of $\pi$ in which $i$ lies. Now the result follows at once. □

Equivalently, for undirected graphs, we can define a partition $\pi$ to be equitable if and only if $F(V, \pi)$ is $\mathcal{A}$-invariant. For directed graphs, there is a choice of the algebra $\langle A \rangle$ or $\langle A, A^T \rangle$ used to define equitable partitions.

If $\pi$ and $\rho$ are partitions of $V$ and each cell of $\rho$ is contained in a cell of $\pi$, we say that $\rho$ is a *refinement* of $\pi$, and write $\rho \leq \pi$. This shows that the set of all partitions of $V$ forms a partially ordered set, but in fact it is even a lattice. The meet $\pi \wedge \rho$ of two partitions $\pi$ and $\rho$ is the partition whose cells are all the nonempty pairwise intersections of cells of $\pi$ with cells of $\rho$. The join $\pi \vee \rho$ is most easily defined as the meet of all the partitions $\sigma$ such that $\pi \leq \sigma$ and $\rho \leq \sigma$. Alternatively, define a graph with vertex set $V$, where two vertices are adjacent if and only if they lie in the same cell of $\pi$ or of $\rho$; the vertex sets of the connected components of this graph are the cells of $\pi \vee \rho$.

The meet of two equitable partitions need not be equitable. An example is given in Figure 4, which is the meet of $\pi_2$ and $\pi_3$ of the Petersen graph in Figure 3. However the join must be. We will derive this from the next result, the proof of which is left as an exercise.

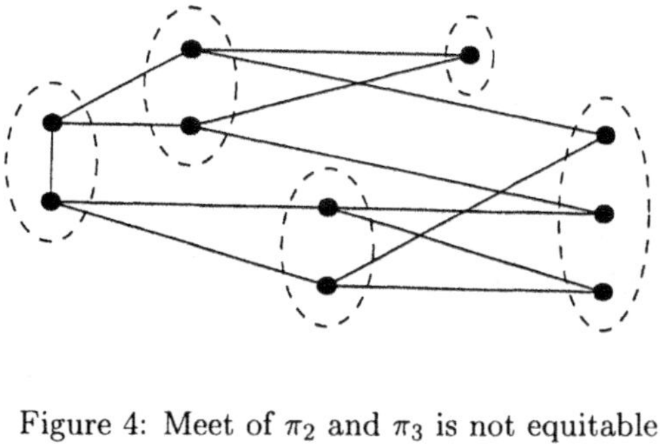

Figure 4: Meet of $\pi_2$ and $\pi_3$ is not equitable

**2.3 Lemma** *Let $\pi$ and $\rho$ be partitions of the set V. Then $F(V, \pi \vee \rho) = F(V, \pi) \cap F(V, \rho)$.* □

**2.4 Corollary (McKay [22, Lemma 5.3])** *Let $\pi$ and $\rho$ be equitable partitions of the graph $X$. Then $\pi \vee \rho$ is equitable.*

**Proof** Assume $V = V(X)$. As $\pi$ and $\rho$ are equitable, both $F(V,\pi)$ and $F(V,\rho)$ are $\mathcal{A}$-invariant. Now the above lemma implies that $F(V,\pi \vee \rho)$ is $\mathcal{A}$-invariant and so $\pi \vee \rho$ is equitable. □

One important consequence is that any partition $\pi$ has a unique maximal equitable refinement—it is the join of all the equitable partitions that refine $\pi$.

If $\pi$ is a partition of $V$ such that each cell is fixed by $\mathrm{Aut}(X)$, then the set of equitable refinements of $\pi$ is fixed by $\mathrm{Aut}(X)$, and therefore the coarsest equitable refinement of $\pi$ is fixed by $\mathrm{Aut}(X)$. (For example, we might choose $\pi$ to be the partition of the vertices of $X$ by valency, which is clearly invariant under $\mathrm{Aut}(X)$. If $\pi'$, the coarsest equitable refinement of $\pi$, is discrete then it follows that $\mathrm{Aut}(X)$ must be the identity group.) If $\pi$ is a partition with cells $C_1, \ldots, C_r$ and $v \in C_i$, define the *profile* of $v$ to be $(n_1, \ldots, n_r; i)$, where $n_j$ is the number of neighbours $v$ has in $C_j$. Let $\pi_1$ be the partition such that two vertices are in the same cell if and only if their profiles (relative to $\pi$) are equal. Then $\pi_1$ is a refinement of $\pi$ which is invariant under any automorphism that fixes each cell of $\pi$. It can be computed from $\pi$ in polynomial time, hence it follows that by repeatedly computing refinements in this manner, we can determine $\pi'$ in polynomial time. So it is not surprising that equitable partitions play an important role in practical graph isomorphism packages such as nauty. Further details will be found in [7].

## 2.2 Quotients

Any subset $S$ of $V$ can be represented by its characteristic function, which takes the value 1 on each element of $S$, and 0 on each element not in $S$. Because a partition of $V$ is a set whose elements are sets, we can represent a partition $\pi$ by its *characteristic matrix*, which is the $|V| \times |\pi|$ matrix with the characteristic functions of the cells of $\pi$ for columns.

**2.5 Lemma** *Let $A$ be the adjacency matrix of $X$ and let $\pi$ be a partition of $V(X)$ with characteristic matrix $D$. Then $\pi$ is equitable if and only if there is a matrix $B$ such that $AD = DB$.*

**Proof** The column space of $D$ can be identified with $F(V,\pi)$. Hence $\pi$ is equitable if and only if the column space of $D$ is $A$-invariant. This holds if and only if, for each column $x$ of $D$, the vector $Ax$ is a linear combination of columns of $D$ and so can be written as $Db$ for some vector $b$. □

The columns of the characteristic matrix of a partition are linearly independent, therefore there is at most one matrix $B$ such that $AD = DB$. Alternatively, note that $D^TD$ is invertible and diagonal with entries being the sizes of the cells, and that $B = (D^TD)^{-1}D^TAD$, thus is determined by $A$ and $D$. If $\pi$ is equitable then $B$ must be the adjacency matrix of $X/\pi$.

**2.6 Theorem** *Let $X$ be a graph with adjacency matrix $A$, and let $\pi$ be an equitable partition of $X$. Then the characteristic polynomial of $A(X/\pi)$ divides the characteristic polynomial of $A$.*

**Proof** Suppose that $v = |V(X)|$ and $m = |\pi|$. Let $x_1, \ldots, x_v$ be a basis for $\mathbb{R}^v$ such that $x_1, \ldots, x_m$ are the distinct columns of the characteristic matrix of $\pi$ and $x_{m+1}, \ldots, x_v$ are each orthogonal to $x_1, \dots, x_m$. Let $D_1 = (x_1, \dots, x_m)$ and $D_2 = (x_{m+1}, \dots, x_v)$. Then the matrix representing $\mathcal{A}$ relative to this basis is

$$\begin{pmatrix} (D_1^T D_1)^{-1} & 0 \\ 0 & (D_2^T D_2)^{-1} \end{pmatrix} \begin{pmatrix} D_1^T \\ D_2^T \end{pmatrix} A \begin{pmatrix} D_1 & D_2 \end{pmatrix}.$$

Since $AD_1 = D_1 B$ and $D_1^T D_2 = 0$, the matrix has the block diagonal form

$$\begin{pmatrix} B & 0 \\ 0 & Y \end{pmatrix} \tag{2.1}$$

where $B$ is the adjacency matrix of $X/\pi$. Since the characteristic polynomial of the matrix in (2.1) is the characteristic polynomial of $A$, the result follows. □

Note that it follows from this that, if the characteristic polynomial of $\mathcal{A}$ is irreducible over the rationals, then the only equitable partition of $X$ is the discrete partition.

Suppose $S \subseteq V(X)$. The *distance partition* of $X$ relative to $S$ is the partition with cells $C_0, \dots, C_r$, where $C_i$ is the set of vertices in $X$ at distance $i$ from $S$. The integer $r$ is the *covering radius* of $S$. We say that $S$ is a *completely regular* subset of $X$ if its distance partition is equitable. For example, the leftmost cell of each of the four partitions in Figure 3 are four different completely regular subsets of the Petersen graph. The *ball of radius* $r$ about a vertex $u$ of $X$ consists of all vertices in $X$ at distance at most $r$ from $u$. The *packing radius* of $S$ is the maximum integer $e$ such that the balls of radius $e$ about the vertices in $S$ are pairwise disjoint. A subset with packing radius $e$ is also called an *e-code*. If $S$ has packing radius $e$ and covering radius $r$ then $e \leq r$. We call $S$ a *perfect code* if $r = e$. (This terminology is consistent with the standard usage in coding theory.)

The *Johnson graph*, $J(n,k)$, has as vertices all the $k$-subsets of a $n$-set, where two $k$-subsets are adjacent if and only if they intersect in $k-1$ elements. Any single vertex in $J(n,k)$ is a perfect code. Furthermore, when $n = 2k$ and $k$ is odd, any pair of disjoint $k$-subsets form a perfect $(k-1)/2$-code. No further perfect codes are known in the Johnson graphs. As a result, Delsarte [8, p. 55] states that "it is tempting to conjecture that such codes do not exist"; this question is still open.

**2.7 Theorem** *Let $X$ be a regular graph. If $X$ has a perfect 1-code, then $-1$ is an eigenvalue of $\mathcal{A}$.*

**Proof** Assume that the valency of $X$ is $k$. Suppose that $C$ is a perfect code in $X$, and that $c = |C|$. Let $\pi$ be the partition with $C$ and $V \backslash C$ as its cells. A vertex in $C$ has no neighbours in $C$ and $k$ neighbours in $V \backslash C$. A vertex in $V \backslash C$ is adjacent to exactly one vertex in $C$, and to $k-1$ vertices in $V \backslash C$. Therefore $\pi$ is equitable, and the adjacency matrix of the quotient is

$$\begin{pmatrix} 0 & k \\ 1 & k-1 \end{pmatrix}.$$

The characteristic polynomial of this matrix is

$$x^2 - (k-1)x - k = (x-k)(x+1),$$

whence Theorem 2.6 implies that $-1$ must be an eigenvalue of $\mathcal{A}$. □

# 3 Algebras

In the next two sections, we derive results concerning some matrix algebras, and apply them to equitable partitions. The following section provides some information about one of these algebras for trees.

## 3.1 Algebras

As we have seen, finding the equitable partitions of a graph is equivalent to finding the $\mathcal{A}$-invariant subalgebras of $F(V)$ that contain the constant functions. If $Y$ is such a subalgebra then $Y$ is invariant under all elements of the algebra generated by $\mathcal{A}$. But there is a larger algebra that we can work with. Let $\mathcal{J}$ denote the square matrix with all entries equal to 1 (and order determined by the context.) A subspace of $F(V)$ is $\mathcal{J}$-invariant if and only if it contains all constant functions, and therefore a subspace $F(V,\pi)$ is invariant under $\mathcal{A}$ and $\mathcal{J}$ if and only if it is $\mathcal{A}$-invariant.

The operator $\mathcal{J}$ is a polynomial in $\mathcal{A}$ if and only if $X$ is a connected regular graph. (This result was first observed by A. J. Hoffman, for a proof see Biggs [3, Prop. 3.2].) Hence, if $X$ is not regular, then $\langle \mathcal{A}, \mathcal{J} \rangle$ is strictly larger than $\langle \mathcal{A} \rangle$. In particular, if $\langle \mathcal{A}, \mathcal{J} \rangle$ is equal to the algebra of all linear operators on $F(V)$, then no proper non-zero subspace of $F(V)$ is $\langle \mathcal{A}, \mathcal{J} \rangle$-invariant, and consequently the only equitable partition of $X$ is the discrete partition. (And $X$ has no non-identity automorphisms.)

In fact, under comparatively mild conditions, $\langle \mathcal{A}, \mathcal{J} \rangle$ is equal to the algebra of all linear operators on $F(V)$. We introduce some machinery to let us justify this claim. If $S \subseteq V(X)$, let $\mathcal{J}_S$ be the operator on $F(V)$ defined by

$$(\mathcal{J}_S f)(i) = \begin{cases} \sum_{j \in S} f(j), & \text{if } i \in S; \\ 0, & \text{otherwise.} \end{cases}$$

**3.1 Theorem** *Let $\mathcal{A}$ be the adjacency operator of the graph $X$ and let $S$ be a subset of $V(X)$ with characteristic function $\sigma$. Let $W$ be the subspace of $F(V)$ spanned by the functions $\mathcal{A}^r\sigma$. Then the algebra generated by $\mathcal{A}$ and $\mathcal{J}_S$ is isomorphic to the direct sum of the algebra of all linear operators on $W$ and the algebra generated by the restriction of $\mathcal{A}$ to $W^\perp$.*

**Proof** The space $F(V)$ is a direct sum of eigenspaces of $\mathcal{A}$; let $\sigma_1, \dots, \sigma_m$ denote the distinct non-zero projections of $\sigma$ on the eigenspaces of $\mathcal{A}$. As the vectors $\sigma_i$ lie in distinct eigenspaces of $\mathcal{A}$, they have distinct eigenvalues and therefore, for each $i$, there is a polynomial $p_i$ such that $p_i(\mathcal{A})\sigma_i = \sigma_i$ and $p_i(\mathcal{A})\sigma_j = 0$ when $j \neq i$. This also shows that the vectors $\sigma_i$ all lie in $W$. Because $\sigma = \sum_i \sigma_i$, we see that $W$ is contained in the span of the vectors $\sigma_i$ and therefore these vectors form a basis for $W$.

Since $\sigma^T\sigma_i = \sigma_i^T\sigma_i^T$ we have

$$p_i(\mathcal{A})\mathcal{J}_S p_j(\mathcal{A})\sigma_r = p_i(\mathcal{A})\,\sigma\sigma^T p_j(\mathcal{A})\sigma_r = \begin{cases} \|\sigma_j\|^2\sigma_i, & \text{if } r = j; \\ 0, & \text{otherwise.} \end{cases}$$

This implies that the $m^2$ operators $p_i(\mathcal{A})\mathcal{J}_S p_j(\mathcal{A})$ span the space of all linear operators on $W$.

As $\sigma \in W$, each vector in $W^\perp$ is orthogonal to $\sigma$ and therefore $\mathcal{J}_S$ acts on $W^\perp$ as the zero operator. □

**3.2 Corollary** *Let $\mathcal{A}$ be the adjacency operator of the graph $X$, let $S$ be a subset of $V(X)$ with characteristic function $\sigma$ and let $\rho$ be the partition of $V$ with $S$ and $V\backslash S$ as its cells. If no eigenvector of $\mathcal{A}$ is orthogonal to $\sigma$ then the discrete partition is the only equitable refinement of $\rho$.*

**Proof** If $W^{\perp}$ is non-zero then it contains an eigenvector for $\mathcal{A}$; this eigenvector is necessarily orthogonal to $\sigma$. Thus the hypothesis implies that $W = F(V)$, and therefore that $\mathcal{A}$ and $\mathcal{J}_S$ generate the algebra of all linear operators on $F(V)$. Consequently no non-zero proper subspace of $F(V)$ is invariant under $\mathcal{A}$ and $\mathcal{J}_S$. This implies that there is no equitable refinement of the partition with cells $S$ and $V \setminus S$. □

If $i \in V(X)$ and $f$ is an eigenvector for $\mathcal{A}$ such that $f(i) = 0$ then the restriction of $f$ to $V\backslash i$ is an eigenvector for $\mathcal{A}(X \setminus i)$, with the same eigenvalue. Consequently Corollary 3.2 implies that, if $\mathcal{A}(X)$ and $\mathcal{A}(X \setminus i)$ have no eigenvalue in common, no non-identity automorphism of $X$ fixes $i$.

If $\mathcal{A}$ has no eigenvector orthogonal to $\mathbf{1}$, then Corollary 3.2 implies that the discrete partition is the only equitable partition of $X$. We leave it as an exercise to show that, if the characteristic polynomial of $\mathcal{A}$ is irreducible over the rationals, then $\mathcal{A}$ and $\mathcal{J}$ have no common eigenvector, and consequently $X$ has no non-identity automorphisms. This is a classical result, first appearing in [25]. Another proof follows from Theorem 2.6.

We present an extension of Theorem 3.1.

**3.3 Theorem** *Let $X$ be a connected graph and let $\sigma_1, \dots, \sigma_m$ be characteristic vectors of subsets $S_1, \dots, S_m$ of $V(X)$. Let $W$ be the space spanned by all vectors of the form $\mathcal{A}^r \sigma_i$. Then the algebra $\mathcal{M}$ generated by $\mathcal{A}$ and $\mathcal{J}_{S_i}$ $(i = 1, \dots, m)$ is isomorphic to the direct sum of the algebra of all linear operators on $W$ and the algebra generated by the restriction of $\mathcal{A}$ to $W^{\perp}$.*

**Proof** Suppose $y$ is an eigenvector for $\mathcal{A}$ in $W$. If $\sigma_i^T y = 0$ for each $i$, then

$$\sigma_i^T \mathcal{A}^r y = 0$$

for each $i$ and so $y \in W^{\perp} \cap W$; hence $y = 0$. Thus we may assume that $\sigma_i^T y \neq 0$ for some $i$.

We now show that $W$ is an irreducible $\mathcal{M}$-module. Suppose that $y$ lies in a submodule $W_1$ of $W$ with $\sigma_i \in W_1$. As $X$ is connected, for any $j$ there is an integer $r$ such that

$$\sigma_j^T \mathcal{A}^r \sigma_i > 0.$$

Consequently $\sigma_j \in W_1$, for all $j$, and therefore $W_1 = W$.

Next, let $L$ be a linear endomorphism of $W$ that commutes with each element of $\mathcal{M}$. We aim to show that $L$ is a scalar multiple of the identity map. For any vector $\sigma$, if

$$L\sigma\sigma^T = \sigma\sigma^T L,$$

then $L\sigma$ and $\sigma$ must be parallel; i.e., $\sigma$ is a right eigenvector for $L$. Similarly $\sigma^T$ and $\sigma^T L$ are parallel and therefore $\sigma^T$ is a left eigenvector. If $L\sigma = c\sigma$ and $\sigma^T L = c\sigma^T$ then

$$c\sigma^T\sigma = \sigma^T L\sigma = d\sigma^T\sigma,$$

whence we see that $\sigma$ and $\sigma^T$ have the same eigenvalue.

Suppose now that $L\sigma_i = 0$. Because $\mathcal{A}$ and $L$ commute we than have $L\mathcal{A}^r\sigma_i = 0$ for all $r$ and therefore, for any $j$, $\sigma_j^T L\mathcal{A}^r\sigma_i = 0$. Since $L$ commutes with $\mathcal{J}_{S_j} = \sigma_j\sigma_j^T$, from above, $\sigma_j^T L = c\sigma_j^T$ for some $c$. It follows that $c\sigma_j^T A^r\sigma_i = 0$ for all $r$, and hence $c = 0$. Thus we have proved that $L\sigma_i = 0$ for all $i$.

Assume next that $L\sigma_i = c_i\sigma_i$ for $i = 1, \ldots, m$, and $c_i \neq 0$ for any $i$. Then

$$c_i\sigma_j^T\mathcal{A}^r\sigma_i = \sigma_j^T\mathcal{A}^r L\sigma_i = c_j\sigma_j^T\mathcal{A}^r\sigma_i,$$

for all values of $r$. This implies that $c_i$ is independent of $i$; denote its common value by $c$.

If $y \in W$ and $\mathcal{A}y = \theta y$ then

$$\theta Ly = L\mathcal{A}y = \mathcal{A}Ly,$$

which shows that each eigenspace of $\mathcal{A}$ in $W$ is fixed by $L$. Suppose that $y_1, \ldots, y_r$ are the non-zero projections of $\sigma_i$ on the eigenspaces of $\mathcal{A}$ in $W$. Then

$$\sigma_i = y_1 + \cdots + y_r$$

and so

$$c\sigma_i = L\sigma_i = Ly_1 + \cdots + Ly_r.$$

From this it follows that $Ly_j = cy_j$ for all $j$.

To complete the proof that $L$ is a scalar operator, it suffices to show that each eigenspace of $\mathcal{A}$ in $W$ is spanned by projections of the vectors $\sigma_i$. Assume by way of contradiction that this is false. Then we would have an eigenspace $U$ and a vector $u$ in $U$ orthogonal to the projection of each vector $\sigma_i$ onto $U$. This implies that $\sigma_i^T u = 0$ for all $i$ and hence $u = 0$.

Summing up, we have shown that $W$ is an irreducible $\mathcal{M}$-module, and that the only linear endomorphisms of $W$ that commute with $\mathcal{M}$ are the scalar operators. By Schur's theorem [19, Theorem 5.3], it follows that $\mathcal{M}$ acts on $W$ as the algebra of all linear endomorphisms. Because $W$ contains each vector $\sigma_i$, we see that $\mathcal{J}_{S_i}$ acts on $W^\perp$ as the zero operator. This completes the proof. □

The condition that $\langle\mathcal{A}, \mathcal{J}\rangle$ is the algebra of all linear operators of $F(V)$ is clearly very strong. Nonetheless it often holds. Some evidence for this is provided by the following table. We believe that the proportion of graphs on $n$ vertices such that $\langle\mathcal{A}, \mathcal{J}\rangle$ is the full matrix algebra $M_{n\times n}(\mathbb{R})$ tends to 1 as $n$ increases. (This belief is based on the results in the table and on experiments with random graphs on up to 20 vertices.)

| $\lvert V\rvert$ | No. of graphs | No. of graphs with $\langle\mathcal{A}, \mathcal{J}\rangle = M_{n\times n}(\mathbb{R})$ | No. of asymmetric graphs | No. of asymmetric & connected graphs |
|---|---|---|---|---|
| 6 | 78 | 4 | 4 | 4 |
| 7 | 522 | 46 | 76 | 72 |
| 8 | 6996 | 1370 | 1848 | 1786 |
| 9 | 154354 | 48457 | 67502 | 65726 |

Figure 5 contains all four graphs of order six with the property that $\langle\mathcal{A}, \mathcal{J}\rangle$ generates the algebra of all linear operators of $F(V)$.

The proofs of the main results in this section (Theorems 3.1 and 3.3) are based on ideas from Laffey [20].

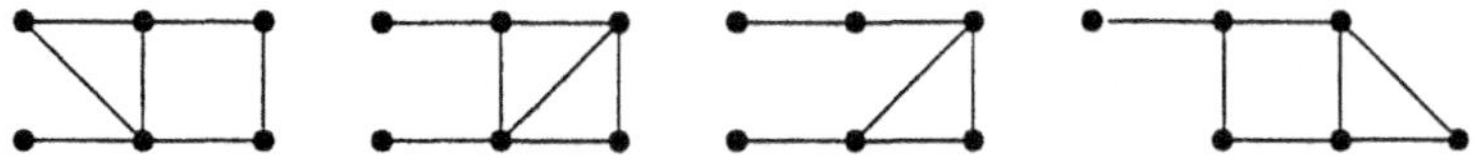

Figure 5: Graphs of order 6 with $\langle \mathcal{A}, \mathcal{J} \rangle = M_{n\times n}(\mathbb{R})$

### 3.2 Walks again

Two graphs $X$ and $Y$ are *cospectral* if $A(X)$ and $A(Y)$ are similar. The following result implies that cospectral graphs have the same closed walk generating function.

**3.4 Lemma** *If $A$ and $B$ are symmetric matrices, then the following are equivalent:*

(i) *$A$ and $B$ are similar.*

(ii) *$A$ and $B$ have the same characteristic polynomial.*

(iii) $\operatorname{tr}(A^n) = \operatorname{tr}(B^n)$, *for all $n \geq 0$.* □

We let $W(X,t)$ denote the generating function for walks in the graph $X$, enumerated by their length. If $S \subseteq V(X)$ then $W_S(X,t)$ is the generating function for the walks in $X$ that start at a vertex in $S$, enumerated by length. If $S$ and $T$ are subsets of $V(X)$, then $W_{S,T}(X,t)$ is the generating function for the walks in $X$ that start at a vertex in $S$ and finish in $T$. In particular, $W_i(X,t)$ is the generating function for the walks in $X$ that start at $i$, and $W_{i,i}(X,t)$ is the generating function for the walks in $X$ that start and finish at the vertex $i$.

The next two results can be derived from identities in [17].

**3.5 Lemma** *Let $X$ and $Y$ be cospectral graphs. Then their complements $\overline{X}$ and $\overline{Y}$ are cospectral if and only if $W(X,t) = W(Y,t)$.* □

**3.6 Lemma** *Let $i$ and $j$ be vertices in the graph $X$. Then $X \setminus i$ and $X \setminus j$ are cospectral if and only if $W_{i,i}(X,t) = W_{j,j}(X,t)$. Their complements are cospectral as well if and only if we also have $W_i(X,t) = W_j(X,t)$.* □

If $\sigma$ is the characteristic vector of $S$ and the space spanned by the vectors $\mathcal{A}^r\sigma$ has dimension $s+1$, we say that $S$ has *dual degree* $s$. As already mentioned, the *covering radius* of $S$ is the least integer $r$ such that each vertex of $X$ is at distance at most $r$ from some vertex in $S$. It follows that $(I+A)^0\sigma, \ldots, (I+A)^r\sigma$ are linearly independent with increasing number of non-zero entries, thus $r \leq s$. As the diameter of $X$ is the maximum value for the covering radius of a vertex, this inequality is a generalization of the well-known fact that the diameter of $X$ is bounded above by the number of distinct eigenvalues of $X$, less 1. The terms covering radius and dual degree come from coding theory, where they refer to the covering radius and dual degree of subsets of the $n$-cube or,more generally, of the Hamming graph $H(n,q)$. For more information see, e.g., [11, Chapter 11].

Let $S$ be a subset of $V(X)$ with characteristic function $\sigma$. Define two vertices $i$ and $j$ of $X$ to be equivalent if

$$(\mathcal{A}^r\sigma)(i) = (\mathcal{A}^r\sigma)(j), \qquad r \geq 0.$$

This determines a partition of $V$, which we call the partition *induced by* $S$. The partition induced by $V$ itself is the *walk partition*, since two vertices $i$ and $j$ are in the same cell if and only the generating function for the walks in $X$ that start at $i$ is equal to the generating function for the walks in $X$ that start at $j$. The walk partition $\omega$ is a refinement of the partition of $V(X)$ by valency. It is refined in turn by the coarsest equitable partition of $X$ that refines the partition with $S$ and its complement as its cells. This shows that the dual degree of $X$ is a lower bound on the number of cells in this partition.

It follows from the proof of Theorem 3.1 that the $\mathcal{A}$-module generated by $\sigma$ is an irreducible module for the algebra $\langle \mathcal{A}, \mathcal{J}_S\rangle$. This module has a basis consisting of vectors $\mathcal{A}^r\sigma$, and so the matrix $M$ representing the action of $\mathcal{A}$ on this module is determined by the inner products

$$\langle \mathcal{A}^r\sigma, \mathcal{A}\,\mathcal{A}^s\sigma\rangle, \qquad r, s \geq 0.$$

As $\langle \mathcal{A}^r\sigma, \mathcal{A}\,\mathcal{A}^s\sigma\rangle = \langle \mathbf{1}, \mathcal{A}^{r+s+1}\sigma\rangle$ it follows that $M$ is determined by the numbers $\sigma^T A^r\sigma$, i.e., by the number of walks in $X$ that start and finish in $S$ and have length $r$, for each non-negative integer $r$.

**3.7 Theorem** *Let $X$ be a graph on $n$ vertices and let $S$ be a subset of $V(X)$ with dual degree $n-1$. Then the walk generating functions $W_{i,S}(X,t)$, for $i \in V(X)$, determine $X$.*

**Proof** Let $\sigma$ be the characteristic function of $S$. As the dual degree of $S$ is $n-1$, it follows from the proof of Theorem 3.1 that the matrices

$$A_{r,s} := A^r\sigma\sigma^T A^s, \quad 1 \leq r, s \leq n$$

form a basis for the algebra of all $n \times n$ matrices. But

$$A^r\sigma\sigma^T A^r = A^r\sigma(A^s\sigma)^T,$$

consequently the matrices $A_{r,s}$ are determined by the given walk generating functions. As these matrices form a basis, we can write $A$ as a linear combination of them. The coefficients in this linear combination are determined by the inner products

$$\langle A, A_{r,s}\rangle = \langle A, A^r\sigma\sigma^T A^s\rangle = \sigma^T A^{r+s+1}\sigma.$$

Hence these coefficients are determined by the walk generating function $W_{S,S}(X,t)$; since this equals $\sum_{i\in S} W_{i,S}(X,t)$, the theorem follows. □

If $W_{i,S} = W_{j,S}$ then

$$\langle e_i - e_j, \mathcal{A}^r\sigma\rangle = 0$$

for all non-negative integers $r$. But if the dual degree of $S$ is $n-1$, then the vectors $\mathcal{A}^r\sigma$ span $\mathbb{R}^n$, consequently in this case $e_i - e_j$ must zero, and therefore $i$ and $j$ are equal. Note also that $W_{i,S}(X,t)$ is determined by its first $n$ coefficients. We can order the set of all real generating functions by writing $F(t) > G(t)$ if the first non-zero coefficient of $F(t) - G(t)$ is positive. The point of Theorem 3.7 is that, under the given hypothesis, the graph $X$ is determined by the ordered set of walk generating functions. Thus it provides an effective isomorphism test.

### 3.3 Trees

Let $D$ be the diagonal matrix whose $i$-th diagonal entry is the valency of the $i$-th vertex of the graph $X$. If $X$ is connected then 0 is a simple eigenvalue of $A - D$ with eigenvector $\mathbf{1}$; it follows from the spectral decomposition that the all-ones matrix $J$ is a polynomial in $A - D$. Hence, when $X$ is connected, the algebra generated by $A$ and $J$ is contained in the algebra generated by $A$ and $D$. If $X$ is a tree we can say more about the latter algebra. What we do say has not been published before.

First we need some more notation. A walk in $X$ is *closed* if its first and last vertices are the same; it is *irreducible* if any two vertices at distance two in the sequence are distinct. If $X$ is a tree, then any two vertices are joined by exactly one irreducible walk, and the length of this walk is the distance between the two vertices.

Let $A$ be the adjacency matrix of $X$, and let $p_r(A)$ denote the matrix with $ij$-entry equal to the number of irreducible walks of length $r$ from $i$ to $j$ in $X$. Define the generating function $\Phi(X,t)$ by

$$\Phi(X,t) = \sum_r p_r(A)t^r.$$

Note that $p_0(A) = I$ and $p_1(A) = A$. As a walk of length two is either closed or irreducible, $p_2(A) = A^2 - D$. Our next result implies that $p_r(A)$ is a polynomial in $A$ and $D$, with degree $r$ in $A$.

**3.8 Theorem** *Let $A$ be the adjacency matrix of $X$. We have*

$$(t^2(D - I) - tA + I)\Phi(X,t) = (1 - t^2)I. \tag{3.1}$$

**Proof** The coefficient of $t^r$ in the left hand side of (3.1) is

$$(D - I)p_{r-2}(A) - Ap_{r-1}(A) + p_r(A);$$

with the understanding that $r \geq 0$ and $p_r(A) = 0$ when $r < 0$.

Suppose that $i$ and $j$ are vertices of $X$. Assume further that $r \geq 3$. Then $(Ap_{r-1}(A))_{i,j}$ equals the number of walks of length $r$ from $i$ to $j$ that are formed by a walk of length 1 from $i$ to one of its neighbours, $\ell$ say, followed by an irreducible walk of length $r-1$ from $\ell$ to $j$. This class of walks includes all irreducible walks of length $r$ from $i$ to $j$. It also contains the walks that start $(i, \ell, i)$, followed by an irreducible walk of length $r-2$ from $i$ to $j$ with second vertex distinct from $\ell$, that is, an irreducible walk of length $r-2$ from $i$ to $j$ with second vertex not equal to $\ell$. The number of these walks is $((D - I)p_{r-2}(A))_{i,j}$.

Accordingly we have proved that the coefficient of $t^r$ in $(t^2(D - I) - tA + I)\Phi(X,t)$ is zero when $r \geq 3$. We have

$$\Phi(X,t) = I + tA + t^2(A^2 - D) + t^3M,$$

for some matrix $M$ (with entries formal power series in $t$). Now

$$(I - tA + t^2(D - I))(I + tA + t^2(A^2 - D) + t^3M) = I + t^2(A^2 - D - A^2 + D - I) + t^3N,$$

where the entries of $N$ are, again, formal power series in $t$. It follows from this that the left and right sides of (3.1) are equal as generating functions. □

**3.9 Corollary** *Let $A$ be the adjacency matrix of the graph $X$, and let $D$ be the diagonal matrix of valencies of $X$. Then $p_r(A)$ is a polynomial in $A$ and $D$ with degree $r$ in $A$.* □

If $X$ has vertex set $V$, let $X_i$ denote the graph with the same vertex set, where two vertices are adjacent in $X_i$ if and only if they are at distance $i$ in $X$. (Thus $X_1 = X$.) Let $A_i$ be the adjacency matrix of $X_i$, and let $A_0$ denote the identity matrix of order $|V| \times |V|$. If $r$ is greater than the diameter of $X$, then $X_r$ has no edges and $A_r = 0$. We call the $A_i$'s the *distance matrices* of $X$.

**3.10 Corollary** *If $X$ is a tree with diameter $d$, then $\langle A, D\rangle$ contains $A_0, \dots, A_d$.*

**Proof** If $X$ is a tree then two vertices are at distance $r$ if and only if there is a unique irreducible walk of length $r$ joining them. Therefore $A_r = p_r(A)$ and

$$\sum_r A_r t^r = (1-t^2)(I - tA + t^2(D-I))^{-1},$$

and therefore the matrices $A_r$ are polynomials in $A$ and $D$. □

The remaining results in this section have no bearing on the main topic of these notes, but may still be interesting.

If $X$ is a graph on $n$ vertices, $W(X,t)$ denotes the $n \times n$ matrix with $ij$-entry equal to the generating function for the walks in $X$ from $i$ to $j$, counted by length. We call $W(X,t)$ the *walk generating function* of $X$. It is not hard to show that, if $A$ is the adjacency matrix of $X$, then

$$W(X,t) = (I - tA)^{-1}.$$

Our next result shows that, for regular graphs, $\Phi(X,t)$ is determined by $W(X,t)$.

**3.11 Corollary** *If $X$ is a regular graph with valency $k$, then*

$$\Phi(X,t) = \frac{1-t^2}{1+(k-1)t^2} W(X, t(1+(k-1)t^2)^{-1}).$$ □

Suppose that $X$ has exactly $c$ connected components. It follows from Theorem 3.8 that each entry of $\Phi(X,t)$ is the product of a formal power series in $t$ with the polynomial $1-t^2$; given this it can be shown that $(1-t^2)^c$ divides $\det \Phi(X,t)$.

**3.12 Lemma** *Let $X$ be a graph. Then*

$$\det(I - tA + t^2(D-I)) = 1 - t^2$$

*if and only if $X$ is a tree.*

**Proof** We first show that the result holds when $X$ is a tree. We proceed by induction on the number of vertices, noting that it can be easily verified when $X$ has two vertices. Suppose $X$ has at least three vertices. Suppose $i$ is a vertex in $X$ with valency 1, adjacent to a vertex $j$ with valency at least 2. Let $Y$ be the graph got by deleting $i$ from $X$. Let $D_1$ and $A_1$ be respectively the valency and adjacency matrices of $Y$. If

$$M := I - tA + t^2(D-I)$$

then $M_{i,i} = 1$, $M_{i,j} = -t$ and $M_{i,r} = 0$ if $r \neq i, j$. If we add $t$ times column $i$ of $M$ to column $j$ and delete row and column $i$, the resulting matrix is $I - tA_1 + t^2(D_1 - I)$, and its determinant is equal to $\det(M)$.

Now we prove the converse. By Theorem 3.8,

$$\Phi(X,t) = (1-t^2)(I - tA + t^2(D-I))^{-1}.$$

If $\det(I-tA+t^2(D-I)) = (1-t^2)$, this implies that the entries of $\Phi(X,t)$ are polynomials in $t$, and therefore there is an upper bound on the length of an irreducible walk in $X$. Consequently $X$ has no cycles. From the first part of the proof, it follows that $\det(I - tA + t^2(D-I))$ is $(1-t^2)^c$, where $c$ is the number of components of $X$; this forces us to conclude that $X$ is a tree. □

It is left an exercise to show that $\det(I - tA + t^2(D-I))$ divides $(1-t^2)^n$ if and only if $X$ is a forest.

Finally, we note that differentiating both sides of (3.1) with respect to $t$ yields, after some manipulation, that

$$(I - tA + t^2(D-I))\Phi'(X,t)(I - tA + t^2(D-I)) = (t^2+1)A - 2tD.$$

Here $\Phi'(X,t)$ is the derivative of $\Phi(X,t)$ with respect to $t$. If $X$ is a forest, then $\Phi'(X,1)$ is the distance matrix of $X$.

# 4 Compact graphs

Compact graphs are a class of graphs with the property that, if a non-identity automorphism exists then we can find one in polynomial time. Equitable partitions provide a useful tool to study them with. Our treatment follows [13], with some adjustments occasioned by interesting work of Evdokimov, M. Karpinski and I. Ponomarenko [9].

## 4.1 A convex polytope

Let $\mathcal{M}$ be a matrix algebra, for example, the adjacency algebra of a graph. We define $\mathcal{S}(\mathcal{M})$ to be the set of all doubly stochastic matrices that commute with each element of $\mathcal{M}$. When $\mathcal{M} = \langle A \rangle$, we will usually write $\mathcal{S}(A)$ rather than $\mathcal{S}(\mathcal{M})$. The set $\mathcal{S}(\mathcal{M})$ is convex and contains the permutation matrices that commute with each element of $\mathcal{M}$. If $G$ and $H$ are two matrices in $\mathcal{S}(\mathcal{M})$, and $F = tG + (1-t)H$ for some $t$ in the interval $[0,1]$, then $F_{i,j} \neq 0$ if either $G_{i,j} \neq 0$ or $H_{i,j} \neq 0$. From this we see that any permutation matrix in $\mathcal{S}(\mathcal{M})$ is an extreme point. Slightly modifying Evdokimov et al., we define $\mathcal{M}$ to be *compact* if all the extreme points of $\mathcal{S}(\mathcal{M})$ are permutation matrices. Note that when $\mathcal{M}$ is finitely generated the set $\mathcal{S}(\mathcal{M})$ is polyhedral—if $\mathcal{M}$ is generated by $A_1, \ldots, A_r$ then $\mathcal{S}(\mathcal{M})$ consists of the matrices $F$ such that

$$\begin{aligned} F &\geq 0, \\ A_iF &= FA_i, \quad i = 1, \ldots, r \\ F\mathbf{1} &= \mathbf{1}, \\ F^T\mathbf{1} &= \mathbf{1}. \end{aligned}$$

As $\mathcal{S}(\mathcal{M})$ is clearly bounded, it will therefore be a convex polytope.

We say that a graph $X$ is *compact* if its adjacency algebra is compact. In this case the terminology, and many of the basic results, are due to Tinhofer [30].

Why is this property significant? One reason is that it is possible to find extreme points of $\mathcal{S}(A)$ in polynomial time. If all extreme points of $\mathcal{S}(A)$ belonged to $\mathrm{Aut}(A)$, then we would be able to decide, in polynomial time, whether a graph admitted an non-identity automorphism. In general, no such algorithm is known.

The complete graphs are compact, a fact which is equivalent to the statement that any doubly stochastic matrix is a convex combination of permutation matrices. Cycles and trees are compact graphs; the disjoint union of two connected non-isomorphic $k$-regular graphs is not compact. (For proofs of these assertions, see [30].) The complement of a compact graph is compact, as is easily shown. The next result provides further evidence that being compact is a strong condition.

**4.1 Lemma** *A compact regular graph is vertex-transitive.*

**Proof** Suppose $X$ is compact and regular, with adjacency matrix $A$. Then $\frac{1}{|V|}\mathcal{J} \in \mathcal{S}(A)$ and, as $X$ is compact, we may assume that

$$\mathcal{J} = \sum_i a_i P_i,$$

where $a_i \geq 0$ and $P_i$ is an automorphism of $X$, for all $i$. Assume $1 \in V(X)$. Since all entries in $J$ are equal to 1 it follows that, for each $r$, there is a permutation matrix in the above sum, $P$ say, such that $P_{1,r} = 1$. Hence, for each vertex $r$ of $X$, there is an automorphism of $X$ that maps 1 to $r$, and thus $\mathrm{Aut}(X)$ acts transitively on $V(X)$ as claimed. □

Lemma 4.1 is not the full truth; we have stated it here because the argument used in its proof is important. We now offer a more precise result, from [13], but with a new proof. Note that a permutation group on a set $V$ is *generously transitive* if, given any two points in $V$, there is a permutation that swaps them.

**4.2 Lemma** *If $X$ is regular, connected and compact, then* $\mathrm{Aut}(X)$ *is generously transitive.*

**Proof** Let $G = \mathrm{Aut}(X)$. As $X$ is compact, the matrices in $G$ span the same subspace as the matrices in $\mathcal{S}(A)$. As $X$ is regular and connected, any matrix which commutes with $A$ commutes with $J$, and therefore its row and column sums are all equal. Hence the real span of the matrices in $\mathcal{S}(A)$ is equal to $C(A)$, and consequently the centralizer of $G$ in $M_{n\times n}(\mathbb{R})$ is $C(C(A)) = \langle A\rangle$, [28, Theorem 39.3], which is commutative. Hence $G$ is multiplicity free, with rank equal to the number of eigenvalues of $A$ [2, Example 2.1]. Also, $\langle A\rangle$ is an association scheme—since it is generated by $A$ and $A = A^T$, all matrices in $\langle A\rangle$ are symmetric and so all orbitals of $G$ are symmetric. Hence $G$ is generously transitive. □

## 4.2 Equitable partitions

Let $\pi$ be an equitable partition of $X$, and let $D$ be the characteristic matrix of $\pi$. Let $P$ be defined by

$$P = D(D^TD)^{-1}D^T.$$

Then $P$ is symmetric, $P^2 = P$, and $P$ and $D$ have the same column space. Hence $P$ is the matrix representing orthogonal projection onto the column space of $D$, i.e., onto $F(V, \pi)$. Note that $D^T D$ is a diagonal matrix, with $i$-th diagonal entry equal to the size of the $i$-th cell of $\pi$, and that $P$ is a block-diagonal matrix with each diagonal block a multiple of the 'all-ones' matrix. In particular, all entries of $P$ are non-negative. Because $\mathbf{1} \in F(V, \pi)$ it follows that $P\mathbf{1} = \mathbf{1}$, which implies that each row of $P$ sums to 1. Similarly each column sums to 1, and therefore $P$ is a doubly stochastic matrix.

**4.3 Lemma** *Let $X$ be a graph and let $\pi$ be a partition of $V(X)$. Then $\pi$ is equitable if and only if the matrix $P$ representing orthogonal projection onto $F(V, \pi)$ lies in $\mathcal{S}(A)$.*

**Proof** The entries of $P$ are non-negative, as we saw above. Because $\mathbf{1} \in F(V, \pi)$, it follows that $P\mathbf{1} = \mathbf{1}$ and, because $P$ is symmetric, it is therefore a doubly stochastic matrix. The column space of $P$ is $F(V, \pi)$, which is $A$-invariant. Hence each column of $AP$ lies in $F(V, \pi)$ and therefore $PAP = AP$. Accordingly

$$PA = (AP)^T = (PAP)^T = PAP = AP,$$

which proves that $A$ and $P$ commute, and therefore $P \in \mathcal{S}(A)$. Conversely, if $P$ and $A$ commute then

$$AP = APP = PAP;$$

since $P$ represents the orthogonal projection onto $F(V, \pi)$, $AP \in F(V, \pi)$. As a result, $F(V, \pi)$ is $A$-invariant, and by Lemma 2.2, $\pi$ is equitable. □

**4.4 Lemma** *If $X$ is a compact graph, each equitable partition is an orbit partition.*

**Proof** Let $\pi$ be an equitable partition of the compact graph $X$, and let $P$ be the matrix representing orthogonal projection on $F(V(X), \pi)$. Then $P \in \mathcal{S}(A(X))$, and hence $P$ is a convex combination of automorphisms of $X$. Arguing as in the proof of Lemma 4.1, it follows that the cells of $\pi$ are the orbits of some group of automorphisms of $X$. □

One consequence of this is that the line graph of $K_n$, $L(K_n)$, is not compact when $n \geq 7$— the spanning subgraph of $K_n$ formed by a 3-cycle and and $(n-3)$-cycle, with its complement, determines an equitable partition of $L(K_n)$ with two cells that is not an orbit partition.

But a stronger statement is true. Evdokimov, M. Karpinski and I. Ponomarenko [9] have recently proved that the Johnson graph $J(v, k)$ is compact if and only if $k = 1$ or $v = 4$ and $k = 2$. This answers a question raised in an earlier draft of this work, where we asked whether the Petersen graph is compact. Their argument is based on the following result.

**4.5 Lemma** *If $X$ is a compact regular graph, then* $\mathrm{Aut}(X)$ *contains automorphisms that map each vertex of $X$ to an adjacent vertex.*

**Proof** Suppose that $X$ has valency $k$ and adjacency matrix $A$. Then $\frac{1}{k}A$ belongs to $\mathcal{S}(A)$, hence is a convex combination of automorphisms. Each of these automorphisms maps any vertex to an adjacent vertex. □

It is a simple exercise using this to determine the compact Johnson graphs. Evdokimov et al. also show that the $n$-cube is compact if and only if $n \geq 4$.

We have shown that each equitable partition of $X$ determines an element of $\mathcal{S}(A)$. There is a converse to this. If $M \in \mathcal{S}(A)$, let $X_M$ denote the directed graph with vertex set $V(X)$ and with an arc from $i$ to $j$ if and only if $M_{i,j} \neq 0$.

**4.6 Theorem** *Let $X$ be a graph and suppose that $M \in \mathcal{S}(A)$. Then the strong components of $X_M$ form an equitable partition of $X$.*

**Proof** Let $Y$ denote $X_M$, and assume that $C$ is a subset of $V(Y)$ such that there is no arc $(u,v)$ with $u \in C$ and $v \notin C$. Then the sum of the entries of $M$ in the rows corresponding to $C$ is $|C|$, whence the sum of the entries in the submatrix of $M$ with rows and columns indexed by $C$ is again $|C|$. But this implies that if $v \notin C$ and $u \in C$ then $(M)_{vu} = 0$, and therefore there are no arcs in $Y$ from a vertex not in $C$ to a vertex in $C$. It follows that, as $M$ is doubly stochastic, we may write it in block-diagonal form as

$$M = \begin{pmatrix} M_1 & & \\ & \ddots & \\ & & M_r \end{pmatrix}$$

where $M_1, \dots, M_r$ are doubly stochastic matrices and the graphs $X_{M_i}$ are strongly connected.

Let $\pi$ be the partition whose cells are the vertex sets of the strong components of $Y$, and let $P$ be the matrix representing orthogonal projection on $F(V(X), \pi)$. We aim to show that $P$ is a polynomial in $M$; since $P$ is doubly stochastic, it will follow that $P \in \mathcal{S}(A)$.

Assume that $Y$ is a strongly connected directed graph on $m$ vertices with adjacency matrix $M_i$. Then the matrix $\frac{1}{2}(M_i + I)$ is doubly stochastic and, for some $s$,

$$N = (\frac{1}{2}(M_i + I))^s > 0.$$

Choose $1 \geq c > 0$ maximal such that $N - \frac{c}{m}J \geq 0$. Now, we want to show by induction that for all $k \geq 1$,

$$N^r \geq (1 - (1-c)^k)\frac{1}{m}J.$$

This is true when $k = 1$. Suppose the inequality holds for all $k \leq t$. Then

$$\begin{aligned} (N - c\frac{1}{m}J)(N^t - (1-(1-c)^t)\frac{1}{m}J) &= N^{t+1} - (1-(1-c)^{t+1})\frac{1}{m}J \\ &\geq 0. \end{aligned}$$

Since $1 \geq c > 0$, we have

$$\lim_{t\to\infty} N^t \geq \lim_{t\to\infty} (1 - (1-c)^t)\frac{1}{m}J = \frac{1}{m}J.$$

Because $N^t$ is doubly stochastic for all $t \geq 0$, we have $\lim_{t\to\infty} N^t = \frac{1}{m}J$. Consequently, any matrices that commute with $M_i$ would commute with $\frac{1}{m}J$, that is, $\frac{1}{m}J \in C(C(M_i))$. In [28], Theorem 39.3 states that any matrix in $C(C(M_i)) = \langle M_i \rangle$; as $\frac{1}{m}J \in C(C(M_i))$ it follows that $J = q_i(M_i)$ for some polynomial $q_i$.

Accordingly, for $i = 1, \dots, r$ there is a polynomial $q_i$ such that

$$q_i(M_i) = \frac{1}{|V(Y_i)|}J.$$

It is now a straightforward exercise to show that, if $q := \prod_i q_i$, then $q(M)$ is the matrix representing orthogonal projection $P$ on $F(V(Y), \pi)$. □

**4.7 Corollary** *The discrete partition is the only equitable partition of $X$ if and only if* $\mathcal{S}(A) = \{I\}$.

To complete this section, we comment further on the recent interesting work by Evdokimov, M. Karpinski and I. Ponomarenko [9]. A *cellular algebra* is a matrix algebra which contains $I$ and $J$ and is closed under Schur multiplication. (Cellular algebras are also known as coherent algebras, and Schur multiplication is often referred to as Hadamard multiplication.) It is not too hard to show that an automorphism of a graph $X$ commutes with each elements of the cellular algebra generated by $A(X)$. Evdokimov et al. define a graph $X$ to be *weakly compact* if the cellular algebra generated by its adjacency matrix is compact. Every compact graph must be weakly compact. However, Evdokimov et al. show that if $\mathrm{Aut}(X)$ is a regular permutation group, or has an abelian regular subgroup of index two, then $X$ is weakly compact. Since the automorphism group of a compact regular graph must be generously transitive, it follows that there are many weakly compact graphs that are not compact. Their works indicates that the class of weakly compact graphs is at least as nice as the class of compact graphs. In particular, they show that the automorphism group of a weakly compact graph can be computed in polynomial time.

We can define a partition $\pi$ of $V(X)$ to be equitable relative to the cellular algebra $\mathcal{C}$ generated by $A(X)$ if $F(V, \pi)$ is invariant under $\mathcal{C}$. With this definition, Lemma 4.3 and Theorem 4.6 can be extended without further work to weakly compact graphs.

# 5 Distance regular graphs

The material in these last three sections follows the treatment from [11, Chapters 11 and 13]. (Fairly closely, which may not be a surprise.)

## 5.1 Definitions

We define a graph $X$ to be *distance regular* if

(a) it is connected,

(b) for each vertex $u$, the distance partition $\partial_u$ of $X$ relative to $u$ is equitable, and

(c) the quotient $X/\partial_u$ is independent of $u$.

In fact, condition (c) can be replaced by the assumption that $X$ is regular, although this is not trivial to prove (see [18]). We consider one family of examples, viz. the *Johnson graphs* $J(n, k)$ (cf. the paragraph preceding Theorem 2.7). The easiest way to see these graphs are distance regular is to show that the distance partition relative to a vertex $u$ is also the orbit partition for the stabilizer of $u$ in $\mathrm{Aut}(J(n, k))$.

A graph $X$ is *distance transitive* if, given two ordered pairs $(u, u')$ and $(v, v')$ such that

$$\mathrm{dist}(u, u') = \mathrm{dist}(v, v')$$

there is an automorphism $a$ of $X$ such that $ua = v$ and $u'a = v'$. Every distance transitive graph is distance regular, but the converse is far from true. We note one class of examples, *Latin square graphs.* Let $L$ be an $n \times n$ Latin square, which we view as a set of $n^2$ ordered triples

$$(i, j, L_{i,j}), \quad 1 \leq i, j \leq n.$$

(Here $L_{i,j}$ denotes the $ij$-entry of $L$.) Let $X(L)$ be the graph with these triples as vertices, where two triples are adjacent if they agree in some coordinate. Note that the definition of a Latin square implies that two triples can agree in at most one coordinate. The graph $X(L)$ has $n^2$ vertices, valency $3(n-1)$ and diameter 2. It is distance regular, but we leave the proof of this for the reader. There are examples of Latin squares $L$ of order as low as 8 such that $X(L)$ has trivial automorphism group. (For details, see [27].)

A *Steiner triple system* is a collection of 3-subsets, called blocks, of a $v$-set such that every 2-subset of the $v$-set occurs in exactly one of the blocks. The line graph of a Steiner triple system has the blocks being its vertices, and two blocks are adjacent if and only if their intersection is non-empty. The line graphs of a Steiner triple system are distance regular with diameter two; they are also called the *strongly regular graphs.* Mendelsohn has shown that every finite group is the automorphism group of a finite Steiner triple system and a strongly regular graph, [23, 24].

If $X$ is a connected graph, let $X_r$ denote the graph with the same vertex set as $X$, but with two vertices adjacent in $X_r$ if and only if they are at distance $r$ in $X$. Let $A_r$ denote the adjacency matrix of $X_r$ and let $A_0$ denote the identity matrix. Let $J$ denote the matrix with all entries equal to 1. If $X$ is connected with diameter $d$, then the following conditions hold (as the reader may verify).

(1) $A_0 = I$,

(2) $\sum_{r=0}^{d} A_r = J$,

(3) $A_r^T = A_r$.

Furthermore, the connected graph $X$ is distance regular if and only if the following holds [11],

(4) $A_r A_s$ belongs to the real span of $A_0, \ldots, A_d$, for all $r$ and $s$.

This shows that the matrices form an *association scheme with d classes.* For background on association schemes see, e.g., [8, 2, 4, 11].

**5.1 Theorem** *If $X$ is a distance regular graph, then the walk generating function $W_{i,j}(X,t)$ is determined by the distance between $i$ and $j$ in $X$.*

**Proof** It follows from Axiom 4 above that $A_1^m$ lies in the span of $A_0, \ldots, A_d$, for all non-negative integers $m$. Hence the $ij$-entry of $A_1^m$ is determined by the distance between $i$ and $j$ in $X$. □

## 5.2 Cosines

Suppose $X$ is a distance regular graph with diameter $d$, and let $u_\theta$ be the representation belonging to the eigenvalue $\theta$ of $\mathcal{A}$. As we noted at the end of the last section, the walk generating function $W_{i,j}(X,t)$ is determined by the distance between $i$ and $j$ in $X$. One consequence of this is that $W_{i,i}(X,t)$ is independent of the choice of the vertex $i$, and therefore $X$ is walk-regular, in the sense used at the end of Section 1.3. This means that all vectors $u_\theta(i)$ have the same length, and hence that the geometry of the points $u_\theta(i)$, for $i$ in $V(X)$, is determined by the inner products $\langle u_\theta(i), u_\theta(j)\rangle$. Hence we define the *cosine* $w_\theta(r)$ to be the cosine of the angle between $u_\theta(i)$ and $u_\theta(j)$, where $i$ and $j$ are two vertices at distance $r$ in $X$.

Let $\partial$ be the distance partition of $X$ relative to some vertex. As $X$ is distance regular, this partition is equitable. Let $\mathcal{B}$ denote the adjacency operator for the quotient $X/\partial$. Then $w_\theta$ is a function on the vertices of $X/\partial$, with the following important property.

**5.2 Theorem** *Let $\theta$ be an eigenvalue of the distance-regular graph $X$ and let $\partial$ be the distance partition of $X$ relative to some vertex. Then $w_\theta$ is an eigenvector for the adjacency operator of $X/\partial$.*

**Proof** Assume $\partial$ is the distance partition relative to the vertex 1 of $X$, with cells $\partial_0, \ldots, \partial_d$, and let $i$ be a vertex at distance $r$ from 1. Let $B_{r,s}$ denote the number of vertices in $\partial_s$ adjacent to a given vertex in $\partial_r$; note that this is 0 if $|r-s| > 1$. Thus the matrix $B$ with $rs$-entry $B_{r,s}$ is tridiagonal. Let $u$ be the representation of $X$ afforded by the eigenspace belonging to $\theta$. Then

$$\theta u(i) = \sum_{j\sim i} u(j).$$

A neighbour of $i$ is at distance $r-1$, $r$ or $r+1$ from 1. Hence, if we take the inner product of each side of the equality with $u(1)$ and then divide by the square of the length of $u(1)$, we find that

$$\theta w_\theta(r) = B_{r,r-1} w_\theta(r-1) + B_{r,r} w_\theta(r) + B_{r,r+1} w_\theta(r+1).$$

This proves our claim. □

One consequence of this theorem is that each eigenvalue of $\mathcal{A}$ must be an eigenvalue of $\mathcal{B}$. (It follows from Theorem 2.6 that each eigenvalue of $\mathcal{B}$ is an eigenvalue of $\mathcal{A}$.) This implies in turn that the eigenvalues of $\mathcal{A}$ are determined by the quotient $X/\partial$. Since the number of vertices in $X/\partial$ is usually much smaller than the number of vertices of $X$, this is somewhat surprising. Theorem 5.2 also implies that the function on $V(X)$ that maps each vertex in $\partial_r$ to $w_\theta(r)$ is an eigenvector for $\mathcal{A}$ contained in $F(V,\partial)$.

The cosines belonging to an eigenvalue of a distance regular graph satisfy a number of useful identities. Assume that $X$ has valency $k$. As

$$\theta u_\theta(i) = \sum_{j\sim i} u_\theta(j),$$

we have

$$\theta = \sum_{j\sim i} \frac{\langle u_\theta(j), u_\theta(i)\rangle}{\langle u_\theta(i), u_\theta(i)\rangle} = k w_\theta(1)$$

and therefore

$$w_\theta(1) = \frac{\theta}{k} \leq 1. \tag{5.1}$$

## 5.3 Feasibility

Let $u$ be the representation belonging to the eigenvalue $\theta$ of the distance regular graph $X$, and let $E$ be the matrix representing orthogonal projection on the eigenspace belonging to $\theta$. If $i$ and $j$ are vertices of $X$ at distance $r$, then

$$E_{i,j} = \langle u(i), u(j) \rangle = w(r)\langle u(i), u(i) \rangle.$$

Let $m$ be the multiplicity of $\theta$. As $E^2 = E$, each eigenvalue of $E$ is 0 or 1, hence $\operatorname{tr} E$ is equal to the rank of $E$, which is $m$. On the other hand,

$$\operatorname{tr} E = \sum_{i \in V} \langle u(i), u(i) \rangle.$$

But $X$ is walk-regular, whence $\langle u(i), u(i) \rangle$ is independent of $i$ and therefore

$$\langle u(i), u(i) \rangle = \frac{m}{|V|}.$$

Consequently $E_{i,j} = w(r)m/|V|$.

As $E^2 = E$ and $E = E^T$, we have

$$\sum_{j \in V} (E_{1,j})^2 = E_{1,1}. \tag{5.2}$$

Let $v_r$ denote the number of vertices at distance $r$ from a fixed vertex in $X$. Then (5.2) can be rewritten as

$$\sum_r v_r \frac{w(r)^2 m^2}{|V|^2} = \frac{m}{|V|},$$

from which it follows that

$$\sum_r v_r w(r)^2 = \frac{|V|}{m}. \tag{5.3}$$

It can be shown that the integers $v_r$ are determined by the quotient of $X$ with respect to the distance partition of a vertex. Given this, (5.3) implies that the multiplicities of the eigenvalues of $X$ can be determined from this quotient.

This leads to the most important feasibility condition for the existence of distance-regular graphs. We do not discuss this here, but simply refer the reader to [4, 11]. Instead we consider a related result, which constrains the possible automorphisms of distance-regular graphs. The result is due to G. Higman, although the first published account is in Cameron [6]. (And our treatment follows Section 13.8 of [11].)

Let $X$ be a distance-regular graph with diameter $d$. If $a \in \operatorname{Aut}(X)$, let $v_r(a)$ denote the number of vertices $i$ of $X$ such that $ia$ is at distance $r$ from $i$. An algebraic integer is a complex number that satisfies a monic polynomial equation with integer coefficients.

**5.3 Theorem (G. Higman)** *Let $X$ be a distance-regular graph with diameter $d$, and let $\theta$ be an eigenvalue of $X$ with multiplicity $m$. If $a \in \mathrm{Aut}(X)$, then*

$$\frac{m}{|V(X)|}\sum_{r=0}^{d} v_r(a)w_\theta(r)$$

*is an algebraic integer.*

**Proof** Let $u = u_\theta$ be the representation belonging to $\theta$. If $a \in \mathrm{Aut}(X)$ then

$$a_\theta = U_\theta^T \hat{a} U_\theta.$$

Given that $\hat{a}$ and $E_\theta = U_\theta U_\theta^T$ commute, it is easy to show that

$$a_\theta^r = U_\theta^T(\hat{a})^r U_\theta.$$

Now $a^m = 1$ for some integer $m$, whence $a_\theta^m = 1$ and hence each eigenvalue of $a_\theta$ is an $m$-th root of unity, and an algebraic integer. Accordingly, $\mathrm{tr}(a_\theta)$ is a sum of algebraic integers, and so is itself an algebraic integer.

To complete the proof, we show that the sum in the statement of the theorem is just the trace of $a_\theta$. We have

$$\mathrm{tr}(a_\theta) = \mathrm{tr}(U_\theta^T \hat{a} U_\theta) = \mathrm{tr}(U_\theta U_\theta^T \hat{a}) = \mathrm{tr}(E_\theta \hat{a}).$$

To compute the last trace, we compute the $ii$-entry of $E_\theta \hat{a}$. This is the $i$-th entry of $E_\theta \hat{a} e_i$. But

$$E_\theta \hat{a} e_i = E_\theta e_{ia^{-1}}$$

and therefore

$$(E_\theta \hat{a})_{i,i} = (E_\theta)_{i,ia^{-1}} = \langle u(i), u(ia^{-1})\rangle.$$

This implies that

$$\mathrm{tr}(a_\theta) = \sum_i \langle u(i), u(ia^{-1})\rangle = \sum_r v_r(a)\frac{m}{|V|}w_\theta(r),$$

from which the theorem follows. □

Following G. Higman, Cameron [6] uses Theorem 5.3 to show that the Moore graph on 3250 vertices cannot be vertex-transitive, if it exists.

Theorem 5.3 can also be used to that show there is no automorphism of the line graph of $K_n$ that maps each vertex to a distinct, non-adjacent vertex. Suppose such an automorphism $a$ exists, that is, $v_0(a) = v_1(a) = 0$ and $v_2(a) = |V(L(K_n))|$. It is sufficient to show that $m_\theta w_\theta(2)$ is not an algebraic integer for some eigenvalue $\theta$ of $L(K_n)$. The graph $L(K_n)$ is a strongly regular graph, from [11, Chapter 5], the eigenvalues of $L(K_n)$ are $k = 2(n-2)$, $\theta = (n-4)$ and $\tau = -2$ with multiplicities $1$, $n-1$ and $n(n-3)/2$, respectively. We have

$$w_\theta(1) = \frac{\theta}{k} = \frac{n-4}{2(n-2)}.$$

The complement of $L(K_n)$ is also strongly regular, therefore,

$$w_\theta(2) = \frac{-\theta-1}{v-1-k} = \frac{-2}{(n-2)},$$

and

$$m_\theta w_\theta(2) = \frac{-2(n-1)}{(n-2)}$$

which is not an algebraic integer when $n \geq 4$.

A similar argument shows that the block graph of a Steiner triple system cannot have an automorphism that maps each vertex to a distinct and non-adjacent vertex.

# References


[1] L. Babai, D. Yu. Grigoryev and D. M. Mount, Isomorphism of graphs with bounded eigenvalue multiplicity, *Proc. 14th ACM STOC* (1982), 310–324.

[2] E. Bannai and T. Ito, *Algebraic Combinatorics I*, Benjamin/Cummings, Menlo Park, CA, 1984.

[3] N. Biggs, *Algebraic Graph Theory*, 2nd edition, Cambridge University Press, Cambridge, 1993.

[4] A. E. Brouwer, A. M. Cohen and A. Neumaier, *Distance-Regular Graphs*, Springer-Verlag, Berlin, 1989.

[5] R. A. Brualdi, Some applications of doubly stochastic matrices, *Linear Algebra Appl.* **107** (1988). 77–100.

[6] P. J. Cameron, Automorphism groups of graphs, in: *Selected Topics in Graph Theory*, Vol. 2 (L.W. Beineke and R.J. Wilson, eds.), Academic Press, London, 1983, 89–127.

[7] D. G. Corneil and C. C. Gottlieb, An efficient algorithm for graph isomorphism, *J. Assoc. Comput. Mach.* **17** (1970), 51-64.

[8] P. Delsarte, An algebraic approach to the association schemes of coding theory, *Philips Res. Rep. Suppl.* **10** (1973).

[9] S. Evdokimov, M. Karpinski and I. Ponomarenko, Compact cellular algebras and permutation groups, preprint, 1996.

[10] C. D. Godsil, Graphs, groups and polytopes, in: *Combinatorial Mathematics* (D. A. Holton and J. Seberry, eds.), Lecture Notes in Math. **686**, Springer-Verlag, Berlin, 1978, 157–164.

[11] C. D. Godsil, *Algebraic Combinatorics*, Chapman and Hall, New York, 1993.

[12] C. D. Godsil, Equitable partitions, in: *Combinatorics, Paul Erdős is Eighty* (Vol. 1) (D. Miklós, V. T. Sós and T. Szőnyi, eds.), János Bolyai Mathematical Society, Budapest, 1993, 173-192.

[13] C. D. Godsil, Compact graphs and equitable partitions, *Linear Algebra Appl.*, to appear.

[14] C. D. Godsil, Eigenpolytopes of distance regular graphs, *Canad. J. Math.*, to appear.

[15] C. D. Godsil and W. J. Martin, Quotients of association schemes, *J. Combin. Theory Ser. A* **69** (1995), 185–199.

[16] C. D. Godsil and B. D. McKay, Feasibility conditions for the existence of walk-regular graphs, *Linear Algebra Appl.* **30** (1980), 51–61.

[17] C. D. Godsil and B. D. McKay, Spectral conditions for the reconstructibility of a graph, *J. Combin. Theory Ser. B* **30** (1981), 285–289.

[18] C. D. Godsil and J. Shawe-Taylor, Distance-regularised graphs are distance-regular or distance biregular, *J. Combin. Theory Ser. B* **43** (1987), 14–24.

[19] D. Gorenstein, *Finite Groups*, Harper & Row, New York, 1968.

[20] T. J. Laffey, A basis theorem for matrix algebras, *Linear and Multilinear Algebra* **8** (1980), 183–187.

[21] P. Lancaster and M. Tismenetsky, *The Theory of Matrices: With Applications*, 2nd edition, Academic Press, New York, 1985.

[22] B. D. McKay, *Backtrack Programming and the Graph Isomorphism Problem*, M. Sc. Thesis, University of Melbourne, 1976.

[23] E. Mendelsohn, Every (finite) group is the group of automorphisms of a (finite) strongly regular graph, *Ars Combin.* **6** (1978), 75–86.

[24] E. Mendelsohn, On the groups of automorphisms of Steiner triple and quadruple systems, *J. Combin. Theory Ser. A* **25** (1978), 97–104.

[25] A. Moshowitz, Graphs, groups and matrices, in: *Proc. 25th Summer Meeting Canad. Math. Congress, Congr. Numer.* **4**, Utilitas Mathematica, Winnipeg, 1971, 509-522.

[26] M. Petersdorf and H. Sachs, Spektrum und Automorphismengruppe eines Graphen, in: *Combinatorial Theory and its Applications, III.*, North-Holland, Amsterdam, 891–907.

[27] K. Phelps, Latin square graphs and their automorphism groups, *Ars Combin.* **7** (1979), 273–299.

[28] V. V. Prasolov, *Problems and Theorems in Linear Algebra*, Transl. Math. Monographs **134**, Amer. Math. Soc., Providence, RI, 1994.

[29] H. Schreck and G. Tinhofer, A note on certain subpolytopes of the assignment polytope associated with circulant graphs, *Linear Algebra Appl.* **111** (1988), 125–134.

[30] G. Tinhofer, Graph isomorphism and theorems of Birkhoff type, *Computing* **36** (1986), 285–300.

[31] G. Tinhofer, A note on compact graphs, *Discrete Appl. Math.* **30** (1991), 253–264.

# Graph homomorphisms: structure and symmetry

Geña HAHN

*Département d'informatique*
*et de recherche opérationnelle*

and

Claude TARDIF

*Département de mathématiques et de statistique*
*Université de Montréal*
*C.P. 6128, succursale Centre-ville*
*Montréal, Québec, H3C 3J7*
*Canada*

**Abstract**

This paper is the first part of an introduction to the subject of graph homomorphism in the mixed form of a course and a survey. We give the basic definitions, examples and uses of graph homomorphisms and mention some results that consider the structure and some parameters of the graphs involved. We discuss vertex-transitive graphs and Cayley graphs and their rather fundamental role in some aspects of graph homomorphisms. Graph colourings are then explored as homomorphisms, followed by a discussion of various graph products.

## 1 Introduction

Homomorphisms provide a way of simplifying the structure of objects one wishes to study while preserving much of it that is of significance. Most mathematicians remember the isomorphism theorems we learn in a first course on group theory, and certainly anyone involved in some way with mathematics or computer science knows about integers modulo some $n$. It is not surprising that homomorphisms also appeared in graph theory, and that they have proven useful in many areas.

We do not claim to provide the definitive survey but only an introductory course. Of necessity, we had to omit most proofs and, sadly, even many results and some aspects of the subject matter. But we do try to provide an extensive bibliography and the interested reader will be able to find the missing pieces with little trouble. For the rest, we will all have to wait for the book that Hell and Nešetřil are reportedly writing. The course/survey is divided into two parts. The present paper is concerned with structure and symmetry, the sequel [56] deals with computational aspects of graph homomorphisms and surveys the computer science roots of the subject.

Graph homomorphisms in the current sense were first studied by Sabidussi in the late fifties and early sixties, with results published in the paper on *Graph derivatives* [109] and



*G. Hahn and G. Sabidussi (eds.), Graph Symmetry,* 107–166.

used by him in, among others, [108]. This was followed by much activity of what has become known as the *Prague school* of category theory. There, people around Hedrlín and Pultr pursued homomorphisms of relational systems in general, and graph homomorphisms in particular, in the sixties and the seventies, with many of the results collected in *Combinatorial, Algebraic and Topological Representations of Groups, Semigroups and Categories* by Pultr and Trnková [102]. Some of this work was continued by Hell in his Ph.D. thesis [65] in 1972 with a study of retracts of graphs. In computer science, graph homomorphisms have been studied as *interpretations*, especially in relation to grammars ([93, 92], for example), and as "general colourings". Welzl's homomorphism density theorem 2.33 (Section 2.5) arose in this context. Others took an algorithmic approach such as *How difficult is it to recognize if a graph is homomorphic to a given one?*, usually – but not always – proving the problem *NP*-complete. This is discussed in more detail in [56].

Graph homomorphisms are mostly used as tools, especially in connection with colouring problems. For example, the chromatic difference sequence of a graph studied by Albertson, Berman, Collins, Tardif, Zhou and others ([3, 5, 119, 132, 134]), relies on them. Since odd cycles must map to odd cycles of at most equal length, people have studied homomorphisms into them, for example [5, 25, 49]. But some of the rare structural results stem from considering graphs which have the property that there are homomorphisms from the cartesian product $G \square G$ into $G$.

Many of the concepts of graph homomorphisms work equally well for finite and for infinite graphs (clearly infinite graphs homomorphically equivalent to finite ones "behave" essentially like them). But the important idea of a core of a graph does not easily generalize to infinite graphs and its considerations lead to undecidability results (these are discussed [56]). This is the main reason for limiting the present paper to finite graphs.

# 2 Basics

## 2.1 Basic definitions

Unless otherwise indicated, graphs in this paper will be simple, loopless and finite. We shall assume the basics of graph theory and, unless otherwise stated, use the notation of [19] for graphs and of [106] for groups. We will denote the vertex set of a graph $G$ by $V(G)$ and its edge set by $E(G)$; the edge between $u$ and $v$ will be denoted by $[u, v]$. The order of a graph $G$, i.e. the number of vertices, will be denoted by $|G|$. Most of the time we shall confuse an equivalence class of graphs under isomorphisms with a particular representative of it. Thus, the complete graph $K_n$ will usually be on the vertex set $\{0, 1, \ldots, n-1\} = [n]$. The complement of a graph $G$ will be denoted by $\overline{G}$. We will denote by $P_n$ and by $C_n$ the path and the cycle on $n$ vertices. Functions will be composed from right to left, $g \circ h(x) = gh(x) = g(h(x))$, and we define $\phi(S) = \{\phi(s) : s \in S\}$.

**Definition 2.1** Let $G$ and $H$ be graphs. A function $\phi : V(G) \longrightarrow V(H)$ is a *homomorphism* from $G$ to $H$ if it preserves edges, that is, if for any edge $[u, v]$ of $G$, $[\phi(u), \phi(v)]$ is an edge of $H$. We write simply $\phi : G \longrightarrow H$.

It is usual to think of graphs as sets of vertices with a binary relation. This is consistent with writing $\phi : G \longrightarrow H$ for a homomorphism which maps vertices to vertices. When

$\phi : G \longrightarrow H$ is a homomorphism, it induces a mapping

$$\phi_E : E(G) \longrightarrow E(H),\ \phi_E([u,v]) = [\phi(u), \phi(v)],$$

and we can define the graph $\phi(G) = (\phi(V(G)), \phi_E(E(G)))$, called the *homomorphic image* of $G$ in $H$.

In group theory, there are trivial homomorphisms from one object to another. Any group can be mapped by a homomorphism into any other by simply sending all its elements to the identity of the target group. In fact, the study of kernels is important in algebra. In the context of graphs without loops, the notion of a homomorphism is far more restrictive. Indeed, there need not be a homomorphism between two graphs, and these cases are as much a part of the theory as those where homomorphisms do exist (there are other categories where homomorphisms do not always exist between two objects, e.g. the category of bounded lattices or that of semigroups, and the interested reader will want to consult [102]).

A surjective homomorphism is often called an *epimorphism*, an injective one a *monomorphism* and a bijective homomorphism is sometimes called a *bimorphism*. Note that unlike in group theory, the inverse of a bijective homomorphism need not be a homomorphism. For example, any bijection from $\overline{K}_n$ to $K_n$ is a bimorphism. A homomorphism from a graph $G$ to itself is called an *endomorphism*. The identity endomorphism on a graph $G$ will be denoted by $id_G$. The set of endomorphisms of a graph $G$ is a semigroup under composition.

**Definition 2.2** A homomorphism $\phi : G \longrightarrow H$ is called *faithful* if $\phi(G)$ is an induced subgraph of $H$. It will be called *full* if $[u,v] \in E(G)$ if and only if $[\phi(u), \phi(v)] \in E(H)$, that is, when $\phi^{-1}(x) \cup \phi^{-1}(y)$ induces a complete bipartite graph whenever $[x,y] \in E(H)$.

In other words, a homomorphism $\phi : G \longrightarrow H$ is faithful when there is an edge between any two pre-images $\phi^{-1}(u)$ and $\phi^{-1}(v)$ such that $[u,v]$ is an edge of $H$. When a faithful homomorphism $\phi$ is bijective, it is full since each $\phi^{-1}(u)$ is a singleton, and we have that $[\phi^{-1}(u), \phi^{-1}(v)]$ is an edge in $G$ if and only if $[u,v]$ is an edge in $H$. Thus a faithful bijective homomorphism is an *isomorphism* and in this case we write $G \cong H$. An isomorphism from $G$ to $G$ is an *automorphism* of $G$; the set of automorphisms of a graph forms a group under composition. We shall denote the group of automorphisms of $G$ by $\mathrm{Aut}(G)$ .

A surjective faithful homomorphisms $\phi$ is sometimes called *complete*. When $\phi : G \longrightarrow H$ is a complete homomorphism, $H$ is a homomorphic image of $G$. Similarly, $G$ is isomorphic to its image under $\phi$ when $\phi$ is injective and faithful.

**Observation 2.3** *Suppose that $\phi : G \longrightarrow H$ is a graph homomorphism. Then $\phi$ is an isomorphism if and only if $\phi$ is bijective and if $\phi^{-1}$ is also a homomorphism. In particular, if $G = H$ then $\phi$ is an automorphism if and only if it is bijective.*

**Example 2.4** Let $P_n$ be the path $u_0 \dots u_{n-1}$ and $K_2$ the complete graph with vertices $0, 1$. It is easy to see that the mappings described below are homomorphisms, and that $\phi_2$ is onto and $\phi_3$ is one-to-one.

- $\phi_1(u_i) = u_i$ when $i = 0, \dots, n-2$, $\phi_1(u_{n-1}) = u_{n-3}$ (from $P_n$ to $P_n$).
- $\phi_2(u_{2k}) = 0$, $\phi_2(u_{2k+1}) = 1$ for $k = 0, \dots, \lfloor \frac{n}{2} \rfloor$ (from $P_n$ to $K_2$).

- $\phi_3(0) = u_i$, $\phi_3(1) = u_{i+1}$, for any $0 \le i < n-1$ (from $K_2$ to $P_n$).

In fact, $K_2$ is the homomorphic image of any bipartite graph. In particular, there always is a homomorphism from the cycle $C_{2k}$ onto $K_2$ for any $k$.

**Example 2.5** Let $C_m = u_0 \dots u_{m-1}$ and $C_n = v_0 \dots v_{n-1}$ be odd cycles, with $m > n$. There is a homomorphism from $C_m$ to $C_n$, for example, $\phi : C_m \longrightarrow C_n$ given by $\phi(u_i) = v_i$ for $0 \le i \le n-1$, and $u_{n+j} = v_{j(\bmod 2)}$. On the other hand, there is no homomorphism from $C_n$ into $C_m$.

**Observation 2.6** *Let $G$ and $H$ be non-bipartite graphs and $\phi : G \longrightarrow H$ a homomorphism. Then*

(1) *the length of a shortest odd cycle in $G$ (the* odd girth *of $G$) is at least the odd girth of $H$;*

(2) *the size $\omega(G)$ of a largest clique in $G$ (the* clique number *of $G$) is at least $\omega(H)$.*

**Example 2.7** A *proper $k$-colouring* of a graph $G$ is an assignment of colours to the vertices of $G$ in such a way that adjacent vertices get different colours. It is useful to think of a proper $k$-colouring as a function $c : V \longrightarrow [k]$, with $[k] = \{0, 1, \dots, k-1\}$. Indeed, it is an easy exercise to see that a graph $G$ has a proper $k$-colouring if and only if there is a homomorphism from $G$ into the complete graph $K_k$. Thus we have a definition of a proper colouring in the language of homomorphisms.

The colours in the above example induce a partition of the vertex set of the graph and the homomorphism into $K_n$ corresponds to identifying the vertices of the same colour. This can be done in general as is explained in the next section.

The reader might have asked whether between any two graphs there is a homomorphism. Proper colourings provide examples of pairs of graphs neither of which maps into the other by a homomorphism. Recall that the chromatic number $\chi(G)$ of a graph $G$ is the least $k$ for which $G$ has a proper $k$-colouring. From the observation made in Example 2.7 it follows that if there is a homomorphism $\phi : G \longrightarrow H$, then $\chi(G) \le \chi(H)$; in other words, a graph with chromatic number $k$ cannot map by a homomorphism into one with a lower chromatic number. Thus the graph $K_3$ and any graph of chromatic number at least four containing no triangles cannot be homomorphic in either direction. The latter kind of graphs exist, for example certain Kneser graphs (defined in Section 3.4; see in particular Proposition 3.14).

## 2.2 Quotients

**Definition 2.8** Let $G$ be a graph and let $\mathcal{P} = \{V_1, \dots, V_k\}$ be a partition of the vertex set of $G$ into non-empty classes. The *quotient* $G/\mathcal{P}$ of $G$ by $\mathcal{P}$ is the graph whose vertices are the sets $V_1, \dots, V_k$ and whose edges are the pairs $[V_i, V_j]$, $i \ne j$, such that there are $u_i \in V_i$, $u_j \in V_j$ with $[u_i, u_j] \in E(G)$. The mapping $\pi_{\mathcal{P}} : V(G) \longrightarrow V(G/\mathcal{P})$ defined by $\pi_{\mathcal{P}}(u) = V_i$ such that $u \in V_i$, is the *natural* map for $\mathcal{P}$.

Quotients often provide a way of deriving the structure of an object from the structure of a larger one; factor groups are a good example. In the present context it is worth our while to investigate when the natural map $\pi_{\mathcal{P}} : V(G) \longrightarrow V(G/\mathcal{P})$ is a homomorphism. Observe that if $\pi_{\mathcal{P}}$ is a homomorphism, then it is automatically faithful.

**Lemma 2.9** *A map $\phi : V(G) \longrightarrow V(H)$ is a graph homomorphism if and only if the pre-image $\phi^{-1}(I)$ of every independent subset $I$ of $V(H)$ is an independent set.*

**Corollary 2.10** *Let $G$ be a graph and $\mathcal{P}$ a partition of the vertex set of $G$. Then $\pi_{\mathcal{P}}$ is a homomorphism if and only if $V_i$ is an independent set for each $i$.*

**Proposition 2.11** *For every homomorphism $\phi : G \longrightarrow H$ there is a partition $\mathcal{P}$ of $V(G)$ into independent sets and a monomorphism $\psi : G/\mathcal{P} \longrightarrow H$ such that $\phi = \psi \circ \pi_{\mathcal{P}}$.*

A homomorphism $\phi$ of a graph $G$ into $H$ gives rise to an equivalence relation $\equiv_\phi$, the *kernel* of $\phi$, defined on $V$ by $u \equiv_\phi v$ if and only if $\phi(u) = \phi(v)$. This, in turn, induces a partition $\mathcal{P}_\phi$ on the vertex set of $G$; it is this partition that works in the Corollary 2.10 and Proposition 2.11. We can then speak of the *quotient of $G$ by $\phi$* and, abusing the notation slightly, we write $G/\phi$ for $G/\mathcal{P}_\phi$. The language of quotients also gives us a more natural definition of a complete homomorphism.

**Proposition 2.12** *A homomorphism $\phi : G \longrightarrow H$ is complete if and only if $\psi : G/\phi \longrightarrow H$ (defined in Proposition 2.11) is an isomorphism.*

One simple but useful consequence is that any homomorphism can be viewed is a sequence of identifications of pairs of vertices by complete homomorphisms.

**Definition 2.13** A complete homomorphism $\phi : G \longrightarrow H$ is *elementary* if there is unique pair of (nonadjacent) vertices $u, v \in V(G)$ which are identified by $\phi$. We call $H$ an *elementary quotient* of $G$.

We then have the very useful observation of Lemma 2.14. It allows us to specify a homomorphism simply by saying which vertices are equivalent under the homomorphism under construction and, further, to say which vertices are identified, pair by pair.

**Lemma 2.14** *Let $G$ and $H$ be graphs and $\phi : G \longrightarrow H$ a homomorphism. Then there is a natural number $k$ and graphs $G = G_0, G_1, \ldots, G_k$ such that $G_{i+1}$ is an elementary quotient of $G_i$ when $i < k$ and $G_k \cong G/\phi$.*

One type of homomorphism which is best defined in terms of partitions although in the literature ([30, 22, 45]) it is not defined in this language, is a folding.

**Definition 2.15** (i) An elementary homomorphism $\phi : G \longrightarrow H$ is a *simple fold* if the two vertices which are identified have a common neighbour.

(ii) A *folding* is a homomorphism obtained as a sequence of simple folds. If $\phi : G \longrightarrow H$ and $\phi$ is a folding, we say that *$G$ folds onto $H$*.

The above-mentioned papers deal with minimal graphs that can fold *onto* complete graphs, a problem raised in [30] and answered in [45].

Usually, a graph cannot fold onto each of its homomorphc images. For example, a disconnected graph cannot fold onto a connected one. A simple connected example is a path of length three which cannot fold onto a triangle but can be mapped to it by an elementary homomorphism. However, we will see in the next section that some important homomorphisms *are* foldings.

## 2.3 Retracts

When $\phi$ is an endomorphism of a graph $G$, the image of $G$ under $\phi$ is a subgraph of $G$ and there is a natural homomorphism back from $\phi(G)$ into $G$, namely the inclusion map. Such relationships between $G$ and $\phi(G)$ are natural objects to consider and lead to the idea of retracts. The concept is not proper to graph theory – retracts have their origin in topology and appear often in category theory. Hell's thesis [65] is concerned with category-theoretic aspects of graph retracts and extends the work of the Prague school ([102]). Our concerns are slightly different and so we shall not elaborate on this.

**Definition 2.16** Let $G$ and $H$ be graphs. Then $H$ is called a *retract* of $G$ if there are homomorphisms $\rho : G \longrightarrow H$ and $\gamma : H \longrightarrow G$ such that $\rho \circ \gamma = id_H$. The homomorphism $\rho$ is called a *retraction* and $\gamma$ a *co-retraction*.

The composition of two retractions is again a retraction and so a retract of a retract of $G$ is a retract of $G$. A co-retraction is always a faithful monomorphism and $\gamma(H)$ an induced subgraph of $G$. Thus retracts of $G$ are (isomorphic to) induced subgraphs of $G$. We usually think of retracts as subgraphs, since an endomorphism $\rho$ of $G$ onto its subgraph $R$ is a retraction whenever $\rho \restriction R = id_R$. In this case the co-retraction is naturally the inclusion map.

**Observation 2.17** Since there exist homomorphisms in both directions between a graph $G$ and any of its retracts $H$, it follows that $G$ and $H$ have the same chromatic number, odd girth and clique number. Note also that $\chi(G) = \omega(G)$ if and only if $K_{\chi(G)}$ is a retract of $G$.

The condition of Definition 2.16 that the composition of a retraction and its associated co-retraction be the identity can be slightly relaxed, providing a more useful characterization of retracts.

**Lemma 2.18** *Let $G$, $H$ be graphs. Then $H$ is a retract of $G$ if and only if there exist homomorphisms $\rho : G \longrightarrow H$ and $\sigma : H \longrightarrow G$ such that $\rho \circ \sigma \in \mathrm{Aut}(H)$.*

*In particular, if $H$ is a subgraph of $G$, then $H$ is a retract of $G$ is and only if there is a homomorphism $\rho : G \longrightarrow H$ whose restriction to $H$ is an automorphism of $H$.*

A retract of a graph $G$ is always a quotient of $G$ (i.e., a homomorphic image), and retractions are complete homomorphisms. In the case of connected graphs we can say more.

**Proposition 2.19** *Any retraction of a connected graph is a folding.*

**Proof** Let $\rho : G \longrightarrow H$ be a retraction, where $G$ is connected and $H$ is a subgraph of $G$. Let $u \in V(G)$ be a vertex not in $H$, adjacent to some vertex $v$ in $H$. Since the retraction $\rho$ which takes $G$ onto $H$ must map $u$ to a neighbour of $v$, the elementary homomorphism $\rho_u$ which identifies $u$ and $\rho(u)$ is a simple fold. The resulting graph still has $H$ as a retract and so a folding can be constructed by repeating the argument. □

In some sense, retracts are easy to find – if we already have an endomorphism, the following lemma allows us to find a retraction.

**Lemma 2.20** *Let $\phi$ be an endomorphism of a graph $G$. Then there is an $n$ such that $R = \phi^n(G)$ is a retract of $G$ (and $\phi^n$ a retraction). Further, $\phi \restriction R$ is an automorphism of $R$.*

Observation 2.17 shows that there are (many) subgraphs of $G$ which are not retracts of $G$. Indeed, no subgraph of $G$ with chromatic number smaller than that of $G$ is a retract of $G$. On the other hand, a graph is always a retract of itself and need not have any other retracts.

## 2.4 Cores

Whenever one looks at substructures with some given property one is naturally tempted to ask whether there is a smallest one that has the property. This is often the case and the property of being a retract is no exception. Minimal retracts are known as cores.

**Definition 2.21** A graph $G$ is a *core* if no proper subgraph of $G$ is a retract of $G$.

Natural though it is, we never work with this definition but replace it at once by an equivalent one which is technically much easier to handle. Note, however, that as it stands, Definition 2.21 applies equally well to finite and infinite graphs, but that this is not the case for its reformulated version:

**Proposition 2.22** *Let $G$ be a graph. Then $G$ is a core if and only if every endomorphism of $G$ is an automorphism of $G$.*

**Proof** Suppose $G$ is a core, and let $\phi : G \longrightarrow G$ be an endomorphism. By Lemma 2.20, $\phi^n(G)$ is a retract of $G$ for some $n$. Thus $\phi^n(G) = G$ by the definition of a core, and hence $\phi \in \mathrm{Aut}(G)$. The converse is trivial. □

The above proposition suggests another name for cores - *automorphic graphs.* This would be consistent with the names for similiar concepts in other branches of mathematics.

**Example 2.23** The following graphs are cores. While some are easily seen to be such, others are justified later.

- $K_n$, $n \geq 1$;
- $C_{2k+1}$ (the odd cycle) for $k \geq 1$;
- $W_{2k+1}$ (the odd wheel, that is, an odd cycle with one extra vertex adjacent to all vertices of the cycle), for $k \geq 1$;
- the Petersen graph;
- any $\chi$-critical graph, that is, a graph for which the chromatic numbers of its proper subgraphs are strictly smaller than its chromatic number.

**Example 2.24** On the other hand, these are not cores.

- $C_{2k}$ for $k > 0$;
- complete graphs with one edge removed;

- a disjoint union of cycles.

**Definition 2.25** A retract $H$ of $G$ is called a *core of G* if it is a core.

Among all subgraphs of $G$ which are retracts of $G$ choose one, say $H$, having the smallest number of vertices. Since retracts of retracts of $G$ are retracts of $G$, it follows that $H$ is a core. Thus:

**Proposition 2.26** *Every finite graph has a core.*

**Proposition 2.27** *If $H_0$ and $H_1$ are cores of a graph $G$ then they are isomorphic.*

**Proof** Consider such a pair of cores of $G$ and consider the retractions $\rho_0$ and $\rho_1$ of $G$ onto $H_0$ and $H_1$, respectively. Without loss of generality we can assume that both $H_0$ and $H_1$ are subgraphs of $G$. Let $\sigma_i : H_i \longrightarrow H_{1-i}$ be the restriction of $\rho_{1-i}$ to $H_i$, $i = 0, 1$. Then $\sigma_{1-i}\sigma_i$ is an endomorphism of $H_i$, and since $H_i$ is a core, $\sigma_{1-i}\sigma_i \in \mathrm{Aut}(H_i)$, by Proposition 2.22. This means that $\sigma_0$, $\sigma_1$ are, in fact, isomorphisms. □

Since all cores of a graph $G$ are isomorphic, it is legitimate to speak of *the* core of a graph $G$; we will denote it by $G^\bullet$. Strictly speaking, $G^\bullet$ is any member of its isomorphism class and we usually take for $G^\bullet$ a subgraph of $G$ (which is necessarily induced). This way of thinking of the core is more intuitive. For example, if $G$ is bipartite then $G^\bullet \cong K_2$, and so any edge of $G$ is its core.

## 2.5 Homomorphic equivalence

It is useful to define a relation on the class of graphs using homomorphisms.

**Definition 2.28** If there is a homomorphism from a graph $G$ to a graph $H$ we say that $G$ *maps to H* and we write $G \to H$.

For a simple example, note that $H \to G$ for any subgraph $H$ of $G$. The relation $\to$ is clearly transitive and reflexive but is not symmetric (an even cycle maps to an odd one but not vice versa). In fact, it does not define a partial order since it is not anti-symmetric either (for example, any two bipartite graphs map one into the other even when they are not isomorphic). Thus the relation $\to$ is only a quasi-order on the class of all graphs.

**Definition 2.29** We say that graphs $G$ and $H$ are *homomorphically equivalent* if $G \to H$ and $H \to G$. When this is the case we write $G \leftrightarrow H$.

For instance, if $R$ is a retract of $G$ then $G \leftrightarrow R$. Indeed, the definition of a retract provides the homomorphisms in both direction, namely the retraction $\rho : G \longrightarrow R$ and the co-retraction $\gamma : R \longrightarrow G$. This is true, in particular, of cores of graphs. Therefore, given any two graphs $G$, $H$, we have that $G \leftrightarrow H$ implies $G^\bullet \leftrightarrow H^\bullet$. However, the following much stronger result holds.

**Proposition 2.30** *Let $G$, $H$ be graphs such that $G \leftrightarrow H$. Then the cores of $G$ and $H$ are isomorphic.*

**Proof** Consider the cores $G^\bullet$ and $H^\bullet$ together with their retractions $\rho_G : G \longrightarrow G^\bullet$ and $\rho_H : H \longrightarrow H^\bullet$ and co-retractions $\gamma_G : G^\bullet \longrightarrow G$ and $\gamma_H : H^\bullet \longrightarrow H$. Let $\phi_G : G \longrightarrow H$ and $\phi_H : H \longrightarrow G$ be homomorphisms. Then the mapping $\psi : G^\bullet \longrightarrow G^\bullet$ defined by $\psi = \rho_G \circ \phi_H \circ \phi_G \circ \gamma_G$ is a homomorphism and hence an automorphism of $G^\bullet$. Therefore $\rho_G \circ \phi_H : H \longrightarrow G^\bullet$ is a retraction of $H$ onto $G^\bullet$ (and $\phi_G \circ \gamma_G : G^\bullet \longrightarrow H$ a co-retraction), thus $G^\bullet$ is a core of $H$ by Lemma 2.18. By Proposition 2.27, $G^\bullet$ is isomorphic to $H^\bullet$. □

Clearly $\leftrightarrow$ is an equivalence relation on the class of all graphs. We shall denote by $\mathcal{H}(G)$ the equivalence class of this relation containing the graph $G$. The following corollary follows from our discussion of retracts, cores and from the definition of homomorphic equivalence.

**Corollary 2.31** *Let $G$ be a graph. Then to within isomorphism, $G^\bullet$ is the unique graph of smallest order in $\mathcal{H}(G)$.*

The above corollary indicates why graph homomorphisms are interesting and important. The various aspects of any property which is preserved under homomorphisms can be studied on graphs best adapted for such investigations, provided they all belong to the same equivalence class (here, as elsewhere, we omit the qualifier *under homomorphisms*). For example, all the graphs in a class $\mathcal{H}(G)$ have the same odd girth, chromatic number, ultimate independence ratio, etc.

The set of equivalence classes of finite graphs can be partially ordered by homomorphisms between their elements.

**Definition 2.32** Let $G$ and $H$ be graphs. Then $\mathcal{H}(G) \prec \mathcal{H}(H)$ if $G \to H$.

This partially ordered set is, in fact, a lattice. We will indicate what the join and the meet are in the next section. Moreover (assuming we consider only graphs having at least one edge), there is a unique equivalence class which is minimal with respect to $\prec$, namely the class $\mathcal{H}(K_2)$ consisting of all bipartite graphs. It was first noticed by Welzl [127] that the part of the lattice above the minimal class $\mathcal{H}(K_2)$ has the remarkable property of being order-dense. Although this result concerns equivalence classes of graphs we shall formulate it in terms of representatives.

**Theorem 2.33 (Welzl)** *Let $G$ and $H$ be graphs such that $G \to H$ and $H \not\to G$. Then there is a graph $K$ such that $G \to K \to H$ and $H \not\to K \not\to G$.*

In terms of the equivalence classes this means that $\mathcal{H}(G) \prec \mathcal{H}(K) \prec \mathcal{H}(H)$, both inequalities being strict. Welzl's proof is rather long, a short one will be given in Section 5.1.

## 2.6 Products

We need to recall some of the various graph products that have been defined and used. We shall use the natural descriptive notation devised by Nešetřil. We will consider four products, two of them in some depth.

**Definition 2.34** Let $G$ and $H$ be graphs. The following products of $G$ and $H$ are defined on the vertex set $V = V(G) \times V(H)$.

- the *cartesian* product $G\Box H$, with

$$E(G\Box H) = \{[(u,x),(v,y)] : \text{ either } u = v, [x,y] \in E(H), \text{ or } [u,v] \in E(G), x = y\};$$

- the *categorical* product $G \times H$, with

$$E(G \times H) = \{[(u,x),(v,y)] : [u,v] \in E(G), [x,y] \in E(H)\};$$

- the *strong* product $G \boxtimes H$, with

$$E(G \boxtimes H) = E(G \times H) \cup E(G\Box H);$$

- the *lexicographic* product $G[H]$, with

$$E(G[H]) = \{[(u,x),(v,y)] : \text{ either } u = v, [x,y] \in E(H), \text{ or } [u,v] \in E(G)\}.$$

All the products can be found in the literature under a variety of names. The notation in the first three products indicates exactly what the result of taking the product of $K_2$ by $K_2$ is; this defines the whole graph.

It is easy to show that all the products are associative and that all but the lexicographic product are commutative. We can then define products of $k$ graphs $G_1, \ldots, G_k$ in the obvious manner and denote them by $\Box_{i=1}^k G_i$ and, analogously, for $\times$ and $\boxtimes$. The meaning of $G^k$ (when $G_i = G$ for all $i$) will be specified as needed.

When a product of graphs $G$ and $H$ is studied, it is often useful to consider the graphs induced by $\{u\} \times V(H)$ or by $V(G) \times \{x\}$. It is easy to see that these graphs, called *fibers*, are isomorphic to $G$ or $H$ in the case of the cartesian, strong, and lexicographic products. Hence the next observation.

**Observation 2.35** *Let $G$ and $H$ be graphs and let $*$ be any of the cartesian, strong or lexicographic product. Then $G \to G * H$ and $H \to G * H$. In fact, in each case $G$ and $H$ are induced subgraphs of the product.*

The converses are not always true. In particular, if both $G$ and $H$ have at least one edge, then $\omega(G \boxtimes H) = \omega(G[H]) \geq \max\{\omega(G), \omega(H)\}$, and so these products do not map into their factors. The cartesian product sometimes does have a homomorphism into its factors, but rarely (see Section 5.2).

Let $X$ and $Y$ be sets and $X \times Y$ their cartesian product. The *projections* of $X \times Y$ onto their factors are the mappings $\mathrm{pr}_X : X \times Y \longrightarrow X$ defined by $\mathrm{pr}_X(u,x) = u$ and $\mathrm{pr}_Y : X \times Y \longrightarrow Y$ defined by $\mathrm{pr}_Y(u,x) = x$. When $G$ and $H$ are graphs, and $X = V(G)$, $Y = V(H)$, we write simply $\mathrm{pr}_G$ and $\mathrm{pr}_H$. Clearly, the projections are not homomorphisms for the cartesian, strong and lexicographic products, and they are homomorphisms for the categorical product. The latter proves the next lemma.

**Lemma 2.36** *Let $G$ and $H$ be graphs. Then $G \times H \to G$ and $G \times H \to H$.*

The converse is not true: $K_2 \times K_3 \cong C_6$, and while the bipartite $C_6$ maps to $K_2$ and can be "wrapped around" $K_3$, the latter does not map into $C_6$

The categorical product allows us to make a lattice of homomorphism equivalence classes and, by extension, of cores. Recall that $\mathcal{H}(G)$ is the equivalence class containing $G$ under homomorphisms.

**Theorem 2.37** *The equivalence classes of graphs under homomorphisms form a lattice under the partial order $\prec$. The join of $\mathcal{H}(G)$ and $\mathcal{H}(H)$ is the equivalence class containing the disjoint union of $G$ and $H$, and the meet of $\mathcal{H}(G)$ and $\mathcal{H}(H)$ is $\mathcal{H}(G \times H)$.*

Products preserve homomorphisms in the following manner. Let $G, H, G', H'$ be graphs with $\phi : G \longrightarrow G'$ and $\psi : H \longrightarrow H'$. Define $\tau : V(G) \times V(H) \longrightarrow V(G') \times V(H')$ by $\tau(u, x) = (\phi(u), \psi(x))$. Then $\tau$ is a homomorphism for each of the products defined above. In particular, $G \Box H \leftrightarrow G^{\bullet} \Box H^{\bullet} \leftrightarrow (G \Box H)^{\bullet}$ and similarly for the other products. It is not true, however, that the product of cores of graphs is the core of the products.

**Example 2.38** Consider $K_2$ and $C_{2k+1}$. Both are cores but $(K_2 \Box C_{2k+1})^{\bullet} = C_{2k+1}$. Similarly, $(K_2 \times C_{2k+1})^{\bullet} = K_2$.

# 3 Vertex-transitive graphs

Bijectivity is a very restrictive condition to impose on a homomorphism. Nonetheless, automorphisms play a significant role in the general study of homomorphisms, in view of some of the facts presented earlier. We have seen in Lemma 2.20 that every endomorphism $\phi : G \longrightarrow G$ acts as an automorphism on some retract of $G$. Also, the core graphs, which represent all classes of graphs, have the property that all of their endomorphisms are automorphisms. By themselves, these observations do not make the subject of vertex-transitive graphs a central theme in the study of homomorphisms, though they emphasize its study, for its own sake, as a fruitful venture. The practical motivation for studying vertex-transitive graphs in the context of graph homomorphisms rather comes from special classes of graphs which are vertex-transitive and play a central role in some aspects of the theory (as complete graphs do in graph colouring).

This section is therefore intended to serve two purposes. First, to present results that are specific to the context of homomorphisms of vertex-transitive graphs, such as Sabidussi's theorem, the No-Homomorphism Lemma and Welzl's theorem on cores of vertex-transitive graphs. Second, to introduce some families of graphs such as Cayley graphs, Kneser graphs and circular graphs, which are of relevance in other sections of this paper.

## 3.1 Cayley graphs

Let $\Gamma$ be a group and $S$ a subset of $\Gamma$ that is closed under inverses and does not contain the identity[1]. The *Cayley graph* $\mathrm{Cay}(\Gamma, S)$ is the graph with $\Gamma$ as its vertex set, two vertices $u$ and $v$ being joined by an edge if and only if $u^{-1}v \in S$. Simple examples of Cayley graphs include the cycles, which are Cayley graphs of cyclic groups, and the complete graphs $K_n$,

[1]Such a set is often called a *Cayley subset* of $\Gamma$.

which are Cayley graphs of any group of order $n$. Specific names are sometimes given to some classes of Cayley graphs; for instance, Cayley graphs of cyclic groups are often called *circulants* (the name comes from the fact that their adjacency matrices are called "circulant matrices"), and a Cayley graph $\mathrm{Cay}(F, S)$, where $F$ is a finite field, $|F| \equiv 1 \bmod 4$, and $S$ is the set of quadratic residues of $F$, is called a *Paley graph.*

Cayley graphs constitute a rich class of vertex-transitive graphs. It is well known that the left translations (that is, bijections $T_a : \Gamma \longrightarrow \Gamma$ defined by $T_a(x) = ax$) of a Cayley graph are automorphisms, thus ensuring vertex transitivity. Since a Cayley graph may also have other automorphisms, the group structure is not encoded in the graph structure.[2] In shifting the focus from automorphisms to homomorphisms, one could expect to see the relative importance of Cayley graphs diminish within the subject of graph symmetry. The topic of graph homomorphisms, however, emphasizes Cayley graphs as a central theme in the study of vertex-transitive graphs for the following reason: *up to homomorphic equivalence, Cayley graphs represent all classes of vertex-transitive graphs.*

Let $G$ be a vertex-transitive graph, and $u_0$ a fixed vertex of $G$. Recall that $\mathrm{Aut}(G)$ denotes the automorphism group of $G$ and $\mathrm{Stab}(u_0)$ the *stabilizer* of $u_0$, i.e., the subgroup of $\mathrm{Aut}(G)$ containing all automorphisms $\phi$ such that $\phi(u_0) = u_0$. Put

$$S = \{\sigma \in \mathrm{Aut}(G) : [u_0, \sigma(u_0)] \in E(G)\}.$$

Thus $S$ is a union of left cosets of $\mathrm{Stab}(u_0)$.

It is easily seen that $S$ contains the inverse of each of its elements, and that the identity does not belong to $S$. Thus we can define the Cayley graph $\mathrm{Cay}(\mathrm{Aut}(G), S)$. By the definition of $S$, for $\phi_1, \phi_2 \in \mathrm{Aut}(G)$, we have $[\phi_1, \phi_2] \in E(\mathrm{Cay}(\mathrm{Aut}(G), S))$ if and only if $[\phi_1(u_0), \phi_2(u_0)] \in E(G)$. With this characterization of adjacency in $\mathrm{Cay}(\mathrm{Aut}(G), S)$, it is easy to see that the maps $\rho$ and $\gamma$ defined below are homomorphisms between $G$ and $\mathrm{Cay}(\mathrm{Aut}(G), S)$.

- $\rho : \mathrm{Cay}(\mathrm{Aut}(G), S) \longrightarrow G$, defined by putting $\rho(\phi) = \phi(u_0)$.
- $\gamma : G \longrightarrow \mathrm{Cay}(\mathrm{Aut}(G), S)$, defined by arbitrarily selecting $\gamma(u) = \phi_u$ such that $\phi_u(u_0) = u$ for each $u \in V(G)$.

Note that $\rho \circ \gamma(u) = u$ for any $u \in V(G)$, so $\rho$ is a retraction. Thus, we have the following.

**Theorem 3.1 (Sabidussi [108])** *Any vertex-transitive graph is a retract of some Cayley graph.* □

It is worthwhile to discuss some additional properties of the retraction $\rho$. Note that for any $u \in V(G)$, the set

$$\rho^{-1}(u) = \{\phi \in \mathrm{Aut}(G) : \phi(u_0) = u\}$$

is a left coset of $\mathrm{Stab}(u_0)$. In fact, $\rho$ is a full homomorphism – if $[u, v]$ is an edge of $G$, then $\rho^{-1}(u) \cup \rho^{-1}(v)$ induces a complete bipartite graph; this accounts for the freedom encountered in defining $\gamma$. But we have more: $\mathrm{Cay}(\mathrm{Aut}(G), S)$ is isomorphic to the lexicographic product

[2] Nonetheless, a result of Babai and Godsil provides evidence that most Cayley graphs have a regular automorphism group, see [7].

$G[\overline{K}_k]$ with $k = |\mathrm{Stab}(u_0)|$. Thus the structure of $\mathrm{Cay}(\mathrm{Aut}(G), S)$ is independent of the choice of $u_0$. A graph isomorphic to $G[\overline{K}_k]$ for some $k$ is called a *multiple* of $G$.

With this in mind, we give an alternative statement of Theorem 3.1 which is closer to its original spirit.

**Theorem 3.1 (Multiples version)** *Let $G$ be a vertex-transitive graph. Then some multiple of $G$ is a Cayley graph.* □

In our approach, a Cayley graph on $\mathrm{Aut}(G)$ is defined first, then the structure of $G$ is recovered by retracting $\mathrm{Cay}(\mathrm{Aut}(G), S)$ onto it. It is possible to interchange the order of these steps; this gives rise to the notion of a Cayley coset graph. Let $\Gamma$ be a group, $\Gamma_0$ a subgroup of $\Gamma$ and $S$ a subset of $\Gamma$ that is closed under inverses and does not contain the identity. The *Cayley coset graph* $\mathrm{Ccg}(\Gamma, S, \Gamma_0)$ is defined by putting

$$\begin{aligned} V(\mathrm{Ccg}(\Gamma, S, \Gamma_0)) &= \{a\Gamma_0 : a \in \Gamma\}, \\ E(\mathrm{Ccg}(\Gamma, S, \Gamma_0)) &= \{[a\Gamma_0, b\Gamma_0] : a\Gamma_0 \neq b\Gamma_0 \text{ and } a\Gamma_0 \cap b\Gamma_0 S \neq \emptyset\}. \end{aligned}$$

The condition $a\Gamma_0 \cap b\Gamma_0 S \neq \emptyset$ means that there exist $c \in a\Gamma_0$, $d \in b\Gamma_0$ such that $d^{-1}c \in S$. In particular, our retraction $\rho$ amounts to collapsing $\mathrm{Cay}(\mathrm{Aut}(G), S)$ onto the coset graph $\mathrm{Ccg}(\mathrm{Aut}(G), S, \mathrm{Stab}(u_0))$. Thus, Theorem 3.1 also admits the following formulation.

**Theorem 3.1 (Cayley coset graphs version)** *Any vertex-transitive graph is isomorphic to a Cayley coset graph.* □

In general, the natural map from a Cayley graph $\mathrm{Cay}(\Gamma, S)$ to a Cayley coset graph $\mathrm{Ccg}(\Gamma, S, \Gamma_0)$ is a homomorphism if and only if $\Gamma_0 \cap S = \emptyset$. However, an element of $\Gamma_0 \cap S$ does not contribute to the structure of $\mathrm{Ccg}(\Gamma, S, \Gamma_0)$ in any way, so the condition $\Gamma_0 \cap S = \emptyset$ is not a severe restriction on a Cayley coset graph.

## 3.2 Indepedemce ratio and the No-Homomorphism Lemma

Other than trial and error, there seems to be no sure-fire way to determine whether there exists a homomorphism from a given graph to another. Parameters such as the odd girth and the chromatic number provide some restrictions, but these are far from being exhaustive. The basic idea behind the No-Homomorphism Lemma is to use symmetry to find new restrictions for the existence of homomorphisms between graphs. Only the simplest version is given here, but it is clear from the proof that many generalizations and variations are possible.

We write $\alpha(G)$ for the *independence number* of a graph $G$, that is, the maximum number of vertices in an independent set of $G$.

**Definition 3.2** The *independence ratio* of a graph $G$ is $i(G) = \alpha(G)/|V(G)|$.

**Lemma 3.3 (No-Homomorphism Lemma, Albertson and Collins [5])** *Let $G$, $H$ be graphs such that $H$ is vertex-transitive and $G \to H$. Then*

$$i(G) \geq i(H).$$

**Proof** Let $\mathcal{S}(H)$ denote the family of independent sets of size $\alpha(H)$ in $H$. By symmetry, every vertex of $H$ is in the same number, say $m$, of members of $\mathcal{S}(H)$. We then have

$$\alpha(H) \cdot |\mathcal{S}(H)| = m \cdot |H|, \tag{3.1}$$

since each expression counts the number of inclusions $u \in I$, with $u \in V(H)$ and $I \in \mathcal{S}(H)$.

Let $\phi : G \longrightarrow H$ be a homomorphism. Then, for each $I \in \mathcal{S}(H)$, we have $|\phi^{-1}(I)| \leq \alpha(G)$. Summing this inequality for all members of $\mathcal{S}(H)$, we get

$$\sum_{I \in \mathcal{S}(H)} |\phi^{-1}(I)| \leq \alpha(G) \cdot |\mathcal{S}(H)|. \tag{3.2}$$

However, each $u \in V(G)$ contributes exactly $m$ to the sum $\sum_{I\in\mathcal{S}(H)} |\phi^{-1}(I)|$, since $\phi(u)$ belongs to exactly $m$ members of $\mathcal{S}(H)$. Thus,

$$\sum_{I \in \mathcal{S}(H)} |\phi^{-1}(I)| = m \cdot |G|. \tag{3.3}$$

Combining (3.1), (3.2), (3.3), we get

$$i(G) = \alpha(G)/|G| \geq m/|\mathcal{S}(H)| = \alpha(H)/|H| = i(H). \tag{3.4}$$

□

The name "No-Homomorphism Lemma" is derived from the contrapositive: *if $H$ is vertex-transitive and $i(G) < i(H)$, then $G \not\to H$.* One trivial consequence of this result is the well-known fact that the chromatic number of a graph $G$ is at least $|G|/\alpha(G)$. In terms of Lemma 3.3, we may rephrase this by saying that a necessary condition for $G$ to admit a homomorphism into $K_n$ is that $i(G) \geq 1/n$.

**Example 3.4** Consider the odd cycles. We have $\alpha(C_{2k+1}) = k$, and the No-Homomorphism Lemma states that if $C_{2k+1} \to C_{2k'+1}$, then $k/(2k+1) \geq k'/(2k'+1)$, i.e., $k \geq k'$.

In fact, there exists a homomorphism from $C_{2k+1}$ to $C_{2k'+1}$ if and only if $k \geq k'$, but sufficiency cannot be deduced from Lemma 3.3.

**Example 3.5** Let $G = \mathrm{Cay}(\mathbb{Z}_8, \{\pm1, \pm2\})$ and $H = \mathrm{Cay}(\mathbb{Z}_7, \{\pm1, \pm2\})$. Then $\alpha(G) = \alpha(H) = 2$. So,

$$i(G) = 2/8 < 2/7 = i(H).$$

Since $H$ is vertex-transitive, this implies that there exists no homomorphism from $G$ to $H$.

In fact, there exists no homomorphism from $H$ to $G$ either for the graphs in the above example, though this cannot be deduced from the No-Homomorphism Lemma. A direct verification of this statement could be messy. However, if we define the parameter $\pi(G)$ as the maximum number of vertices in an induced subgraph of $G$ that does not contain a triangle, we get $\pi(G) = 5$, $\pi(H) = 4$, so $\pi(H)/|H| < \pi(G)/|G|$. The reader might want to check how the proof of the No-Homomorphism Lemma can be adapted to deduce from this that $H \not\to G$.

The critical situation in the No-Homomorphism Lemma is the case of homomorphisms between graphs with the same independence ratio. The next result shows how the behaviour of homomorphisms is restricted in that case.

**Lemma 3.6** *Let $G$, $H$ be graphs such that $H$ is vertex-transitive, $i(G) = i(H)$ and $G \to H$. Then for any independent set $I$ of cardinality $\alpha(H)$ in $H$ and any homomorphism $\phi : G \longrightarrow H$ we have*

$$|\phi^{-1}(I)| = \alpha(G).$$

**Proof** With the notation of the proof of Lemma 3.3, our hypotheses imply that equality holds in (3.4). Thus, $m \cdot |G| = \alpha(G) \cdot |\mathcal{S}(H)|$. By (3.3), we then have $\sum_{I \in \mathcal{S}(H)} |\phi^{-1}(I)| = \alpha(G) \cdot |\mathcal{S}(H)|$. Since $|\phi^{-1}(I)| \leq \alpha(G)$ for all $I \in \mathcal{S}(H)$, we must have $|\phi^{-1}(I)| = \alpha(G)$ for all $I \in \mathcal{S}(H)$. □

In particular, the situation described in Lemma 3.6 arises in the case of homomorphically equivalent graphs. If $G$ and $H$ are homomorphically equivalent vertex-transitive graphs, then the No-Homomorphism Lemma states that $i(G) \geq i(H)$ and $i(H) \geq i(G)$. Thus, $i(G) = i(H)$, and Lemma 3.6 says that any homomorphism between $G$ and $H$ induces a map between $\mathcal{S}(H)$ and $\mathcal{S}(G)$. Sometimes this information helps to characterize such homomorphisms (see Proposition 3.15).

## 3.3 Cores of vertex-transitive graphs

Theorem 3.1 tells us that any vertex-transitive graph is a retract of a Cayley graph. As retractions often provide a way of simplifying the structure of an object, it would be worthwhile to investigate the retracts *of* vertex-transitive graphs as well. However, retractions do not generally preserve vertex transitivity; for instance, any even cycle retracts onto a path of the same diameter. With respect to symmetry, the most interesting retract of a vertex-transitive graph is its core. Recall that core graphs have the important property that all of their endomorphisms are automorphisms, and this is the basis of many useful properties of cores of vertex-transitive graphs, starting with the following (also independently proved by MacGillivray):

**Theorem 3.7 (Welzl [128])** *Let $G$ be a vertex-transitive graph. Then its core $G^\bullet$ is vertex-transitive.*

**Proof** Fix a retraction $\rho : G \longrightarrow G^\bullet$ with co-retraction $\gamma : G^\bullet \longrightarrow G$. For any $u, v \in V(G^\bullet)$, there exists an automorphism $\phi$ of $G$ mapping $\gamma(u)$ to $\gamma(v)$. We then have $\rho \circ \phi \circ \gamma(u) = v$ (since $\rho \circ \gamma$ is the identity map on $G^\bullet$). But since $G^\bullet$ is a core, the map $\rho \circ \phi \circ \gamma : G^\bullet \longrightarrow G^\bullet$ is an automorphism of $G^\bullet$. Thus, $\mathrm{Aut}(G^\bullet)$ acts transitively on $G^\bullet$. □

Along the same lines, it is easy to show that the core of an edge-transitive graph is again edge-transitive and the same idea applies to many other types of symmetry.

Theorem 3.7 allows us to apply the results of the preceding section. A vertex-transitive graph and its core are homomorphically equivalent vertex-transitive graphs, so Lemma 3.3 states that they have the same independence ratio. Since this ratio is defined with integer parameters on graphs, we deduce a sufficient condition for a graph to be a core.

**Corollary 3.8** *Let $G$ be a vertex-transitive graph. If $\alpha(G)$ and $|G|$ are relatively prime, then $G$ is a core.*

In particular, this occurs for vertex-transitive graphs on a prime number of vertices. It is known that such a graph is isomorphic to a circulant $\mathrm{Cay}(\mathbb{Z}_p, S)$, whose automorphism group is generated by the left translations and the automorphisms of $\mathbb{Z}_p$ that preserve the connection set $S$ (see Turner [122]). By the preceding result, these graphs are also cores, and thus have no endomorphisms other than their automorphisms.

In general, the result of Lemma 3.6 cannot be formulated for vertices rather than maximum independent sets. For instance, consider the homomorphisms between the 6-cycle and the 4-cycle. Note that the cardinalities of inverse images of individual vertices vary, while the sum over maximal independent sets is constant. The next result show that the situation is different in the case of cores.

**Theorem 3.9** *Let $\phi$ be a homomorphism of a vertex-transitive graph $G$ onto $G^\bullet$. Then all inverse images $\phi^{-1}(u), u \in V(G^\bullet)$, have the same cardinality, namely $|G|/|G^\bullet|$.*

**Proof** Fix a homomorphism $\gamma : G^\bullet \longrightarrow G$ and a vertex $u$ of $G^\bullet$. Put

$$X = \{(v, \psi) : v \in V(G^\bullet), \psi \in \mathrm{Aut}(G) \text{ and } \phi \circ \psi \circ \gamma(v) = u\}.$$

The cardinality of $X$ can be evaluated in two ways: First, for any $\psi \in \mathrm{Aut}(G)$, the map $\phi\circ\psi\circ\gamma : G^\bullet \longrightarrow G^\bullet$ is an automorphism of $G^\bullet$, so it must be bijective. Therefore there exists a unique $v \in V(G^\bullet)$ such that $\phi\circ\psi\circ\gamma(v) = u$, i.e., $(v, \psi) \in X$. Thus $|X| = |\mathrm{Aut}(G)|$. Second, for any $v \in V(G^\bullet)$ and $x \in \phi^{-1}(u)$, the set of automorphisms of $G$ mapping $\gamma(v)$ to $x$ is a left-coset of $\mathrm{Stab}(\gamma(v))$, and has cardinality $|\mathrm{Aut}(G)|/|G|$. Thus $|X| = |G^\bullet| \cdot |\phi^{-1}(u)| \cdot (|\mathrm{Aut}(G)|/|G|)$. Combining these two expressions for $|X|$, we have $|\phi^{-1}(u)| = |G|/|G^\bullet|$. □

This last result is an analogue for cores of vertex-transitive graphs of Lagrange's theorem: the order of $G^\bullet$ divides that of $G$.

Any fixed retraction $\rho$ and co-retraction $\gamma$ between a graph $G$ and its core $G^\bullet$ induce a map between $\mathrm{Aut}(G)$ and $\mathrm{Aut}(G^\bullet)$, defined by mapping $\phi \in \mathrm{Aut}(G)$ to $\rho \circ \phi \circ \gamma \in \mathrm{Aut}(G^\bullet)$. Some results of this section are applications of this fact, but notice that this induced mapping between groups is devoid of algebraic significance. A case in point is the fact that the core of a Cayley graph is not necessarily a Cayley graph. The Petersen graph is a well-known example of a vertex-transitive graph which is not a Cayley graph. It will be shown in the next section that the Petersen graph is a core. However, by Theorem 3.1, the Petersen graph is a retract of a Cayley graph, so it is the core of a Cayley graph.

## 3.4 Kneser graphs

Let $r, s$ be integers such that $1 \le r < s/2$. The *Kneser graph* $K(r, s)$ is the graph whose vertices are the $r$-subsets of $[s]$, two vertices being joined by an edge if and only if they are disjoint. Kneser graphs are a combinatorial structure which arises naturally in different contexts. The graphs $K(r, 2r+1), r \ge 1$ are also called *odd graphs*; amongst them we find $K(2, 5)$, which is the Petersen graph. Also, $K(1, s)$ is the complete graph $K_s$. We will see below a connection between Kneser graphs and the Erdős-Ko-Rado theorem, and other sections of this paper will show their relation with the lexicographic product of graphs and some parameters such as the fractional chromatic number.

It will also be shown that Kneser graphs provide examples of graphs with arbitrarily large odd girth and chromatic number.

The name 'Kneser graph' is derived from Kneser's conjecture, which is the following statement.

> *Let $r, s$ be integers such that $2 \leq r < s/2$. If the $r$-subsets of a $s$-set are partitioned into $s - 2r + 1$ families, then one family contains two disjoint sets.*

Note that a partition of the $r$-subsets of $[s]$ into families of pairwise intersecting subsets amounts to a proper vertex colouring of the Kneser graph $K(r,s)$. Thus Kneser's conjecture says that $\chi(K(r,s)) \geq s - 2r + 2$. On the other hand, it is easily seen that $s - 2r + 2$ colours are sufficient: For $k = 0, \ldots, s - 2r$, assign the colour $k$ to the set

$$I_k = \{A \in V(K(r,s)) : \min A = k\},$$

and assign the colour $s - 2r + 1$ to the set

$$I_{s-2r+1} = \{A \in V(K(r,s)) : \min A \geq s - 2r + 1\}.$$

This graph-theoretic interpretation of Kneser's conjecture is due to Lovász, who gave the first proof of it.

**Theorem 3.10 (Lovász [91])** *The chromatic number of $K(r,s)$ is $s - 2r + 2$.*

Lovász used a surprising approach; simplifications and generalizations were later made by Bárány [14] and Walker [126], but the crux remains the basic idea of Lovász, which is, to transpose the problem to some topological space associated with the graph, and then use some powerful tools of algebraic topology, notably the Borsuk-Ulam antipodal theorem. It is somewhat unsettling that there is as yet no genuine graph-theoretic argument that explains why $K(r,s)$ cannot be coloured with $s - 2r + 1$ colours; on the other hand, Lovász's result gives an interesting topological flavour to the problem of determining whether there exists a homomorphism between two given graphs.

The independence number of Kneser graphs is related to the following classical inequality.

**Theorem 3.11 (Erdős, Ko, Rado [42])** *Let $r, s$ be integers such that $1 \leq r < s/2$, and $\mathcal{F}$ a family of pairwise intersecting $r$-subsets of $[s]$. Then*

$$|\mathcal{F}| \leq \binom{s-1}{r-1}.$$

As the families of pairwise intersecting $r$-subsets of $[s]$ are precisely the independent sets of $K(r,s)$, this result gives a bound for $\alpha(K(r,s))$. This bound is easily seen to be sharp, as it coincides with the cardinality of each set

$$J_k = \{A \in V(K(r,s)) : k \in A\}, \quad k = 0, \ldots, s-1.$$

One other combinatorial inequality provides further information on the structure of Kneser graphs:

**Theorem 3.12 (Hilton, Milner [81])** *Let $r, s$ be integers such that $1 \leq r < s/2$, and $\mathcal{F}$ a family of pairwise intersecting $r$-subsets of $[s]$. If $\bigcap \mathcal{F} = \emptyset$, then*

$$|\mathcal{F}| \leq \binom{s-1}{r-1} - \binom{s-r-1}{r-1} + 1.$$

This result shows that the only independent sets of maximal cardinality in $K(r,s)$ are the sets $J_0, \ldots, J_{s-1}$ defined above. We see from this that the automorphism group of $K(r,s)$ is isomorphic to the symmetric group $S_s$: Any permutation of the base set $[s]$ induces an automorphism of $K(r,s)$ in a natural way. Conversely, any automorphism of $K(r,s)$ must interchange the sets $J_1, \ldots, J_s$ (since these are the only independent sets of maximal size) thus induces a permutation of $[s]$. With the help of Lemma 3.6, it is possible to complete the characterization of endomorphisms of Kneser graphs.

**Proposition 3.13** *All Kneser graphs are cores.*

**Proof** Let $\phi : K(r,s) \longrightarrow K(r,s)$ be a homomorphism. Then for $k = 0, \ldots, s-1$, $|\phi^{-1}(J_k)| = \alpha(K(r,s))$, so there exists an index $\psi(k) \in [s]$ such that $\phi^{-1}(J_k) = J_{\psi(k)}$. By definition of the sets $J_k, k = 0, \ldots, s-1$, this means that for any $A \in V(K(r,s))$, we have $k \in \phi(A)$ if and only if $\psi(k) \in A$. Therefore, $\psi$ must be bijective, $\phi$ is the automorphism induced by $\psi^{-1} : [s] \longrightarrow [s]$, and $K(r,s)$ is a core. □

The Kneser graphs with a given chromatic number $k$ are the graphs $K(n, 2n-2+k)$, $n \geq 1$, and the odd girth of these can be made arbitrarily large. Thus, we get a constructive proof of the well-known result that there exist graphs with arbitrarily large odd girth and chromatic number (see [41]). The following statement provides specific examples for any given odd girth and chromatic number.

**Proposition 3.14** *Given integers $k$, $n$ such that $n \geq 3$, let $r = k(n-2)$, $s = (2k+1)(n-2)$. Then $K(r,s)$ has chromatic number $n$ and odd girth $2k+1$.*

**Proof** By Theorem 3.10, $\chi(K(r,s)) = n$. Also, since Kneser graphs are vertex-transitive, Lemma 3.3 states that $C_{2i+1} \rightarrow K(r,s)$ only if $i/(2i+1) \geq k/(2k+1)$, that is, $i \geq k$. Thus, the odd girth of $K(r,s)$ is at least $2k+1$. The interested reader may want to verify that $K(r,s)$ indeed contains a $(2k+1)$-cycle, but this also follows from our discussion of circular graphs in the next section. □

At this point it is worthwhile to reflect upon the condition $r < s/2$ that we imposed in the definition of Kneser graphs. If we allow $r = s/2$, the resulting graph $K(r, 2r)$ is a perfect matching. The formulas above still provide the correct chromatic number and independence number, but the subsequent results on the structure of independent sets and automorphisms groups are no longer valid. If $s/2 < r \leq s$, the resulting graph $K(r,s)$ has no edge, and the formulas for the chromatic number and the independence number are wrong in most cases. However, for some applications of Kneser graphs such as subset colourings and the fractional chromatic number, it is convenient to consider these marginal graphs as Kneser graphs. The context will make clear what restrictions on $r$ and $s$ are to be imposed.

The remainder of this section is devoted to homomorphisms between Kneser graphs. A question raised by Hell asks for which integers $r, s, r', s'$ we have $K(r,s) \rightarrow K(r', s')$ (see Godsil [50]). Lovász's theorem and the No-Homomorphism Lemma provide some restrictions. If $K(r,s) \rightarrow K(r',s')$, then

$$\chi(K(r,s)) = s - 2r + 2 \leq s' - 2r' + 2 = \chi(K(r',s'))$$

and

$$i(K(r,s)) = r/s \geq r'/s' = i(K(r',s')).$$

Also, the proof of Proposition 3.13 can easily be adapted to the following characterization of homomorphisms between Kneser graphs with the same independence ratio.

**Proposition 3.15 (Stahl [113])** *Let $r, s$ be relatively prime numbers such that $1 \leq r < s/2$. Then $K(mr, ms) \to K(nr, ns)$ for integers $m, n$ if and only if $n$ is a multiple of $m$.*

We now turn our attention to a wide list of homomorphisms between Kneser graphs. Note that $K(r,s)$ is an induced subgraph of $K(r, s+1)$. Also, it is easily seen that there exists a homomorphism from $K(r+1, s+1)$ to $K(r,s)$ defined by removing the maximal element from each $A \in V(K(r+1, s+1))$. A refinement of this idea is used in the proof of the following result:

**Proposition 3.16 (Stahl [113])** *For any two integers such that $1 \leq r < s/2$ we have $K(r+1, s+2) \to K(r,s)$.*

**Proof** Define a map $\phi$ from $K(r+1, s+2)$ to $K(r,s)$ by

$$\phi(A) = \begin{cases} A \setminus \{\max(A)\} & \text{if } |A \cap \{s, s+1\}| \leq 1 \\ (A \setminus \{s, s+1\}) \cup \{\max(\overline{A})\} & \text{if } \{s, s+1\} \subseteq A \end{cases}$$

(where $\overline{A}$ denotes the complement of $A$). We show that $\phi$ is a homomorphism. Take $A, B \in V(K(r+1, s+2))$ such that $[A, B] \in E(K(r+1, s+2))$. Then $A \cap B = \emptyset$, so $A$ and $B$ cannot both contain $\{s, s+1\}$. We may therefore assume that $|A \cap \{s, s+1\}| \leq 1$. We then have $\phi(A) \subseteq A$. If $|B \cap \{s, s+1\}| \leq 1$, then $\phi(B) \subseteq B$, thus $\phi(A) \cap \phi(B) = \emptyset$ and $[\phi(A), \phi(B)] \in E(K(r,s))$. Otherwise, we have $\phi(B) \setminus B = \{k\}$, where $k \leq s-1$ and $\{k+1, \ldots, s+1\} \subseteq B$. Since $A \cap B = \emptyset$, we have $\max(A) \leq k$, thus $k \notin \phi(A)$. But $\phi(B) \setminus \{k\} \subseteq B$, so $\phi(A) \cap \phi(B) = \emptyset$ and $[\phi(A), \phi(B)] \in E(K(r,s))$. □

Note that the graphs $K(r, k-2+2r), r \in \mathbb{N}$, are precisely the Kneser graphs with chromatic number $k$. By the preceding result, these are linearly ordered by the relation $\to$. Combining this with inclusions, we get the following.

**Corollary 3.17** *Let $r, r', s, s'$ be integers such that $r \geq r'$, $1 \leq r < s/2$ and $1 \leq r' < s'/2$. Then $K(r,s) \to K(r', s')$ if and only if $s - 2r + 2 \leq s' - 2r' + 2$.* □

It remains to investigate homomorphisms from $K(r,s)$ to $K(r',s')$, with $r < r'$. The next result allows us to make use of the homomorphisms we have already defined.

**Proposition 3.18** *Let $G$ be a graph, and let $r, r', s, s'$ be integers such that $1 \leq r < s/2$ and $1 \leq r' < s'/2$. If $G \to K(r,s)$ and $G \to K(r', s')$ then $G \to K(r + r', s + s')$.*

**Proof** Let $\phi_1 : G \longrightarrow K(r,s)$ and $\phi_2 : G \longrightarrow K(r', s')$ be homomorphisms. Define $\phi : G \longrightarrow K(r + r', s + s')$ by putting $\phi(u) = \phi_1(u) \cup \{i + s : i \in \phi_2(u)\}$. Then $\phi$ is a homomorphism. □

By Theorem 3.16, $K(r,s) \to K(r-k, s-2k)$ for all $k \in \{1,\ldots,r-1\}$, so this last result shows that $K(r,s) \to K(nr-k, ns-2k)$ for all $n \in \mathbb{N}$ and $k \in \{1,\ldots,r-1\}$. This is where the question stands today. Stahl [113] conjectures that $K(r,s) \to K(r',s')$ if and only if $s' \geq ns - 2k$, where $r' = nr - k$. This conjecture is known to be true for low values of $r$, and for all odd graphs $K(r, 2r+1)$ (see Stahl [114]).

## 3.5 Circular graphs

This section presents a class of graphs that behaves with respect to $\to$ like the set of rationals with respect to its usual order. Let $r,s$ be integers such that $1 \leq r \leq s/2$. We define the *circular graph* $G_s^r$ as the Cayley graph $\mathrm{Cay}(\mathbb{Z}_s, \{r, r+1, \ldots, s-r\})$. These graphs play an important role in the definition of the circular chromatic number (see Section 4) and some essential properties of this parameter are derived from the results presented here. It is worthwhile to warn the reader about some notation used in the proofs. Formally, the elements of $\mathbb{Z}_s$ are congruence classes modulo $s$, but are usually written as integers from $[s]$. This convention becomes inconvenient and confusing when defining maps between groups $\mathbb{Z}_s$ and $\mathbb{Z}_{s'}$. Consequently some proofs will be given in more detail than usual in order to avoid any possible misinterpretation.

**Lemma 3.19** *Let $r,s$ be positive integers such that $r \leq s/2$. Then $\alpha(G_s^r) = r$.*

**Proof** Obviously, $\alpha(G_s^r) \geq r$, since $\{0,1,\ldots,r-1\}$ is an independent set. Let $I$ be an independent set of $G_s^r$ and $u$ an element of $I$. Then $I \subseteq \{u-(r-1),\ldots,u,\ldots,u+r-1\}$. However, $I$ can contain only one vertex of each edge $[u-i, u-i+r]$, so $|I| \leq r$. □

This simple result shows that the independence ratio of $G_s^r$ is $r/s$, and allows us to use the No-Homomorphism Lemma to investigate the existence of homomorphisms between circular graphs. Also note that the Erdős-Ko-Rado inequality (Theorem 3.11) can be deduced from Lemma 3.19 (see [120]). Define $\phi : G_s^r \longrightarrow K(r,s)$ by putting $\phi(u) = \{u, u+1, \ldots, u+r-1\}$ (we identify a congruence class in $\mathbb{Z}_s$ with its unique representative in $[s]$). Clearly, $\phi$ is a homomorphism, and with Lemma 3.3, the independence ratio of $G_s^r$ provides an upper bound for $\alpha(K(r,s))$ which coincides with the Erdős-Ko-Rado bound.

**Lemma 3.20 (Bondy, Hell [18])** *Let $r,s,k$ be integers such that $r \leq s/2$. Then $G_s^r \leftrightarrow G_{ks}^{kr}$.*

**Proof** Let $\phi$ be the unique group homomorphism mapping the generator 1 of $\mathbb{Z}_s$ to the element $k$ of $\mathbb{Z}_{ks}$. It is easily seen that $\phi$ is a homomorphism from $G_s^r$ to $G_{ks}^{kr}$. We can also define a homomorphism $\psi : G_{ks}^{kr} \longrightarrow G_s^r$ by putting $\psi(v) = u$ if and only if $v \in \{\phi(u), \phi(u)+1, \ldots, \phi(u)+k-1\}$. □

**Proposition 3.21 (Bondy, Hell [18])** *Let $r,s,m,n$ be integers such that $r \leq s/2$ and $m \leq n/2$. Then $G_s^r \to G_n^m$ if and only if $r/s \geq m/n$.*

**Proof** By Lemma 3.3 and Lemma 3.19, if there exists a homomorphism from $G_s^r$ to $G_n^m$, then $r/s \geq m/n$. Conversely, suppose that $r/s \geq m/n$. By Lemma 3.20, $G_s^r \leftrightarrow G_{ns}^{nr}$

and $G_n^m \not\leftrightarrow G_{sn}^{sm}$. But $nr \geq sm$, so $G_{ns}^{nr}$ is a subgraph of $G_{sn}^{sm}$. Therefore, there exists a homomorphism from $G_s^r$ to $G_n^m$. □

This last result shows that with respect to the relation $\rightarrow$, the class of circular graphs behaves like the rationals in $]0, 1/2]$ with respect to the relation $\geq$. This is an example of a family of graphs which is 'dense' in the sense of Theorem 2.33: between any two rationals $r/s < r'/s'$ in $]0, 1/2]$, there exists a rational $r''/s''$ which is strictly in between them; we then have $G_{s'}^{r'} \rightarrow G_{s''}^{r''} \rightarrow G_s^r$ but no two of the three graphs are homomorphically equivalent. Recall that all circular graphs are Cayley graphs and so are vertex-transitive. It is natural to ask whether Welzl's density theorem 2.33 can be restricted to vertex-transitive graphs. This question was raised by Welzl [128]. Albertson and Booth [4] have shown that every vertex-transitive graph is an endpoint of an interval containing no other vertex-transitive graph. They also proved a special case of the following theorem.

**Theorem 3.22 (Tardif [118])** *If $G$ and $H$ are vertex-transitive graphs such that $G$ maps strictly into $H$ then there is a vertex-transitive graph $K$ such that $G \rightarrow K \rightarrow H$ and no two of the three graphs are homomorphically equivalent.*

By Proposition 3.21, $G_s^r \not\rightarrow G_{s'}^{r'}$ for any integers $r', s'$ such that $r/s < r'/s'$. We conclude this section with a result showing that $G_s^r$ is critical with respect to this property. Recall that $G - u$ is the graph obtained from $G$ by the removal of the vertex $u$ and all edges incident with it.

**Proposition 3.23 (Bondy, Hell [18])** *Let $r, s$ be relatively prime integers such that $r < s/2$. Then there exists integers $m, n$ such that $r/s < m/n$, and for any $u \in V(G_s^r)$, $G_s^r - u \rightarrow G_n^m$.*

**Proof** Since $r$ and $s$ are relatively prime, there exist integers $a < s$ and $b < r$ such that $ar = bs + 1$. Put $m = r - b$ and $n = s - a$. Then, $r/s < m/n$. We show that for any $u \in V(G_s^r)$ there exists a homomorphism from $G_s^r - u$ to $G_n^m$.

By vertex transitivity, it suffices to consider the case $u = 0$. Also, by Lemma 3.20, it suffices to find a homomorphism from $G_s^r - \{0\}$ to $G_{rn}^{rm}$. To do this, we identify the vertices of $G_s^r - \{0\}$ with their representatives $1, \ldots, s-1$ in $\mathbb{N}$, and the vertices of $G_{rn}^{rm}$ with their representatives $0, \ldots, rn-1$ in $\mathbb{N}$ (to recover multiplication in $\mathbb{N}$). We then define $\phi : G_s^r - \{0\} \longrightarrow G_{rn}^{rm}$ by putting $\phi(i) = im$. This map is well defined since $1 \leq i \leq s-1$ implies $0 \leq im \leq rn-1$. Take $1 \leq i < j \leq s-1$ such that $[i, j]$ is an edge of $G_s^r - \{0\}$. Then $j - i \in \{r, \ldots, s-r\}$, so $\phi(j) - \phi(i) = (j-i)m \in \{rm, \ldots, r(n-m)\}$ (since $(s-r)m \leq r(n-m)$). Thus, $[\phi(i), \phi(j)]$ is an edge of $G_{rn}^{rm}$, and $\phi$ is a homomorphism. □

# 4 Graph colourings and variations

One of the sources of the subject of graph homomorphisms is the theory of graph colourings. We have seen (Example 2.7) that a proper colouring of a graph $G$ with $k$ colours can be viewed as a homomorphism from $G$ into $K_k$. In this section we will explore the homomorphism approach to graph colourings, beginning with definitions phrased in the language of homomorphisms. We begin with proper colourings and the chromatic number. We will be brief most of the time since there are recent publications surveying much of the literature on graph colourings; see [82, 121, 142].

A $k$-colouring of a graph $G$ is a homomorphism into the complete graph $K_k$. Such a homomorphism is often wasteful in that not all the vertices (or edges) of $K_k$ are images of vertices (or edges) of $G$. It is, therefore, of interest to restrict our attention only to colourings that use all of the target graph.

**Definition 4.1** Given a natural number $k$, a *complete $k$-colouring* of a graph $G$ is a complete homomorphism $\phi : G \longrightarrow K_k$.

That is, a $k$-colouring of $G$ is complete if it is proper and every pair of colours appears on the endpoints of some edge of $G$. The definition can be rephrased by saying that a complete colouring of a graph $G$ is a partition $\mathcal{P}$ of $V(G)$ into independent sets such that $G/\mathcal{P}$ is a complete graph. When the partition has $k$ classes, we speak of a *complete $k$-colouring.*

We will prove (Proposition 4.12) that if there is a complete homomorphism from $G$ onto $K_r$ and onto $K_t$ then there also is one onto $K_s$ for any $s$ between $r$ and $t$. Since clearly $r$ is bounded below (by 1) and $t$ is bounded above (by $|G|$), it is reasonable to ask for tight bounds. These have been studied extensively, especially the former, and we consider them briefly in the next two sections.

## 4.1 Chromatic number

There is a large number of papers dealing with the minimum value of the $k$ for which there is a complete $k$-colouring of a graph, most not concerned with homomorphisms. We will only mention, in the language of homomorphisms, a small number of very basic results.

**Lemma 4.2** *Let $G$ and $H$ be graphs and assume that $G \to H$. Then $G$ has a proper $k$-colouring whenever $H$ does.*

**Proof** Let $\phi : G \longrightarrow H$ and $\psi : H \longrightarrow K_k$ be homomorphisms. Then the composition of $\psi$ and $\phi$ is a $k$-colouring of $G$. □

The chromatic number of a graph is usually defined as the least $k$ so that the graph has a proper $k$-colouring. We do the same thing in the language of homomorphisms.

**Definition 4.3** Let $G$ be a graph. The *chromatic number* of $G$, denoted by $\chi(G)$, is the least $n$ such that there is a homomorphism from $G$ onto $K_n$.

We can immediately observe the following.

**Lemma 4.4** *Let $G$ be a graph with $\chi(G) = n$. Then every homomorphism $G \longrightarrow K_n$ is complete.*

Lemma 4.2 then has a few corollaries.

**Corollary 4.5** *If $G \to H$ then $\chi(G) \leq \chi(H)$.*

**Corollary 4.6** *If $H$ is the quotient of $G$ by an elementary homomorphism then* $\chi(G) \leq \chi(H) \leq \chi(G)+1$.

**Proof** Let $\phi : G \longrightarrow H$ be an elementary homomorphism and let $u, v$ be the only vertices of $G$ such that $\phi(u) = \phi(v)$. Let $\psi : G \longrightarrow K_n$ (recall that $V(K_n) = [n]$). By Lemma 4.5 we only need to prove the upper bound. Define a complete homomorphism $\psi' : H \longrightarrow K_{n+1}$ by $\psi'(z) = \psi(\phi^{-1}(z))$ for $z \neq \{u, v\}$ (this is well defined since $\phi$ is elementary) and $\psi'(u) = \psi'(v) = n$. □

**Corollary 4.7** *The chromatic number of a graph and its core are the same.*

This corollary provides a sufficient condition for a graph to be a core. If we define a graph $G$ to be $\chi$-*critical* whenever each of its proper induced subgraphs has chromatic number strictly smaller than $\chi(G)$, we can immediately deduce that all $\chi$-critical graphs are cores.

Another way to define the chromatic number of a graph is described by X.Zhu in [138]. Let $I$ be an interval of length 1 and $r \geq 1$ any real number. Let $I(r)$ be the graph whose vertices are the open intervals in $I$ of length $1/r$, two of which are joined by an edge if and only if their intersection is empty (the non-intersection graph of these intervals).

**Theorem 4.8 (X.Zhu [138])** *For any graph $G$,* $\chi(G) = \inf\{r : G \rightarrow I(r)\}$.

**Proof** Let $r \geq 1$ be a real number. Define $\phi : I(r) \longrightarrow K_{\lfloor r \rfloor}$ by $\phi((a, a+1/r)) = i$ if and only if $i/r \leq a < (i+1)/r$. This is a homomorphism. Since $I$ clearly contains $\lfloor r \rfloor$ pairwise disjoint open intervals of length $1/r$, $I(r) \leftrightarrow K_{\lfloor r \rfloor}$. □

This way of defining a parameter related to graph homomorphism is often useful and will appear throughout the sections on colourings.

## 4.2 Achromatic number

The maximum $k$ for which a graph has a complete $k$ colouring is a more recent idea than the chromatic number. It stems partly from Proposition 4.12.

**Definition 4.9** The *achromatic number* of a graph $G$, denoted by $\mathrm{achr}(G)$, is the largest $s$ such that there is a complete homomorphism from $G$ onto $K_s$.

For example, $\mathrm{achr}(K_n) = n$, $\mathrm{achr}(P_7) = 3$, $\mathrm{achr}(P_8) = 4$, $\mathrm{achr}(P_{11}) = 5$, etc. The next lemma follows easily from the definition.

**Lemma 4.10 ([59])** *If $\phi : G \longrightarrow H$ is complete then* $\mathrm{achr}(G) \geq \mathrm{achr}(H)$.

As with the chromatic number, the achromatic number of a quotient of a graph $G$ by an elementary homomorphism cannot differ much from that of $G$.

**Proposition 4.11 ([59])** *If there is an elementary homomorphism $\phi : G \longrightarrow H$, then*

$$\mathrm{achr}(G) - 2 \leq \mathrm{achr}(H) \leq \mathrm{achr}(G).$$

The lower bound is attained by graphs constructed from complete bipartite graphs with at least three vertices in each class of the bipartition by removing two non-adjacent edges. The elementary homomorphism identifying the endpoints of the removed edges in one of the classes of the bipartition is onto a complete bipartite graph whose achromatic number is 2 while the achromatic number of the original graph is 4.

The next result, known as the Interpolation Theorem for complete homomorphisms, is due to Harary, Hedetniemi and Prins [60].

**Proposition 4.12** *Let $G$ be a graph. There is a complete homomorphism from $G$ onto $K_s$ for each $s$ such that $\chi(G) \leq s \leq \mathrm{achr}(G)$.*

**Proof** We give the proof in terms of elementary quotients. By definition of the achromatic number there is a quotient $G/\phi$ which is the complete graph on achr($G$) vertices and, by Lemma 2.14 there are graphs $G_0, \ldots, G_k$ such that $G_0 = G$, $G_k \cong K_{\mathrm{achr}(G)}$, and $G_{i+1}$ is an elementary quotient of $G_i$, $i = 0, \ldots, k-1$. Now $G \to G_i$ and $G_i \to K_{s_i}$ for each $i$, with $s_i = \chi(G_i)$. Since, by Lemma 4.6, the sequence $s_0, s_1, \ldots, s_k = s$ is non-decreasing and $s_i \leq s_{i-1} + 1$, there is, for each $\chi(G) \leq s \leq \mathrm{achr}(G)$, a least $i_s$ such that $\chi(G_{i_s}) = s$. Thus, the composition of the natural map from $G$ to $G_{i_s}$ with any $s$-colouring of $G_{i_s}$ is a complete homomorphism from $G$ to $K_s$. □

The achromatic number has been studied by many. For example, in their work on graphs with high achromatic number, Hell and Miller ([69]) use quotients obtained by defining $u$ and $v$ to be equivalent if and only if $u$ and $v$ have the same neighbourhood. Graphs in which distinct vertices have distinct neighbourhoods are called *irreducible*. The important theorem then says than the number of irreducible graphs with a given achromatic number is finite.

**Theorem 4.13 ([69])** *For every $k$ there is a $K$ such that $|V(G)| \leq K$ for any irreducible graph $G$ with* $\mathrm{achr}(G) = k$.

In [68] Hell and Miller also study the achromatic number of paths and cycles by exploring the relationship between eulerian trails and complete homomorphisms.

**Lemma 4.14 ([68])** *Let $G$ be a graph, $n$ an integer.*

(1) *There is a complete homomorphism from $P_n$ onto $G$ if and only if $G$ is the underlying graph of some multigraph $G'$ with $n-1$ edges and an eulerian trail.*

(2) *There is a complete homomorphism from $C_n$ onto $G$ if and only if $G$ is the underlying graph of some multigraph $G'$ with $n$ edges and an eulerian tour.*

Paths, of course, are much easier to analyze than general graphs since, as the lemma indicates, they only need to be "wrapped around" a complete graph in order to get a lower bound on their achromatic number. It is an indication of the difficulty of the general problem that, except in a few special cases, the achromatic number is not known even for trees. For a recent survey of results on the achromatic number see [82].

Hedrlín, Hell and Ko consider another kind of interpolation in [63] by defining the class $\mathcal{H}_{n+\frac{1}{2}}$ of graphs obtained form $K_n$ by adding two new vertices, each adjacent to some but not all the vertices of $K_n$, each of which is adjacent to at least one of the two new vertices. Note that the graphs in $\mathcal{H}_{n+\frac{1}{2}}$ are homomorphically equivalent to $K_n$.

**Theorem 4.15 (Hedrlín, Hell, Ko)** *For every graph $G$ for which there are complete homomorphisms onto $K_n$ and $K_{n+1}$ there is a graph $K \in \mathcal{H}_{n+\frac{1}{2}}$ and a complete homomorphism of $G$ onto $K$.*

## 4.3 Kneser colourings

From the point of view of homomorphisms, a proper colouring is a homomorphism of a graph into a complete graph. The minimum and the maximum order of the complete graphs for which there is a surjective homomorphism from $G$ define numerical parameters which give us information about the graphs by implying the existence of homomorphisms. It is then natural to realize that any homomorphism could be called a 'colouring' if we do not restrict the image to complete graphs. This point of view is adopted by Hell and Nešetřil in [72] (see also [56]). This, however, destroys the very desirable property of having a link between homomorphisms and numerical parameters. In the case of the chromatic and the achromatic numbers, this link is a consequence of the structure of the images allowed (complete graphs), and of our knowledge of homomorphisms between the members of this family. Many parameters that have been studied in graphs can be presented within the same framework. Start with a given family $\mathcal{F}$ of graphs. For a graph $G$, the homomorphisms from $G$ to members of $\mathcal{F}$ are called 'colourings' by $\mathcal{F}$. Then if some numerical parameter $\pi$ can be naturally associated to all members of $\mathcal{F}$, we may expand the definition of this parameter to the class of all graphs, by putting

$$\pi(G) = \inf\{\pi(H) : H \in \mathcal{F} \text{ and } G \to H\} \tag{4.1}$$

and hope that the infimum will turn out to be minimum. In some cases this happens naturally, as it does for the chromatic number. The parameters defined in this manner share some fundamental properties, the principal one being that $\pi(G) \leq \pi(H)$ whenever $G \to H$.

Recall that the Kneser graph $K(r, s)$, $r < s/2$, is the non-intersection graph of the $r$-element subsets of $[s]$.

**Definition 4.16** Let $r, s$ be positive integers, $r < s/2$. A *Kneser $(r, s)$-colouring* of a graph $G$ is a homomorphism from $G$ to the Kneser graph $K(r, s)$.

This concept was introduced by Stahl [113]. The "normal" definition is clearly the following. An $(r, s)$*-subset colouring*, $r < s/2$, of a graph $G$ is a map $\phi$ from the vertex set of $G$ to the $r$-subsets of the set $\{1, \ldots, s\}$ such that adjacent vertices are mapped to disjoint sets.

Notice that we have $K(r, s) \to K(r, s+1)$ for all $s \in \mathbb{N}$. This allows a definition of the $r$-chromatic number.

**Definition 4.17** Given a natural number $r$, the *$r$-chromatic number* $\chi_r(G)$ of a graph $G$ is the least $s$ such that $G \to K(r, s)$.

In particular, a Kneser $(1, s)$-colouring is a homomorphism into a complete graph on $s$ vertices and so $\chi_1(G) = \chi(G)$ for any graph $G$. The chromatic number $\chi(G)$ of a graph $G$ is thus generalized to a sequence of numerical parameters $\{\chi_r(G)\}_{r\in\mathbb{N}}$. Such a sequence tells us more about the graph than the chromatic number alone. For example, there is an infinite sequence of Kneser graphs with a given chromatic number but the sequence of $r$-chromatic numbers determines a Kneser graph uniquely.

Observe that for a graph $G$, $\chi_r(G) = r$ if and only if $G$ has no edges, and that $\chi_r(G) = 2r$ if and only if $G$ is bipartite with at least one edge. From the No-Homomorphism Lemma 3.3 we have that if there is a homomorphism from $G$ into $K(r,s)$, then $\alpha(G)/|G| \geq r/2$ and so $\chi_r(G) \geq r|G|/\alpha(G)$.

Obviously, homomorphisms between Kneser graphs can be translated as relationships between members of the sequence of $r$-chromatic numbers. For instance, from Propositions 3.16 and 3.18 we have that $K(r+1, s+2) \to K(r,s)$ and $\chi_{r+1}(G) \geq \chi_r(G) + 2$, as well as $\chi_{r+s}(G) \leq \chi_r(G) + \chi_s(G)$, see [113].

The latter is related to homomorphisms between categorical products of Kneser graphs (see Section 5). Some additional properties of the $r$-chromatic numbers are linked to the fractional chromatic number defined below. Also, the $r$-chromatic numbers of graphs are related to the chromatic number of lexicographic products of graphs (see Section 5.3).

## 4.4 Circular colourings

The next family of graphs does not immediately provide a minimum in the equation (4.1) defining $\pi$, but it can be shown that the infimum is, in fact, a minimum.

Let $1 \leq r \leq s/2$. Recall that $G_s^r$ is the Cayley graph $\mathrm{Cay}(\mathbb{Z}_s, \{r, r+1, \ldots, s-r\})$.

**Definition 4.18** Let $1 \leq r \leq s/2$. A *circular $(r,s)$-colouring* of a graph $G$ is a homomorphism $\phi : G \longrightarrow G_s^r$

This concept was introduced by Vince [125], without the name, with the next (equivalent) definition.

**Definition 4.19** A *circular $(r,s)$-colouring* of a graph $G$ is a map $\phi : V(G) \to \{0, \ldots s-1\}$ such that

$$[u,v] \in E(G) \Rightarrow r \leq |\phi(u) - \phi(v)| \leq s - r.$$

Our presentation follows the combinatorial approach of Bondy and Hell [18] rather than that of Vince who used continuous methods. The name "circular" is suggested by the circulant target graphs of the homomorphisms.

The *circular chromatic number*[3] $\chi_c(G)$ of a graph $G$ is defined by putting

$$\chi_c(G) = \inf\{s/r : G \to G_s^r\}.$$

The range of this new parameter is not restricted to the integers, so we have to define it as an infimum instead of a minimum. From Proposition 3.21 we have that $G \to G_s^r$ if and only if $s/r \geq \chi_c(G)$. Further, since by Proposition 3.23 we only need to consider surjective homomorphisms, these results imply that the infimum is actually a minimum.

**Theorem 4.20 (Bondy and Hell [18])** *The circular chromatic number of a graph $G$ is*

$$\chi_c(G) = \min\{s/r : G \to G_s^r \text{ and } s \leq |G|\}.$$

[3]The name *star chromatic number* usually given to this parameter only relates to the current notation $\chi*$. The name *circular* is not only justified by the homomorphisms into circulants but also by the characterization of the parameter due to X.Zhu given later.

Thus, the circular chromatic number of a finite graph $G$ is always rational, and completely determines the pairs of integers $r, s$ such that $G$ has a circular $(r, s)$-colouring: $G \to G^r_s$ if and only if $s/r \geq \chi_c(G)$.

Also note that since $G^s_1$ is the complete graph $K_s$, we have the inequalities

$$\chi(G) - 1 < \chi_c(G) \leq \chi(G).$$

The following characterization of circular colourings and of the circular chromatic number is due to X.Zhu [138]. Let $C$ be a circle in $\mathbb{R}^2$ of length 1, and let $r \geq 1$ be any real number. Let $C(r)$ be the non-intersection graph of the open arcs on $C$ of length $1/r$.

**Theorem 4.21** *For any graph $G$, $\chi_c(G) = \inf\{r : G \to C(r)\}$.*

From the above theorem together with Theorem 4.8 we can obtain a sufficient condition for graphs for to have the same the chromatic and the circular chromatic number.

**Theorem 4.22** *If $G$ has a vertex $v$ which is adjacent to every other vertex of $G$, then $\chi_c(G) = \chi(G) = n$.*

**Proof** A vertex adjacent to all other vertices in $G$ must map to an interval $I$ which intersects no other. Thus an $r$-circle colouring exists if and only if an $r$-interval colouring exists since the circle colouring corresponds to an interval one by "cutting" the circle at one of the endpoints of $I$ (the other direction is trivial). □

By refining somewhat the proof of the theorem we obtain a slightly more general result.

**Corollary 4.23** *Suppose that $G$ has chromatic number $s$ and a vertex whose neighbours induce a subgraph of chromatic number $s-1$. Then $\chi_c(G) = \chi(G)$. In particular, if $\chi(G) = s$ and $G$ contains (a copy of) $K_s$ then $\chi_c(G) = \chi(G)$.*

The corollary can, of course, be obtained simply by considering what the circular graph which realizes the circular chromatic number must look like in this case.

Abbott and B.Zhou [1], and, independently, G.Gao, Mendelsohn and H.Zhou [47], have found a more general sufficient condition for the circular chromatic number to be equal to the chromatic number, again by using $r$-circle colourings.

**Theorem 4.24** *If $G$ is a graph whose complement is disconnected, then $\chi_c(G) = \chi(G)$.*

The circular chromatic number has been studied by Gao, Hahn, Hell, H.Zhou, X.Zhu [18, 26, 46, 48, 115, 138] and others. We shall not list here the many results obtained for particular classes of graphs, for more see the survey [142]. Deuber and X.Zhu [34] have generalized the circular chromatic number further (incidentally again justifying the change of name). They start with a graph $G$ and a weight function which assigns to each vertex of $G$ a non-negative real weight $w(u)$. Then they consider a circle of circumference $t$ in the plane and a mapping $\phi$ assigning to each vertex $u$ of a graph $G$ an arc of length $w(u)$ on the circle in such a way that the arcs assigned to adjacent vertices do not intersect. The *weighted circular chromatic number* $\chi_c(G, w)$ is then the infimum of the $t$ for which such a $\phi$ exists.

Clearly $\chi_c(G,1) = \chi_c(G)$. Other weighted colourings are considered as well, but the methods do not use homomorphism.

The interest of the circular chromatic number comes in part from the bound it provides on the ultimate independence ratio discussed in Section 4.7. The fractional chromatic number considered next also gives a bound on the ultimate independence ratio.

## 4.5 Fractional chromatic number

**Definition 4.25** Let $\mathcal{I}(G)$ denote the set of all independent sets of a graph $G$. A *fractional colouring* of $G$ is a weight function $\mu : \mathcal{I}(G) \to [0,1]$ such that the constraints

$$\sum_{u \in I \in \mathcal{I}(G)} \mu(I) \geq 1$$

are satisfied for all $u \in V(G)$.

While the concepts of both subset colouring and circular colouring remain close to our intuitive notion of colouring, that is, the affectation of colours to vertices, the idea of a fractional colouring is somewhat different. It can be motivated by our definitions of colourings through homomorphisms: we assign vertices to colours rather than the other way around. Thus, the usual vertex colourings are constructed by selecting disjoint independent set which are assigned to different colours. The corresponding weight function has value 1 for all selected independent sets, and 0 for all other independent sets. With fractional colourings, we allow independent sets to be 'partially' selected. Examples of such fractional colourings can be found by looking at the Kneser $(r,s)$ colourings of a graph. If $\phi : G \longrightarrow K(r,s)$ is a homomorphism, we may define a fractional colouring $\mu : \mathcal{I}(G) \to [0,1]$ by putting $\mu(I) = k_I/r$, where $k_I$ is the number of indices $i \in [s]$ such that $I = \{u \in V(G) : i \in \phi(u)\}$.

The *fractional chromatic number* $\chi_f(G)$ of a graph $G$ is defined by putting

$$\chi_f(G) = \inf\{ \sum_{I \in \mathcal{I}(G)} \mu(I) : \mu \text{ is a fractional colouring of } G\}.$$

Thus, the fractional chromatic number of a graph is the solution of a linear program with $|\mathcal{I}(G)|$ variables and $|G|$ constraints. Therefore, the infimum is always attained and is truly a 'fraction'. Note that this definition of the fractional chromatic number does not seem to relate to homomorphisms. However, we can express the fractional chromatic number of a graph in terms of graph homomorphisms in at least two ways, the second of which uses the independence number $\alpha(G)$ of the graph $G$.

**Proposition 4.26** *Let $G$ be a graph. Then*

$$\chi_f(G) = \inf\{s/r : G \to K(r,s)\} = \sup\{|H|/\alpha(H) : H \to G\}.$$

Some relations between the fractional chromatic number, the circular chromatic number and the subset chromatic numbers of a graph can be derived from the first of these equalities. First of all we easily have $\chi_f(G) \leq \chi_r(G)/r$ for all $r \in \mathbb{N}$ by the definition of these parameters (in fact, $\chi_f(G) = \lim_{r\to\infty} \chi_r(G)/r$). We also have $G_r^s \to K(r,s)$, so that $\chi_c(G) \leq \chi_f(G)$. Unlike the circular chromatic number, the fractional chromatic number of a graph can differ arbitrarily from its chromatic number, the simplest example being that of the Kneser graphs themselves. Finally, note that the last characterization of the fractional chromatic number strays from our intended mold. This may be explained by the dual version of the linear program defining the fractional chromatic number of a graph, which can be expressed as follows.

$$\max \textstyle\sum_{u \in V(G)} \nu(u)$$
$$\text{subject to } \textstyle\sum_{u \in I} \nu(u) \leq 1 \text{ for all } I \in \mathcal{I}(G).$$

Observe also that the $0-1$ weight functions $\nu$ on $V(G)$ that satisfy the above constraints are obtained by assigning the value 1 to all vertices of some complete subgraph of $G$ and 0 to all the other vertices. Hence the above expression can be thought of as a *fractional clique number* of $G$. We can then note that while the chromatic number and the clique number of a graph can differ arbitrarily, the fractional versions always coincide.

From the primal and the dual formulations of the fractional chromatic number we can prove something for vertex-transitive graphs.

**Theorem 4.27** *If $G$ is a vertex-transitive graph then $\chi_f(G) = |G|/\alpha(G)$.*

The No-Homomorphism Lemma 3.3 can be viewed as a consequence of this and Proposition 4.26.

An alternate description of the fractional chromatic number is given by X.Zhu in [48] and [34]. Let $I$ be an interval of length 1 and $r \geq 1$ any real number. Let $M(r)$ be the non-intersection graph of the measurable subsets of $I$ of measure $1/r$.

**Theorem 4.28 (X.Zhu [48])** *For any graph $G$, $\chi_f(G) = \inf\{r : G \to M(r)\}$.*

## 4.6 Chromatic difference sequence

There is but a small step from the chromatic number to asking *How many vertices of a graph can be coloured with a given number of colours?* More precisely, following Albertson and Berman [3], we define, for a given graph $G$ and a positive integer $k$, $\alpha_k(G)$ to be the maximum number of vertices in a $k$-colourable induced subgraph of $G$. We also define the differences of successive $\alpha_k$'s by setting $\beta_k(G) = \alpha_k(G) - \alpha_{k-1}(G)$, for $k > 0$ and $\alpha_0(G) = 0$. Thus $\sum_{i=1}^{k} \beta_i(G) = \alpha_k(G)$.

**Definition 4.29** Let $G$ be a graph and let $n = |G|$. The *chromatic difference sequence* of $G$ is the sequence $(\beta_1(G), \ldots, \beta_{\chi(G)}(G))$. The *normalized* chromatic difference sequence of $G$ is the sequence $(\beta_1(G)/n, \ldots, \beta_{\chi(G)}(G)/n)$.

We will say that a sequence $a_1, \ldots, a_n$ *dominates* the sequence $b_1, \ldots, b_n$, if $\sum_{i=1}^{k} a_i \geq \sum_{i=1}^{k} b_i$ for all $k$, with equality when $k = n$ (this can be extended to sequences of different lengths, see [132]).

Chromatic difference sequences have been used to prove that graphs do not have homomorphisms into other graphs. The No-Homomorphism Lemma 3.3 was first proved in this context in the following form.

**Lemma 4.30** **([5])** *If there is a homomorphism from a graph $G$ to a graph $H$ and $H$ is vertex-transitive, then the normalized chromatic sequence of $G$ dominates that of $H$.*

It was then used to show that for the Petersen graph $P$, $P^r \not\to P^s$ for any $r > s$, where $P^n$ is the $n$-th cartesian power of $P$. Others, notably H. Zhou in his thesis [135] and subsequent papers, considered the behaviour of the chromatic difference sequences under homomorphisms and in relation to graph products.

**Theorem 4.31** *The normalized chromatic difference sequence of $G^k$ is equal to that of $G$ for any Cayley graph $G$ of an abelian group, with the cartesian product.*

In general, H.Zhou has shown the following.

**Theorem 4.32** **([132])** *Let $G$ and $H$ be graphs. Then:*

(1) *The normalized chromatic difference sequence of $G \times H$ dominates those of $G$ and $H$.*

(2) *The normalized chromatic difference sequence of $G \Box H$ is dominated by those of $G$ and $H$, and so the normalized chromatic difference sequence of the cartesian power $G^k$ is dominated by that of $G$.*

Albertson and Collins [5] conjectured that the chromatic difference sequence of a vertex-transitive graph was always non-increasing. In a similar direction, H.Zhou asked if a Cayley graph $G$ has a non-increasing *achievable* chromatic difference sequence (that is, one with colour classes of sizes $\beta_k$ for $k = 1, \dots, \chi(G)$). Tardif [119] constructed a circulant whose chromatic difference sequence is not monotonic.

**Example 4.33** The chromatic difference sequence of $(C_7[C_5]) \times K_6$ is $(36, 36, 36, 36, 31, 35)$.

While the chromatic difference sequence does not lead to a numerical parameter, a concept whose roots are in the study of the chromatic difference sequence does. It is the ultimate independence ratio of a graph considered next.

## 4.7 Ultimate independence ratio

Recall that the independence ratio of a graph $G$ is $i(G) = \alpha(G)/|G|$ (Definition 3.2). Since in this section we will be concerned with the cartesian product, we will have $G^1 = G$ and $G^k = G^{k-1} \Box G$.

We begin by asking about $\alpha(G \Box H)$. Clearly, $\alpha(G \Box H) \leq |G|\alpha(H)$ since each fiber $F_u = \{u\} \times V(H)$ of $G \Box H$ can contribute at most $\alpha(G)$ vertices to a maximum independent set of $G \Box H$. This is achievable if and only if we can choose, in each fiber $F_u$, a maximum independent set $\{u\} \times I_u$ of size $\alpha(G)$ so that $I_u \cap I_v = \emptyset$ whenever $[u, v]$ is an edge of $G$. This can be expressed in a setting using homomorphisms.

Let $G$ be a graph and define the *maximum independent set graph* of $G$, $\text{Ind}(G)$, to be the non-intersection graph the maximum independent sets of $G$.

**Definition 4.34** An *independent set cover* of a graph $G$ by a graph $H$ is a homomorphism $\iota : G \longrightarrow \text{Ind}(H)$. When $G$ has an independent set cover by $G$, we say simply that $G$ *has an independent set cover.*

For example, $\mathrm{Ind}(C_{2k+1})$ is the odd cycle $C_{2k+1}$; the maximum independent set graph of the Petersen graph consists of five isolated vertices. Thus each odd cycle has an independent set cover but the Petersen graph does not.

The following results are direct consequences of the above discussion.

**Lemma 4.35** *Let $G$ and $H$ be graphs. Then*

$$i(G \Box H) \leq \min\{i(G), i(H)\},$$

*with equality if and only if $G \to \mathrm{Ind}(H)$ or $H \to \mathrm{Ind}(G)$, depending on whether $i(H) \leq i(G)$ or $i(G) \leq i(H)$.*

In particular this says that a graph $G$ has an independent set cover if and only if $i(G^2) = i(G)$. In the language of Section 5.2 this can be restated by saying that the graph $\mathrm{Hom}(G, \mathrm{Ind}(G))$ (see Definition 5.18) is non-empty. In fact we have the stronger statement that $i(G^2) = i(G)$ if and only if the graphs $\mathrm{Hom}(G, \mathrm{Ind}(G))$ and $\mathrm{Ind}(G^2)$ are isomorphic.[4]

Given a graph $G$, consider the powers $G^k$. By Lemma 4.35 we have that $i(G^k) \geq i(G^{k+1}) > 0$ for all $k$. Hence the sequence $i(G^k)$, $k = 1, 2, \ldots$, has a limit $\geq 0$. In fact the sequence is bounded away from zero since by Lemma 5.13, $\chi(G^k) = \chi(G)$ for all $k$, so that $i(G^k) \geq 1/\chi(G^k) = 1/\chi(G)$. Hell, Yu and H. Zhou introduced the concept of ultimate independence ratio in [75].

**Definition 4.36** The *ultimate independence ratio* of a graph $G$, denoted by $I(G)$, is

$$I(G) = \lim_{k \longrightarrow \infty} i(G^k).$$

It is not known if the limit is always rational.

There exist graphs for which $I(G)$ is strictly less than $i(G^k)$ for any $k$. For example, if $G$ is the Petersen graph, then $I(G) = 1/3$ whereas $i(G^k) = \alpha(G^k)/10^k$ (two proper fractions with relatively prime denominators). At the other extreme are graphs whose ultimate independence ratio is already equal to $i(G)$. A class of such graphs is provided by the following result of Hell, Yu and H. Zhou [75] which is based on Lemma 4.35.

**Theorem 4.37** *If $G^2 \to G \to \mathrm{Ind}(G)$ then $I(G) = i(G)$.*

Thus the powers of a graph satisfying $G^2 \to G$ and having an independent set cover all have the same independence ratio. Conversely, if $G$ is any graph for which $I(G) = i(G)$, then in particular $i(G^2) = i(G)$, hence by the remark after Lemma 4.35 we have that the condition $G \to \mathrm{Ind}(G)$ in Theorem 4.37 is necessary. We do not know whether the same is true for the condition $G^2 \to G$. In general, Theorem 4.37 is false without it (see the end of this section).

We can use Theorem 4.37 to prove the following result.

**Theorem 4.38** *Let $G$ be a Cayley graph $\mathrm{Cay}(\Gamma, S)$ on an abelian group $\Gamma$. Then $I(G) = i(G)$.*

---

[4]This was noticed independently by B. Larose, F. Laviolette and G. Sabidussi.

**Proof** For an abelian group $\Gamma$ the mapping $\phi : G^2 \longrightarrow G$ defined by $\phi(g,h) = gh$ is a homomorphism. Let $I$ be any maximum independent set of $G$ and for any $g \in \Gamma$ set $I_g = gI$. Since left translations are automorphisms, $I_g$ is a maximum independent set. If now $gh$ is an edge of $G$, $I_g \cap I_h = \emptyset$ since otherwise $gk = hk'$ for some $k, k' \in I$ and so $k^{-1}k' = h^{-1}g \in S$, contradicting the independence of $I$. Thus the map $g \mapsto I_g$ is a homomorphism $G \to \mathrm{Ind}(G)$. □

The above theorem can be generalized to normal Cayley graphs, see Section 5. For arbitrary Cayley graphs very little is known, and results even for Cayley graphs of dihedral groups are nonexistent. We do know, however, that the tempting conjecture that $I(G) = i(G)$ for Cayley graphs fails for infinitely many Cayley graphs, see [57].

**Corollary 4.39** *If $G$ is a circulant, then $I(G) = i(G)$. In particular, $I(C_{2k}) = 1/2$ and $I(C_{2k+1}) = k/(2k+1)$ for $k > 0$.*

With the ultimate independence ratio we can prove an analogue of the No-Homomorphism Lemma 3.3. It does not require that the target graph $H$ be transitive, but we pay a price by having to deal with limits. The proof uses multiples of graphs mentioned in Section 3.1.

**Theorem 4.40 (Hahn, Hell, Poljak [55])** *If $G \to H$ then $I(H) \leq I(G)$. In particular, if $H$ is a retract of $G$, then $I(H) = I(G)$.*

This gives, as corollaries, that $I(G) \leq 1/\omega(G)$ ($\omega(G)$ is the size of a largest clique in $G$) and that $I(G) = 1/\chi(G)$ for perfect graphs.

Homomorphisms also allow us to prove:

**Theorem 4.41 (Hahn, Hell, Poljak [55])** *For any graph $G$,*

$$\frac{1}{\chi(G)} \leq I(G) \leq \frac{1}{\chi_f(G)}.$$

X.Zhu gives more bounds related to homomorphisms. The first was also proved independently by Favaron[44].

**Theorem 4.42 (X.Zhu [137])** *For any graph $G$, the sequence $1/\chi_f(G^k)$, $k = 1, 2, \ldots$, is non-increasing and*

$$I(G) = \lim_{k\to\infty} \frac{1}{\chi_f(G^k)}.$$

**Theorem 4.43 (X.Zhu [137])** *For any graph $G$, $I(G) \geq 1/\chi_c(G)$.*

In general, the problem of determining the ultimate independence ratio of a graph is open. The conjectures which would equate $I(G)$ to one of the bounds have proven wrong (for example, X.Zhu constructs a graph with $1/\chi(G) < I(G) < 1/\chi_f(G)$ in [137]). The simplest case of interest – because of the methods which must be invented – is that of the odd wheels in general and $W_5$ in particular. From solving the linear program for the fractional chromatic number of the square of $W_5$ we get $1/4 \leq I(W_5) \leq 11/41$. The working hypothesis is that $1/4$ is the value for any odd wheel.

As an application of these remarks and in order to illustrate the usefulness of Theorem 4.40, let us show that in general Theorem 4.37 fails if we only assume that $G \to \mathrm{Ind}(G)$. In other words, it is not true that if $i(G^2) = i(G)$, then all the higher powers of $G$ also have the same independence ratio. Let $W_5'$ be a 5-wheel with a "double hub", i.e. a pentagon with two additional vertices that are joined to all vertices of the pentagon but not to each other. Clearly the wheel $W_5$ is a retract of $W_5'$. Hence by Theorem 4.40,

$$I(W_5') = I(W_5) \leq 11/41 < 2/7 = i(W_5').$$

On the other hand, it is easy to check that $\mathrm{Ind}(W_5') = W_5$, and hence the condition $W_5' \to \mathrm{Ind}(W_5')$ is satisfied.

# 5 Graph products

The main constructions known as 'graph products' have already been introduced in Section 2.6. In this section, we take a closer look at some of the themes linking graph homomorphisms and products. The principal motivation of this subject is the investigation of the behaviour of some parameters, such as the chromatic number, the clique number or the independence number, under some product. In terms of homomorphisms, these questions often amount to finding criteria for the existence of a homomorphism from a product of graphs to a given graph, or from a given graph to a product of graphs.

## 5.1 The categorical product and Hedetniemi's conjecture

Recall that the categorical product $G \times H$ of two graphs $G$ and $H$ is the graph with vertex set $V(G) \times V(H)$, where two vertices $(u_1, u_2)$, $(v_1, v_2)$ are adjacent if and only if $u_1$, $v_1$ are adjacent in $G$, and $u_2$, $v_2$ are adjacent in $H$. This is the product with the universal property in the category of graphs. Indeed, both projections $\mathrm{pr}_G$ and $\mathrm{pr}_H$ of $G \times H$ onto its factors are homomorphisms, and a graph $K$ admits homomorphisms $\phi_1 : K \longrightarrow G$ and $\phi_2 : K \longrightarrow H$ if and only if it admits a homomorphism $\phi : K \longrightarrow G \times H$ such that $\phi_1 = \mathrm{pr}_1 \circ \phi$ and $\phi_2 = \mathrm{pr}_2 \circ \phi$. This homomorphism $\phi$ is defined by putting $\phi(u) = (\phi_1(u), \phi_2(u))$. In general, there need not exist a homomorphism from $G$ to $G \times H$; such a homomorphism exists if and only if $G \to H$. However, two homomorphisms $\phi_1 : G \longrightarrow G'$ and $\phi_2 : H \longrightarrow H'$ can be used coordinatewise to define a homomorphism $\phi : G \times H \longrightarrow G' \times H'$. In terms of the quasi-order $\to$, the universal property of the categorical product of graphs says that for graphs $G$, $H$, $K$, we have $K \to G \times H$ if and only if both $K \to G$ and $K \to H$ hold. Thus, the class $\mathcal{H}(G \times H)$ is the infimum of the classes $\mathcal{H}(G)$ and $\mathcal{H}(H)$ with respect to the partial order $\prec$.

On a more concrete level, note that both the odd girth and the clique number of a graph are defined by homomorphisms from odd cycles or complete graphs into this graph. With this in mind, the following facts are direct applications of the universal property of the categorical product.

**Observation 5.1** *Let $G$, $H$ be graphs. Then the odd girth of $G \times H$ is the maximum of the odd girths of $G$ and $H$, and the clique number of $G \times H$ is the minimum of the clique numbers of $G$ and $H$.*

It is very enlightening to be able to express the value of some parameters in a product as a function of the same parameters in the factors in this fashion. We may wonder if we might be able to do the same with other parameters. In particular, note that the chromatic number of $G \times H$ is defined by homomorphisms *from* $G \times H$ *to* complete graphs, and its characterization does not fall within the framework of the defining property of the categorical product. Thus, the colourings of $G \times H$ are much less predictable. Some 'canonical' colourings can be defined by projecting $G \times H$ onto one of its factors and then colouring this factor, but other colourings also exist. However, we do not know any instances where less than $\min\{\chi(G), \chi(H)\}$ colours are used. It is believed that such colourings are impossible, and this is the essence of the following.

**Conjecture 5.2 (Product conjecture, Hedetniemi [62])** Let $G$, $H$ be graphs. Then

$$\chi(G \times H) = \min\{\chi(G), \chi(H)\}. \tag{5.1}$$

It is easy to see that the identity (5.1) holds for $\min\{\chi(G), \chi(H)\} \leq 3$. In terms of homomorphisms, Conjecture 5.2 admits many equivalent formulations, which take into account the obvious inequality $\chi(G \times H) \leq \min\{\chi(G), \chi(H)\}$. Each of the following statements is equivalent to the product conjecture.

- Let $n$ be a positive integer. If for two graphs $G$, $H$ we have $G \not\to K_n$ and $H \not\to K_n$, then $G \times H \not\to K_n$.
- Let $G$, $H$ be $m$-chromatic graphs. Then there exists an $m$-chromatic graph $K$ such that $K \to G$ and $K \to H$.

The first of these statements amounts to saying that $K_n$ is *multiplicative* (see Definition 5.9 below). Note that in the second statement, no restriction is made on the graph $K$. Of course, if the product conjecture is true, we can put $K = G \times H$. However, it is sometimes convenient to consider other possible choices for $K$. For instance, we can try to find a graph $K$ that is $\chi$-critical. In particular, the only 3-colourable graphs that are $\chi$-critical are the graphs $K_1$, $K_2$ and the odd cycles. This simple family of graphs accounts for the validity of the product conjecture for $\min\{\chi(G), \chi(H)\} \leq 3$.

The next step towards a proof of the product conjecture would be to settle the case $\min\{\chi(G), \chi(H)\} = 4$. However the family of 4-critical graphs is much more complex than the family of odd cycles, and there is no hope of generalizing the arguments that worked before. Nonetheless, progress has indeed been made with a new approach due to El-Zahar and Sauer, and it is now known that the identity (5.1) holds for $\min\{\chi(G), \chi(H)\} \leq 4$.

For a graph $G$ and an integer $n$, put

$$\mathcal{F} = \{H : G \times H \text{ is } n\text{-colourable}\}.$$

Then, $H_1 \to H_2$ and $H_2 \in \mathcal{F}$ implies $H_1 \in \mathcal{F}$. Thus, $\mathcal{F}$ is an ideal with respect to the quasi-order $\to$. If $G$ is $n$-colourable, then $\mathcal{F}$ contains all graphs; the interesting situation arises when $\chi(G) > n$. Then, we find a counterexample to the product conjecture if there exists a graph $H \in \mathcal{F}$ such that $\chi(H) > n$. The following construction, due to El-Zahar and Sauer, allows us to find a maximal element in $\mathcal{F}$.

**Definition 5.3** For a graph $G$ and an integer $n$, we define the *$n$-colouring graph* $\mathcal{C}_n(G)$ of $G$ by putting

$$\begin{aligned} V(\mathcal{C}_n(G)) &= \{f : V(G) \longrightarrow [n]\}, \\ E(\mathcal{C}_n(G)) &= \{[f, g] : \text{ for all } (u, v) \in V(G)^2, [u, v] \in E(G), f(u) \neq g(v)\}. \end{aligned}$$

The vertices of $\mathcal{C}_n(G)$ are arbitrary functions, not proper colourings in general. Thus, $|\mathcal{C}_n(G)| = n^{|G|}$, and the structure of $G$ only determines the structure of $E(\mathcal{C}_n(G))$. By the definition of $E(\mathcal{C}_n(G))$, a function $f$ is adjacent to itself (i.e., $[f, f]$ is a loop) if and only if it is a proper colouring of $G$.[5] Thus, $\mathcal{C}_n(G)$ has no loops if and only if $\chi(G) > n$.

For $H \in \mathcal{F}$ and a proper $n$-colouring $\phi$ of $G \times H$, we define a map $\psi : H \longrightarrow \mathcal{C}_n(G)$ by putting $\psi(x) = f_x$, where $f_x : V(G) \longrightarrow \{0, \dots, n-1\}$ is the function defined by $f_x(u) = \phi(u, x)$. By the definition of $\mathcal{C}_n(G)$, it is straightforward to verify that $\psi$ is a homomorphism. Conversely, we can define a proper $n$-colouring $\phi$ of $G \times \mathcal{C}_n(G)$ by putting $\phi(u, f) = f(u)$. Thus, $\mathcal{C}_n(G) \in \mathcal{F}$, and $\mathcal{F} = \{H : H \to \mathcal{C}_n(G)\}$.

Now assume that $\chi(G) > n$. If $\chi(\mathcal{C}_n(G)) \leq n$, then $\chi(H) \leq n$ for all $H \in \mathcal{F}$, and there does not exist any graph $H$ such that $\chi(H) > n$ and $\chi(G \times H) \leq n$. On the other hand, if $\chi(\mathcal{C}_n(G)) > n$, we find a counterexample to the product conjecture since $\chi(G \times \mathcal{C}_n(G)) \leq n$. Therefore, the product conjecture is equivalent to the statement that for any graph $G$ and for any $n < \chi(G)$, we have $\chi(\mathcal{C}_n(G)) \leq n$. The main results of El-Zahar and Sauer are the following.

**Theorem 5.4** **(El-Zahar, Sauer [40])** *Let $G$ be a graph such that $\chi(G) \geq 4$. Then $\chi(\mathcal{C}_3(G)) = 3$.*

**Corollary 5.5** *If $\chi(G) \geq 4$ and $\chi(H) \geq 4$, then $\chi(G \times H) \geq 4$.*

The proof of Theorem 5.4 is intricate and, once again, it is not clear how it can be generalized to higher values of $n$. However, El-Zahar and Sauer's approach provides a fruitful insight into colourings of a categorical product of graphs. For instance, for a graph $G$ and $n = \chi(G) - 1$, we at least know that $\chi(\mathcal{C}_n(G))$ is bounded, so there exists an integer $m \geq \chi(G)$ such that if $\chi(H) \geq m$, then $\chi(G \times H) = \chi(G)$. The situation is different if we do not fix a graph $G$ and consider the function $a : \mathbb{N} \longrightarrow \mathbb{N}$ defined by putting

$$a(m) = \min\{\chi(G \times H) : \chi(G) = \chi(H) = m\}.$$

Though $a$ is non decreasing, it is not even clear that $a$ is not bounded. Poljak and Rödl [101] obtained the surprising result that *if* $a$ is bounded, then $a(m) \leq 16$ for all $m \in \mathbb{N}$ (Poljak [100] later improved this bound to 9).

The next three results show how our knowledge of the general structure of the graph $\mathcal{C}_n(G)$ helps to solve further instances of the product conjecture.

**Lemma 5.6 (El-Zahar, Sauer [40])** *Let $G$ be a connected graph and $n$ an integer such that $n < \chi(G)$. Then $\mathcal{C}_n(G)$ contains a unique complete subgraph of cardinality $n$, namely, the subgraph induced by the constant functions.*

**Proof** Let $f_1, \dots, f_n$ denote the vertices of a complete subgraph in $\mathcal{C}_n(G)$. Suppose that $G$ contains a vertex $u$ such that the values $f_1(u), \dots, f_n(u)$ are all distinct. Then for any neighbour $v$ of $u$ and for any $i \in \{1, \dots, n\}$, we have $f_i(v) \neq f_j(u)$ for all $j \neq i$. Thus, $f_i(v) = f_i(u)$ for all $i \in \{1, \dots, n\}$, and the values $f_1(v), \dots, f_n(v)$ are all distinct. Since $G$ is

[5] Technically, we do not allow loops in this paper, so we should always specify that $\chi(G) > n$. However, the validity of the approach remains unaltered even if we disregard the possible presence of loops.

connected, we can repeat this argument to show that $f_i(u) = f_i(v)$ for all $u, v \in V(G)$ and $i \in \{1, \dots, n\}$. Therefore, $\{f_1, \dots, f_n\}$ are the constant functions.

It remains to consider the case where for all $u \in V(G)$, the sequence $f_1(u), \dots, f_n(u)$ contains at least one repetition. We show that if this were the case, then $G$ would be $n$-colourable. Define a map $\phi : G \longrightarrow \{0, \dots, n-1\}$ by selecting, for each $u \in V(G)$, a value $\phi(u)$ that appears at least twice in the sequence $f_1(u), \dots, f_n(u)$. It is easy to verify that $\phi$ is a proper colouring of $G$. Since $\chi(G) > n$, such a colouring is impossible. Therefore, $\mathcal{C}_n(G)$ contains a unique complete subgraph of cardinality $n$. □

**Theorem 5.7 (Burr, Erdős, Lovász [23])** *Let $H$ be a graph such that each vertex of $H$ is contained in a complete subgraph of cardinality $n$, and $\chi(H) > n$. Then for any graph $G$ such that $\chi(G) > n$ we have $\chi(G \times H) > n$.*

**Proof** We may assume that $G$ is connected. Suppose that $\chi(G \times H) \leq n$. Then $H \to \mathcal{C}_n(G)$, and any homomorphism $\phi : H \longrightarrow \mathcal{C}_n(G)$ must map all vertices of $H$ to the unique $n$-clique of $\mathcal{C}_n(G)$. This implies that $H$ is $n$-colourable, which contradicts our hypothesis. □

**Theorem 5.8 (Duffus, Sands, Woodrow [38])** *Let $G$, $H$ be two connected graphs, both containing a complete subgraph of cardinality $n$, and such that $\chi(G), \chi(H) > n$. Then $\chi(G \times H) > n$.*

**Proof** Suppose that $\chi(G \times H) \leq n$. Then there exists a homomorphism $\phi : H \longrightarrow \mathcal{C}_n(G)$, and since $H$ is connected, $\phi$ must map all vertices of $H$ into the same connected component of $\mathcal{C}_n(G)$, namely, the component $C$ that contains the constant functions. It remains to show that if $G$ contains a complete subgraph of cardinality $n$, then $C$ is $n$-colourable.

Let $u_1, \dots, u_n$ denote the vertices of a complete subgraph of $G$. Suppose that $C$ contains a function $f$ such that the values $f(u_1), \dots, f(u_n)$ are all distinct. For any neighbour $g$ of $f$, we have $f(u_i) \neq g(u_j)$ for all $i \neq j$, so $f(u_i) = g(u_i)$ for $i = 1, \dots, n$. Thus, the values $g(u_1), \dots, g(u_n)$ are all distinct, and since $C$ is connected, any function $h$ in $C$ has the property the values $h(u_1), \dots, h(u_n)$ are all distinct. However, this is impossible since $C$ contains the constant functions.

Therefore, for any function $f$ in $C$, the values $f(u_1), \dots, f(u_n)$ are not all distinct. We can then define a map $\phi : C \longrightarrow \{0, \dots, n-1\}$ by selecting, for each $f \in V(C)$, a value $\phi(f)$ that appears at least twice in the sequence $f(u_1), \dots, f(u_n)$. It is easy to verify that $\phi$ is a proper $n$-colouring of $C$. □

Note that since the categorical product of graphs is commutative, we have $H \to \mathcal{C}_n(G)$ if and only if $G \to \mathcal{C}_n(H)$. This accounts for the similarity of the proofs of Lemma 5.6 and Theorem 5.8. The reader may have noticed that connectedness plays an important role in these proofs. Of course, if the product conjecture is false, then there exist two *connected* graphs $G$ and $H$ such that $\chi(G \times H) < \min\{\chi(G), \chi(H)\}$. Also, it is an easy exercise to verify that if $C_1, \dots, C_m$ are the connected components of $G$, then $\mathcal{C}_n(G)$ is isomorphic to $\times_{k=1}^{m} \mathcal{C}_n(C_k)$. Hence, everything seems to indicate that difficulties related to connectedness are avoidable. However, a closer inspection of the order-theoretic properties of the construction $\mathcal{C}_n$ with respect to the quasi-order $\to$ reveals that connectedness is indeed an important issue.[6]

Given a homomorphism $\phi : G \longrightarrow H$ one can define a homomorphism $\psi : \mathcal{C}_n(H) \longrightarrow \mathcal{C}_n(G)$ by putting $\psi(f) = f'$, where $f'(u) = f(\phi(u))$. Thus, the construction $\mathcal{C}_n$ 'reverses arrows' in the sense that $G \to H$ implies $\mathcal{C}_n(H) \to \mathcal{C}_n(G)$. Suppose that neither $G$ nor $\mathcal{C}_n(G)$ is $n$-colourable. Then we may iterate the construction and consider $\mathcal{C}_n(\mathcal{C}_n(G))$. Since $\mathcal{C}_n(G) \times G$ is $n$-colourable, we have $G \to \mathcal{C}_n(\mathcal{C}_n(G))$. Therefore, $\chi(\mathcal{C}_n(\mathcal{C}_n(G))) \geq \chi(G)$, and $\mathcal{C}_n(\mathcal{C}_n(G))$ is not $n$-colourable. Again, $\mathcal{C}_n(\mathcal{C}_n(G)) \times \mathcal{C}_n(G)$ is $n$-colourable, and $\mathcal{C}_n(G) \to \mathcal{C}_n(\mathcal{C}_n(\mathcal{C}_n(G)))$. However, since the construction $\mathcal{C}_n$ reverses arrows and $G \to \mathcal{C}_n(\mathcal{C}_n(G))$, we also have $\mathcal{C}_n(\mathcal{C}_n(\mathcal{C}_n(G))) \to \mathcal{C}_n(G)$, that is, $\mathcal{C}_n(\mathcal{C}_n(\mathcal{C}_n(G)))$ and $\mathcal{C}_n(G)$ are homomorphically equivalent. Let $H_1$, $H_2$ denote the cores of $\mathcal{C}_n(G)$ and $\mathcal{C}_n(\mathcal{C}_n(G))$, respectively. We then have $H_1 \leftrightarrow \mathcal{C}_n(H_2)$ and $H_2 \leftrightarrow \mathcal{C}_n(H_1)$.

Hence, if the product conjecture is false, then there exist an integer $n$ and two core graphs $G$ and $H$ such that $G \leftrightarrow \mathcal{C}_n(H)$ and $H \leftrightarrow \mathcal{C}_n(G)$.[7] In particular, $G$ and $H$ must contain complete subgraphs of cardinality $n$. Thus, if $G$ and $H$ were connected, we would have a contradiction to Corollary 5.8. So, we have to assume that $G$ and $H$ are disconnected.

An essential element of the proofs of Lemma 5.6 and Corollary 5.8 is the recognition of 'invariants' of a connected component of a $n$-colouring graph. In the proof of Theorem 5.4, the role of these invariants is played by a parameter defined as the 'parity' of a 3-colouring of an odd cycle. It seems that more invariants of this type need to be discovered if any progress is to be made with this approach to the product conjecture. However, the product conjecture admits interesting variations, and the methods of El-Zahar and Sauer have been applied successfully to some of these.

**Definition 5.9** A graph $K$ is called *multiplicative* if for any graphs $G$ and $H$ such that $G \not\to K$ and $H \not\to K$, we have $G \times H \not\to K$.

With this terminology, the product conjecture states that all complete graphs are multiplicative, and falls within the general problem of characterizing all multiplicative graphs. In this respect, a circular chromatic number analogue of Hedetniemi's conjecture has been formulated by X.Zhu in [138]: *all circular graphs are multiplicative.*

**Conjecture 5.10** Let $G$ and $H$ be graphs. Then $\chi_c(G \times H) = \min\{\chi_c(G), \chi_c(H)\}$.

Note however that there exist graphs which are not multiplicative. For instance, if $G, H$ are graphs such that $G \not\to H$ and $H \not\to G$, then $K = G \times H$ is not multiplicative.

By analogy with the definition of $\mathcal{C}_n(G)$, we define, for any graph $K$, the *$K$-colouring graph* $\mathcal{C}_K(G)$ of a graph $G$. The vertices of $\mathcal{C}_K(G)$ are all functions $f : V(G) \longrightarrow V(K)$, and two functions $f$, $g$ are joined by an edge if and only if for all $[u, v] \in E(G)$, we have $[f(u), g(v)] \in E(K)$. In particular, $\mathcal{C}_{K_n}(G) = \mathcal{C}_n(G)$. Note that a function $f \in V(\mathcal{C}_K(G))$ is adjacent to itself if and only if it is a homomorphism form $G$ to $K$. Hence, $\mathcal{C}_K(G)$ is a genuine (irreflexive) graph if and only if $G$ has *no* "$K$-colourings"; in this respect, the name "$K$-colouring graph" is somewhat misleading[8].

---

[6]These order-theoretic properties are investigated by Duffus and Sauer [39].

[7]If we allow loops, this relationship holds between the complete graph $K_n$ and the one-vertex graph with a loop. If the product conjecture is true, this is the only instance.

[8]Also note that the categorical product is sometimes called the tensor product, although it is not a tensor product from the point of view of category theory.

The essential properties of $K$-colouring graphs are summarized in the following statement.

**Lemma 5.11** *Let $G$, $H$, $K$ be graphs. Then $G \times H \to K$ if and only if $H \to \mathcal{C}_K(G)$.*

**Proof** Let $\phi : G \times H \longrightarrow K$ be a homomorphism. We define a map $\psi : V(H) \longrightarrow V(\mathcal{C}_K(G))$ by putting $\psi(x) = f_x$, where $f_x(u) = \phi(u, x)$. By the definition of $\mathcal{C}_K(G)$, it is straightforward to verify that $\psi$ is a homomorphism.

Conversely, for a homomorphism $\psi : H \longrightarrow \mathcal{C}_K(G)$, we can define a homomorphism $\phi : G \times H \longrightarrow K$ by $\phi(u, x) = f_x(u)$, where $f_x = \psi(x)$. □

In particular, the relation $K \to \mathcal{C}_K(G)$ holds for any graphs $G$ and $K$. Also, a graph $K$ is multiplicative if and only if for every graph $G$ such that $G \not\to K$, we have $\mathcal{C}_K(G) \to K$. In this connection, Theorem 5.4 has been generalized to the following.

**Theorem 5.12 (Häggkvist, Hell, Miller, Neumann-Lara, [54])** *All odd cycles are multiplicative.*

In addition, the categorical product and the concept of multiplicative graphs is also investigated in the context of directed graphs. There, the approach of El-Zahar and Sauer has some noteworthy competition, namely the concept of 'complete sets of obstructions', which generalizes the role of the odd cycles in the characterization of bipartite graphs (see [54]). In another direction, infinite graphs and infinite chromatic numbers have also been considered with respect to the product conjecture. The categorical product of two graphs with countably infinite chromatic number has countably infinite chromatic number, but Hajnal [58] has shown that the product conjecture is false for higher cardinalities.

We conclude this section with an elegant proof of Welzl's density theorem 2.33 that uses the concepts presented above. This proof (to our knowledge unpublished) is due to M. Perles [99] and is very different form Welzl's original approach.

**Proof of Theorem 2.33** Let $G$, $H$ be graphs with at least one edge, and such that $G \to H$ and $H \not\to G$. Note that $H$ need not be connected. However, we may assume that every connected component of $H$ is nonbipartite. Let $m$ denote the maximum of the odd girths of the connected components of $H$, and put $n = \chi(\mathcal{C}_G(H))$ (since $H \not\to G$, $\mathcal{C}_G(H)$ has no loops and hence its chromatic number is well defined). It is well known that there exists a graph $L$ with odd girth greater than $m$ and chromatic number greater than $n$ (for instance, some Kneser graphs satisfy this property, see Proposition 3.14). We then have $L \not\to \mathcal{C}_G(H)$, and $H' \not\to L$ for every connected component $H'$ of $H$. Let $K$ be the disjoint union of the graphs $G$ and $H \times L$. Obviously, we have $G \to K \to H$; we show that $H \not\to K \not\to G$. Suppose that $K \to G$. Then $H \times L \to G$, thus $L \to \mathcal{C}_G(H)$ by Lemma 5.11; this contradicts our choice of $L$. Similarly, suppose that $H \to K = G \cup H \times L$. Since $H \not\to G$, we must have $H' \to H \times L$ for some connected component $H'$ of $H$. We then have $H' \to l$, which again contradicts our choice of $L$. Therefore, we have $G \to K \to H$ and $H \not\to K \not\to G$; that is, $K$ is homomorphically "strictly in between" $G$ and $H$, and this concludes the proof. □

## 5.2 The cartesian product and normal Cayley graphs

Recall that the cartesian product $G \Box H$ of two graphs $G$ and $H$ has vertex set $V(G) \times V(H)$, and two vertices $(u_1, v_1), (u_2, v_2)$ are joined by an edge if either $u_1 = u_2$ and $[v_1, v_2] \in E(H)$, or $[u_1, u_2] \in E(G)$ and $v_1 = v_2$. Hence the projections of $G \Box H$ onto its factors are not homomorphisms (except in the case of factors with no edges). However, for any $(u, v) \in V(G) \times V(H)$, the fibers induced by the sets $\{u\} \times V(H)$ and $V(G) \times \{v\}$ are isomorphic copies of $H$ and $G$, respectively. This shows that both relations $G \to G \Box H$ and $H \to G \Box H$ hold. Thus, in shifting the focus from the categorical product to the cartesian product, we find that the respective roles of the product and the factors are reversed, in the sense that the factors admit natural maps into the product instead of the opposite. In particular, the inequality relating the respective chromatic numbers of the product now becomes $\chi(G \Box H) \geq \max\{\chi(G), \chi(H)\}$. However, in the case of the cartesian product, the analogue of the product conjecture admits a simple proof.

**Lemma 5.13** ([107]) *For graphs $G$ and $H$, $\chi(G \Box H) = \max\{\chi(G), \chi(H)\}$.*

**Proof** Recall our convention that for any $n \in \mathbb{N}$, the graph $K_n$ has as its vertices the integers $\{0, 1, \ldots, n-1\}$. Let $m = \max\{\chi(G), \chi(H)\}$. Given homomorphisms $\phi : G \longrightarrow K_{\chi(G)}$ and $\psi : H \longrightarrow K_{\chi(H)}$, we define $\nu : G \Box H \longrightarrow K_m$ by $\nu(u, x) = \phi(u) + \psi(x) \bmod m$. This is a homomorphism and so $\chi(G \Box H) \leq m$. Since both $G$ and $H$ map into $G \Box H$, we also have the inverse inequality. □

From the point of view of homomorphisms, Lemma 5.13 admits the following interpretation. For any graphs $G$, $H$ such that $G \to K_n$ and $H \to K_n$, we have $G \Box H \to K_n$. By analogy with the concept of multiplicative graphs (Definition 5.9), this property suggests the following problem.

> *Characterize the graphs $K$ with the property that for any graphs $G, H$ such that $G \to K$ and $H \to K$, we have $G \Box H \to K$.*

In particular, for any graph $K$, we have $K \to K$, so this property implies $K \Box K \to K$. Conversely, any pair of homomorphisms $\phi_1 : G \longrightarrow K$ and $\phi_2 : H \longrightarrow K$ can be used coordinatewise to define a homomorphism $\phi : G \Box H \longrightarrow K \Box K$. Therefore, a graph $K$ has the property described above if and only if $K \Box K \to K$.

**Definition 5.14** A graph $K$ is called *hom-idempotent* if $K \Box K \leftrightarrow K$.

As any graph $K$ satisfies $K \to K \Box K$, a graph $K$ is hom-idempotent if and only if $K \Box K \to K$. This property seems very restrictive, since the projections are not homomorphisms. The complete graphs are hom-idempotent, but a close inspection of the proof of Lemma 5.13 hints at a construction of other examples.

**Definition 5.15** A *normal Cayley graph* is a Cayley graph $\mathrm{Cay}(\Gamma, S)$ such that $x^{-1}sx \in S$ for any $x \in \Gamma$ and $s \in S$.

Note that the condition $x^{-1}sx \in S$ is satisfied whenever the group $\Gamma$ is abelian. Thus, all Cayley graphs of abelian groups, and in particular all circulants, are normal Cayley graphs. The defining property of normal Cayley graphs has the following interpretation.

**Lemma 5.16** *A Cayley graph* $\mathrm{Cay}(\Gamma, S)$ *is normal if and only if both the left translations and the right translations of* $\Gamma$ *are automorphisms of* $\mathrm{Cay}(\Gamma, S)$.

**Proof** Since the left translations are always automorphisms of Cayley graphs, it suffices to consider the right translations. Suppose that $\mathrm{Cay}(\Gamma, S)$ is normal. For $w \in \Gamma$, let $T : \Gamma \longrightarrow \Gamma$ be the right translation $T(u) = uw$. For $[u, v] \in E(\mathrm{Cay}(\Gamma, S))$, we have $u^{-1}v \in S$, so $T(u)^{-1}T(v) = w^{-1}u^{-1}vw \in S$ and $[T(u), T(v)] \in E(\mathrm{Cay}(\Gamma, S))$. Therefore, $T$ is an automorphism of $\mathrm{Cay}(\Gamma, S)$. Conversely, suppose that all right translations of a Cayley graph $\mathrm{Cay}(\Gamma, S)$ are automorphisms. Then, for $s \in S$ and $w \in \Gamma$, we have $[1_\Gamma, s] \in E(\mathrm{Cay}(\Gamma, S))$, so $[w, sw] \in E(\mathrm{Cay}(\Gamma, S))$, that is, $w^{-1}sw \in S$. □

**Corollary 5.17 (Hahn, Hell, Poljak [55])** *Let* $\mathrm{Cay}(\Gamma, S)$ *be a normal Cayley graph. Then the map* $\phi : \mathrm{Cay}(\Gamma, S)\Box\mathrm{Cay}(\Gamma, S) \longrightarrow \mathrm{Cay}(\Gamma, S)$ *defined by putting* $\phi(x, y) = xy$ *is a homomorphism. Therefore,* $\mathrm{Cay}(\Gamma, S)$ *is hom-idempotent.*

**Proof** The map $\phi$ acts as a right translation on each fiber of the first factor and as a left translation on each fiber of the second factor. Thus, by Lemma 5.16, $\phi$ is a homomorphism. □

Two classes of circulants appear respectively in the definition of the chromatic number and the circular chromatic number, namely the complete graphs and the circular graphs. By Corollary 5.17, these are hom-idempotent graphs, and this accounts for the identities $\chi(G\Box H) = \max\{\chi(G), \chi(H)\}$ and $\chi_c(G\Box H) = \max\{\chi_c(G), \chi_c(H)\}$. In contrast, the fractional chromatic number is defined using homomorphisms into Kneser graphs, which are not hom-idempotent. It follows that in general, $\chi_f(G\Box H)$ cannot be expressed as a function of $\chi_f(G)$ and $\chi_f(H)$. For instance, we have $\chi_f(C_5) = \chi_f(K(2,5)) = 5/2$. However, since $C_5$ is hom-idempotent, we have $\chi_f(C_5\Box C_5) = \chi_f(C_5) = 5/2$, while $\chi_f(K(2,5)\Box K(2,5)) = 50/17$, as shown by Albertson and Collins [5].

These examples show how hom-idempotent graphs are related to parameters which are well behaved with respect to the cartesian product. The characterization of hom-idempotent graphs is closely related to the definition of the cartesian product of graphs as a tensor product in the category of graphs.

**Definition 5.18** Let $G$ and $K$ be graphs. We define the *homomorphism graph* $\mathrm{Hom}(G, K)$ by putting

$$\begin{aligned} V(\mathrm{Hom}(G,K)) &= \{\phi : G \longrightarrow K : \ \phi \text{ is a homomorphism}\}, \\ E(\mathrm{Hom}(G,K)) &= \{[\phi, \psi] : \ \text{for all } u \in V(G), [\phi(u), \psi(u)] \in E(K)\}. \end{aligned}$$

It is interesting to compare this definition to that of the graph $\mathcal{C}_K(G)$ defined in the preceding section. Note that $\mathcal{C}_K(G)$ has $|K|^{|G|}$ vertices regardless of the structure of $G$ and $K$, whereas the vertices of $\mathrm{Hom}(G, K)$ are homomorphisms, and hence their number depends heavily on the structure of both $G$ and $K$. In particular, the number of vertices of $\mathrm{Hom}(G, K_n)$ is the value $p_G(n)$, where $p_G$ is the chromatic polynomial of $G$. The relative importance of structure is reversed when we consider the edge sets of $\mathcal{C}_K(G)$ and $\mathrm{Hom}(G, K)$. The definition of $E(\mathcal{C}_K(G))$ uses both definitions of $E(G)$ and $E(K)$, while the definition of

$E(\mathrm{Hom}(G,K))$ omits that of $E(G)$. In fact, $\mathrm{Hom}(G,K)$ can be viewed as a subgraph of the categorical product of $|G|$ copies of $K$. This brings forth the last point of comparison between these graphs. The relation $\mathrm{Hom}(G,K) \to K$ trivially holds in all cases, as for any $u \in V(G)$, the *evaluation map* $\epsilon_u : \mathrm{Hom}(G,K) \to K$ defined by $\epsilon_u(\phi) = \phi(u)$ is a homomorphism. However, to decide in which cases the relation $\mathcal{C}_K(G) \to K$ holds, amounts to characterizing multiplicative graphs. Note, however, that the relation $K \to \mathcal{C}_K(G)$ always holds.

Any homomorphism $\phi : G\Box H \longrightarrow K$ acts as a homomorphism on each of the fibers of $H$. It is the kind of interplay needed between these that dictates the definition of $\mathrm{Hom}(H,K)$. Formally, this means that the homomorphism graphs can be used in a characterization of the cartesian product of graphs as a tensor product. The essential points of this characterization are summarized in the following technical statement.

**Proposition 5.19** *Let $G$, $H$, $K$ be graphs.*

(i) *Let $\phi : G\Box H \longrightarrow K$ be a homomorphism. Then the map $\psi : G \longrightarrow \mathrm{Hom}(H,K)$ defined by setting $\psi(u) = \phi_u$ is a homomorphism, where $\phi_u : H \longrightarrow K$ is defined by $\phi_u(v) = \phi(u,v)$.*

(ii) *Let $\psi : G \longrightarrow \mathrm{Hom}(H,K)$ be a homomorphism. Then the map $\phi : G\Box H \longrightarrow K$ defined by $\phi(u,v) = \phi_u(v)$ is a homomorphism, where $\phi_u = \psi(u)$.*

We omit the proof which is a straightforward application of the definition. Our main interest is the hom-idempotent graphs, that is, the situation where $G = H = K$. Proposition 5.19 states that for a graph $K$, we have $K\Box K \to K$ if and only if $K \to \mathrm{Hom}(K,K)$. Since the relation $\mathrm{Hom}(K,K) \to K$ always holds, we have the following.

**Corollary 5.20** *A graph $K$ is hom-idempotent if and only if $K \leftrightarrow \mathrm{Hom}(K,K)$.*

We can make this characterization more precise by restricting it to core graphs.

**Proposition 5.21** *Let $K$ be a core. Then $\mathrm{Hom}(K,K)$ is a normal Cayley graph.*

**Proof** If $K$ is a core, then $V(\mathrm{Hom}(K,K)) = \mathrm{Aut}(K)$. Two automorphisms $\phi, \psi$ of $K$ are adjacent if and only for all $u \in V(K)$, we have $[\phi(u), \psi(u)] \in E(K)$. Applying $\phi^{-1}$ yields $[u, \phi^{-1} \circ \psi(u)] \in E(K)$ for all $u \in V(K)$. Therefore, $\mathrm{Hom}(K,K)$ is the Cayley graph $\mathrm{Cay}(\mathrm{Aut}(K), S_K)$, where

$$S_K = \{\sigma \in \mathrm{Aut}(K) : [u, \sigma(u)] \in \mathrm{Aut}(K) \text{ for all } u \in V(K)\}.$$

The elements of $S_K$ are called the *shifts* of $K$. It remains to show that for any $\sigma \in S_K$ and any $\phi \in \mathrm{Aut}(K)$, we have $\phi^{-1} \circ \sigma \circ \phi \in S_K$. For any $u \in V(K)$, we have $[\phi(u), \sigma \circ \phi(u)] \in E(K)$ by the definition of shifts. Applying $\phi^{-1}$ yields $[u, \phi^{-1} \circ \sigma \circ \phi(u)] \in E(K)$. Thus, $\phi^{-1} \circ \sigma \circ \phi \in S_K$, and $\mathrm{Hom}(K,K) = \mathrm{Cay}(\mathrm{Aut}(K), S_K)$ is a normal Cayley graph. □

Since homomorphisms can be used coordinatewise, hom-idempotency of a graph $K$ is equivalent to hom-idempotency of its core. Therefore, combining the previous results with Corollary 5.17, we get the following.

**Theorem 5.22 (Larose, Laviolette, Tardif [90])** *A graph is hom-idempotent if and only if it is homomorphically equivalent to a normal Cayley graph.* □

Thus, up to homomorphic equivalence, the examples described in Corollary 5.17 represent all hom-idempotent graphs. This is surprising in view of the fact that even for a core graph $K$, the homomorphisms $\phi : K \Box K \longrightarrow K$ are usually not group operations. For instance, any latin square on the vertex set of $K_n$ defines a homomorphism $\phi : K_n \Box K_n \longrightarrow K_n$, and only a few of these are group operations. Of course, some always are, and $K_n$ is indeed a normal Cayley graph. However, it is conceivable that there exists a hom-idempotent core $K$ such that none of the homomorphisms $\phi : K \Box K \longrightarrow K$ are group operations, though no such example is known.

Since the fibers are isomorphic copies of the factors in a cartesian product of graph, we have for any graph $K$ and any $n \leq m$ that $K^n \to K^m$ (where $K^n$ is short for $\Box_{i=1}^n K$). In fact, it is quite common that the relation $\to$ defines a strict order on the cartesian powers of a graph. At the other extreme lie the hom-idempotent graphs, which have the property that all of their cartesian powers are homomorphically equivalent. Between these two extremes, there exist graphs which satisfy some nontrivial relations $K^m \to K^n$ but not all of them. The next result shows that normal Cayley graphs also help to characterize all of these graphs.

**Theorem 5.23 (Larose, Laviolette, Tardif [90])** *Let $K$ be a graph and $m > n$ integers. Then $K^m \to K^n$ if and only if there exist normal Cayley graphs* $\mathrm{Cay}(\Gamma_1, S_1), \ldots, \mathrm{Cay}(\Gamma_n, S_n)$ *such that* $\mathrm{Cay}(\Gamma_k, S_k) \to K$ *for $k = 1, \ldots, n$, and* $K \to \Box_{k=1}^n \mathrm{Cay}(\Gamma_k, S_k)$.

We have seen in Example 3.5 that $G = \mathrm{Cay}(\mathbb{Z}_8, \{\pm 1, \pm 2\})$ and $H = \mathrm{Cay}(\mathbb{Z}_7, \{\pm 1, \pm 2\})$ are two circulants such that $G \not\to H$ and $H \not\to G$. Both of these are normal Cayley graphs. Let $K$ be the graph obtained from the disjoint union of these two graphs by adding an edge between the element 0 of $\mathbb{Z}_8$ and the element 0 of $\mathbb{Z}_7$. Then we have $G \to K$ and $H \to K$. Also, a homomorphism $\phi : K \longrightarrow G \Box H$ can be defined by putting $\phi(u) = (u, 0)$ if $u \in \mathbb{Z}_8$ and $\phi(u) = (1, u)$ if $u \in \mathbb{Z}_7$. Therefore, we also have $K \to G \Box H$.

Theorem 5.23 then states that for any $m > 2$, we have $K^m \to K^2$. A homomorphism from $K^m$ to $K^2$ can be constructed as follows. First, use $\phi : K \longrightarrow G \Box H$ coordinatewise to define a homomorphism from $K^m$ to $(G \Box H)^m$. Then, note that $G \Box H$ is a normal Cayley graph (the cartesian product of normal Cayley graphs always is), so the multiplication of the coordinates is a homomorphism from $(G \Box H)^m$ to $G \Box H$. Finally, use any homomorphisms $\psi_1 : G \longrightarrow K$ and $\psi_2 : H \longrightarrow K$ coordinatewise to define a homomorphism from $G \Box H$ to $K^2$.

However, there is no homomorphism from $K^2$ to $K$. Indeed, we have $K^2 \leftrightarrow G \Box H$, and every edge of $G \Box H$ is contained in a triangle. Hence, no edge of $G \Box H$ can be mapped by homomorphism to the edge of $K$ joining the element 0 of $\mathbb{Z}_8$ and the element 0 of $\mathbb{Z}_7$. Therefore, the image of a homomorphism from $G \Box H$ to $K$ would either be contained in $G$ or in $H$. However, this is impossible since $G \not\to H$ and $H \not\to G$.

We now turn our attention to homomorphisms *into* a cartesian product of graphs. Let $\phi : G \longrightarrow \Box_{k=1}^n G_k$ be a homomorphism. Then the maps $\mathrm{pr}_k \circ \phi : G \longrightarrow G_k, k = 1, \ldots, n$, need not be homomorphisms. However, for any $k$, the partition

$$\mathcal{P}_k = \{(pr_k \circ \phi)^{-1}(u) : u \in V(G_k)\}$$

of $V(G)$ has the property that $G/\mathcal{P}_k \to G_k$, and the least common refinement of $\mathcal{P}_1, \ldots, \mathcal{P}_n$ is the partition

$$\mathcal{P} = \{\phi^{-1}(u) : u \in V(\Box_{k=1}^n G_k)\}$$

of $V(G)$. Hence, the essential information concerning $\phi$ is encoded in the family $\mathcal{P}_1, \ldots, \mathcal{P}_n$ of partitions of $V(G)$. This representation is still cumbersome, but we will be able to refine it by using the structure of the cartesian product of graphs.

**Definition 5.24**

- Let $\phi : G \longrightarrow \square_{k=1}^{n} G_k$ be a homomorphism. The *edge-labeling* induced by $\phi$ is the map

$$\ell_\phi : E(G) \longrightarrow \{1, \ldots, n\},$$

  where $\ell_\phi([u,v])$ is the unique index $k$ such that $\mathrm{pr}_k \circ \phi(u) \neq \mathrm{pr}_k \circ \phi(v)$.

- Let $\ell : E(G) \longrightarrow \{1, \ldots, n\}$ be a map. Then for $k \in \{1, \ldots, n\}$, let $G/\ell^{-1}(k)$ denote the quotient $G/\mathcal{Q}_k$, where $\mathcal{Q}_k$ is the partition of $V(G)$ whose cells are the (vertex-sets of) connected components of $G - \ell^{-1}(k)$.

Let $\phi : G \longrightarrow \square_{k=1}^{n} G_k$ be a homomorphism with induced edge labeling $\ell_\phi$. Then $\mathrm{pr}_k \circ \phi$ is constant on each connected component of $G - \ell_\phi^{-1}(k)$, and there exists a homomorphism $\phi_k : G/\ell_\phi^{-1}(k) \longrightarrow G_k$ (not necessarily injective) such that $\mathrm{pr}_k \circ \phi = \phi_k \circ \pi_k$ (where $\pi_k : G \longrightarrow G/\ell_\phi^{-1}(k)$ is the natural map) for $k = 1, \ldots, n$. In addition, we can use the natural maps $\pi_k : G \longrightarrow G/\ell_\phi^{-1}(k)$ coordinatewise to define a homomorphism $\pi : G \longrightarrow \square_{k=1}^{n} G/\ell_\phi^{-1}(k)$. To some extent, this shows how some essential information concerning homomorphisms of $G$ into a cartesian product is encoded in maps from $E(G)$ to $\{1, \ldots, n\}$ for some integer $n$, in other words, in edge partitions. The following is a characterization of the maps which give rise to a homomorphism in this fashion.

**Observation 5.25** *Let $G$ be a graph, $n$ an integer, and $\ell : E(G) \longrightarrow \{1, \ldots, n\}$ a map. Then the following conditions are equivalent.*

(i) *There exists a homomorphism $\phi : G \longrightarrow \square_{k=1}^{n} G_k$ such that $\ell_\phi = \ell$.*

(ii) *The natural map $\pi : G \longrightarrow \square_{k=1}^{n} G/\ell^{-1}(k)$ is a homomorphism.*

(iii) *For any edge $e = [u,v]$ of $G$ and for any path $P$ from $u$ to $v$, there exists an edge $e'$ of $P$ such that $\ell(e') = \ell(e)$.*

Therefore, the problem of finding a homomorphism from a graph to a cartesian product of graphs often amounts to finding a suitable edge partition. In that vein, we find the long standing problem of characterizing the induced subgraphs of cubes (a *cube* is a cartesian product of copies of $K_2$). This problem was brought forth by Shapiro [112] along with the first definition of the cartesian product of graphs, and has since been studied by many (see [27, 31]). As is to be expected, the problems concerning homomorphisms into cartesian product of graphs turn out to be as intricate as many other problems concerning graph homomorphisms. However, a recent wave of interest in some particular homomorphisms called isometric embeddings has given rise to interesting results concerning embeddings into cartesian products. These and other aspects of isometric embeddings are investigated in Section 5.4.

## 5.3 The strong product and the lexicographic product

The strong product $G \boxtimes H$ and the lexicographic product $G[H]$ bear a close resemblance to each other. The edges of $G \boxtimes H$ are the pairs $[(u_1, u_2), (v_1, v_2)]$ such that $u_i$ and $v_i$ are adjacent or equal for $i = 1, 2$. All these edges are contained in $G[H]$, in addition to all other edges $[(u_1, u_2), (v_1, v_2)]$ such that $[u_1, v_1] \in E(G)$. In particular, $G \boxtimes K_n = G[K_n]$ for any graph $G$ and any integer $n$. This identity summarizes the relation between the colouring problems between both of these products. The problem of finding the chromatic number of a lexicographic product of graphs reduces to that of finding the chromatic number of the graphs $G[K_n]$, which in turn provides bounds on the chromatic number of a strong product of graphs.

In section 4.3, we defined the $n$-chromatic number of a graph $G$ as the least integer $m = \chi_n(G)$ such that $G$ admits a Kneser $(n, m)$-colouring, that is, a homomorphism $\phi : G \longrightarrow K(n, m)$.

**Lemma 5.26** *For any graph $G$, and any integer $n$, $\chi(G[K_n]) = \chi_n(G)$.*

**Proof** Let $\phi : \chi(G[K_n]) \longrightarrow [m]$ be a colouring. Then, we find a $(n, m)$-subset colouring of $G$ by assigning to each vertex $u$ the set of colours used on the fiber $\{(u, v) : v \in K_n\}$. Conversely, if $\phi : G \longrightarrow K(n, m)$ is a homomorphism, then we can $m$-colour $G[K_n]$ by colouring the fiber $\{(u, v) : v \in K_n\}$ with colours in $\phi(u)$ for each $u \in V(G)$. □

**Theorem 5.27 (Stahl [113])** *For any graphs $G$ and $H$, we have $\chi(G[H]) = \chi_n(G)$, where $n = \chi(H)$.*

**Proof** By Lemma 5.26, we have $\chi(G[H]) \leq \chi_n(G)$ since $G[H] \to G[K_n]$. On the other hand, for any $m$-colouring $\phi : G[H] \longrightarrow [m]$, each fiber $\{(u, v) : v \in V(H)\}$ uses at least $n = \chi(H)$ colours, and the fibers corresponding to adjacent vertices of $G$ are coloured with disjoint sets of colours. Thus, we can define a Kneser $(n, m)$-colouring of $G$ by independently selecting, for each $u$ in $V(G)$, a $n$-subset of the colours used for its fiber. □

Thus, the chromatic number of $G[H]$ does not depend on the structure of $H$, only on its chromatic number. A similar phenomenon occurs when we consider the circular chromatic number of a lexicographic product of graphs.

**Theorem 5.28 (Zhu [138])** *For any graphs $G$ and $H$, we have $\chi_c(G[H]) = \chi_c(G[K_n])$, where $n = \chi(H)$.*

In the case of the strong product, Lemma 5.26 can be used to find lower bounds for the chromatic number.

**Lemma 5.29** *For any graphs $G$ and $H$,*

$$\chi(G \boxtimes H) \geq \max\{\chi_{\omega(H)}(G), \chi_{\omega(G)}(H)\}.$$

**Proof** Since $G[K_{\omega(H)}] \simeq G \boxtimes K_{\omega(H)} \to G \boxtimes H$ and $H[K_{\omega(G)}] \simeq H \boxtimes K_{\omega(G)} \to H \boxtimes G \simeq G \boxtimes H$, the result follows from Lemma 5.26. □

It is also possible to express bounds for the chromatic number of a strong product of graphs in terms of the chromatic numbers and clique numbers of the factors alone. If $G$ has at least one edge, then for any $n$, $\chi_n(G) \geq 2n$. For $m = \chi_n(G)$, we have $G \to K(n,m)$. By Theorem 3.10, $\chi(K(n,m)) = m - 2n + 2$, and $\chi(G) + 2n - 2 \leq m = \chi_n(G)$. Therefore, we have the following.

**Theorem 5.30 (Klavžar, Milutinović [85])** *If $G$ and $H$ are graphs with at least one edge, then*

$$\chi(G \boxtimes H) \geq \max\{\chi(G) + 2\omega(H) - 2, \chi(H) + 2\omega(G) - 2\}. \qquad \square$$

Note that this bound is obtained by considering subgraphs of the product which are isomorphic to lexicographic products. Vesztergombi [124] proposed a different approach which takes into account the full structure of the strong product of graphs. For a graph $G$ and an integer $m$, define the graph $\mathcal{C}_m^{\boxtimes}(G)$ by putting

$$\begin{aligned} V(\mathcal{C}_m^{\boxtimes}(G)) &= \{\phi : G \longrightarrow [m] : \phi \text{ is a proper colouring}\}, \\ E(\mathcal{C}_m^{\boxtimes}(G)) &= \{[\phi,\psi] : \phi(u) \neq \psi(u) \text{ for all } u \in V(G) \text{ and} \\ &\qquad \phi(u) \neq \psi(v) \text{ for all } [u,v] \in E(G)\}. \end{aligned}$$

This construction bears the same relation to the colourings of a strong product of graphs as do the $n$-colouring graphs of Definition 5.3 to the colourings of a categorical product of graphs.[9] Namely, the following holds.

**Theorem 5.31 (Vesztergombi [124])** *For two graphs $G$ and $H$, $G \boxtimes H$ is $m$-colourable if and only if $H \to \mathcal{C}_m^{\boxtimes}(G)$.*

It is clear that the chromatic number of a strong product of graphs cannot be expressed as a function of the chromatic numbers of the factors. Thus, there is no strong product-analogue of the product conjecture. Nonetheless, it would be interesting to know more about the general structure of $\mathcal{C}_m^{\boxtimes}(G)^\bullet$. In particular, Vesztergombi [124] has shown that $\mathcal{C}_m^{\boxtimes}(K_n)^\bullet = K(n,m)$, and $\mathcal{C}_m^{\boxtimes}(C_5)^\bullet = C_5$.

We now turn our attention to retracts of strong products and lexicographic products of graphs. These are more manageable than the retracts of products we have investigated so far. We will be particularly interested in finding out when the product of two cores is a core. Since homomorphisms can be used coordinatewise, it is of course necessary that both factors be cores. However, this condition is not sufficient for either of the products. Consider the graphs $K(3,9)$ and $K_3$. Both are cores; however, their product $K(3,9) \boxtimes K_3 \simeq K(3,9)[K_3]$ is a 9-chromatic graph with a 9-clique, so its core is $K_9$.

**Proposition 5.32** *For any connected graphs $G$, $H$, the core of $G[H]$ is $G'[H^\bullet]$, where $G'$ is a subgraph of $G$ which is itself a core.*

**Proof** Clearly, $G[H] \leftrightarrow G[H^\bullet]$. Let $\rho : G[H^\bullet] \longrightarrow G[H^\bullet]^\bullet$ be a retraction, and identify $G[H^\bullet]^\bullet$ with a subgraph of $G[H^\bullet]$. Then for any $u \in V(G)$ and $[v,w] \in E(H^\bullet)$ such that $(u,v) \in V(G[H^\bullet]^\bullet)$, $\rho(u,w)$ must be adjacent to $(u,v)$ and cannot be adjacent to $(u,w)$.

[9]Note that the definition of $\mathcal{C}_m^{\boxtimes}(G)$ predates that of $\mathcal{C}_m(G)$.

Since $H^\bullet$ is connected, it follows that $\rho$ maps the fiber $F_u = \{(u,v) : v \in V(H^\bullet)\}$ into itself. Since $H^\bullet$ is a core, this means that $G[H^\bullet]^\bullet = G'[H^\bullet]$, where $G'$ is a subgraph of $G$. □

However note that not all retracts of $G[H]$ are lexicographic products, since retractions can be performed independently on the fibers of $H$. The situation is different in the case of the strong product, as the following result shows.

**Theorem 5.33 (Imrich, Klavžar [84])** *Let $G$, $H$ be connected graphs, and $R$ a retract of $G \boxtimes H$. Then there exist subgraphs $G'$ of $G$ and $H'$ of $H$ such that $R$ is isomorphic to $G' \boxtimes H'$.*

In particular, the core of $G \boxtimes H$ is a strong product of subgraphs of $G$ and $H$. The example of $K(3,9)[K_3]$ shows that neither Proposition 5.32 nor Theorem 5.33 can be refined to the statement that these subgraphs are retracts of the respective factors. The reason lies in the fact that the presence of triangles in the factors allows some interplay between the factors. This is made clear by the following results.

**Proposition 5.34** *Let $G$ and $H$ be connected graphs such that $G$ does not contain any triangles. Then $G[H]^\bullet = G^\bullet[H^\bullet]$.*

**Proof** By Proposition 5.32, $G[H]^\bullet = G'[H^\bullet]$, where $G'$ is a core, and it remains to show that it is the core of $G$. Let $\rho : G[H^\bullet] \longrightarrow G'[H^\bullet]$ be a retraction. Put

$$S = \{u \in V(G) : \rho(F_u) = F_{\psi(u)} \text{ for some } \psi(u) \in V(H^\bullet)\}.$$

We show that $S = V(G)$. Clearly, $S$ is not empty since $V(G') \subseteq S$. If $u'$ is adjacent to some $u \in S$, then $\rho(F_{u'})$ cannot intersect $F_{\psi(u)}$ since every vertex of $F_{u'}$ is adjacent to all vertices of $F_u$. Also, since $H^\bullet$ is connected, $\mathrm{pr}_G(F_{u'})$ induces a connected subgraph of $G$, whose vertices are all adjacent to $\psi(u)$. However, $G$ does not contain any triangles, so $\mathrm{pr}_G(F_{u'})$ consists of a single vertex $\psi(u')$. Then, $\rho$ must map $F_{u'}$ to $F_{\psi(u')}$ bijectively, since all fibers are isomorphic to the core $H^\bullet$. Thus, $u' \in S$. $G$ being connected, this implies that $S = V(G)$. Therefore, $\psi : V(G) \longrightarrow G'$ is a well-defined retraction, and $G' = G^\bullet$. □

**Theorem 5.35 (Imrich, Klavžar [84])** *Let $G$, $H$ be connected triangle-free graphs, and $R$ a retract of $G \boxtimes H$. Then there exist retracts $G'$ of $G$ and $H'$ of $H$ such that $R$ is isomorphic to $G' \boxtimes H'$. In particular, $(G \boxtimes H)^\bullet = G^\bullet \boxtimes H^\bullet$.*

In the case of these two products, retracts and cores can be expressed as products of subgraphs or retracts. The situation is different with other products. In particular, the identity $(G \times G)^\bullet = G^\bullet$ holds for all graphs, and $G^\bullet$ cannot be expressed as a categorical product of subgraphs of $G$. Also, the hom-idempotent graphs are the graphs $G$ that satisfy $(G \Box G)^\bullet = G^\bullet$, and $G^\bullet$ is a cartesian product of subgraphs of $G$. In general, it is known (see Tardif [117]) that if $(\Box_{k=1}^n G_k)^\bullet$ is vertex-transitive, then it is a cartesian product of subgraphs of the factors. However, it is not known if the same holds for all cores of connected cartesian products of graphs.

## 5.4 Isometric embeddings and retracts

The *distance* $d_G(u,v)$ between two vertices $u,v$ of a connected graph $G$ is the length of a shortest path joining them. Such a path is called a *uv-geodesic.* The function $d_G$ itself is often called the 'geodesic metric' or 'shortest path metric' of $G$. It is indeed a metric in the usual sense, as it satisfies the axioms $d_G(u,v) = 0$ if and only if $u = v$, $d_G(u,v) = d_G(v,u)$ and the triangle inequality $d_G(u,w) \leq d_G(u,v) + d_G(v,w)$.

The interplay between graph homomorphisms and the shortest path metric is illustrated by the following observation. Let $G$, $H$ be connected graphs, and $\phi : G \longrightarrow H$ a homomorphism. Then, for any $u,v \in V(G)$ and any $uv$-geodesic $P$, $\phi(P)$ is a $\phi(u)\phi(v)$-trail, and we see from this that $d_H(\phi(u),\phi(v)) \leq d_G(u,v)$. This motivates the following definitions.

**Definition 5.36** Let $G$, $H$ be connected graphs, and $\phi : V(G) \longrightarrow V(H)$ a map.

(i) $\phi$ is a *contraction* if $d_H(\phi(u),\phi(v)) \leq d_G(u,v)$ for any two vertices $u,v$ of $G$.

(ii) $\phi$ is an *isometric embedding* if $d_H(\phi(u),\phi(v)) = d_G(u,v)$ for any two vertices $u,v$ of $G$.

Any homomorphism between connected graphs is a contraction, but the converse fails since contractions can map adjacent vertices to the same vertex. However, isometric embeddings are homomorphisms. The term 'embedding' is normally reserved for injective maps and indeed, it is easy to see that any isometric embedding is a full monomorphism. The following result shows that some important homomorphisms that we have already encountered are in fact isometric embeddings.

**Lemma 5.37** *Let $R$ be a retract of a connected graph $G$. Then any co-retraction $\gamma : R \longrightarrow G$ is an isometric embedding.*

**Proof** Let $\rho : G \longrightarrow R$ be a retraction such that $\rho \circ \gamma = \mathrm{id}_R$. Then for any $u,v \in V(R)$, we have

$$d_R(u,v) \geq d_G(\gamma(u),\gamma(v)) \geq d_R(\rho\circ\gamma(u),\rho\circ\gamma(v)) = d_R(u,v),$$

hence $\gamma$ is an isometric embedding. □

Therefore, metric considerations turn out to play an important role in the characterization of retracts of graphs, though we have managed to avoid the subject up to this point. Part of the reasons for avoiding it lie in the fact that when considering discrete metric spaces, the natural morphisms are the contractions rather than graph homomorphisms.

Isometric embeddings into products of graphs often provide an insight into the metric structure of a graph; this has become a flourishing topic in view of some recent applications of graph theory to computer science. Szamkołowicz [116] has shown that there exist precisely two products of graphs in which the distance can be expressed as a reasonably 'nice' function of the coordinate distances. These are the cartesian product, where the distance is the sum of coordinate distances, and the strong product, where the distance is the maximum of coordinate distances. Each of these products gives rise to an interesting theory regarding isometric embeddings and retracts. We shall present both in some detail.

Recall that Observation 5.25 shows how the analysis of homomorphisms into cartesian products is reduced to the study of edge labelings and natural maps into products of the quotients introduced in Definition 5.24

**Theorem 5.38 (Graham, Winkler [52])** *Let $G$ be a graph, $n$ an integer, and $\ell : E(G) \longrightarrow \{1, \dots, n\}$ a map. Then the following conditions are equivalent.*

(i) *There exists an isometric embedding $\phi : G \longrightarrow \Box_{k=1}^{n} G_k$ such that $\ell_\phi = \ell$.*

(ii) *The natural map $\pi : G \longrightarrow \Box_{k=1}^{n} G/\ell^{-1}(k)$ is an isometric embedding.*

(iii) *For all $u, v \in V(G)$, for all $k \in \{1, \dots, n\}$ and for all $uv$-geodesic $P$, we have*

$$|E(P) \cap \ell^{-1}(k)| = \min\{|E(Q) \cap \ell^{-1}(k)| : Q \textit{ is a } uv\textit{-path } \}.$$

(iv) *If $[u, v], [u', v']$ are edges of $G$ such that $\ell^{-1}([u, v]) \neq \ell^{-1}([u', v'])$, then*

$$d_G(u, u') - d_G(u, v') = d_G(v, u') - d_G(v, v').$$

In particular, condition (iv) suggests the following. Define a relation $\theta$ on the edge set of a connected graph $G$ by putting

$$[u, v]\theta[u', v'] \equiv d_G(u, u') + d_G(v, v') \neq d_G(u, v') + d_G(v, u').$$

This relation was introduced by Djoković for the purpose of characterizing the graphs which admit an isometric embedding in a cube. Note that $\theta$ is well defined, reflexive and symmetric. Let $\hat{\theta}$ denote the transitive closure of $\theta$. Then $\hat{\theta}$ is an equivalence relation and partitions $E(G)$ into classes $E_1, \dots, E_n$. Define $\ell : E(G) \longrightarrow \{1, \dots, n\}$ by putting $\ell(e) = k$ if $e \in E_k$. Then the natural map $\pi : G \longrightarrow \Box_{k=1}^{n} G/\ell^{-1}(k)$ is an isometric embedding. This map is optimal in the sense that if $\phi : G \longrightarrow \Box_{k=1}^{m} G_k$ is an isometric embedding that is 'irredundant' (that is, $\mathrm{pr}_k \circ \phi(G) = G_k$ for all $k$ and none of the factors $G_k$ is isomorphic to $K_1$), then $\phi$ can be refined to $\pi$.

Thus, isometric embeddings into cartesian products of graphs admit an efficient characterization, contrary to homomorphisms in general. A similar treatment applies to isomorphisms into cartesian products, which can be viewed as a particular kind of isometric embeddings (see [83]). However, we shall not go into this subject and turn our attention to the retracts of a cartesian product of graphs.

Recall that the cubes are the cartesian powers of $K_2$. These are the simplest instances of cartesian product of graphs. No general characterization of the retracts of cartesian products of graphs is known, but such a characterization exists for simple structures such as cubes. We define the *diameter* of a graph $G$ as the maximum distance between any two vertices of $G$. For instance, the cube $Q_n = \Box_{k=1}^{n} K_2$ has diameter $n$. Also, the *covering graph* (or *unoriented Hasse diagram*) of a partially ordered set $(P, \leq)$ is the graph $G$ which has $P$ as its vertex set, and whose edges are the pairs $[p, q]$ such that $p$ *covers* $q$, that is, $p < q$ and no element $r$ of $P$ satisfies $p < r < q$.

**Theorem 5.39 (Duffus, Rival [37])** *The diameter-preserving retracts of cubes are the covering graphs of distributive lattices.*

This result may come as a surprise, since it is not clear how lattice operations can be related to a graph structure, let alone to graph retractions. The following concept helps to clarify these matters. A *median* of three vertices of a connected graph is a vertex that lies simultaneously on geodesics between any two of them. It often happens that three given vertices of a graph have no median, or more than one. A graph $G$ is called a *median graph* if any triple of (not necessarily distinct) vertices of $G$ has a unique median. For instance, trees are simple examples of median graphs. Other examples include the cubes and the covering graphs of distributive lattices. In fact, if $G$ is the covering graph of a distributive lattice $L$, then the median of any three vertices $u, v, w$ of $G$ coincides with their order-theoretic median $(u \wedge v) \vee (u \wedge w) \vee (v \wedge w)$ in L.

Let $R$ be a retract of a median graph $G$. By Lemma 5.37, any co-retraction $\gamma : R \longrightarrow G$ is an isometric embedding. Let $\rho : G \longrightarrow R$ be a retraction such that $\rho \circ \gamma = \mathrm{id}_R$. Let $u, v, w$ be vertices of $R$, and $x \in V(G)$ the median of $\gamma(u)$, $\gamma(v)$ and $\gamma(w)$. Then $\rho(x)$ is a median of $u$, $v$ and $w$, which is easily seen to be unique since $\gamma$ is an isometric embedding. Thus, $R$ is a median graph, and we have shown that the class of median graphs is closed under taking retracts. In fact, Theorem 5.39 generalizes to the following characterization.

**Theorem 5.40 (Bandelt [10])** *The median graphs are the retracts of cubes.*

The reader may have noticed that the essential feature of retractions that is used in this characterization is the fact that they are contractions, not that they are edge-preserving. Indeed, we can define a *weak retract* of a graph $G$ as a graph $R$ such that there exist contractions $\rho : G \longrightarrow R$ and $\gamma : R \longrightarrow G$ satisfying $\rho \circ \gamma = \mathrm{id}_R$. The metric properties defining median graphs can then be relaxed to a definition of *quasi-median graphs*, which are the weak retracts of Hamming graphs (that is, cartesian products of complete graphs, see Chung, Graham, Saks [28] and Wilkeit [129]).

In view of these observations, the median graphs can also be characterized as the weak retracts of cubes. Thus, Theorem 5.40 not only provides a characterization of the retracts of cubes, but also shows that no other graphs are found if we allow contractions to serve as retractions. This is a bit surprising in view of the freedom gained when considering contractions rather than homomorphisms. However, note that median graphs are bipartite, and in the context of bipartite graphs, contractions and homomorphisms bear a special relationship to each other, as is shown by the following result.

**Proposition 5.41** *Let $\phi : G \longrightarrow H$ be a contraction, where $G$ and $H$ are bipartite graphs. Then there exists a bipartite graph $G'$ and a homomorphism $\psi : G \longrightarrow H \square G'$ such that $\phi = \mathrm{pr}_H \circ \psi$.*

**Proof** Put $F = \{[u, v] \in E(G) : \phi(u) = \phi(v)\}$, and let $\mathcal{P}$ be the partition of $V(G)$ whose cells are the connected components of $G - F$. Put $G' = G/\mathcal{P}$, and let the map $\psi : G \longrightarrow H \square G'$ be defined by $\psi(u) = (\phi(u), \pi(u))$, where $\pi : G \longrightarrow G'$ is the natural map. We then have $\phi = \mathrm{pr}_H \circ \psi$.

We show that $\psi$ is a homomorphism. Let $[u, v]$ be an edge of $G$. If $\phi(u) \neq \phi(v)$, then $[\phi(u), \phi(v)] \in E(H)$ and $\pi(u) = \pi(v)$, since $[u, v] \notin F$. Thus, $[\psi(u), \psi(v)] \in E(H \square G')$. Now suppose that $\phi(u) = \phi(v)$. Since $[u, v] \in E(G)$ and $G$ is bipartite, any $uv$-path $P$ has odd length. If $E(P) \cap F$ were empty, then $\phi(P)$ would be an odd closed trail in $H$, which is impossible since $H$ is bipartite. Therefore, $u$ and $v$ belong to different connected components of $G - F$, and $[\pi(u), \pi(v)] \in E(G')$. Thus, $[\psi(u), \psi(v)] \in E(H \square H')$.

Along the the same lines, it is easy to show that any cycle of $G'$ must have even length. Thus, $G'$ is also bipartite. □

We now turn our attention to isometric embeddings into strong products of paths. We begin with a pleasing result that is both general and simple.

**Proposition 5.42 (Quilliot [104], Nowakowski, Rival [97])** *Every connected graph admits an isometric embedding into a strong product of paths.*

**Proof** Let $G$ be a connected graph with diameter $n$, and $P = x_0, x_1, \ldots, x_n$ a path. For any $u \in V(G)$, we can define a map $\phi_u : G \longrightarrow P$ by putting $\phi_u(v) = x_k$, where $k = d_G(u,v)$. By the triangle inequality, $\phi_u$ is a contraction. We can then use the maps $\phi_u, u \in V(G)$, coordinatewise to define a contraction $\phi : G \longrightarrow \boxtimes_{u \in V(G)} P$. For any $v, w \in V(G)$, we then have

$$d_{\boxtimes_{u\in V(G)} P}(\phi(u), \phi(v)) = \max\{|d_G(u,v) - d_G(u,w)| : u \in V(G)\}.$$

This maximum is attained by putting $u = v$ (or $u = w$). Thus,

$$d_{\boxtimes_{u\in V(G)} P}(\phi(u), \phi(v)) = d_G(v, w),$$

and $\phi$ is an isometric embedding. □

As was the case for the cubes, a characterization of the weak retracts of strong products of paths is heavily based on some special metric properties. We use the following concepts.

**Definition 5.43** Let $G$ be a connected graph.

(i) For $u \in V(G)$ and an integer $r$, the *ball* of center $u$ and radius $r$ is the set

$$B(u,r) = \{v \in V(G) : d_G(u,v) \leq r\}.$$

(ii) $G$ is a *Helly graph* if the family of its balls has the Helly property, that is, for any family $B(u_1, r_1), \ldots, B(u_n, r_n)$ of pairwise intersecting balls, we have $\bigcap_{k=1}^{n} B(u_k, r_k) \neq \emptyset$.

In particular, all trees are Helly graphs, since their balls are subtrees, and these satisfy the Helly property. The balls of a strong product of graphs are products of balls, and a strong product of Helly graphs is a Helly graph. In particular, the strong products of paths are Helly graphs. Also, weak retractions preserve intersecting balls, so the weak retracts of Helly graphs are Helly graphs. All in all, the situation is very similar to the case of cubes and median graphs, and indeed, we get the following:

**Proposition 5.44 (Quilliot [104], Nowakowski, Rival [98])** *The Helly graphs are the weak retracts of strong product of paths.*

The proof of Proposition 5.44 is based on the following property.

**Theorem 5.45 (Extension Theorem, Quilliot [103])** *Let $G$ be a connected graph and $H$ a Helly graph. Let $S$ be a subset of $V(G)$ and $\phi_0 : S \longrightarrow V(H)$ a map such that $d_H(\phi_0(u), \phi_0(v)) \leq d_G(u,v)$ for all $u, v \in S$. Then there exists a contraction $\phi : G \longrightarrow H$ whose restriction to $S$ is $\phi_0$.*

**Proof** We construct $\phi$ by gradually expanding the domain $S$ of $\phi_0$. If $S = V(G)$, we are done. Otherwise, there exists $u \in V(G)\backslash S$. Let $S' = S \cup \{u\}$ and define $\phi_0' : S' \longrightarrow H$ as follows. Put $\phi_0'(v) = \phi_0(v)$ if $v \in S$, and notice that the balls $B(\phi_0(v), d_G(u,v)), v \in S$, are pairwise intersecting. Indeed, the condition $B(\phi_0(v_1), d_G(u,v_1)) \cap B(\phi_0(v_2), d_G(u,v_2)) \neq \emptyset$ is equivalent to the condition $d_H(\phi_0(v_1), \phi_0(v_2)) \leq d_G(u,v_1) + d_G(u,v_2)$, and the latter is fulfilled since $\phi_0$ does not increase distances. Since $H$ is a Helly graph, there exists $w \in \bigcap_{v\in S} B(\phi_0(v), d_G(u,v))$; put $\phi_0'(v) = w$. Then, $S'$ and $\phi_0'$ still satisfy the conditions of the theorem, so we can iterate our construction. Eventually, we get $S = V(G)$. □

We see that Proposition 5.42 and Theorem 5.45 imply Proposition 5.44. If G is a Helly graph, and $\psi : G \longrightarrow \boxtimes_{k=1}^n P$ an isometric embedding, then it is a full monomorphism, and we can define a partial inverse $\phi_0 : S = \phi(V(G)) \longrightarrow V(G)$ such that $\phi_0 \circ \phi(u) = u$ for all $u \in V(G)$. Since $G$ is a Helly graph, $\phi_0$ extends to a contraction $\psi : \boxtimes_{k=1}^n P \longrightarrow G$, so $G$ is a weak retract of $\boxtimes_{k=1}^n P$.

These results on Helly graphs show how the context of contractions and weak retractions strays from the subject of homomorphisms. By contrast, Theorem 5.33 states that the retracts of strong products of paths are strong products of (possibly shorter) paths. However, this topic also brought us close to one of the earliest results concerning the retracts of graphs.

**Theorem 5.46 (Hell [65])** *Let $T$ be a tree, $G$ a bipartite graph, and $\gamma : T \longrightarrow G$ an isometric embedding. Then $T$ is a retract of $G$.*

**Proof** Since any tree is a Helly graph, the partial inverse $\rho_0 : \gamma(V(T)) \longrightarrow T$ of $\gamma$ extends to a contraction $\rho : G \longrightarrow T$. Both $G$ and $T$ are bipartite, so by Proposition 5.41, there exists a bipartite graph $H$ and a homomorphism $\phi : G \longrightarrow T\square H$ such that $\rho = \mathrm{pr}_T \circ \phi$. We can then properly 2-colour $H$ to obtain a homomorphism $\psi : G \longrightarrow T\square K_2$. Since $\rho \circ \gamma = \mathrm{id}_T$, the vertices of $\psi \circ \gamma(T)$ have all the same second coordinate, say 0. We then define a map $\rho' : T\square K_2 \longrightarrow T$ as follows. Fix a vertex $u \in T$ with a neighbour $v$, and define $\sigma : T \longrightarrow T$ by taking $\sigma(w)$ as the neighbour of $w$ on the unique $uw$-geodesic, if $u \neq w$, and putting $\sigma(u) = v$. The map $\rho' : T\square K_2 \longrightarrow T$ is then defined by putting $\rho'((w,i)) = w$ if $i = 0$, and $\rho'((w,i)) = \sigma(w)$ if $i = 1$. Thus, $\rho' \circ \psi : G \longrightarrow T$ is a homomorphism such that $\rho' \circ \psi \circ \gamma = \mathrm{id}_T$, so $T$ is a retract of $G$. □

A graph $G$ is called an *absolute retract* of the bipartite graphs if it is a retract of any bipartite graph into which it admits an isometric embedding. For instance, Theorem 5.46 states that all trees are absolute retracts of the bipartite graphs. By analogy, Theorem 5.45 is a characterization of the Helly graphs as the 'absolute weak retracts' of the connected graphs. In fact, the characterization of the absolute retracts of the bipartite graphs is similar to our presentation of Helly graphs (see [11, 66]). Note that a categorical product of paths is not connected; however, the connected components of categorical products of paths play the same role in characterizing the absolute retracts of the bipartite graphs as do the strong products of paths in that of the 'absolute weak retracts' of the connected graphs: Every bipartite graph is an isometric subgraph of a connected component of a categorical product of paths, and the absolute retracts of the bipartite graphs are precisely the retracts of connected components of categorical products of paths.

# 6 Remarks

We have concentrated on finite graphs two reasons. The first is that when an infinite graph has finite homomorphic images, compactness comes to play. For example, if $H$ is finite, then an arbitrary graph $G$ admits a homomorphism to $H$ if and only if all of its finite subgraphs do (see [21, 51]).

The second reason for not considering infinite graphs in this paper is the work Bauslaugh ([15, 16, 17]) and the fact that cores are fundamental to the study of homomorphisms. It turns out that not only does Proposition 2.22 not generalize to infinite graphs, but many problems related to cores and homomorphisms are undecidable. We survey some of this work in [56]. Bauslaugh's work treats directed graphs, but the cores present a problem even for undirected ones. For an example, consider the ray whose vertices are the positive integers and each of whose edges connects two consecutive integers. For each $k > 1$, let $G_k$ be a distinct copy of the complete graph on $k$ vertices $\{v_1, \ldots, v_k\}$. Construct a graph $H$ by identifying $k$ with $v_k$ for $k \in \mathbb{N}$. We claim this graph has no core in the sense that there is no minimal retract. Indeed, any retraction can map $G_i$ to $G_j$ only if $i < j$ and so the retracts of $H$ are the infinite components of $H - k$, $k \in \mathbb{N}$. Clearly there is no minimal one.

We have also skipped some other interesting work. For example, the concept of a *rigid* graph, studied by Hell, Nešetřil, Babai and others (see, for example, [2, 64, 29, 8, 9], and also [102] for many more references), is a natural one: consider a graph whose only endomorphism is the identity. Brewster and MacGillivray [20] have looked at graphs with the property that every homomorphic image is a subgraph. Also, all the work, some quite recent by Hell, MacGillivray, Bang-Jensen ([12, 13]) and others, on homomorphisms of directed graphs is omitted, as is the large body of results on the complexity of what has become known as *$H$-colourings*, that is, problems of the existence of homomorphisms into given graphs $H$. Both the complexity and the undecidability considerations, as well as the related topics of homomorphisms into directed graphs are treated in [56].

**Acknowledgment.** The authors would like to express their gratitude to Gert Sabidussi not only for helpful comments concerning this manuscript, but also for the knowledge he has shared with them throughout the years.

# References

[1] H. L. Abbott, B. Zhou, The star chromatic number of a graph, *J. Graph Theory* **17** (1993), 349 – 360.

[2] M. E. Adams, J. Nešetřil, J. Sichler, Quotients of rigid graphs, *J. Combin. Theory Ser. B* **30** (1981), 351 – 359.

[3] M. O. Albertson, D. M. Berman, The chromatic difference sequence of a graph, *J. Combin. Theory Ser. B* **29** (1980), 1 – 12.

[4] M. O. Albertson, V. Booth, Homomorphisms of symmetric graphs, *Congr. Numer.* **53** (1986), 79 – 86.

[5] M. O. Albertson, K. L. Collins, Homomorphisms of 3-chromatic graphs, *Discrete Math.* **54** (1985), 127–132.

[6] M. O. Albertson, P. A. Catlin, L. Gibbob, Homomorphisms of 3-chromatic graphs II, *Congr. Numer.* **47** (1985), 19 – 28.

[7] L. Babai, C. Godsil, On the automorphism group of almost all Cayley graphs, *European J. Comb.* **3** (1982), 9 – 15.

[8] L. Babai, J. Nešetřil, High chromatic rigid graphs I, in: *Combinatorics* (A. Hajnal, V. T. Sós, eds.), Colloq. Math. Soc. János Bolyai **18**, North-Holland, Amsterdam, 1978, 53 – 60.

[9] L. Babai, J. Nešetřil, High chromatic rigid graphs II, in: *Algebraic and Geometric Combinatorics* (E. Mendelsohn, ed.), *Ann. Discrete Math.* **15** (1982), 55 – 61.

[10] H. J. Bandelt, Retracts of hypercubes, *J. Graph Theory* **8** (1984), 501–510.

[11] H. J. Bandelt, A. Dählmann, H. Schütte, Absolute retracts of bipartite graphs, *Discrete Appl. Math.* **16** (1987), 191–215.

[12] J. Bang-Jensen, P. Hell, and G. MacGillivray, The complexity of colouring by semi-complete digraphs, *SIAM J. Discrete Math.* **1** (1988), 281-298.

[13] J. Bang-Jensen, P. Hell, and G. MacGillivray, Hereditarily hard $H$-colouring problems, *Discrete Math.* **138** (1995), 75–92.

[14] I. Bárány, A short proof of Kneser's conjecture, *J. Combin. Theory Ser. A* **25** (1978), 325–326.

[15] B. Bauslaugh, Core-like properties of infinite graphs and structures, *Discrete Math.* **138** (1995), 101 – 111.

[16] B. Bauslaugh, Homomorphisms in infinite structures, Ph. D. Thesis, Simon Fraser University, 1994.

[17] B. Bauslaugh, The complexity of infinite $H$-colouring, *J. Combin. Theory Ser. B* **61** (1994), 141–154.

[18] J. A. Bondy, P. Hell, A note on the star chromatic number, *J. Graph Theory* **14** (1990), 479–482.

[19] J. A. Bondy, U. S. R. Murty, *Graph Theory With Applications*, Macmillan, London, 1976.

[20] R. Brewster, G. MacGillivray, Homomorphically full graphs, *Discrete Appl. Math.* **66** (1996), 23-31.

[21] N. de Bruijn, P. Erdős, A colour problem for infinite graphs and a problem in the theory of relations, *Nederl. Akad. Wetensch. Proc. Ser. A* **54** (1951), 371–373 (*Indag. Math.* **13**).

[22] F. Buckley, L. Superville, The $a$-jointed number and graph homomorphism problems, in: *The Theory and Applications of Graphs* (G. Chartrand et al., eds.), Wiley, New York, 1981, 149 – 158.

[23] S. A. Burr, P. Erdős, L Lovász, On graphs of Ramsey type, *Ars Combin.* **1** (1976), 167–190.

[24] P. Catlin, Graph homomorphisms onto the five-cycle, *J. Combin. Theory Ser. B* **45** (1988), 199 – 211.

[25] P. Catlin, Homomorphisms as generalizations of graph colouring, *Congr. Numer.* **50** (1985), 179 – 186.

[26] G. Chang, L. Huang, X. Zhu, The star-chromatic number of Mycielski's graphs, preprint, National Sun Yat-sen University, Taiwan, 1996.

[27] F. R. K. Chung, Z. Füredi, R. L. Graham, P. Seymour, On induced subgraphs of the cube, *J. Combin. Theory Ser. A* **49** (1988), 180–187.

[28] F. R. K. Chung, R. L. Graham, M. E. Saks, A dynamic location problem for graphs, *Combinatorica* **9** (1989), 111–131.

[29] V. Chvátal, P. Hell, L. Kučera, J. Nešetřil, Every finite graph is a full subgraph of a rigid graph, *J. Combin. Theory* **11** (1971), 284 – 286.

[30] C. R. Cook, A. B. Evans, Graph folding, *Congr. Numer.* **23** (1979), 305 – 314.

[31] D. G. Corneil, A. Wagner, On the complexity of the embedding problem for hypercube related graphs, *Discrete Appl. Math.* **43** (1993), 75–95 .

[32] W. Deuber, X. Zhu, Chromatic numbers of distance graphs, to appear in *Discrete Math.*

[33] W. Deuber, X. Zhu, Relaxed coloring of a graph, to appear in *Graphs and Combin.*

[34] W. Deuber, X. Zhu, Circular coloring of weighted graphs, *J. Graph Theory* **23** (1996), 365–376.

[35] G. A. Dirac, Homomorphism theorems for graphs, *Math. Ann.* **153** (1964), 69 - 80.

[36] D. Ž. Djoković, Distance-preserving subgraphs of hypercubes, *J. Combin. Theory Ser. B* **14** (1973), 263–267.

[37] D. Duffus, I. Rival, Graphs orientable as distributive lattices, *Proc. Amer. Math. Soc.* **88** (1983), 197–200.

[38] D. Duffus, B. Sands, R. E. Woodrow, On the chromatic number of the product of graphs, *J. Graph Theory* **9** (1985), 487–495.

[39] D. Duffus, N. Sauer, Lattices arising in categorial investigations of Hedetniemi's conjecture, *Discrete Math.* **152** (1996), 125–139.

[40] M. El-Zahar, N. Sauer, The chromatic number of the product of two 4-chromatic graphs is 4, *Combinatorica* **5** (1985), 121–126.

[41] P. Erdős, Graph theory and probability, *Canad. J. Math.* **11** (1959), 34 – 38.

[42] P. Erdős, C. Ko, R. Rado, Intersection theorems for systems of finite sets, *Quart. J. Math.* **12** (1961), 313–320,

[43] P. Erdős, J. Spencer, *Probabilistic Methods in Combinatorics*, Academic Press, New York, 1974.

[44] O. Favaron, personal communication.

[45] G. Gao, G. Hahn, Minimal graphs that fold onto $K_n$, *Discrete Math.* **142** (1995), 277 – 280.

[46] G. Gao, G. Hahn, H. Zhou, Star chromatic number of flower graphs, preprint, Université de Montréal, 1992.

[47] G. Gao, E. Mendelsohn, H. Zhou, Computing star chromatic number from related graph invariants, *J. Combin. Math. Combin. Comput.* **16** (1994), 87 – 95.

[48] G. Gao, X. Zhu, Star-extremal graphs and lexicographic product. *Discrete Math.* **152** (1996), 147–156.

[49] A. H. M. Gerards, Homomorphisms of graphs into odd cycles, *J. Graph Theory* **12** (1988), 73 – 83.

[50] C. D. Godsil, Problems in algebraic combinatorics, *Electron. J. Combin.* **2** (1995), Feature 1, approx. 20 pp. (electronic).

[51] W. H. Gottschalk, Choice functions and Tychonoff's theorem, *Proc. Amer. Math. Soc.* **2** (1951), 172.

[52] R. L. Graham, P. M. Winkler, On isometric embeddings of graphs, *Trans. Amer. Math. Soc.* **288** (1985), 527–536.

[53] D. Guan, X. Zhu, A coloring problem for weighted graphs, to appear in *Inform. Process. Lett.*

[54] R. Häggkvist, P. Hell, D. J. Miller, V. Neumann Lara, On multiplicative graphs and the product conjecture, *Combinatorica* **8** (1988), 63–74.

[55] G. Hahn, P. Hell, S. Poljak, On the ultimate independence ratio of a graph, *European J. Combin.* **16** (1995), 253–261.

[56] G. Hahn, G. MacGillivray, Graph homomorphisms II: Computational aspects and infinite graphs, preprint, Université de Montréal, 1997.

[57] G. Hahn, J. Širáň, A note on intersecting cliques, *J. Combin. Math. Combin. Comput.* **18** (1995), 57 – 63.

[58] A. Hajnal, The chromatic number of the product of two $\aleph_1$-chromatic graphs can be countable, *Combinatorica* **5** (1985), 137–139.

[59] F. Harary, S. Hedetniemi, Achromatic number of a graph, *J. Combin. Theory* **8** (1970), 154 - 161.

[60] F. Harary, S. Hedetniemi, G. Prins, An interpolation theorem for graphical homomorphisms, *Portugal. Math.* **26** (1967), 453–462.

[61] F. Harary. D. Hsu, Z. Miller, The bichromaticity of a tree, in: *Theory and Applications of Graphs* (Y. Alavi, D. R. Lick, eds.), Lecture Notes in Math. **642**, Springer-Verlag, Berlin, 1978, 236 - 246.

[62] S. Hedetniemi, Homomorphisms of graphs and automata, University of Michigan Technical Report 03105–44–T, 1966.

[63] Z. Hedrlín, P. Hell, C. S. Ko, Homomorphism interpolation and approximation, in: *Algebraic and Geometric Combinatorics* (E. Mendelsohn, ed.), *Ann. Discrete Math.* **15** (1982), 213 - 227.

[64] P. Hell, Rigid undirected graphs with given number of vertices, *Comment. Math. Univ. Carolinae* **9** (1968), 51 - 69.

[65] P. Hell, Rétractions de graphes, Ph. D. Thesis, Université de Montréal, 1972.

[66] P. Hell, Absolute retracts in graphs, in: *Graphs and Combinatorics* (R. A. Bari, F. Harary, eds.), Lecture Notes in Math. **406** (1974), 291–301.

[67] P. Hell, An introduction to the category of graphs, *Ann. N. Y. Acad. Sciences* **328** (1979), 120 - 136.

[68] P. Hell, D. Miller, Graphs with given achromatic number, *Discrete Math.* **16** (1976), 195 - 207.

[69] P. Hell, D. Miller, On forbidden quotients and the achromatic number, in *Congr. Numer.* **15** (1976), 283 - 292.

[70] P. Hell, D. Miller, Achromatic number and graph operations, *Discrete Math.* **108** (1992), 297 - 305.

[71] P. Hell, D. Miller, Graphs with forbidden homomorphic images, *Ann. N. Y. Acad. Sci.* **319** (1979), 270 - 280.

[72] P. Hell, J. Nešetřil, On the complexity of $H$-colourings, *J. Combin. Theory Ser. B* **48** (1990), 92 - 110.

[73] P. Hell, J. Nešetřil, The core of a graph, *Discrete Math.* **109** (1992), 117 - 126.

[74] P. Hell, J. Nešetřil, X. Zhu, Duality of graph homomorphisms, in: *Combinatorics, Paul Erdős is Eighty*, vol. 2, Bolyai Society Mathematical Studies, Budapest, 1996, 271–282.

[75] P. Hell, X. Yu, H. Zhou, Independence ratios of graph powers, *Discrete Math.* **127** (1994), 213–220.

[76] P. Hell, H. Zhou, X. Zhu, Homomorphisms to oriented cycles, *Combinatorica* **13** (1993), 421–433.

[77] P. Hell, H. Zhou, X. Zhu, Multiplicative oriented cycles, *J. Combin. Theory Ser. B* **60** (1994), 239–253.

[78] P. Hell, X. Zhu, Homomorphisms to oriented paths, *Discrete Math.* **132** (1994), 107–114.

[79] P. Hell, X. Zhu, The existence of homomorphisms to oriented cycles, *SIAM J. Discrete Math.* **8** (1995), 208–222.

[80] P. Hell, H. Zhou, X. Zhu, A note on homomorphisms to acyclic local tournaments, *J. Graph Theory* **20** (1995), 467–471.

[81] A. J. W. Hilton, E. C. Milner, Some intersection theorems for systems of finite sets, *Quart. J. Math. Oxford Ser. 2* **18** (1967), 369–384.

[82] F. Hughes, G. MacGillivray, The achromatic number of graphs: A survey and some new results, *Bull. Inst. Combin. Appl.* **19** (1997), 27-56.

[83] W. Imrich, J. Žerovnik, Factoring Cartesian-product graphs, *J. Graph Theory* **18** (1994), 557–567.

[84] W. Imrich, S. Klavžar, Retracts of strong product of graphs, *Discrete Math.* **109** (1992), 147–154.

[85] S. Klavžar, U. Milutinović, Strong products of Kneser graphs, *Discrete Math.* **133** (1994), 297–300.

[86] A. Kostochka, E. Sopena, X. Zhu, Acyclic and oriented chromatic numbers of graphs, to appear in *J. Graph Theory.*

[87] P. Křivka, On homomorphism perfect graphs, *Comment. Math. Univ. Carolinae* **12** (1971), 619 – 626.

[88] H. -J. Lai, Unique graph homomorphisms onto odd cycles, *Utilitas Math.* **31** (1987), 199 – 208.

[89] H. -J. Lai, Unique graph homomorphisms onto odd cycles II, *J. Combin. Theory Ser. B* **46** (1989), 363 – 376.

[90] B. Larose, F. Laviolette, C. Tardif, Normal Cayley graphs and homomorphisms of cartesian powers of graphs, preprint, Université de Montréal, 1994, submitted to *European J. Combin.*

[91] L. Lovász, Kneser's conjecture, chromatic number, and homotopy, *J. Combin. Theory Ser. A* **25** (1978), 319–324.

[92] H. A. Maurer, A. Salomaa, E. Welzl, On the complexity of the general colouring problem, *Inform. and Control* **51** (1981), 128 – 145.

[93] H. A. Maurer, A. Salomaa, D. Wood, Colorings and interpretations: a connection between graphs and grammar forms, *Discrete Appl. Math.* **3** (1981), 119 – 135.

[94] J. Nešetřil, Homomorphisms of derivative graphs, *Discrete Math.* **1** (1971), 257 – 268.

[95] J. Nešetřil, X. Zhu, Path homomorphisms, *Math. Proc. Cambridge Philos. Soc.* **120** (1996), 207–220.

[96] J. Nešetřil, X. Zhu, On bounded treewidth duality of graphs, *J. Graph Theory* **23** (1996), 151–162.

[97] R. Nowakowski, I. Rival, On a class of isometric subgraphs of a graph, *Combinatorica* **2** (1982), 79–90.

[98] R. Nowakowski, I, Rival, The smallest graph variety containing all paths, *Discrete Math.* **43** (1983), 235–239.

[99] M. Perles, personal communication via P. Hell.

[100] S. Poljak, Coloring digraphs by iterated antichains, *Comment. Math. Univ. Carolinae* **32** (1992), 209–212.

[101] S. Poljak, V. Rödl, On the arc-chromatic number of a digraph, *J. Combin. Theory Ser. B* **31** (1981), 190–198.

[102] A. Pultr, V. Trnková, *Combinatorial, Algebraic and Topological Representations of Groups, Semigroups and Categories*, North-Holland, Amsterdam, 1980.

[103] A. Quilliot, Un problème de point fixe sur les graphes, *Ann. Rev. Inst. Mat. Mexico*, 1981.

[104] A. Quilliot, On the Helly property working as a compactness criterion on graphs, *J. Combin. Theory Ser. A* **40** (1985), 186–193.

[105] I. Rival, Maximal sublattices of finite distributive lattices, *Proc. Amer. Math. Soc.* **37** (1973), 417–420.

[106] J. Rotman, *The Theory of Groups*, Allyn and Bacon, Boston, 1973.

[107] G. Sabidussi, Graphs with given group and given graph-theoretical properties, *Canad. J. Math.* **9** (1957), 515–525.

[108] G. Sabidussi, Vertex-transitive graphs, Monatsh. Math. **68** (1964), 426–438.

[109] G. Sabidussi, Graph derivatives, *Math. Z.* **76** (1961), 385 – 401.

[110] N. Sauer, X. Zhu, An approach to Hedetniemi's conjecture, *J. Graph Theory* **16** (1992), 423–436.

[111] N. Sauer, X. Zhu, Multiplicative posets, *Order* **8** (1992), 349–358.

[112] H. Shapiro, The embedding of graphs in cubes and the design of sequential relay circuits, Bell Telephone Laboratories Memorandum, 1953.

[113] S. Stahl, $n$-tuple colorings and associated graphs, *J. Combin. Theory Ser. B* **20** (1976), 185–203.

[114] S. Stahl, The multichromatic numbers of some Kneser graphs, preprint, University of Kansas, 1996.

[115] E. Steffen X. Zhu, Star chromatic numbers of graphs, *Combinatorica* **16** (1996), 439–448.

[116] L. Szamkołowicz, Remarks on the Cartesian product of two graphs, *Colloq. Math.* **9** (1962), 43–47.

[117] C. Tardif, A fixed box theorem for the cartesian product of graphs and metric spaces, to appear in *Discrete Math.*

[118] C. Tardif, Fractional multiples of graphs and the density of vertex-transitive graphs, preprint, Université de Montréal, 1996.

[119] C. Tardif, Graph products and the chromatic difference sequence of vertex transitive graphs, preprint, Université de Montréal, 1996.

[120] C. Tardif, Homomorphismes du graphe de Petersen et inégalités combinatoires, preprint, Université de Montréal, 1996.

[121] B. Toft, *Graph Colouring Problems*, Wiley-Intersci. Ser. Discrete Math. Optim., Wiley, New York, 1995.

[122] J. Turner, Point-symmetric graphs with a prime number of points, *J. Combin. Theory Ser. B* **3** (1967), 136–145.

[123] D. Turzik, A note on chromatic number of direct product of graphs, *Comment. Math. Univ. Carolinae* **24** (1983), 461 – 463.

[124] K. Vesztergombi, Chromatic number of strong products of graphs, in: *Algebraic Methods in Graph Theory*, vol. 2 (L. Lovász, V. T. Sós, eds.), Colloq. Math. Soc. János Bolyai **25**, North-Holland, Amsterdam, 1981, 819–825.

[125] A. Vince, Star chromatic number, *J. Graph Theory* **12** (1988), 551-559.

[126] J. Walker, From graphs to ortholattices and equivariant maps, *J. Combin. Theory Ser. B* **35** (1983), 171–192.

[127] E. Welzl, Color-families are dense, *J. Theoret. Comput. Sci.* **17** (1982), 29–41.

[128] E. Welzl, Symmetric graphs and interpretations, *J. Combin. Theory Ser. B* **37** (1984), 235 – 244.

[129] E. Wilkeit, The retracts of Hamming graphs, *Discrete Math.* **102** (1992), 197–218.

[130] B. Zelinka, Homomorphisms of finite bipartite graphs onto complete bipartite graphs, *Math. Slovaca* **33** (1983), 545 – 547.

[131] B. Zelinka, Homomorphisms of infinite bipartite graphs onto complete bipartite graphs, *Czechoslovak Math. J.* **32** (1982), 361 - 366.

[132] H. Zhou, Chromatic difference sequences and homomorphisms, *Discrete Math.* **113** (1993), 285 - 292.

[133] H. Zhou, The chromatic difference sequence of the cartesian product of graphs, *Discrete Math.* **90** (1991), 297 - 311.

[134] H. Zhou, The chromatic difference sequence of the cartesian product of graphs: Part II, *Discrete Appl. Math.* **41** (1993), 263 - 267.

[135] H. Zhou, Homomorphism properties of graphs, Ph. D. Thesis, Simon Fraser University, 1988.

[136] H. Zhou, X. Zhu, On the multiplicativity of acyclic local tournaments, to appear in *Combinatorica.*

[137] X. Zhu, On the bounds for the ultimate independence ratio of graphs, *Discrete Math.* **156** (1996), 207 - 220.

[138] X. Zhu, Star chromatic numbers and products of graphs, *J. Graph Theory* **16** (1992), 557–569.

[139] X. Zhu, On the chromatic number of the products of hypergraphs *Ars Combin.* **34** (1992), 25–31.

[140] X. Zhu, A simple proof of the multiplicativity of directed cycles of prime power length, *Discrete Appl. Math.* **36** (1992), 313–315.

[141] X. Zhu, A note on graph reconstruction, to appear in *Ars Combinatoria.*

[142] X. Zhu, Circular chromatic number: a survey, preprint, National Sun Yat-sen University, Taiwan, 1997.

# Cayley graphs and interconnection networks

Marie-Claude HEYDEMANN

*Laboratoire de Recherche en Informatique*[*]
*Bâtiment 490*
*Université de Paris-Sud,*
*91405 Orsay*
*France*

Based on notes by
Bertrand DUCOURTHIAL

**Abstract**

Due to recent developments of parallel and distributed computing, the design and analysis of various interconnection networks has been a main topic of research for the past few years and is still stimulated by the new technologies of communication networks such as optic fibers. There are many advantages in using Cayley (di)graphs as models for interconnection networks. This work first surveys some classes of Cayley graphs which are well studied as models of interconnection networks. Results and problems related to routings in networks are then presented, with emphasis on loads of nodes and links in routings.

## 1 Introduction

As defined by D.F. Hsu in the introduction of a special issue of the journal *Networks* ([97]), "an interconnection network consists of hardware and software entities that are interconnected to facilitate efficient computation and communication. These entities can be in the form of processors, processes, memory modules or computer systems." Due to recent developments in parallel and distributed computing, the design and analysis of various interconnection networks has been a main topic of research for the past few years. An essential component of a supercomputer based on large-scale parallel processing is the interconnection network. It provides communication among the processors and the memories. Interconnection networks also play a key role in the design and implementation of communication networks and the recent advent of optic technology adds more design problems. This explains that a growing number of research articles and congresses (such as IWIN every two years since 1991) are devoted to interconnection networks, and that special issues of journals have already been published on the subject (for example [97], [17] and [107]).

Interconnection networks are often modeled by finite graphs (or digraphs). The vertices of the graph represent the nodes of the network, that is, processing elements, memory modules or switches, and the edges correspond to communication lines. What qualities are required in order for a graph to be a good model for interconnection networks ? A basic constraint

[*]UA 410, CNRS.



*G. Hahn and G. Sabidussi (eds.), Graph Symmetry,* 167–224.

in many network design problems is that a bound on the maximum node degree is imposed by cost and fundamental engineering limitations. That is, the nodes of the network can be connected by at most a fixed number of communication lines to other nodes. At the same time other properties are crucial for many applications which need efficient network communications. This leads to some well studied design problems. We can distinguish at least three categories of interconnection network problems which lead to well studied problems in graph theory:

(a) network topologies and analysis : find graphs or digraphs which model good interconnection networks with regard to quality measures such as diameter and connectivity. For example, consider the diameter (i.e. the maximum of the distance between two vertices in the graph) as a measure of the maximum communication delay. The *degree/diameter problem* is then to construct the largest possible graphs satisfying bounds on the maximum vertex degree and diameter (see Section 4).

(b) communication : efficiently implement distributed algorithms such as those for broadcasting, gossiping or accumulation on a specific network (a graph with a communication hypothesis).

(c) mapping and simulation : in order to efficiently emulate a network A on a network B, study mappings from a graph into another one (in particular homomorphisms which are arc-preserving mappings). This problem has received increasing attention from researchers in the last years, but the results concerning symmetric graphs are not as developped as for the first two problems.

Symmetry is important for all of these problems since it simplifies their study and allows the use of algebra, which is an efficient tool in proofs. In particular, there are many advantages in using Cayley (di)graphs. They have the property of node symmetry (vertex transitivity) which makes it possible to implement the same routing and communication schemes at each node of the network they model. Furthermore, some of them have other advantages such as line symmetry (edge transitivity), hierarchical structure (allowing recursive constructions), high fault tolerance and so on, as we will see in the following. This explains why the topology chosen for some parallel machines is or was a hypercube (for example the nCUBE2 is based on a hypercube of dimension 13) or a torus (like the parallel machine Fujitsu AP1000) despite their drawbacks, the degree for the former, the diameter for the latter. It also explains why hypercubes have probably been the most studied graphs for interconnection networks in the past decade, and why star graphs, toruses and other graphs generalizing hypercubes, are much studied in recent publications.

Most of the time, researchers first study some problems in a particular Cayley graph, for example one kind of communication problem in the hypercube (like [95] or [73]) or the star graph (like [72]). Then the same problem is studied in other Cayley graphs and, finally, the result is obtained in a more general class of Cayley graphs which includes the graphs studied to begin with (like [21] or [74]). Therefore, it seems important to specify and study large classes of Cayley graphs having good properties for given interconnection problems.

We cannot survey here all the problems of interconnection networks modeled by Cayley graphs. Good references on this subject are the first two published surveys [3] and [103]. In this course, we will focus on routing problems including connectivity, diameter and loads of routings. We place the emphasis on load problems, since new classes of vertex-transitive graphs (see quasi-Cayley graphs in Section 5.3) or edge-transitive graphs (see regular orbital graphs in Section 5.6) were studied in order to enlarge the class of graphs behaving well for loads of routings. More precisely, after recalling some basic definitions and properties in Section 2, we survey particular classes of Cayley graphs which are often studied as models of interconnection networks. First, we study Cayley graphs generated by transpositions (Section 3.1) and those with a special automorphism which rotates the generators and is called a complete rotation (Section 3.2). In Section 3.3, we present hierarchical Cayley digraphs and results about fault tolerance. Finally, multistage Cayley digraphs and their quotients are studied in Section 3.4. In a second part of this course (Section 4), we consider diameter and path problems. The last part is devoted to the specific problem of routing. The notion of load of routings was first studied in graph theory in [35] which introduced the parameter called (vertex-)forwarding index. This notion becomes more and more important, and Section 5 introduces the subject. One of the reasons why Cayley graphs behave well as models for networks is that they allow routings of shortest paths with uniform loads on the nodes. Generalizations of Cayley graphs with the same behaviour are then presented. A similar notion for load of edges instead of vertices is considered in Section 5.5 with the edge-forwarding index. Multicommodity flow theory can be useful here as shown in Section 5.7. As a conclusion, Section 6 gives some trends of research in this rich field. Finally, an Appendix contains the definitions and drawings of Cayley graphs not given in the text.

# 2 Definitions and basic properties

In this section, some basic definitions are given, and notation is introduced. We also recall some well-known properties of groups and Cayley graphs that we will use later. For more definitions and basic results on groups, we refer the reader for example to [117], [7], [32], for graph theory to [16], [28], [27], and for interconnection networks to [106], [47].

## 2.1 Groups and permutations

We only consider finite groups. Therefore the inverse of any element is a power of this element. The identity (unit element) of a group $G$ will be denoted by $I$ and the operation will be written as multiplication, except for abelian groups, where the operation is denoted additively and the unit element is 0. By abuse of notation, we use the same letter to denote a group and the subset of its elements and specify the law of the group only when a confusion can arise. Throughout $\mathbb{Z}_d$ will denote the group of integers modulo $d$. The elements of a subset $S$ of a group $G$ are called *generators of* $G$, and $S$ is said to be a *generating set*, if every element of $G$ can be expressed as a finite product of their powers. We will also say that $G$ is generated by $S$. We will denote by $Gr(S)$[1] the subgroup of a group $G$ generated by a subset $S$ of $G$. A subset $S$ of $G$ is said to be *unit free* if $I \notin S$. It is said to be *symmetric* (or closed under inverses) if $s \in S$ implies $s^{-1} \in S$.

The direct product of two groups $G$ and $G'$ is denoted by $G \times G'$.

A *permutation* $\sigma$ on the set $\{1, \dots, n\}$ is a bijective mapping (i.e. one-to-one and surjective) from $V$ to $V$. As usual, it is denoted by the images of the elements $(\sigma(1), \dots, \sigma(n))$. In a

[1] We do not use the classical notation $\langle S \rangle$ since we need it for transpositions and cyclic permutations.

product of permutations, $\sigma\tau$ signifies that we apply first mapping $\tau$ on the set $\{1,\ldots,n\}$ and then the mapping $\sigma$, i.e. $\sigma\tau = (\sigma(\tau(1)),\ldots,\sigma(\tau(n)))$. We denote by $\mathfrak{S}_V$ the group of all permutations on a set $V$, and by $\mathfrak{S}_n$ the group of all permutations on $\{1,\ldots,n\}$, also called the *symmetric group*. A *permutation group* on $\{1,\ldots,n\}$ (we will also say a group of permutations) is some subgroup $G$ of $\mathfrak{S}_n$.

For a permutation $\sigma$ on a set $X$, $Supp(\sigma)$ is the set of elements $i$ of $X$ such that $\sigma(i) \neq i$. The *degree of a permutation* $\sigma$ is the cardinality of $Supp(\sigma)$. If the permutation $\sigma$ is a *cycle* of length $\ell \geq 2$, we denote it by $\langle\sigma(a),\sigma^2(a),\ldots,\sigma^\ell(a)\rangle$, where $a$ is any element of $X$ such that $\sigma(a) \neq a$. For example, $\langle 3,5\rangle$ denotes the transposition of the elements 3 and 5.

The *stabilizer* of an element $x \in X$ in a permutation group $G$ on $X$ is the subgroup of $G$, defined by $Stab_G(x) = \{g \in G : g(x) = x\}$.

Let $G$ be any subgroup of $\mathfrak{S}_V$. The group $G$ *acts transitively* on $V$ if for any $u, v \in V$, there exists $f \in G$ such that $f(u) = v$. The permutation group $G$ is said to be *semi-regular* if the stabilizer of every vertex in $G$ is reduced to the identity. Finally, $G$ is said to be *regular* (or *acts regularly* on $V$) if $G$ acts transitively and is semi-regular, or equivalently, if for any two elements $u$ and $v$ of $V$, there exists a unique permutation $f \in G$ such that $f(u) = v$.

The following well-known theorem of Cayley allows one to consider only groups of permutations.

**Theorem 1** *Every finite group is isomorphic to a group of permutations.*

Thus, except for commutative groups, groups are mainly defined as groups of permutations in articles on interconnection networks as in [3], [103]. But it should be noticed that it can also be convenient to use a definition by generators and relators. This is the case in recent articles like [69], [53], [92].

## 2.2 Graphs and Cayley graphs

A *digraph* $\Gamma = (V, E)$ is defined by a set of *vertices* $V$ and a set $E$ of *arcs* or *oriented edges* which is a subset of elements $(u, v)$ of $V \times V$. If the subset $E$ is symmetric ($(u,v) \in E$ implies $(v,u) \in E$), we can identify two opposite arcs $(u,v)$ and $(v,u)$ to a single edge denoted by $uv$, or $[u,v]$ if $u$ and $v$ are composite. We then obtain a *graph*. The *out-degree* (respectively *in-degree*) of a vertex $u$ of a digraph $\Gamma$ is the number of arcs $(u,v)$ (respectively $(v,u)$) of $\Gamma$ and is denoted by $d^+(u)$ (respectively $d^-(u)$). A digraph $\Gamma$ is said to be *regular of out-degree* $d$ if $d^+(u) = d^-(u) = d$ for every vertex $u$ of $\Gamma$.

A *directed path* or *dipath* from vertex $u$ to vertex $v$ in a digraph $\Gamma$ is a sequence of vertices $x_0, x_1, \ldots, x_l$ with $x_0 = u$, $x_l = v$, such that $(x_i, x_{i+1})$ is an arc of the digraph for any $i$, $0 \leq i \leq l-1$. In such a case, $l$ is the *length* of the dipath, that is, the number of arcs it contains. The *distance* from vertex $u$ to $v$, denoted by $d(u,v)$, is the length of a shortest dipath from $u$ to $v$. A digraph is *strongly connected* if there is a dipath from $u$ to $v$ for all $u \neq v$ in $\Gamma$. The *diameter* of $\Gamma$ is $Diam(\Gamma) = \max_{(u,v)\in V\times V} d(u,v)$.

Let $G$ be a group and let $S$ be a subset of $G$. Then $Cay(G,S)$ denotes the Cayley[2] digraph of the group $G$ and the subset $S$. The vertices of $Cay(G,S)$ are the elements of $G$ and the

[2] Arthur Cayley (1821-1895) lived in his youth in St-Petersburg, studied in Cambridge, and became professor of Pure Mathematics at Cambridge in 1863 after he had been a lawyer during 14 years. He published over 900 papers and notes covering nearly every aspect of modern mathematics. The most important of his work is in developing the algebra of matrices. He also worked on non-euclidean geometry and n-dimensional geometry. He introduced the drawing in groups in 1878. At this time groups were considered mainly related to geometry.

arcs are all ordered pairs $(g, gs)$ for $g \in G, s \in S$. The arc $(g, gs)$ is said to be *labeled by* $s$. If $S$ is a generating set of $G$, then $Cay(G, S)$ is strongly connected. In this case, we will say that $Cay(G, S)$ is the *Cayley digraph of G generated by S*. If $S$ is unit free and closed under inverses, then $Cay(G, S)$ is a simple graph. The Appendix in Section 8 gives examples of well-known Cayley graphs. Notice that the definitions of a Cayley digraph given in the literature can differ. Many authors consider only the case where $S$ is a symmetric generating set of $G$. Furthermore, the terminology can also differ. For example, [27] speaks of *Cayley diagrams* instead of Cayley digraphs.

## 2.3 Homomorphisms

An *automorphism* of a digraph $\Gamma = (V, E)$ is a permutation $f$ of $V$ such that $(u, v)$ is an arc of $\Gamma$ if and only if $(f(u), f(v))$ is an arc of $\Gamma$. The automorphisms of $\Gamma$ form a subgroup of $\mathfrak{S}_V$ denoted by $Aut(\Gamma)$.

If $Aut(\Gamma)$ acts transitively on $V$, the digraph $\Gamma$ is said to be *vertex transitive*. In a Cayley digraph, the automorphisms $f_a : x \mapsto ax$, $a \in G$ (also called *translations*), form a subgroup of $Aut(\Gamma)$ which is isomorphic to the group $G$. This subgroup acts regularly on $V(= G)$ since for any two vertices $u$ and $v$ of $V$, there exists a unique element $a$ of $G$, $a = vu^{-1}$, such that $f_a(u) = v$. Therefore a Cayley digraph $\Gamma(V, E) = Cay(G, S)$ is vertex transitive. Furthermore, the group of translations allows a characterization of Cayley digraphs.

**Theorem 2 ([124])** *Let $\Gamma$ be a strongly connected digraph. Then $\Gamma$ is a Cayley digraph if and only if $Aut(\Gamma)$ contains a subgroup which acts regularly on $V$.*

The following results on automorphisms of Cayley graphs will be used in the next section.

**Proposition 3** *Let $G = Gr(S)$ and $G' = Gr(S')$ be two groups generated by $S$ and $S'$ respectively, and with the same cardinality $|G| = |G'|$. Let $h$ be a group homomorphism from $G$ into $G'$ such that $h(S) = S'$. Then $h$ is an isomorphism from the digraph $Cay(G, S)$ onto the digraph $Cay(G', S')$.*

**Proof** Let $x'$ be any element of $G'$. Then $x'$ can be written as a product of elements of $S'$, $x' = s'_1 s'_2 \cdots s'_k$, and since $h(S) = S'$, every element $s'_j$, $1 \le j \le k$, can be written as $s'_j = h(s_j)$, so that $x' = h(s_1)h(s_2)\cdots h(s_k) = h(x)$ with $x \in G$. Thus, the mapping $h$ is surjective. Since $|G| = |G'|$, $h$ is bijective, whence it follows from $h(S) = S'$ that $|S| = |S'|$, showing that the Cayley graphs have the same degree and consequently the same number of edges. Thus, it is sufficient to prove that $h$ is edge-preserving to finish the proof. For this, consider any two adjacent vertices $x$, $y$ of $Cay(G, S)$, with $y = xs$ for some element $s$ of $S$. By $h(y) = h(xs) = h(x)h(s)$, with $h(s) \in S'$, $h(y)$ and $h(x)$ are adjacent. □

By applying Proposition 3 to the case $S = S'$, we get the following well-known result.

**Corollary 4 ([146, 24])** *Let $h$ be a homomorphism of the group $G$ generated by $S$ into itself such that $h(S) = S$. Then $h$ is a graph automorphism of $Cay(G, S)$.*

A graph $\Gamma$ is said to be *edge transitive* if for any given pair of edges $uv, u'v'$, there exists an automorphism $f \in Aut(\Gamma)$ such that $f(\{u, v\}) = \{u', v'\}$. A digraph (symmetric or not)

Its definition used in 1878 the left product by a generator, but it used the right product in 1889... The first drawing of a Cayley graph concerned the group of the isometries of the tetrahedron. The usual form of Cayley digraphs is more that of Max Dehn (1878-1951) than that of Arthur Cayley.

is *arc transistive* (1-*arc-transitive* for some authors, *symmetric* in [24]) if for any given pair of arcs $(u,v),(u',v')$ there exists an automorphism $f \in Aut(\Gamma)$ such that $f(u) = u'$ and $f(v) = v'$. Moreover, a digraph $\Gamma$ is said to be *distance transitive* if for all vertices $u, v, u', v'$ of $V$ such that $d(u,v) = d(u',v')$, there exists an automorphism $f$ of $\Gamma$ such that $f(u) = u'$ and $f(v) = v'$. Figure 1 shows the implications between these definitions in the case of graphs. An arrow from one property to another one indicates that the former implies the latter. If an implication is not true, the name of a graph which is a counterexample is given besides a crossed arrow.

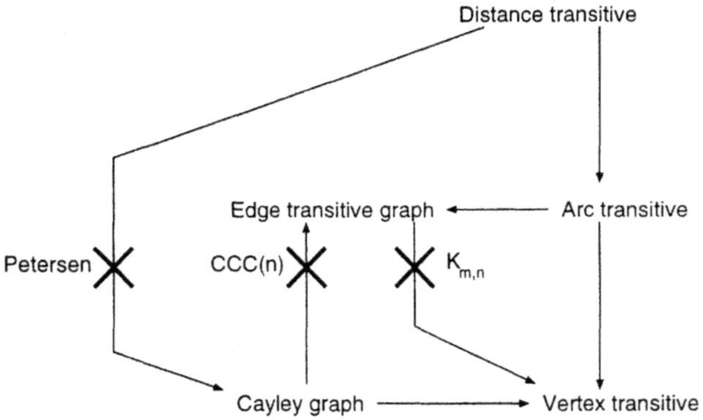


Figure 1: Relations between definitions

We now recall a condition which is sufficient for a Cayley digraph to be arc transitive.

**Proposition 5** **([3])** *A Cayley digraph $Cay(Gr(S),S)$ such that, for every pair $s, s'$ of $S$, there exists a group homomorphism $h$ of $Gr(S)$ satisfying $h(S) = S$ and $h(s) = s'$, is arc transitive.*

**Proof** Consider two arcs $(u,v)$ and $(u',v')$ of $Cay(Gr(S),S)$ with $v = us$, $s \in S$, and $v' = u's'$, $s' \in S$. Let $h$ be a group homomorphism satisfying $h(S) = S$ and $h(s) = s'$. By Corollary 4, $h$ is a graph automorphism. Consider $\phi$ defined by $\phi(x) = u'h(u^{-1}x)$. Then $\phi$ is a graph automorphism since it is the composition of the three automorphisms $x \mapsto u^{-1}x$, $h$ and $y \mapsto u'y$. Furthermore, $\phi(u) = u'$, $\phi(v) = u'h(u^{-1}v) = u'h(s) = u's' = v'$. □

**Remark 6** *The converse of Proposition* 5 *is false.*

**Proof** (An error in [3]) Consider the complete graph $K_n$ which is arc transitive as the Cayley graph on the additive group $\mathbb{Z}_n$ generated by $S = \mathbb{Z}_n \setminus \{0\}$. Notice that any group homomorphism $h$ on $\mathbb{Z}_n$ satisfies $h(x) = h(\underbrace{1 + \cdots + 1}_{x}) = xh(1)$. If $h(S) = S$, then $h(1) \neq 0$ and $h(x) = ax$ with $a = h(1) \neq 0$. If for every pair $s, s'$ of $\mathbb{Z}_n \setminus \{0\}$, there exists $a \neq 0$, and $s' = as$, then in particular $1 = as$ for some $a \neq 0$. But if $n$ is not prime, not every element has a multiplicative inverse in $\mathbb{Z}_n \setminus \{0\}$, and therefore the condition of Proposition 5 cannot be fulfilled. This proves that the converse of Proposition 5 is not true for $K_n$ if $n$ is not prime. □

## 2.4 Products

We recall the definitions and notation of the products of graphs we will use later.

- The *cartesian product* of two digraphs $\Gamma = (V, E)$ and $\Gamma' = (V', E')$, denoted by $\Gamma \square \Gamma'$, is the digraph with vertex set $V \times V'$ and arc set

  $$\{((x,x'),(y,x')) : x' \in V',\ xy \in E\} \cup \{((x,x'),(x,y')) : x \in V,\ x'y' \in E'\}.$$

  For example, the hypercube $H(n)$ and the $n$-dimensional torus $MT_{n,k}$ (see definitions in Section 8) are cartesian products of $n$ complete graphs $K_2$ and of $n$ cycles $C_k$, respectively.

- The *weak product* (also known as the *categorical* or *cardinal* product) of two digraphs $\Gamma = (V, E)$ and $\Gamma' = (V', E')$, denoted by $\Gamma \times \Gamma'$, is the digraph with vertex set $V \times V'$ and arc set

  $$\{((x,x'),(y,y')) :\ xy \in E,\ x'y' \in E'\}.$$

- The *strong product* of two digraphs $\Gamma = (V, E)$ and $\Gamma' = (V', E')$, denoted by $\Gamma \boxtimes \Gamma'$, is the digraph with vertex set $V \times V'$ and arc set the union of the arc sets of $\Gamma \square \Gamma'$ and of $\Gamma \times \Gamma'$.

- The *lexicographic product* of two digraphs $\Gamma = (V, E)$ and $\Gamma' = (V', E')$, denoted by $\Gamma[\Gamma']$, is the digraph with vertex set $V \times V'$ and arc set

  $$\{((x,x'),(x,y')) : x \in V,\ x'y' \in E'\} \cup \{((x,x'),(y,y')) :\ x' \in V', y' \in V',\ xy \in E\}.$$

A digraph produced by any of the three products: cartesian product, weak product, lexicographic product, is vertex transitive if and only if each factor is vertex transitive [125], [126]. For factors which are Cayley graphs, the following results are easy to obtain.

**Proposition 7** *Let $\Gamma = Cay(G, S)$ and $\Gamma' = Cay(G', S')$ be two Cayley digraphs. Then*

- *the cartesian product $\Gamma \square \Gamma'$ is the Cayley digraph of the direct product $G \times G'$ and of the subset $(S \times I) \cup (I \times S')$;*

- *the weak product $\Gamma \times \Gamma'$ is the Cayley digraph of the direct product $G \times G'$ and of the subset $S \times S'$;*

- *the strong product $\Gamma \boxtimes \Gamma'$ is the Cayley digraph of the direct product $G \times G'$ and of the subset $(S \times I) \cup (I \times S') \cup (S \times S')$;*

- *the lexicographic product $\Gamma[\Gamma']$ is the Cayley digraph of the direct product $G \times G'$ and of the subset $(S \times G') \cup (I \times S')$.*

In Section 3.4.2 we will consider some Cayley digraphs of wreath products of groups.

### 2.4.1 Quotients

We recall here some classical definitions and fix our notation.

For any group $G$ and any subgroup $H$ of $G$, the *left cosets of* $H$ in $G$ are the subsets $gH$ of $G$ defined by $gH = \{gh : h \in H\}$. We denote by $(G/H)_l$ the set of the left cosets of $H$ in $G$. Similarly, the *right cosets of* $H$ in $G$ are the subsets $Hg$ of $G$ defined by $Hg = \{hg : h \in H\}$. We denote by $(G/H)_r$ the set of the right cosets of $H$ in $G$.

A *set of left coset representatives* of $H$ in $G$, also called *left transversal*, is a set of elements $\{g_i : 1 \leq i \leq r\}$ such that $G$ is the disjoint union of the left cosets $g_iH$, $1 \leq i \leq r$. A set of *right coset representatives* or *right transversal* is defined similarly. Given a Cayley digraph $Cay(G,S)$ and a subgroup $H$ of $G$, it is possible to construct two natural quotient graphs of $Cay(G,S)$.

The *left coset digraph of* $G$ with respect to a subgroup $H$ and to a subset $S$, denoted by $Cos_l(G,H,S)$, is the digraph with vertex set $(G/H)_l$, such that there exists an arc $(gH, g'H)$ if and only if there exists an element $s \in S$ and $ghs = g'h'$ for some $h, h' \in H$. This digraph can be seen as obtained from $Cay(G,S)$ by contracting the vertices of each left coset and deleting multiple arcs. For $H = \{I\}$, $Cos_l(G,H,S) = Cay(G,S)$.

The Sabidussi representation theorem [127] asserts that a (di)graph is vertex transitive if and only if it is a left coset (di)graph.

The *right coset digraph of* $G$ with respect to a subgroup $H$ and to a subset $S$, denoted by $Cos_r(G,H,S)$, is the digraph with vertex set $(G/H)_r$, such that there exists an arc $(Hg, Hg')$ if and only if there exists $s \in S$ and $Hgs = Hg'$. This digraph is also called a *Schreier coset graph* or a *Schreier diagram of* $G(\text{mod}H)$ [27]. For $H = \{I\}$, $Cos_r(G,H,S) = Cay(G,S)$. As we will see in Section 3.4.3 a right coset digraph may not be vertex transitive.

# 3 Special classes of Cayley graphs

## 3.1 Cayley graphs generated by transpositions

In this part we only consider Cayley digraphs $Cay(G,S)$, where $S$ is a set of transpositions on $\{1,2,\ldots,n\}$. This case is simpler than the general case since $s = s^{-1}$ for any $s \in S$. Thus $Cay(G,S)$ is a graph. It is called a *Cayley graph generated by transpositions.* This includes many of the frequently studied interconnection Cayley graphs such as hypercubes, (generalized) star graphs, complete transposition graphs, (modified) bubble sort graphs (see the definitions in Section 8 and other examples in [103]). Notice that a Cayley graph generated by transpositions is bipartite, each class of the bipartition being defined by the parity of the permutations.

Following [103], we define the transposition graph associated to $S$, and hence to the Cayley graph $Cay(Gr(S),S)$, as follows.

**Definition 8** Let $S$ be a set of transpositions of $\{1,2,\ldots,n\}$. The *transposition graph* of $S$, denoted by $T(S)$, is the graph with vertex set $\{1,2,\ldots,n\}$ and edges $ij$ for all $\langle i,j\rangle \in S$.

For example, the transposition graph associated to the star graph $ST(n)$ is the star $K_{1,n-1}$ (see Figures 10 and 11 which display the star graph $ST(4)$ and its associated transposition graph). The transposition graph associated to the bubble sort graph is a hamiltonian path on $n$ vertices. Other examples of transpositions graphs are given in the Appendix.

If $T(S)$ is a tree, it is called *a transposition tree*. This case is studied in [3] (see Theorem 31 in Section 3.3). For this case a useful result is the following lemma ([15], p.118).

**Lemma 9** *If the set of transpositions $S$ has cardinality $n-1$, then $Gr(S) = \mathfrak{S}_n$ if and only if $T(S)$ is a tree.*

Lemma 9 implies that if $T(S)$ contains a spanning tree, in other words, if $T(S)$ is connected, then $Gr(S) = \mathfrak{S}_n$. This is in particular the case for the generalized star graph defined in Section 3.2.

The problem is to get some properties of the Cayley graphs $Cay(G, S)$ generated by transpositions, using the characteristics of the associated transposition graphs $T(S)$. Following the results of [103], we will see below that in fact $T(S)$ characterizes $Cay(G, S)$, and that if $T(S)$ is edge transitive, then $Cay(G, S)$ is arc transitive. Our proofs differ from the proofs of [103].

We will use the following lemma to prove that $T(S)$ characterizes $Cay(G, S)$.

**Lemma 10** *Let $S$ and $S'$ be two sets of transpositions of $\{1, 2, \ldots, n\}$, with $s = \langle i, j\rangle \in S$ and $s' = \langle i', j'\rangle \in S'$. Assume that $f$ is an isomorphism from $T(S)$ onto $T(S')$ such that $\{f(i), f(j)\} = \{i', j'\}$. Then there exists a group automorphism $h$ of $\mathfrak{S}_n$ mapping $Gr(S)$ onto $Gr(S')$ such that $h(s) = s'$ and $h(S) = S'$.*

**Proof** Let $f$ be an isomorphism from $T(S)$ onto $T(S')$ satisfying the hypothesis. We can consider $f$ as a permutation of $\{1, 2, \ldots, n\}$ and define a mapping $h$ by $h(g) = fgf^{-1}$, for any permutation $g$ of $\mathfrak{S}_n$. Then $h$ is a group automorphism of $\mathfrak{S}_n$. Notice that for $t = \langle k, l\rangle$, we have $h(t) = \langle f(k), f(l)\rangle$. In particular, $h(s) = s'$. Since $f$ preserves edges, if $\langle k, l\rangle$ belongs to $S$, then $\langle f(k), f(l)\rangle$ belongs to $S'$. Thus $h(S)$ is a subset of $S'$ and $h$ is a group homomorphism from $Gr(S)$ into $Gr(S')$. Furthermore $h(S) = S'$, since $f$ is an isomorphism from $T(S)$ onto $T(S')$. □

**Corollary 11** **([103])** *Let $S$ and $S'$ be two sets of transpositions of $\{1, 2, \ldots, n\}$ such that the two graphs $T(S)$ and $T(S')$ are isomorphic. Then the Cayley graphs $Cay(Gr(S), S)$ and $Cay(Gr(S'), S')$ are also isomorphic.*

**Proof** Since the two graphs $T(S)$ and $T(S')$ are isomorphic, the groups $Gr(S)$ and $Gr(S')$ are also isomorphic. By Proposition 3, the group homomorphism $h$ defined in Lemma 10 is an isomorphism from $Cay(Gr(S), S)$ onto $Cay(Gr(S'), S')$. □

Recently C. Delorme [52] and J. Fournier [70] have proved the converse of Corollary 11.

**Theorem 12** **([52, 70])** *Let $S$, $S'$ be two sets of transpositions of $\{1, \ldots, n\}$. Then the graphs $T(S)$ and $T(S')$ are isomorphic if and only if the Cayley graphs $Cay(Gr(S), S)$ and $Cay(Gr(S'), S')$ are isomorphic.*

**Corollary 13** **([103])** *If the transposition graph $T(S)$ is edge transitive, then the Cayley graph $Cay(G, S)$ is arc transitive.*

**Proof** Apply Lemma 10 with $S = S'$ and then Proposition 5. □

Finally, let us recall that the conjecture affirming that any Cayley graph $Cay(G, S)$ is hamiltonian is proved if $G = \mathfrak{S}_n$ and $S$ is a set of transpositions, but open if $G = \mathfrak{S}_n$ and every element of $S$ is of order 2 (see the survey [42]).

### 3.2 Cayley digraphs with the rotation property

In order to get more properties of symmetry on Cayley graphs and easier communication schemes in networks, a class of graphs with special graph automorphisms called *rotations* have been considered independently by P. Fragopoulou and S. Akl in [71], [74] and by J.-C. Bermond, T. Kodate and S. Perennes in [21]. The notion of rotation in graph theory was first used in the context of imbeddings (see for example [25], [147]). In this context, a *rotation* of a graph $\Gamma = (V, E)$ is a cyclic permutation of all the vertices adjacent to a vertex, and a *rotation scheme* is a collection $\{r_i, i \in V\}$, where $r_i$ is a rotation at the vertex $i$. It is used to imbed the graph $\Gamma$ into a surface. For a Cayley graph, any cyclic permutation $r$ of the generators allows one to define a rotation scheme by $r_i(j) = ir(i^{-1}j)$ for any edge $ij$ (see [25], page 117).

The notion of *complete rotation* we now define is related, but different.

**Definition 14 ([21])** Let $S = \{s_i, 0 \leq i \leq d-1\}$ be a subset of a group $G$ with $G = Gr(S)$. A mapping $\omega : G \to G$ is a *complete rotation* of $(G, S)$ if it is bijective and satisfies, for any $x \in G$ and any $i \in \mathbb{Z}_d$, the following two properties:

$$\omega(I) = I, \quad (1)$$

$$\omega(xs_i) = \omega(x)s_{i+1}. \quad (2)$$

Some properties of complete rotations are listed here. Some of them are used in [21],[71] and [74]. A study of complete rotations can also be found in [92].

**Lemma 15** *A mapping $\omega$ is a complete rotation of $(G, S)$ if and only if it is a group automorphism of $Gr(S)$ which cyclically permutes the elements of $S$.*

**Proof** Let $\omega$ be a complete rotation of $(G, S)$. By taking $x = I$ in the equation (2) of Definition 14 and using the equation (1), we get $\omega(s_i) = s_{i+1}$. By induction on the number of terms in the product of generators which gives an element of $G$, we get from condition (2) of Definition 14 that $\omega(xy) = \omega(x)\omega(y)$. Thus $\omega$ is a bijective group homomorphism. Conversely, it is obvious that a group automorphism of $Gr(S)$ which cyclically permutes the elements of $S$ is a complete rotation of $(G, S)$. □

By applying Corollary 4, we get that a complete rotation $\omega$ of $(G, S)$ induces a graph homomorphism and therefore a graph automorphism of $Cay(Gr(S), S)$. In the following, we will also say that the graph $Cay(Gr(S), S)$ *has a complete rotation.*

A digraph $\Gamma$ such that there exists a group $G$ and a subset $S$ with $\Gamma = Cay(G, S)$ and such that $(G, S)$ has a complete rotation is said to be *rotational.* Most popular Cayley digraphs used for interconnection networks such as the hypercube, the generalized hypercube, the star graph, the modified bubble sort graph, the $n$-dimensional torus with the same number of vertices for each dimension (see definitions given in Section 8) are rotational. In fact, for these Cayley digraphs, the set of generators $S = \{s_i, i \in \mathbb{Z}_d\}$ of the group $G$ (considered as a permutation group on a set $X$) satisfies $s_{i+1} = \sigma s_i \sigma^{-1}$ for some permutation $\sigma$ on $X$ (depending on the digraph) and any $i \in \mathbb{Z}_d$, or equivalently, $s_i = \sigma^i s_0 \sigma^{-i}$. In such a case $\omega(x) = \sigma x \sigma^{-1}$ is a complete rotation as noticed in [21] and [74].

We use this property to define a complete rotation for a new class of Cayley graphs, the generalized star graphs. Other examples can be found in the Appendix and in [71] and [74].

**Definition 16** Let $k$ and $n$ be two integers such that $1 \le k < n$. The *generalized star graph* $GST(n,k)$ is the Cayley graph $Cay(Gr(S),S)$, where $S$ is the set of the transpositions $\langle i,j\rangle, i \in \{1,\ldots,k\}, j \in \{k+1,\ldots,n\}$.

The transposition graph $T(S)$ associated to $GST(n,k)$ is the complete bipartite graph $K_{k,n-k}$. In the case $k=1$, $GST(n,1)$ is isomorphic to the star graph $ST(n)$ and $GST(2,1)$ to the graph $K_2$. Notice that $GST(4,2)$, which is depicted in Figure 12, is isomorphic to the modified bubble sort graph $MBS(4)$, since in both cases $T(S)$ is a cycle of length 4.

The graph $GST(n,k)$ is related to the *arrangement graph* $A(n,k)$ studied in [44]. The graph $A(n,k)$ is the graph with vertex set the set of arrangements of $k$ elements chosen out of $n$ elements in which an edge connects two vertices if and only if they correspond to arrangements differing in exactly one position. Thus $A(n,k)$ is obtained from $GST(n,k)$ by contracting into one vertex all the vertices of $GST(n,k)$ which are permutations of $\mathfrak{S}_n$ which fix $i$, $1 \le i \le k$ (and deleting loops and multiple edges). In other words, $A(n,k)$ is the left coset graph of $\mathfrak{S}_n$ with respect to the subgroup $H$ and the subset $S$, where $H = \{\sigma \in \mathfrak{S}_n : Supp(\sigma) = \{k+1,\ldots,n\}\}$.

**Proposition 17 ([92])** *If $k$ and $n-k$ are relatively prime, then the generalized star graph $GST(n,k)$ is rotational.*

**Proof** Consider the permutation $\sigma$ defined on $\{1,\ldots,n\}$ by

$$\sigma = \langle 1,2,\ldots,k\rangle\ \langle k+1,k+2,\ldots,n\rangle.$$

Then $\sigma^{-1} = \langle k, k-1,\ldots,1\rangle\ \langle n,n-1,\ldots,k+1\rangle$. For $1 \le i \le k$, $k+1 \le j \le n$,

$$\sigma\langle i,j\rangle\sigma^{-1} = \langle 1+i \pmod k,\ k+1+(j-k) \pmod{n-k}\rangle.$$

So let $\omega$ be the group automorphism defined by $\omega(x) = \sigma x \sigma^{-1}$. Since $n-k$ and $k$ are relatively prime, the order of $\sigma$ is the least common multiple of $k$ and $n-k$, that is $k(n-k)$, and condition (2) of Definition 14 is fulfilled. Thus $\omega$ is a complete rotation. □

Not all pairs $(Gr(S),S)$ have a complete rotation, as can be seen in the following example.

**Example 18** *Let $(G,S)$ be $(\mathbb{Z}_n, \mathbb{Z}_n\backslash\{0\})$. Then $(G,S)$ has a complete rotation if and only if $n$ is prime.*

**Proof** A nonzero group homomorphism $\omega$ of the additive group $\mathbb{Z}_n$ satisfies $\omega(x) = x\omega(1)$, so that if $\omega(1) = a$, then by induction $\omega(x) = ax$, and the set of generators $\mathbb{Z}_n \setminus \{0\}$ is equal to $\{1, a, a^2, \ldots a^{n-2}\}$. But this is possible if and only if $n$ is prime. □

Note that this example does not give all the values of $n$ for which the complete graph $K_n$ is rotational. For this, we have to consider $K_n = Cay(G, G \setminus \{I\})$, where $G$ is any group.

**Proposition 19 ([112])** *The complete graph $K_n$ is rotational if and only if $n$ is a power of a prime.*

This result shows that a graph can have a complete rotation for some, but not for every structure as a Cayley graph.

By applying Proposition 5, we get the following corollary.

**Corollary 20** *Any rotational Cayley digraph is arc transitive.*

**Remark 21** Example 18 shows that there exist arc transitive Cayley digraphs which are not rotational.

If $G$ and $H$ are two rotational graphs, is the cartesian product $G \times H$ also rotational ? The following proposition was found independently by several people, in particular [11] and [92].

**Proposition 22** *Let $\Gamma$ be a rotational Cayley graph. Then, for any integer $n \geq 1$, the cartesian product of $n$ graphs isomorphic to $\Gamma$, $\Gamma_n = \underbrace{\Gamma \square \cdots \square \Gamma}_{n}$, is also rotational.*

**Proof** Assume $\omega$ is a complete rotation of $(G, S)$, with $S = \{s_i, i \in \mathbb{Z}_d\}$ and $\Gamma = Cay(G, S)$. We denote the vertices of $\Gamma_n$ by $(x_0, x_1, \ldots, x_{n-1})$. The $n \times d$ generators of $\Gamma_n$ are:

$$\begin{array}{rclrclcrcl}
t_0 &=& (s_0, I, \ldots, I) & t_1 &=& (I, s_0, \ldots, I) & \ldots & t_{n-1} &=& (I, \ldots, I, s_0) \\
t_n &=& (s_1, I, \ldots, I) & t_{n+1} &=& (I, s_1, \ldots, I) & \ldots & t_{2n-1} &=& (I, \ldots, I, s_1) \\
\vdots &&& \vdots &&&& \vdots && \\
t_{(d-1)n} &=& (s_{d-1}, I, \ldots, I) & & \ldots & & \ldots & t_{dn-1} &=& (I, \ldots, I, s_{d-1})
\end{array}$$

The rotation on $\Gamma_n$ is given by $\rho(x_0, x_1, \ldots, x_{n-1}) = (\omega(x_{n-1}), x_0, x_1 \ldots, x_{n-2})$. The mapping $\rho$ is a group homomorphism since

$$\begin{aligned}
\rho((x_0, x_1, \ldots, x_{n-1})(y_0, y_1, \ldots, y_{n-1})) &= \rho(x_0y_0, x_1y_1, \ldots, x_{n-1}y_{n-1}) \\
&= (\omega(x_{n-1}y_{n-1}), x_0y_0, \ldots, x_{n-2}y_{n-2}) \\
&= (\omega(x_{n-1})\omega(y_{n-1}), x_0y_0, \ldots, x_{n-2}y_{n-2}) \\
&= \rho(x_0, x_1, \ldots, x_{n-1})\rho(y_0, y_1, \ldots, y_{n-1}).
\end{aligned}$$

Furthermore, $\rho(t_i) = t_{i+1}$ for $0 \leq i \leq dn - 1$. □

Proposition 22 implies that cartesian products of isomorphic modified bubblesort graphs and cartesian products of isomorphic generalized star graphs are rotational. Recall that the hypercube is a particular case of a cartesian product of isomorphic (generalized) star graphs $GST(2, 1)$, so that Proposition 22 implies also that the hypercube is rotational.

The following results characterizes Cayley graphs generated by a set of transpositions which have a complete rotation.

**Theorem 23 ([92])** *The only Cayley digraphs $Cay(G, S)$, with $G$ a permutation group generated by a set $S$ of transpositions, such that $(G, S)$ has a complete rotation are cartesian products of isomorphic modified bubblesort graphs and cartesian products of isomorphic generalized stargraphs $GST(n, k)$ with $k$ and $n - k$ relatively prime.*

The proof is based on the following proposition.

**Proposition 24 ([92])** *Consider a set $S$ of $d$ transpositions and let $G$ be the permutation group generated by $S$. Then $(G, S)$ has a complete rotation if and only if the transposition graph $T(S)$ is the vertex disjoint union of isomorphic cycles $C_q$ or the vertex disjoint union of isomorphic complete bipartite graphs $K_{t,q}$ with $t$ and $q$ relatively prime.*

### 3.3 Hierarchical Cayley digraphs and fault tolerance

One important problem in parallel computers, and more generally in distributed systems, is the problem of fault tolerance. We only consider here the basic notion corresponding to connectivity of the graph or the digraph (see the survey of R. J. Faudree [67] and references in [58] for advanced concepts). Other notions of fault tolerance will be considered in Section 4.5.

The *vertex connectivity* of a non-complete (di)graph is the minimal number of vertices that must be removed in order to disconnect the (di)graph, while the *edge connectivity* of a (di)graph is the minimal number of edges (arcs) needed to break connectivity. A vertex transitive graph is maximally edge connected ([145], [110]), that is, its edge connectivity equals its degree. Similarly, every strongly connected vertex transitive digraph of out-degree $d$ is $d$-arc connected ([85]).

The *fault tolerance* of a (strongly) connected (di)graph $\Gamma$ is the maximum number $k$ such that, if any $k$ vertices are removed from $\Gamma$, the resulting subgraph is still (strongly) connected. The fault tolerance is equal to one less than the vertex connectivity of the (di)graph unless the (di)graph is complete, in which case it is equal to the vertex connectivity. A regular (di)graph $\Gamma$ of out-degree $d$ and fault tolerance $d-1$ is said to be *optimally fault tolerant.* We only present here results about undirected graphs.

**Proposition 25** **([4])** *Every connected vertex transitive graph of degree* 2, 3, 4 *or* 6 *is optimally fault tolerant.*

The following example gives a Cayley graph of degree 5 having vertex connectivity equal to one less than the degree, i.e. fault tolerance equal to two less than the degree.

**Example 26** **([2])** Consider the additive commutative group $\mathbb{Z}_4 \times \mathbb{Z}_2$ and the ordered set of generators $S = \{(2,0),(1,0),(3,0),(1,1),(3,1)\}$ (see Figure 2). The corresponding graph $Cay(G,S)$ has 8 vertices and is of degree 5. It is composed of two complete graphs $K_4$, one with vertices $(x,0)$, the other with vertices $(x,1)$, joined by edges all incident to one vertex of the set $\{(1,0),(1,1),(3,0),(3,1)\}$, which is therefore a cutset.

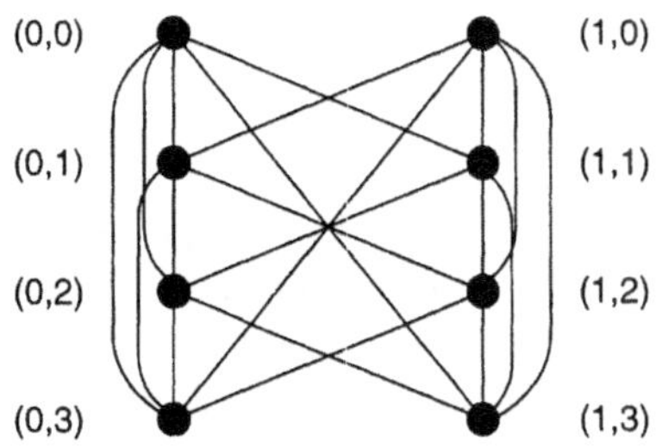


Figure 2: Example of graph having fault tolerance equal to two less than the degree

This example was generalized by B. Alspach ([4], see Theorem 32) to a family of graphs that was more precisely studied in [114]. In order to introduce a class of Cayley graphs which are optimally fault tolerant, we need the following definition.

**Definition 27** A *minimal generating set* $S$ of a group $G = Gr(S)$ is a unit free generating set such that $Gr(S \setminus \{s\})$ is a proper subgroup[3] of $G$ for any $s \in S$.

A *minimal symmetric generating set* $S$ of a group $G = Gr(S)$ is a unit free symmetric generating set such that $Gr(S \setminus \{s, s^{-1}\})$ is a proper subgroup of $G$ for any $s \in S$.

A Cayley digraph (resp. Cayley graph) $Cay(G, S)$ is said to be *strong hierarchical* if $S$ is a minimal generating set (resp. minimal symmetric generating set).

For example, the hypercube and the star graph are strong hierarchical Cayley graphs. C. Godsil proved that if $S$ is a minimal generating set, then the graph $Cay(G, S)$ is optimally fault tolerant [81]. The result of Godsil was extended to a larger class of Cayley graphs as we explain now.

**Definition 28** An ordered subset $S = \{s_1, s_2, \ldots, s_d\}$ of $G$ is said to be *quasiminimal* if $s_i \notin Gr(s_1, \ldots, s_{i-1})$, $2 \leq i \leq d$.

In other words, a minimal generating set is a quasiminimal generating set for any order of the generators.

We give the definition used in [4].

**Definition 29** A *quasiminimal symmetric generating set* is a unit free and symmetric generating set such that there exists a numbering of the generators in $S$, $s_1, s_2, \ldots, s_d$, such that

- if the order of $s_i$ is greater than 2, then $s_i^{-1}$ is either $s_{i-1}$ or $s_{i+1}$;
- if the order of $s_i$ is 2, then $Gr(s_1, s_2, \ldots, s_{i-1})$ is a proper subgroup of $Gr(s_1, s_2, \ldots, s_i)$;
- if the order of $s_i$ is greater than 2 and $s_i^{-1} = s_{i-1}$, then $Gr(s_1, s_2, \ldots, s_{i-2})$ is a proper subgroup of $Gr(s_1, s_2, \ldots, s_i)$.

For example, the generating set $S = \{(2,0), (1,0), (3,0), (1,1), (3,1)\}$ with the given order is a quasiminimal symmetric generating set of the group of Example 26. But it is not a minimal symmetric generating set since $Gr(S) = Gr(S \setminus (2,0))$.

This notion is related to the notion of *hierarchical* Cayley graph introduced in [1].

**Definition 30 ([1])** A Cayley graph $Cay(G, S)$ is said to be *hierarchical* if $S$ is a quasiminimal symmetric generating set of $G$.

By definition, a strong hierarchical Cayley graph is hierarchical, but the converse is false. For example, the pancake graph $PR(n)$ (defined in Section 8.9 and depicted in Figure 15) is a hierarchical Cayley graph, but not a strong hierarchical Cayley graph. Other examples of hierarchical Cayley graphs are given by Cayley graphs associated to a transposition tree, as shows the following theorem.

**Theorem 31 ([3])** *If the transposition graph $T(S)$ is a tree, then $\Gamma = Cay(G, S)$ satisfies the following properties.*

- *$\Gamma$ has $n!$ vertices and is bipartite;*

[3] A proper subgroup is a subgroup different from the group itself.

- *the connectivity of* $\Gamma$ *is equal to its degree;*
- *the edges of* $T(S)$ *can be ordered to have* $\Gamma$ *as a hierarchical Cayley graph. In this case,* $\Gamma$ *is obtained by interconnecting* $n$ *copies of a Cayley graph defined by a transposition tree of order* $n-1$.

Theorem 31 applies in particular to the star graph and the bubble sort graph.

Hierarchical Cayley (di)graphs are fault tolerant since their connectivity is equal to their degree, except for a particular family. Many authors have studied their connectivity. The strongest result is the following theorem found independently (and by different methods) by B. Alspach and M. Baumslag. It generalizes results of [81] on minimal generating sets, and of [1] on some hierarchical Cayley graphs.

**Theorem 32** **([4, 14])** *If* $S$ *is a quasiminimal symmetric generating set of the group* $G$*, then the graph* $Cay(G,S)$ *has connectivity* $|S|$*, unless* $Cay(G,S)$ *belongs to a particular family of graphs of connectivity* $|S|-1$.

This particular family contains graphs for which the quasiminimal generating set $S$ satisfies the following conditions:

- $S = \{s_1, s_2, s_2^{-1}, \ldots, s_k, s_k^{-1}\}$ (thus the degree is $2k+1$) is quasiminimal;
- $s_1$ is of order 2;
- for $2 \le i \le k$, $s_i$ is of order 4 with $s_i^2 = s_1$ (and therefore $s_i s_1 = s_1 s_i$).

These graphs are studied in [114].

The analogue of Theorem 32 for digraphs has been proved by Y. Hamidoune, A. Lladó and O. Serra [87].

Notice that the connectivity of circulant digraphs has been studied by Van Doorn [144]. Maximally connected circulant graphs have been characterized by F. Boesch and R. Tindell [26]. This article considered furthermore superconnected circulant graphs in the following sense. A *superconnected digraph* is a digraph in which every disconnecting set has cardinality at least $d+1$ (where $d = \min_{u \in V}\{d^+(u), d^-(u)\}$), unless it is the set of out-neighbours or the set of in-neighbours of a vertex. Y. Hamidoune, A. Lladó and O. Serra have also studied the superconnectivity of abelian Cayley digraphs [86] and the superconnectivity of hierarchical Cayley graphs [88]. In particular, they have proved the following theorem.

**Theorem 33** **([88])** *Let* $G$ *be a group containing a quasiminimal generating set* $M$ *such that* $|M| \ge 4$ *and, for* $S = M \cup M^{-1}$*, the connectivity of the Cayley digraph* $Cay(G,S)$ *is* $|S|$*. Then either* $Cay(G,S)$ *is superconnected or there exists a hierarchical ordering of* $M$ *with a minimal element* $u \in M$ *and an element* $v \in M$ *such that* $u^2 = v^2 = I$ *and* $s^2 = u$ *for any* $s \in S, s \neq u, s \neq v$.

Other results on hierarchical Cayley graphs improving the preceding ones can be found in [89] and [90].

The interest of hierarchical Cayley graphs is that they often allow inductive proofs by decomposing (stripping) the graph into smaller members of the same family. These graphs are *scalable* in the sense that they recursively consist of copies of smaller Cayley graphs of the same variety. For example, the pancake graph of dimension 4, $PR(4)$, with generators $S = \{(2,1,3,4), (3,2,1,4), (4,3,2,1)\}$ (see Section 8.9), is composed of 4 pancake graphs of dimension 3 connected together by edges labeled by the generator $(4,3,2,1)$.

Numerous results about communications (as in [83], [22], [38]) and embeddings (as in [29]) in the star graph $ST(n)$ or the pancake graph, $PR(n)$ have been proved by induction on $n$ using the recursive construction of these graphs. In fact, the proofs work for a class of hierarchical Cayley graphs which generalizes both the star graph and the pancake graph, and is defined as follows.

**Definition 34 ([83])** A Cayley graph on the symmetric group, $Cay(\mathfrak{S}_n, S)$, is *recursively decomposable* if $S$ is composed of $n-1$ permutations each of order 2, which can be arranged in a sequence $s_2, s_3, \ldots, s_n$, such that for every $s_i \in S$, $s_i(1) = i, s_i(i) = 1$, $s_i(j) = j$ for $i < j \leq n-1$.

To emphasize the importance of hierarchical Cayley graphs, we recall that J. Liu proves in [108] that there exists a hamiltonian decomposition of strong hierarchical Cayley graphs on abelian groups of odd order.

## 3.4 Multistage Cayley digraphs and quotients

### 3.4.1 Motivation

The hypercube is a model of interconnection networks with many good properties which are most of the time a consequence of the fact that it is a Cayley graph and a cartesian product. But it has a major drawback: its degree is $\log_2 N$, where $N$ is the number of vertices. So other graphs with the same number of vertices and the same diameter, but a constant degree, have been studied, such as the binary de Bruijn graph. In fact, a family of graphs, called *hypercubic graphs*, has been intensively studied in the literature with respect to the relations between them. Some of these graphs are called multistage graphs since they model dynamic networks in which nodes are commutators displayed in different levels (see [47] for a short introduction to multistage interconnection networks). We present here some of these hypercubic graphs and their relations, following [13] and [6].

### 3.4.2 Cube-connected cycles and butterfly graphs

Binary strings are used as vertex labels or components of vertex labels for all of the classes of graphs studied in this part. We use $x(i)$ or $x_0x_1\ldots\overline{x_i}\ldots x_{n-1}$ to denote the binary string obtained from $x = x_0x_1\ldots x_{n-1}$ by complementing the bit in position $i$. In this section, all arithmetic on indices and levels is assumed to be mod $n$, and all arithmetic on bits is assumed to be mod 2. For example, the hypercube $H(n)$ is here considered as a Cayley graph on the additive group $(\mathbb{Z}_2)^n$ with vertices denoted by $x = x_0x_1\ldots x_{n-1}$ with $x_i \in \mathbb{Z}_2, 0 \leq i \leq n-1$.

**Definition 35** The *cube-connected cycles graph* of dimension $n$, denoted by $CCC(n)$, is the graph with vertex set $\mathbb{Z}_n \times (\mathbb{Z}_2)^n$ and with edges defined as follows. Any vertex $(l, x)$, where $l \in \{0, \ldots, n-1\}$, $x = x_0x_1\ldots x_{n-1}$, $x_i \in \mathbb{Z}_2, 0 \leq i \leq n-1$, is adjacent to the vertex $(l+1, x)$ and to the vertex $(l, x(l))$, where $x(l) = x_0x_1\ldots x_{l-1}\overline{x_l}x_{l+1}\ldots x_{n-1}$.

For $n \geq 3$, the graph $CCC(n)$ is derived from $H(n)$ by replacing each vertex $x$ of $H(n)$ by a cycle of length $n$ (see $CCC(3)$ in Figure 16). Such a cycle is composed of the vertices labelled $(l, x)$, $0 \le l \le n-1$, where $l$ is called the *level* of the vertex. Each of the $n$ vertices $(l, x)$ of $CCC(n)$ that replaces the vertex $x$ of $H(n)$ inherits one of the $n$ edges that were incident to $x$ in $H(n)$, namely the edge $[(l, x), (l, x(l))]$, which is called an $H$-edge.

This graph has been introduced by Preparata and Vuillemin [116] as a model of very efficient network for algorithms of the "Ascend-Descend" class, which contains classical algorithms (sorting algorithms, Fast-Fourier-Transform algorithms, permutations, convolutions, matrix operations, ...). Furthermore, it has a good VLSI layout area. It also contains cycles of almost every length [80].

A directed version of $CCC(n)$, called in what follows the *digraph* $CCC(n)$, can be considered by replacing each $H$-edge by two opposite arcs and each edge $[(l, x), (l+1, x)]$ by the arc $((l, x), (l+1, x))$. The resulting digraph is regular of out- and in-degree 2.

The *butterfly* graph of dimension $n$ (also called *wrapped butterfly* by some authors), $BF(n)$, is derived from the hypercube in a similar way to $CCC(n)$. The difference is that $BF(n)$ has two $H$-edges corresponding to each former hypercube edge instead of one (Figure 18).

**Definition 36** The *butterfly graph* of dimension $n$, denoted by $BF(n)$, is the graph with vertex set $\mathbb{Z}_n \times (\mathbb{Z}_2)^n$ and with edges defined as follows. Any vertex $(l, x)$, where $l \in \{0, \dots, n-1\}$, $x = x_0 x_1 \dots x_{n-1}$, $x_i \in \mathbb{Z}_2, 0 \le i \le n-1$, is adjacent to the vertex $(l+1, x)$ and to the vertex $(l+1, x(l))$, where $x(l) = x_0 x_1 \dots x_{l-1} \overline{x_l} x_{l+1} \dots x_{n-1}$.

The edges incident with the vertex $(l, x)$ in $BF(n)$ and inherited from hypercube edges are $[(l, x), (l+1, x(l))]$ and $[(l, x), (l-1, x(l-1))]$. Similarly as for $CCC(n)$, the graph $BF(n)$ can be oriented as a regular digraph of out- and in-degree 2. Each edge $[(l, x), (l+1, y)]$ is replaced by an arc $((l, x), (l+1, y))$.

Both oriented versions of $BF(n)$ and $CCC(n)$ are Cayley digraphs on the group which is the *wreath product* $\mathbb{Z}_2 \wr \mathbb{Z}_n$, i.e. a particular semi-direct product of $\mathbb{Z}_n$ by $\mathbb{Z}_2^n$. The elements of this group $G$ are $(l, x)$, where $l \in \mathbb{Z}_n$ and $x = x_0 x_1 \dots x_n$ with $x_i \in \mathbb{Z}_2$ for $0 \le i \le n-1$. The identity of the group is $(0, 0^n)$. The homomorphism $h$ from the additive group $\mathbb{Z}_n$ into the automorphisms of the additive group $\mathbb{Z}_2^n$ used in the semi-direct product is defined by $l \mapsto \{y \mapsto \sigma^l(y)\}$, where $\sigma(y_0 y_1 \dots y_{n-1}) = (y_{n-1} y_0 \dots y_{n-2})$, and therefore the product is given by:

$$(l, x)(l', x') = \left(l + l', x + \sigma^l(x')\right) = (l + l', (x_0 + x'_{n-l})(x_1 + x'_{n-l+1}) \dots (x_{n-1} + x'_{n-l-1}))\,.$$

It is easy to prove the following result.

**Lemma 37** **([5])** *The digraph $BF(n)$ is a Cayley digraph $Cay(G, S)$ for $S = \{(1, \underbrace{0 \dots 0}_{n}),$ $(1, 1\underbrace{0 \dots 0}_{n-1})\}$.*
*The digraph $CCC(n)$ is a Cayley digraph $Cay(G, S')$ for $S' = \{(1, \underbrace{0 \dots 0}_{n}), (0, 1\underbrace{0 \dots 0}_{n-1})\}$.*

By definition, the hypercube $H(n)$, considered as a symmetric digraph, is a quotient of both graphs $BF(n)$ and $CCC(n)$, obtained by contraction using the mapping $(l, x) \mapsto x$ and deleting loops. $H(n)$ can also be viewed as the left coset digraph of $G$ with respect to the subgroup $H = \{(l, 0^n) : l \in \mathbb{Z}_n\}$ and to $S$ or $S'$,

$$H(n) = Cos_l(G, H, S) = Cos_l(G, H, S').$$

We will see in the next section what is the right coset graph of $G$ with respect to $H$ and $S$.

We finish this presentation by a few remarks on these digraphs.

Despite the similarity of the digraphs $CCC(n)$ and $BF(n)$, it has been proved only recently that $CCC(n)$ is a subgraph of $BF(n)$ [68].

A generalization of these Cayley digraphs is defined similarly on the wreath product $\mathbb{Z}_d \wr \mathbb{Z}_n$ (see [6] and [13]).

There exist other multistage graphs defined in the literature, like the *baseline graph* and the *omega graph*, but they are isomorphic to the butterfly graph (see [47] and the references given there).

### 3.4.3 Group action graphs

We present here mainly the results of [6].

**Definition 38 ([6])** Given a set $V$ and a set $\Pi$ of permutations on $V$, the *group action graph* ($GAG$ for short) defined by $V$ and $\Pi$, and denoted by $GAG(V, \Pi)$, is the digraph with vertex set $V$ and arcs $(x, \pi(x)), x \in V, \pi \in \Pi$.

In such a case, we will say that the arc $(x, \pi(x))$ is labelled by $\pi$. To illustrate this definition, we consider the de Bruijn digraph.

**Definition 39** The *de Bruijn digraph* $B(d, n)$ is the digraph with vertex set $V = \mathbb{Z}_d^n$ and arc set given by $E(B(d,n)) = \{(x_0x_1 \dots x_{n-1}, x_1 \dots x_{n-1}x_n) : x_i \in \mathbb{Z}_d, 0 \le i \le n\}$.

This digraph is regular of out-degree $d$, but not vertex transitive (only $d$ vertices, $i^n, 0 \le i \le d-1$, are incident with loops). It has been intensively studied and proposed as a model of parallel architecture. Many of its good properties come from its being an iterated line-digraph (see [47] and references cited here).

For the sake of simplicity here we only consider the binary de Bruijn digraph $B(2, n)$ (see Figure 3).

**Example 40** The *binary de Bruijn digraph* of order $n$, $B(2, n)$, is a GAG, by taking $V = \mathbb{Z}_2^n$ and $\Pi = \{\pi_0, \pi_1\}$ with

$$\pi_0(x_0x_1 \dots x_{n-1}) = x_1 \dots x_{n-1}x_0 \text{ and } \pi_1(x_0x_1 \dots x_{n-1}) = x_1 \dots x_{n-1}\overline{x_0}.$$

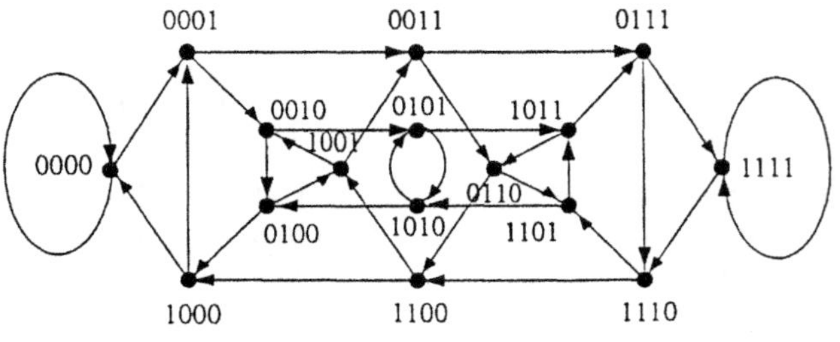


Figure 3: Binary de Bruijn digraph: an example of a GAG

**Lemma 41** ([6])

(i) *Every GAG is regular of out-degree* $|\Pi|$.

(ii) *Every regular digraph is a GAG.*

(iii) *Every connected GAG is strongly connected.*

**Proof** (i) is evident by definition. (ii) results from Proposition 74 in Section 5.3. (iii) can be proved by using the fact that a connected GAG has an eulerian circuit. □

Let $\Pi$ be a set of permutations on $V$. We now consider $Gr(\Pi)$, the group of permutations on $V$ generated by $\Pi$, and the Cayley digraph $Cay(Gr(\Pi),\Pi)$. The digraph $Cay(Gr(\Pi),\Pi)$ is said to be the *Cayley digraph associated to* $GAG(V,\Pi)$ .

The following theorem shows how GAGs and Cayley digraphs are related. Recall that the stabilizer of a vertex $v \in V$, $Stab_{Gr(\Pi)}(v) = \{\sigma \in Gr(\Pi) : \sigma(v) = v\}$, is a subgroup of $Gr(\Pi)$.

**Theorem 42** ([6]) *Every connected* $GAG(V,\Pi)$ *is isomorphic to the right coset digraph of* $G = Gr(\Pi)$ *with respect to* $Stab_{Gr(\Pi)}(v_0)$ *and* $\Pi$, *where* $v_0$ *is any vertex of* $V$.

**Proof** Following [6], for convenience of notation, we adopt here a different sense for products: $\pi_1 \cdot \pi_2$ is the permutation such that $\pi_1 \cdot \pi_2(u) = \pi_2(\pi_1(u))$.

If $GAG(V,\Pi)$ is connected, then $GAG(V,\Pi)$ is strongly connected by Lemma 41. This implies that $Gr(\Pi)$ acts transitively on $V$ since for any vertices $u$, $v$ of $V$ there exists a path from $u$ to $v$. If the arcs of this path are labelled from $u$ to $v$ successively by $\pi_1, \pi_2, \ldots, \pi_l$, where $\pi_i \in \Pi$, $1 \le i \le l$, then $v = \pi_l(\pi_{l-1}(\cdots(\pi_1(u))\cdots))$. Therefore, $v = g(u)$ with $g = \pi_1 \cdot \pi_2 \cdot \ldots \cdot \pi_l$, $g \in Gr(\Pi)$ (notice here the order of the product).

Let $v_0$ be any vertex of $V$ and $H = Stab_{Gr(\Pi)}(v_0)$. An isomorphism from $(G/H)_r$ to $GAG(V,\Pi)$ results from the mapping $\phi$ defined as follows :

$$\begin{array}{ccl} G & \longrightarrow & V \\ g & \longmapsto & v_0 g = g(v_0). \end{array}$$

In fact, $\phi(g) = \phi(g')$ if and only if $v_0 g = v_0 g'$, or $gg'^{-1} \in H$, i.e. $Hg = Hg'$. Furthermore, $\phi$ is surjective since $Gr(\Pi)$ acts transitively on $V$. Thus the mapping $Hg \in Cos_r(G,H,\Pi) \mapsto g(v_0) \in GAG(V,\Pi)$ is bijective. It is arc preserving since if $Hg' = Hg\pi$, then $\pi(g(h(v_0))) = \pi(g(v_0)) = g'(v_0)$ and $(g(v_0), g'(v_0))$ is an arc in $GAG(V,\Pi)$. Conversely we have that any arc $(g(v_0), \pi(g(v_0)))$ is the image of an arc $(Hg, Hg \cdot \pi)$ of $Cos_r(G,H,\Pi)$. □

**Example 43** ([6]) The binary butterfly digraph $BF(n)$ of dimension $n$ is the Cayley digraph associated to the de Bruijn digraph $B(2,n)$ considered as the GAG of Example 40.

Taking $v_0 = 0^n$, the mapping $\phi$ of the proof of Theorem 42 is defined in the case of $BF(n)$ by

$$(l, x_0 x_1 \cdots x_{n-1}) \mapsto x_l x_{l+1} \cdots x_{l-1}.$$

Thus, keeping the notation of Lemma 37, the right coset digraph $Cos_r(G,H,S)$, which is a quotient of $BF(n)$, is $B(2,n)$ which is not vertex transitive. Similarly, $Cos_r(G,H,S')$ is a non-transitive digraph, different from $B(2,n)$, and called the *shuffle-exchange digraph* of dimension $n$ (see [6]). This digraph is a subgraph of $B(2,n)$ [68].

To illustrate the interest of the relation between GAGs and Cayley digraphs and the power of algebraic tools, we give an application of the results cited in this section.

Notice that, since $Gr(\Pi)$ acts transitively on $V$, all the stabilizers of elements of $V$ are conjugate subgroups of $Gr(\Pi)$ and therefore are all isomorphic.

**Theorem 44** **([6])** *If the group action graph $GAG(V,\Pi)$ has a subgraph $T$ which is a directed tree, then the Cayley digraph $Cay(Gr(\Pi),\Pi)$ contains $|Stab_{Gr(\Pi)}(v_0)|$ vertex-disjoint directed trees isomorphic to $T$, where $v_0$ is any vertex of $GAG(V,\Pi)$.*

**Corollary 45** **([6])** *The butterfly digraph $BF(n)$ contains $n$ vertex disjoint copies of a complete directed binary tree of height $n-1$.*

**Proof** This is an easy consequence of Theorem 44, Theorem 42, and Example 40, since the de Bruijn graph $B(2,n)$ contains a complete binary tree of height $n-1$. This tree can be easily constructed taking as the root $0^{n-1}1$, and adding, for each vertex $x_0x_1\ldots x_{n-1}$ already put in the tree, the two children $x_1\ldots x_{n-1}0$ and $x_1\ldots x_{n-1}1$. □

Corollary 45 was first proved by a complicated purely combinatorial proof (see [6]).

We conclude this part with an open problem on group action graphs.

**Definition 46** **([100])** The *Kautz digraph* $K(d,n)$ is the digraph with vertex set

$$V=\{x_0x_1\ldots x_{n-1}\in\mathbb{Z}_{d+1}^n : x_i\neq x_{i+1}, 0\leq i\leq n-2\}$$

and arc set given by

$$\{(x_0x_1\ldots x_{n-1},x_1\ldots x_{n-1}x_n) : x_0x_1\ldots x_{n-1}\in V, x_n\in\mathbb{Z}_{d+1}, x_{n-1}\neq x_n\}.$$

The Kautz digraph $K(d,n)$ is a Cayley digraph if and only if $n=1$ or $n=2$ and $d+1$ is a prime power [30]. For any integers $d,n$, $K(d,n)$ is regular of out- and in-degree $d$ and strongly $d$-connected, thus it is a GAG and a coset digraph.

**Problem 47** Give an explicit construction of the Cayley digraph $Cay(Gr(\Pi),\Pi)$ associated to $K(d,n)$ considered as $GAG(V,\Pi)$.

# 4 Routing and diameter problems

## 4.1 Paths and factoring

Let $Cay(G,S)$ be a Cayley digraph and $g,g'\in G$ two of its vertices. If $g'=gs_1s_2\ldots s_t$ with $s_i\in S, 1\leq i\leq t$, then $s_1,s_2,\ldots,s_t$ defines a path from vertex $g$ to vertex $g'$ with arcs labelled by $s_1,s_2,\ldots,s_t$. It is equivalent to finding a path from $g^{-1}g'=s_1s_2\cdots s_t$ to $I$. Thus, as noticed in [132], to represent an element $g$ in the group $Gr(S)$ as a word written with generators of $S$, is equivalent to finding a path in the Cayley digraph $Cay(G,S)$ from the vertex $I$ to the vertex $g$.

Rewriting techniques can be used to find a shortest path between two vertices if the presentation of the group is known [69]. Anyway, a minimal presentation of a finite group can be constructed starting from fundamental cycles in the Cayley graph (see [34] or [140]). Notice that the minimal word problem in groups (for a given element $g$, find $s_1, s_2, \dots, s_t$ with $s_i \in S, 1 \leq i \leq t$, and $t$ minimal) is known to be NP-hard [64] if the set of generators is not fixed in advance (more precisely, this problem is PSPACE-complete [99]).

In the case of permutation groups, the problem of factoring has been studied extensively. Algorithms that have been developed by group theorists for permutation groups can be used. One tool is the *Sim's factorisation* ([31], [132]). Let $G$ be a group of permutations on $X$. A *base* $B$ is a subset of $X$ such that if $g \in G$ satisfies $g(\beta) = \beta$ for every $\beta \in B$, then $g = I$. Thus any permutation of $G$ is defined by its action on $B$. Assume $B = \{\beta_1, \beta_2, \cdots, \beta_b\}$ and consider the subgroups $G_i$ of $G$,

$$G_i = \{g \in G : g(\beta_j) = \beta_j, 1 \leq j \leq i+1\}, i = 0, \dots, b-1.$$

$G_i$ is a subgroup of $G_{i-1}$. $S$ is said to be a *strong generating set* relative to the base $B$ if $S \cap G_i$ generates $G_i$ for any $0 \leq i \leq b-1$.

**Example 48 ([31])** Take for $G$ the group of the symmetries of the square generated by $S = \{s_1, s_2\}$ with $s_1 = \langle 1,2,3,4 \rangle$ and $s_2 = \langle 2,4 \rangle$. The order of $G$ is 8. A base is $B = \{1,2\}$, $G_0 = Stab_G(1) = Gr(s_2)$, $G_1 = Stab_G(1) \cap Stab_G(2) = I$, and $S$ is a strong generating set relative to $B$.

**Proposition 49 ([132])** *With the notation given above, let $U_i$ be a set of right coset representatives of $G_i$ in $G$. Then every element of $G$ has a unique representation as a product $u_1 u_2 \cdots u_b$, with $u_i \in U_i, 1 \leq i \leq b$.*

Sim's algorithm gives a factorisation of each group element as a unique product of coset representatives This property is used in [132] to find efficient routing algorithms for Cayley graphs $Cay(G,S)$ such that $S$ is a strong generating set, in particular, for the star graph and the pancake graph.

## 4.2 Diameter calculation

Let $\Gamma = Cay(G,S)$ be a Cayley digraph of order $n$ and out-degree $d$. Since $\Gamma$ is vertex transitive its diameter can be obtained by constructing a breadth-first search tree from any vertex of the graph, in particular $I$. Thus, the diameter is the maximum distance $d(I,g)$ from $I$ to a vertex $g$ of the graph. It can be computed in time in $O(dn)$ instead of $O(dn^2)$ for a general digraph.

Let $D = Diam(Cay(G,S))$ be the diameter of $Cay(G,S)$. Then we can express $D$ as follows:

$$D = \max_{g \in G} \{\min \{t : \exists s_1, s_2, \dots, s_t, g = s_1 s_2 \dots s_t, s_i \in S, \forall i,\ 1 \leq i \leq t\}\}.$$

In the case of a Cayley digraph $Cay(G,S)$, where $G$ is a group of permutations, the calculation of $d(\sigma, I)$ is equivalent to sorting the permutation $\sigma$ of $G$ using the generators in the minimum number of operations.

According to Cooperman *et al.* [40] it is important to observe that the computation of the diameter for large groups is more often limited by memory resources rather than by the CPU time. They give the example of the calculation of the diameter of the Cayley graph of the cube $2 \times 2 \times 2$, with 9 generators, which needs $3,674,160$ nodes to be stored in semiconductor memory. They succeed with 8 megabytes using techniques from [39] against 80 megabytes in a method using simple breadth-first search to construct a spanning tree for the graph.

## 4.3 Diameter of particular graphs

For transposition graphs we have the following theorem.

**Theorem 50 ([3])** *If the transposition graph $T(S)$ is a tree, then $\Gamma = Cay(Gr(S), S)$ satisfies*

$$Diam(\Gamma) \leq \max_{\pi \in Gr(S)} \{c(\pi) - n + \sum_{i=1}^{n} \delta(i, \pi(i))\},$$

*where $\delta$ is the distance in the tree $T(S)$, and $c(\pi)$ is the number of cycles in the permutation $\pi$ on $\{1, 2, \ldots, n\}$.*

This can be made precise for the star graph and the bubble sort graph.

**Proposition 51 ([1, 103])** *The diameter of the star graph $ST(n)$ is $\lceil 3(n-1)/2 \rceil$ for $n \geq 3$. The diameter of the bubble sort graph $BS(n)$ is $n(n-1)/2$.*

Despite the similarity between the star and pancake graphs, the diameter of the pancake graph $PR(n)$ is unknown for $n > 13$ (for example, the diameter of $PR(13)$ is 15). This problem is known as the *pancake flipping problem.* There exist bounds given by Gates and Papadimitriou.

**Proposition 52 ([76])** *The diameter of the n-pancake graph $PR(n)$ satisfies*

$$Diam(PR(n)) \leq \frac{5n+5}{3}$$

*for all $n$, and*

$$Diam(PR(n)) \geq \frac{17n}{16}$$

*for infinitely many $n$.*

A generalization of this problem to *burnt pancakes* has been studied in [91]. A vertex in the burnt pancake graph is a signed permutation $(a_1, a_2, \ldots, a_n)$, where $a_i \in X$ or $-a_i \in X$. It is conjectured that the furthest vertex from $I$ (or equivalently the hardest to sort permutation) is $-I$ with $-I = (-1, -2, \ldots, -n)$. If this is true, results of [91] give an upper bound equal to $3(n+1)/2$ for all $n$ congruent to 3 (mod 12) which is better than the one of proposition 52. D. Cohen and M. Blum have shown that $-I(n)$ can be sorted in $\frac{23}{14}n = 1.63n$ steps. The problem is also to find the furthest vertex from $I$ in the $n$-pancake graph. M. Heydari and H. Sudborough also showed that the problem of deciding, given two vertices $x$ and $y$ in the pancake graph and an integer $k$, whether there is a path of length $k$ connecting $x$ and $y$, is NP-complete [91].

It should be noted that it can even be difficult to compute the diameter of Cayley (di)graphs on additive groups, as proves the study of *circulant (di)graphs* (also called *multiple loop networks*), which are Cayley (di)graphs defined on $\mathbb{Z}_n$ (see the survey [18] and references given here).

In [13] and [5], relationships are established between the diameter of a Cayley graph and the diameter of a subgraph and of a coset graph.

## 4.4 $(\Delta, D)$-graphs

In this part we only consider graphs. The design of interconnection networks comes up against two fundamental constraints: the number of connections for every node is limited, and the number of intermediate nodes on a communication path must be small, in order for the time to broadcast information to be acceptable. This is translated in graph theory as follows.

**Problem 53 ($(\Delta, D)$-problem)** [63] Find graphs with the maximal number of vertices with given constraints on the maximum degree $\Delta$ and the diameter $D$.

This problem has been intensively studied in the last fifteen years (see, for example, [19], [56] and the references given there).

A $(\Delta, D)$-graph is defined as a graph of maximum degree $\Delta$ and diameter $D$. Following [47], we denote by $N(\Delta, D)$ the maximum order of a $(\Delta, D)$-graph. The only known bound on $N(\Delta, D)$ is the Moore bound. This bound is obtained by calculating the maximum number of vertices at distance $0, 1, \dots, D$ from one vertex.

**Proposition 54 ([113])**

$$N(\Delta, D) \leq 1 + \Delta + \Delta(\Delta - 1) + \cdots + \Delta(\Delta - 1)^{D-1}.$$

*Thus* $N(\Delta, D) = (\Delta(\Delta - 1)^D - 2)/(\Delta - 2)$ *for* $\Delta \geq 3$, *and* $N(2, D) \leq 1 + 2D$.

Graphs which attain the bound are called *Moore graphs.*

**Theorem 55 ([10, 96, 43])** *Moore graphs exist only for:* $\Delta = 2$ : *cycles* $C_{2D+1}$;
$D = 1$ : *complete graphs* $K_{\Delta+1}$;
$D = 2$ *and* $\Delta = 3$ : *Petersen graph;*
$D = 2$ *and* $\Delta = 7$ : *Hoffman-Singleton graph (a vertex-transitive graph with* 50 *vertices);*
*and, perhaps, for* $D = 2$ *and* $\Delta = 57$ *(if such a graph exists it has* 3250 *vertices and is not vertex transitive).*

Furthermore, apart from Moore graphs, the only graphs known to be optimal are $(3,3)$-, $(4,2)$- and $(5,2)$-graphs with 20, 15, and 24 vertices, respectively. Except for the cycle of length 4, $N(\Delta, D)$ is not equal to the Moore bound minus 1. In the absence of good bounds, researchers are running a race to construct ever "better" $(\Delta, D)$-graphs. A table of large $(\Delta, D)$ graphs is regularly updated by C. Delorme ([20], see Table given in [56]). The constructions are mostly algebraic ones, using Cayley graphs, coset graphs and compounds (see [51]). For some values of $(\Delta, D)$, the best graphs presently known are Cayley graphs. In this case the problem splits into two parts: choice of the group and then choice of the generating set.

With experimental computing the nine authors of [33] have found small diameter Cayley graphs constructed from linear groups. Their experiment gives no method except the fact that abelian groups are to be avoided in this kind of problem. They use a particular kind of matrix groups and fix the cardinality of the generating set. The diameter is then computed by repeated random choices of the generating set using the expression for the diameter given in Section 4.2. The bounds given in [33] have been improved by constructions of M. Dinneen and P. R. Hafner [56] which use semidirect products of cyclic groups $\mathbb{Z}_n$. For example, they obtain more than 30 entries of Delorme's table by taking a semidirect product of $\mathbb{Z}_m$ with $\mathbb{Z}_n$ of the type

$$[x, y] \times_a [u, v] = [(x + u) \bmod m, (ya^u + v) \bmod n].$$

In particular, with $m = 15$, $n = 77$, $a = 4$ and $S = \{[6, 2], [9, 5], [10, 9], [5, 24]\}$, they construct a $(4, 7)$-graph with 1155 vertices. Recently, M. Sampels obtained better constructions by using abstract presentations of groups. In the article [131], the use of abstract presentations of groups is explained, and presentations of (new) graphs are given with the following values of $(\Delta, D, N)$ : $(4, 10, 17316)$, $(6, 5, 1344)$ and $(4, 8, 3080)$. The value $(4, 10, 17316)$ is now updated to $(4, 10, 17525)$ by O. Delmas [48] ; F. Comellas and M. Mitjana have also a construction for the value $(8, 3, 253)$ [37].

The results of a simulation study about the practical use of degree/diameter-graphs as communication networks are presented in [129]. Since then, using the technique above and genetic algorithms on metacyclic groups, the following values of $(\Delta, D, N)$ were obtained by M. Sampels [130]: $(8, 3, 253)$, $(5, 4, 210)$, $(7, 4, 672)$, $(5, 5, 546)$, $(6, 5, 1395)$, $(5, 8, 15990)$, $(4, 10, 17535)$.

A table on this problem with details on all the graphs can be found on the WWW at

`http://maite71.upc.es/grup_de_grafs/table_g.html.`

Let us note that similar work exists for bipartite graphs and digraphs. The Moore bounds for these classes of graphs are only reached for small degrees or small diameters (see [47]); F. Comellas and M.A. Fiol found some vertex-symmetric digraphs with small diameter (see [36]). Also, some good non-symmetric $(\Delta, D)$ digraphs are known, like Kautz digraphs (see Definition 46 and results cited in [47]) or Cyclic prefix digraphs which are based on coset digraphs [65].

Cayley graphs of non-solvable groups, in particular non-abelian simple groups (groups without non-trivial normal subgroup) are considered by many authors as producing good interconnection networks. This is a field which is of interest also to group theorists.

**Theorem 56 ([8])** *There exists a constant $c$ such that every non-abelian simple group has a set $S$ of at most seven generators for which the diameter of the Cayley graph $Cay(G, S)$ is at most $c \log_2(|G|)$.*

It is proved in [5] that this property is not true for Cayley graphs of abelian groups. On the other hand, it is conjectured that a general bound exists in the non-abelian case:

**Conjecture 57 ([9])** There exists a constant $c$ such that for every non-abelian finite simple group $G$ and any generating set $S$,

$$Diam(Cay(G, S)) \leq (\log_2(|G|))^c.$$

## 4.5 Diameter vulnerability

It is of practical importance to know what is the diameter of the resulting digraph after faulty vertices (or edges) have been deleted from a digraph. A first result of this kind concerns the hypercube.

**Theorem 58 ([47])** *Every pair of vertices of the hypercube $H(n)$ is connected by $n-1$ paths of length at most $n$ and a path of length at most $n+1$, all these paths being internally vertex disjoint.*

Thus after deleting at most $n-1$ vertices, the diameter of the resulting graph is at most $n+1$, i.e. increases by at most 1. But Theorem 58 gives more than the value of the diameter of the graph obtained after deleting faulty vertices, since it bounds the length of parallel paths between any two vertices.

Different generalizations of the diameter can be found in the literature in order to measure the diameter after deleting vertices, such as the notions of *fault-diameter* and *wide-diameter* (see [101] and the survey of Hsu [98]). We give here some definitions and results from [98].

**Definition 59** Let $k$ and $\kappa$ be two integers, $k \leq \kappa$. Let $\Gamma$ be a $\kappa$-connected graph, and $x$ and $y$ two distinct vertices.

- The *k-fault diameter* of $\Gamma$, denoted by $D_k(\Gamma)$, is the maximum of the diameters of the graphs obtained from $\Gamma$ after deleting any set of $k-1$ vertices.
- The *k-distance* $d_k(x,y)$ between $x$ and $y$ is equal to the least number $l$ such that there exist $k$ vertex-disjoint paths between $x$ and $y$ whose lengths are at most $l$.
- The *k-diameter* of $\Gamma$, $Diam_k(\Gamma)$, is the maximum of the $k$-distances $d_k(x,y)$ taken over all distinct pairs $x$, $y$.

Notice that $D_1(\Gamma) = Diam_1(\Gamma) = Diam(\Gamma)$. By Menger's theorem, if a graph $\Gamma$ is $\kappa$-connected then, for $k \leq \kappa$, its $k$-diameter and its $k$-fault diameter are well defined and finite.

If $\Gamma$ is a $k$-connected graph, its $k$-diameter is also called its *wide-diameter* and is denoted by $wdiam(\Gamma)$, and its $k$-fault diameter is simply called *fault diameter*. It is easy to see that

$$Diam(\Gamma) \leq D_k(\Gamma) \leq wdiam(\Gamma).$$

Graphs with small wide-diameter and large connectivity are desirable because they enable fast communication using parallel paths between two vertices. A graph with connectivity $k$ is said to be *strongly resilient* if $wdiam(\Gamma) = Diam(\Gamma) + O(1)$. For a $k$-regular, $k$-connected graph $\Gamma$ it is known that

$$wdiam(\Gamma) \geq Diam(\Gamma) + 1.$$

In this case $wdiam(\Gamma) = Diam(\Gamma) + 1$ is optimal. Thus by Theorem 58, the hypercube $H(n)$ which is $n$-regular, $n$-connected and has wide-diameter equal to $n+1$, is optimal in this sense. Notice that this is not the case for every Cayley graph, since, for the cycle $C_n$, which is 2-regular and 2-connected, the wide-diameter is equal to $n-1$ and the fault diameter is equal to $n-2$. Thus it is interesting to know which graphs, and in particular which Cayley graphs, are strongly resilient.

The star graph is strongly resilient like the hypercube. The diameter of the resulting graph after deleting at most $n-1$ vertices from the star graph $ST(n)$ increases by at most one for $n \geq 7$ (see [121] and [104]). Furthermore, between two vertices $x$ and $y$ of $ST(n)$, there exist $n-1$ paths of length $d(x, y)$ or $d(x, y)+2$ or $d(x, y)+4$ [45].

The generalized hypercubes (defined as cartesian products of complete graphs) are also strongly resilient [60].

The same notion of wide-diameter can be defined for digraphs and some results have been obtained for particular Cayley digraphs, proving that there are optimal. For example, Hsu proved that if $\Gamma$ is the circulant digraph $Cay(\mathbb{Z}_n, \{1, a, a^2, \ldots, a^{n-1}\})$, then $wdiam(\Gamma) \leq diam(\Gamma)+1$. It can be noticed that in the case of the multidimensional torus, the undirected case is more complicated than the directed one [98].

To finish this section we give some related questions and results.

- The results on the wide-diameter of de Bruijn and Kautz digraphs (which are quotients of Cayley digraphs, see Section 3.4.3) extend to iterated line digraphs (see [115] and references given there).

- It can be interesting to know the connectivity under forbidden faulty sets (for example, the neighborhood of any vertex) so that more faults are supported without increasing the diameter too much. In [120] it is proved that any set of $2n-5$ vertices not containing all the neighbors of some vertex can be deleted from the star graph $ST(n)$, and the resulting diameter is at most increased by 2. The same property is true for hypercubes [123].

- One can relax the condition on the diameter in order to find better algorithms. For example, Q. Gu et S. Peng [84] give linear time algorithms for finding paths in the faulty hypercube $H(n)$ obtained by deleting $(n-1)$ arbitrary vertices, and in the faulty star graph $ST(n)$ after deleting $(n-2)$ arbitrary vertices. They obtain in particular a path of length at most $Diam(ST(n))+3$ between any two non-faulty vertices of $ST(n)$ in $O(n)$ time.

- Given a strongly connected digraph $\Gamma$, a routing $R$ (see definition in Section 5) and a set $F$ of faults (arcs and vertices), the *surviving routing digraph* $S(\Gamma, R, F)$ is defined as the digraph with vertex set the non-faulty vertices of $\Gamma$, and in which $(x, y)$ is an arc if and only if the path $R(x, y)$ does not contain any element from $F$ [57]. The diameter of $S(\Gamma, R, F)$ gives the maximum number of retransmissions needed to insure the communication between two nodes of the network after failure. It is interesting to find routings $R$ such that $Diam(S(\Gamma, R, F))$ is small. Some results are known for particular Cayley graphs, like hypercubes and some circulant digraphs [62].

# 5 Indices of routings

In this section, we study loads on vertices and arcs induced by routings in digraphs. Following [35], we define a routing as follows.

For a given (strongly) connected (di)graph $\Gamma$ of order $n$, a *routing* $R$ is defined as a set of $n(n-1)$ simple (di)paths $R(x, y)$, specified for every ordered pair $(x, y)$ of vertices in $\Gamma$, such that $R(x, y)$ joins $x$ to $y$. A *partial routing* is any subset of a routing.

## 5.1 Types of routings

Different types of routings can be considered. We define some of them below. These types of routings have interesting properties and are often used in practice.

A routing $R$ is said to be:

(a) *of shortest paths* or *minimal* if, for any two vertices $x$ and $y$, the path $R(x,y)$ is a path of shortest length between $x$ and $y$;

(b) *symmetric* if, for any two vertices $x$ and $y$, the path $R(y,x)$ is the reverse of the path $R(x,y)$;

(c) *consistent* if, for any two vertices $x$ and $y$ and for each vertex $z$ belonging to the dipath $R(x,y)$, the dipath $R(x,y)$ is the concatenation of the dipaths $R(x,z)$ and $R(z,y)$.

**Example 60** Let $\Gamma$ be the cycle $C_4$ with vertices labeled by $\mathbb{Z}_4$. We illustrate the above definitions by considering some routings on $\Gamma$.

A routing $R$ such that $R(0,3) = (0,1,2,3)$ and $R(0,2) = (0,3,2)$ is not consistent. Notice that since the diameter is equal to 2, any routing of shortest paths is consistent.

Consider $R_1$ such that $R_1(i,j)$ is the path from $i$ to $j$ following a chosen direction of the cycle. More precisely, $R_1(0,1) = (0,1)$, $R_1(0,2) = (0,1,2)$, $R_1(0,3) = (0,1,2,3)$. $R(i,j)$ is obtained from $R(0,j-i)$ by applying the translation $x \mapsto x + i(\text{mod}4)$. $R_1$ is not a routing of shortest paths and is not symmetric, but it is consistent.

We define $R_2$ such that $R_2(0,2) = (0,1,2)$, $R_2(2,0) = (2,1,0)$, $R_2(1,3) = (1,2,3)$, $R_2(3,1) = (3,2,1)$, and if $i$ and $j$ are adjacent, $R_2(i,j) = (i,j)$. $R_2$ is a symmetric routing of shortest paths.

We define $R_3$ such that $R_3(0,2) = (0,1,2)$, $R_3(2,0) = (2,3,0)$, $R_3(1,3) = (1,2,3)$, $R_3(3,1) = (3,0,1)$, and if $i$ and $j$ are adjacent, $R_3(i,j) = (i,j)$. $R_3$ is a routing of shortest paths but it is not symmetric.

## 5.2 Vertex-forwarding index

For a given digraph $\Gamma$, a vertex $x$ of $\Gamma$ and a specified routing $R$, the *load* of $x$, denoted by $\xi(\Gamma,R,x)$, is the number of dipaths of $R$ going through $x$. The *vertex-forwarding index* $\xi(\Gamma,R)$ is the maximum number of dipaths of $R$ going through a vertex of $\Gamma$. The minimum of $\xi(\Gamma,R)$ over all possible routings $R$ of $\Gamma$ is denoted by $\xi(\Gamma)$ and is called the *vertex-forwarding index of* $\Gamma$ or for short the *forwarding index of* $\Gamma$:

$$\xi(\Gamma) = \min_R \xi(\Gamma,R).$$

The minimum of $\xi(\Gamma,R)$ over all possible routings of *shortest paths* $R$ of a strongly connected digraph $\Gamma$ is called the *minimal forwarding index* of $\Gamma$ and is denoted by $\xi_m(\Gamma)$.

**Example 61** Let us consider the routings defined in Example 60 on the graph $C_4$. We have:
$\xi(\Gamma,R_1,x) = 3$ for any vertex $x$, and $\xi(\Gamma,R_1) = 3$;
$\xi(\Gamma,R_2,1) = \xi(\Gamma,R_2,2) = 2$, $\xi(\Gamma,R_2,0) = \xi(\Gamma,R_2,3) = 0$, $\xi(\Gamma,R_2) = 2$;
$\xi(\Gamma,R_3,x) = 1$ for any vertex $x$, and $\xi(\Gamma,R_3) = 1$;
thus $\xi(\Gamma) = \xi_m(\Gamma) = \xi(\Gamma,R_3) = 1$.

Since forwarding indices were first defined in graphs [35], most of the results in the literature are given for graphs instead of digraphs, but they can often be easily extended to digraphs [111]. Nevertheless, we give here most of the results for graphs as they appear in the literature.

To compute $\xi(G)$ for an arbitrary graph is difficult as shows the complexity of the following problem.

**Problem 62** FI (Forwarding Index problem)
*Instance: A graph* $\Gamma$ *and an integer* $k$.
*Question:* $\xi(\Gamma) \leq k$ ?

**Theorem 63** **([122])** *FI is NP-complete.*

It has been shown that there are graphs for which the forwarding index is obtained neither for a minimal nor for a symmetric routing. For example, the forwarding index of wheels cannot be obtained for minimal routings [35], and the forwarding index of cycles of even length cannot be obtained for symmetric routings [94].

Similarly, we define the problems **SFI** (symmetric FI), **MFI** (minimal FI), as the above problems when the considered routings are restricted to symmetric routings, and routings of shortest paths, respectively.

**Theorem 64** **([122])** *SFI is NP-complete.*

The problem FI is still NP-complete for graphs of diameter 2. By contrast:

**Proposition 65** **([93])** *MFI is polynomial for graphs of diameter* 2.

The preceding results show how important it is to find classes of graphs for which the forwarding index can be easily calculated.

The following proposition gives bounds on $\xi(\Gamma)$ and characterizes the routings for which the lower bound is attained.

**Proposition 66** **([35])** *Let* $\Gamma$ *be a simple connected graph of order* $n$. *Then*

(i) $\frac{1}{n}\sum_u \sum_{v \neq u}(d(u,v)-1) \leq \xi(\Gamma) \leq \xi_m(\Gamma) \leq (n-1)(n-2)$.

(ii) *The equalities* $\xi(\Gamma) = \xi_m(\Gamma) = \frac{1}{n}\sum_u \sum_{v \neq u}(d(u,v)-1)$ *hold if and only if there exists a routing of shortest paths in* $\Gamma$ *for which the load of all vertices is the same.*

The proof given in [35] works also for digraphs. This proposition shows how interesting are the graphs in which there exists a routing of shortest paths which loads all vertices of $\Gamma$ uniformly, in which case $\xi(\Gamma)$ has the lowest possible value, given by $\frac{1}{n}\sum_u \sum_{v \neq u}(d(u,v)-1)$. The next theorem proves that Cayley digraphs form such a class of digraphs.

**Theorem 67** **([93])** *In any Cayley digraph* $\Gamma = (V, E)$, *there exists a consistent routing of shortest paths in which the load of every vertex* $u$, *and therefore also the vertex forwarding index, is equal to*

$$\sum_{v \in V, v \neq u} d(u,v) - (n-1).$$

**Proof** Let $\Gamma = Cay(G, S)$ be a Cayley digraph and consider an order of $S$, $S = s_1, s_2, \ldots, s_d$, with $s_1 < s_2 < \ldots < s_d$. This induces a lexicographic order on the dipaths of shortest length from $x$ to $y$, for any two distinct vertices $x$ and $y$ of $\Gamma$. The dipath $\rho = x_0 x_1 \ldots x_l$ is smaller than the dipath $\rho' = y_0 y_1 \ldots y_l$ ($x_0 = y_0 = x$ and $x_l = y_l = y$) if there exists $i$, $1 \leq i < l$, such that

$$x_k = y_k \text{ for } 0 \leq k < i, \text{ and } x_{i-1}^{-1} x_i < y_{i-1}^{-1} y_i.$$

For any two vertices $x$ and $y$ of $\Gamma$, define $R(x, y)$ to be the smallest dipath (for the induced order) of shortest length among all dipaths of shortest length from $x$ to $y$.

It is easy to verify that the routing $R$ is consistent. If $R(x, y) = x x_1 \ldots x_{l-1} y$, the path $x x_1 \ldots x_i$, $1 \leq i < l$, has length $d(x, x_i)$ since $R(x, y)$ is of shortest length. Moreover, it is a smallest path for the order, otherwise $x_0 x_1 \ldots x_l$ would not be a smallest path. Thus $R(x, x_i) = x_0 x_1 \ldots x_i$. Similarly, $R(x_i, y) = x_i \ldots x_{l-1} y$ and $R(x, y)$ is the concatenation of $R(x, x_i)$ and $R(x_i, y)$.

Let us now prove that the number of paths of the routing going through any vertex is a constant. Consider the regular subgroup of translations of $Aut(\Gamma)$, $F = \{f_a : a \in G\}$, as defined before Theorem 2. For any two vertices $x$ and $y$ of $\Gamma$, there exists a unique $a \in G$ such that $f_a(x) = y$. Notice that an arc labeled by $s \in S$ is transformed by $f_a$ into an arc with the same label, since $xs = y$ implies $axs = ay$ (the associativity of the law is here important). Thus any path $R(u, v)$ going through $x$ is transformed by $f_a$ into a path of shortest length from $au$ to $av$ which is, moreover, a smallest path, i.e., the path $R(au, av)$, which goes through $y$. Therefore, by symmetry, $\xi(\Gamma, R, x) = \xi(\Gamma, R, y)$. We can then finish by Proposition 66. □

In order to appreciate the strength of this property of Cayley graphs, observe that a class of graphs is constructed in [93] for which the vertex forwarding index is not given by a consistent routing of shortest paths. Thus it is worth considering the following problems.

**Problem 68** (**[93]**) For which digraphs $\Gamma$ does there exist a consistent routing $R$ such that $\xi(\Gamma) = \xi(\Gamma, R)$ ?

**Problem 69** (**[93]**) For which digraphs $\Gamma$ does there exist a consistent routing of shortest paths $R$ such that $\xi_m(\Gamma) = \xi(\Gamma, R)$ ?

The following conjecture has attracted many researchers for the last ten years without a complete success.

**Conjecture 70** (**[94]**) In any vertex-transitive graph $\Gamma = (V, E)$, there exists a routing of shortest paths in which the load of every vertex $u$, and therefore also the vertex-forwarding index, is equal to

$$\sum_{v \neq u} d(u, v) - (n - 1).$$

The next two sections present the major results obtained so far on this conjecture.

## 5.3 Quasi-Cayley digraphs

We now present the theory of quasi-Cayley (di)graphs as given by G. Gauyacq in [78] and [79]. The class of quasi-Cayley graphs is a new class of vertex-transitive graphs which contains the class of Cayley graphs, and which satisfies Conjecture 70.

**Definition 71** A *quasigroup* $Q$ is a set of elements and a binary operation denoted as a product such that

(i) for every ordered pair $(a, b)$ of elements of $Q$ the product $a \cdot b$ exists and belongs to Q;

(ii) for every ordered pair $(a, b)$ of $Q$ there exists a unique element $x \in Q$ such that $a \cdot x = b$;

(iii) for every ordered pair $(a, b)$ of $Q$ there exists a unique element $y \in Q$ such that $y \cdot a = b$.

By definition, the multiplication table of the binary operation of a quasigroup is a latin square. Notice that an associative quasigroup is a group [54].

**Example 72 ([78])** Take $\mathbb{Z}_2 \times \mathbb{Z}_5$ and the binary operation defined for any $a, c \in \mathbb{Z}_2$ and $b, d \in \mathbb{Z}_5$ by $(a, b) \cdot (c, d) = (a + c, b + 2^a d)$, where $2^0 = 1$ and $2^1 = 2$. Then $(\mathbb{Z}_2 \times \mathbb{Z}_5, \cdot)$ is a quasigroup.

In the following, we denote for short by $Q$ the quasigroup $(Q, .)$ and omit the dot $\cdot$ in the products if no confusion can arise.

**Definition 73** Let $Q$ be a quasigroup and let $S$ be a subset of $Q$. The *quasigroup digraph* $QD(Q, S)$ is defined as the digraph with vertex set $Q$ and arcs $(g, g \cdot s)$, $g \in Q$, $s \in S$.

The digraph $QD(Q, S)$ is regular since for every vertex $g \in Q$, $d^+(g) = |S|$, and by property (iii) of the definition $d^-(g) = |S|$. By property (ii) of the definition, a quasigroup digraph has no multiple arc.

**Proposition 74 ([54])** *A digraph $\Gamma = (V, E)$ is regular if and only if it is a quasigroup digraph.*

This notion is similar to the notion of group action graph (GAG) defined in Section 3.4.3. Every quasigroup digraph $QD(Q, S)$ is a $GAG(Q, \Pi)$ with $\Pi = \{x \mapsto x \cdot s, s \in S\}$. Conversely, every GAG without multiple arc is a quasigroup digraph [78].

Quasi-Cayley digraphs are particular quasigroup digraphs defined as follows.

**Definition 75** A subset $S$ of $Q$ is said to be *right associative* if for any two elements $a$ and $b$ of $Q$, $a \cdot (b \cdot S) = (a \cdot b) \cdot S$.

**Definition 76** A subset $S$ of $Q$ is said to *generate* $Q$ if, for any element $g$ of $Q$, there exist elements $s_1, s_2, \ldots, s_t$ of $S$ such that $g = (\ldots((s_1 \cdot s_2) \cdot s_3) \ldots) \cdot s_t$.

**Definition 77** An element $I$ of $Q$ is a *right unit* if $a \cdot I = a$ for any $a \in Q$.

**Definition 78** Given a quasigroup $Q$ with a right unit and a subset $S$ of $Q$ which is right associative, the quasigroup digraph $QD(Q, S)$ is called a *quasi-Cayley digraph.*

We list below some easily obtained properties of any quasi-Cayley digraph $QD(Q, S)$.

• If $S$ generates $Q$, then $QD(Q, S)$ is strongly connected. For this, one can construct a path from $I$ to every element $g$, using the facts that $g$ is a product of elements of $S$ and that

$S$ is right associative. Similarly, a path from every element $g$ to $I$ can be obtained, using the existence of $g'$ such that $gg' = I$, with $g'$ a product of elements of $S$.

• If the right unit $I$ does not belong to $S$, $QD(Q, S)$ has no loop. If furthermore $Q$ is a group, then the quasi-Cayley digraph $QD(Q, S)$ is the same as the Cayley digraph $Cay(Q, S)$. Thus any Cayley digraph is a quasi-Cayley digraph.

• If $S$ has the property that for any $s \in S$, the element $s'$ which satisfies $ss' = I$ (see property (ii) of Definition 71) belongs to $S$, then $QD(Q, S)$ is a graph. For this, assume $(u, v)$ is an arc of $QD(Q, S)$, then $v = us$. Consider $s' \in S$ such that $ss' = I$. Since $S$ is right associative, there exists $s''$ such that $u(ss') = (us)s''$. This implies $u = uI = u(ss') = (us)s'' = vs''$, therefore $(v, u)$ is an arc of $QD(Q, S)$.

**Example 79 ([78])** The Petersen graph is a quasi-Cayley graph which is not a Cayley graph. Indeed, taking $Q$ as defined in Example 72 and $S = \{(0,1),\ (1,0),\ (0,4)\}$, $QD(Q, S)$ is the Petersen graph Since it is well known that the Petersen graph is not a Cayley graph, this proves that not every quasi-Cayley digraph is a Cayley digraph.

G. Gauyacq also proved that the Hoffman-Singleton graph and the metacirculant graphs are quasi-Cayley graphs, but that some Kneser graphs and some Johnson graphs are not. In particular the line-graph of $K_n$ for $n$ even and $n > 5$, is not a quasi-Cayley graph [78].

In the following we assume that $S$ generates $Q$, and we give some properties of quasi-Cayley digraphs which generalize properties of Cayley digraphs.

**Proposition 80 ([78])** *Any quasi-Cayley digraph is vertex-transitive.*

**Proof** Let $QD(Q, S)$ be a quasi-Cayley digraph. As for Cayley digraphs, every element $a \in Q$ defines an automorphism $f_a$ of $QD(Q, S)$ by $f_a(x) = ax$. This a consequence of property (ii) of Definition 71 and the fact that $S$ is right associative. Furthermore, the set $F = \{f_a, a \in Q\}$ acts regularly on $Q$ by property (iii) since, for any two elements $x$ and $y$ of $Q$, there exists a unique element $a$ of $Q$ such that $ax = y$, i.e. $f_a(x) = y$. □

As for Cayley digraphs, regular automorphisms characterize quasi-Cayley digraphs as is shown by Theorem 82 which generalizes Theorem 2.

**Definition 81 ([78])** A *regular family* of a digraph $\Gamma = (V, E)$ is a subset $F$ of the automorphism group $Aut(\Gamma)$ which acts regularly on $V$, i.e. for any two vertices $x$ and $y$ of $\Gamma$ there exists a unique automorphism $f \in F$ such that $f(x) = y$.

**Theorem 82 ([78, 79])** *A strongly connected digraph $\Gamma = (V, E)$ is a quasi-Cayley digraph if and only if $Aut(\Gamma)$ contains a regular family on $V$.*

**Proof** We only give a sketch of the if part of the proof. Assume $\Gamma = (V, E)$ is a strongly connected digraph and $F$ a regular family of $\Gamma$. Let $V = \{1, 2, \ldots, n\}$ and define $f_i$ as the unique element of $F$ such that $f_i(1) = i$, $1 \leq i \leq n$. A binary operation $\cdot$ is defined on $V$ by

$$i \cdot j = f_i(j).$$

It can be easily verified that $(V, \cdot)$ is a quasigroup. Since by definition $f_i(1) = i \cdot 1 = i$, 1 is a right unit.

Let $S$ be the set of the out-neighbours of 1, i.e. the vertices $u$ of $V$ such that $(1,u) \in E$. One can verify that $S$ is right associative and generates $V$ (using the existence of a dipath from 1 to every element of $V$). Finally,

$$\begin{aligned}(u,v) \in E &\Leftrightarrow (f_u^{-1}(u), f_u^{-1}(v)) \in E \\ &\Leftrightarrow (1, f_u^{-1}(v)) \in E \Leftrightarrow f_u^{-1}(v) \in S \Leftrightarrow v = f_u(s) = u \cdot s, s \in S. \qquad \square\end{aligned}$$

As for Cayley graphs, products of quasi-Cayley graphs are quasi-Cayley graphs.

**Proposition 83 ([78])** *If $\Gamma_1$ and $\Gamma_2$ are two quasi-Cayley digraphs, then the cartesian product, the weak product and the lexicographic product of $\Gamma_1$ and $\Gamma_2$ are also quasi-Cayley digraphs.*

**Proof** Consider a regular family $F_i \subseteq Aut(\Gamma_i)$ of $\Gamma_i$, $i = 1,2$. Given $f_i \in F_i$, $i = 1,2$, define the mapping $f$ on $V_1 \times V_2$ by $f(x_1,x_2) = (f_1(x_1), f_2(x_2))$. It is easy to see that $f$ is an automorphism for any of the three products and that $\{f : f_1 \in F_1, f_2 \in F_2\}$ is a regular family of the product. □

In [78], G. Gauyacq characterizes the quasi-Cayley graphs with the notion of quasi-normal subgroup.

**Definition 84 ([78])** A subgroup $H$ of a group $G$ is said to be *quasi-normal* if there exists a common system of left coset representatives of $H$ and all its conjugate subgroups $gHg^{-1}$ for $g \in G$.

**Theorem 85 ([78])** *Let $\Gamma$ be a vertex transitive graph, and $H$ the subgroup of $Aut(\Gamma)$ which is the stabilizer $Stab_{Aut(\Gamma)}(u)$ of any fixed vertex $u$ of $\Gamma$. Then $\Gamma$ is a quasi-Cayley graph if and only if $H$ is a quasi-normal subgroup of $Aut(\Gamma)$.*

Other properties of quasi-normal subgroups can be found in [78].

Finally, we give the theorem which motivates all this study.

**Theorem 86 ([78, 79])** *In any quasi-Cayley digraph there exists a routing of shortest paths in which the load on every vertex is the same.*

**Proof** We only give the idea of the proof. It is similar to that of Theorem 67 except that we cannot define the routing so easily and insure that it is consistent.

Let $\Gamma = QD(Q,S)$ be a quasi-Cayley digraph. The idea is to consider a partial routing $R'$ consisting of paths of shortest length from the vertex $I$ (the right unit) to any vertex $u$ of $\Gamma$, $\{R'(I,u) : u \in Q, u \neq I\}$ . Consider the regular family $F$ of $Aut(\Gamma)$, $F = \{f_x, x \in Q\}$, as defined after Proposition 80. In particular, $f_x(I) = xI = x$ for any $x \in Q$. Since the image by $f_x$ of a dipath of shortest length $P$ is a dipath of shortest length, denoted by $f_x(P)$, a routing $R$ can be defined by

$$R(x,y) = f_x(R'(I, f_x^{-1}(y)))$$

for any two vertices $x$ and $y$ (notice that $R(I,y) = f_I(R'(I, f_I^{-1}(y)))$ can be different from $R'(I,y)$ since $I$ is not assumed to be a left unit). Then $R = \{f_x(R'(I,u)) : x \in Q, u \in Q, u \neq I\}$. It can then be proved that the number of paths of the routing $R$ going through any vertex is a constant. □

## 5.4 Digraphs with regular orbits

G. Gauyacq has defined another class of vertex transitive graphs which strictly contains quasi-Cayley graphs and satisfies Conjecture 70. Our presentation differs a little from [78] (stating results for digraphs), but all the results and ideas come from it. To understand the definition of this class, recall that in the proof of Theorem 86, the key point is the uniform load induced on a vertex by a well chosen subset of paths obtained by automorphisms. Thus the idea is to consider a subset of $R$ consisting of paths between vertices at the same distance and such that these paths can be obtained from one of them by applying automorphisms. This leads to the notion of $V \times V$ orbit.

**Definition 87** Let $\Gamma = (V, E)$ be a strongly connected vertex-transitive digraph, and $k \leq Diam(\Gamma)$. A *$k$-orbit* $\Omega$ of $\Gamma$ is a subset of $V \times V$ such that

(i) there exists $(u, v) \in \Omega$ with $d(u, v) = k$;

(ii) for every $(u', v') \in V \times V$, $(u', v') \in \Omega$ if and only if there exists $g \in Aut(\Gamma)$ such that $g(u) = u'$ and $g(v) = v'$.

Since the distance between vertices is unchanged by automorphisms, $d(u', v') = d(u, v) = k$ for every $(u', v')$ in a $k$-orbit. Thus a $k$-orbit is an orbit for the action of the group $Aut(\Gamma)$ on $V \times V$ consisting of ordered pairs of vertices $(u', v')$ at distance $k$. Therefore $V \times V$ is partitioned into $k$-orbits.

**Lemma 88** *Let $\Omega$ be a $k$-orbit. Then:*

(i) *For any vertex $u \in V$, there exists $v \in V$ such that $(u, v) \in \Omega$.*

(ii) *For any vertex $u \in V$, the number of vertices $v \in V$ such that $(u, v) \in \Omega$ is independent of the choice of $u$; let $\alpha(\Omega)$ be this number.*

(iii) $|\Omega| = |V| \times \alpha(\Omega)$.

**Proof** (i) Since $\Gamma$ is vertex-transitive, for any $u' \in V$ there exists an automorphism $f$ and a vertex $v'$ such that $f(u) = u'$, $f(v) = v'$ and $(u', v') \in \Omega$.

(ii) Let $(u, x_1), (u, x_2), \ldots, (u, x_p)$ be the elements of $\Omega$ whose first coordinate is $u$. Given any $(u', v') \in \Omega$, there exists $g \in Aut(\Gamma)$ such that $g(u) = u'$ and $(u', g(x_1)), (u', g(x_2)), \ldots, (u', g(x_p))$ are all in $\Omega$. Thus $|\{v : (u, v) \in \Omega\}| \leq |\{v' : (u', v') \in \Omega\}|$ and by symmetry, $|\{v : (u, v) \in \Omega\}| = |\{v' : (u', v') \in \Omega\}|$.

(iii) The equality $|\Omega| = |V|\alpha(\Omega)$ comes from facts (i) and (ii). □

**Proposition 89 ([78])** *Let $\Gamma$ be a vertex-transitive digraph and $\Omega$ a 2-orbit of $\Gamma$. Then there exists a partial routing $\{R(u, v) : (u, v) \in \Omega\}$, consisting of shortest paths between the couples of vertices of $\Omega$, which loads uniformly all vertices of $\Gamma$.*

**Proof** Consider a bipartite graph $B$ in which one class consists of $\Omega$ and the other, denoted by $Y$, consists of the ordered pairs $(x, i)$, where $x \in V$ and $i$ is any integer $\in \{1, \ldots, \alpha(\Omega)\}$. Notice that $|\Omega| = |V|\alpha(\Omega) = |Y|$. An edge connects $(u, v) \in \Omega$ with $(x, i)$ if and only if $d(u, x) = d(x, v) = 1$. $B$ is regular, thus it contains a perfect matching. Let $R(u, v) = (u, s, v)$

if $(u, v)$ is matched with $(s, i)$. The load of any vertex $s$ is then equal to $\alpha(\Omega)$, since $s$ belongs to the paths connecting the $\alpha(\Omega)$ ordered pairs $(u, v)$ matched respectively with $(s, 1), (s, 2), \ldots, (s, \alpha(\Omega))$. □

The following corollary shows that Conjecture 70 is true for vertex-transitive digraphs of diameter 2.

**Corollary 90** **([78])** *If* $\Gamma$ *is a vertex-transitive digraph of diameter* 2, *then there exists a routing of shortest paths which loads uniformly all vertices of* $\Gamma$.

We turn now to the general case.

**Definition 91** **([78])** A $k$-orbit $\Omega$ is said to be *regular* if, for every $(u, v) \in \Omega$, there exists a shortest path $u, s_1, s_2, \ldots, s_k = v$ and a subset $F$ of $Aut(\Gamma)$ such that

(i) for every $(u', v') \in \Omega$, there exists a unique $f \in F$ such that $f(u) = u'$ and $f(v) = v'$;

(ii) for any $i, 1 \leq i \leq k-1$, and for any $x \in V$, the number of $f \in F$ such that $f(s_i) = x$ depends only on $\Omega$.

Let $\Omega$ be a regular $k$-orbit. By property (i) of Definition 91, $|\Omega| = |F|$. By the property (ii) of the definition, let $\beta(\Omega)$ be the number of $f \in F$ such that $f(s_i) = x$ for any $i, 1 \leq i \leq k-1$, and for any $x \in V$. Let $(u, x_1), (u, x_2), \ldots, (u, x_p)$, with $p = \alpha(\Omega)$, be the elements of $\Omega$ with first coordinate $u$. Consider $t$ such that $(s_1, t) \in \Omega$. There exist $\beta(\Omega)$ automorphisms $f \in F$ such that $f(s_1) = u$ so that $f(t) \in \{x_1, x_2, \ldots, x_p\}$ and by property (i), $\{f(t), f \in F\} = \{x_1, x_2, \ldots, x_p\}$. Thus $\beta(\Omega) = \alpha(\Omega)$, i.e., the number of $v \in V$ such that $(u, v) \in \Omega$ for any $u \in V$.

We first give the result which justifies Definition 91, before defining a new class of vertex-transitive digraphs.

**Lemma 92** **([78])** *If* $\Omega$ *is a regular* $k$*-orbit, then there exists a partial routing* $R_\Omega$ *defined on* $\Omega$ *for which the load of all the vertices of the digraph* $\Gamma$ *is the same.*

**Proof** Define $R_\Omega$ by $R_\Omega(u, v) = u, s_1, s_2, \ldots, s_k$ with $s_k = v$, and for any $(u', v') \in \Omega$ $R_\Omega(u', v') = f(R_\Omega(u, v))$, where $f$ is the unique automorphism of $F$ such that $f(u) = u'$ and $f(v) = v'$. Thus the partial routing $R_\Omega$ is the collection of paths $f(R_\Omega(u, v))$, $f \in F$. To calculate the load of any vertex $x$ by $R_\Omega$, consider the number of times $x$ is in position $i$ on a path $f(R_\Omega(u, v))$. By property (ii) of Definition 91, this is exactly $\alpha(\Omega)$. Thus the load of the vertex $x$ by $R_\Omega$ is equal to $(k-1)$ times $\alpha(\Omega)$ and is independent of $x$.

□

From Lemma 92, it is easy to get :

**Corollary 93** **([78])** *If* $\Gamma$ *is a vertex-transitive digraph in which all* $k$*-orbits are regular for* $1 \leq k \leq Diam(\Gamma)$, *then there exists a routing of shortest paths which loads uniformly all vertices of* $\Gamma$.

**Proposition 94** **([78])** *In any quasi-Cayley digraph, all* $k$*-orbits are regular.*

G. Gauyacq gives some examples of vertex-transitive digraphs in which all $k$-orbits are regular and which are not quasi-Cayley digraphs, such as the odd graphs $O_4, O_5, O_6, O_8$ [78]. Notice that the proof of this involves far from evident combinatorics.

**Problem 95** (**[78]**) Find an example of vertex-transitive digraph in which not all $k$-orbits are regular.

Similar to Figure 1, Figure 4 summarizes the implications between the different classes of vertex transitive graphs introduced in this part.

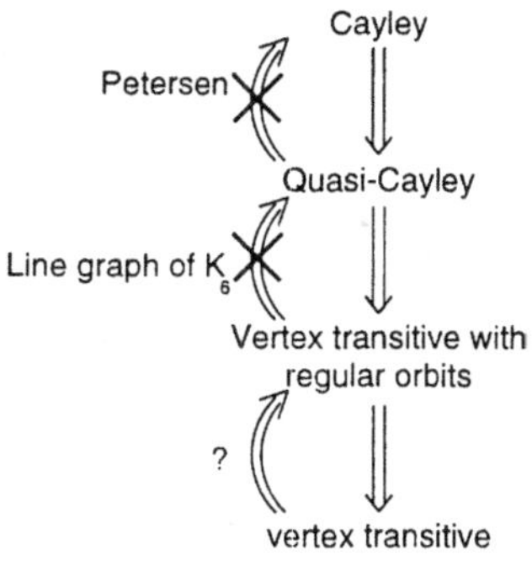


Figure 4: Relations between classes of vertex transitive graphs

## 5.5 Edge-forwarding index

The vertex-forwarding index of a routing can be considered as a measure of the load of the vertices. Similarly, one can consider the *load of any arc* $e$ in a given routing $R$ of a strongly connected digraph $\Gamma = (V, E)$, that is, the number of dipaths of $R$ going through the arc $e$. If $\Gamma$ is a graph, the original definition of [94] considers for each edge $uv$ the sum of the loads of the two arcs $(u, v)$ and $(v, u)$ of the equivalent symmetric digraph. In the following, to avoid confusion, we only consider graphs and give results as they appear in the literature. But the reader can keep in mind that the results can also be considered for digraphs.

For a graph $\Gamma = (V, E)$ and a routing $R$, let $\pi(\Gamma, R, e)$ be the number of paths of $R$ going through the edge $e$. The *edge-forwarding index* of $(\Gamma, R)$, denoted by $\pi(\Gamma, R)$, is the maximum number of paths of $R$ going through any edge of $\Gamma$, that is :

$$\pi(\Gamma, R) = \max_{e \in E} \pi(\Gamma, R, e).$$

The *edge-forwarding index* of $\Gamma$ [94] is defined as

$$\pi(\Gamma) = \min_{R} \pi(\Gamma, R).$$

Similarly, the *minimal edge-forwarding index*, $\pi_m(\Gamma)$, is defined by replacing paths with paths of shortest length in the definitions.

As for the vertex-forwarding index, the following inequality is easily obtained from the definition.

**Proposition 96** **([94])** *Let $\Gamma = (V, E)$ be a connected graph. Then*

$$\pi_m(\Gamma) \geq \pi(\Gamma) \geq \frac{1}{|E|} \sum_{(u,v) \in V \times V} d(u,v)$$

*The equalities*

$$\pi_m(\Gamma) = \pi(\Gamma) = \frac{1}{|E|} \sum_{(u,v) \in V \times V} d(u,v)$$

*hold if and only if there exists a routing of dipaths of shortest length in $\Gamma$ for which the load of all the edges is the same.*

Despite the similarity of the two notions of forwarding indices, there does not exist an edge-uniform routing of shortest paths in every Cayley graph (consider for example the cartesian product of the two complete graphs $K_2$ and $K_p$ for $p \geq 3$). On the other hand, there exist infinite families of Cayley graphs in which such routings exist [50], [138]. In [77] it is proved that the edge-forwarding index of star graphs and complete transposition graphs achieves (within one) the lower bound given above.

We finish this section by giving two conjectures in order to find sufficient conditions on graphs insuring the existence of routings with uniform loads for the edges or almost uniform loads (only different by one) for the vertices and the edges.

**Conjecture 97** **([94])** In any distance-transitive graph $\Gamma = (V, E)$, there exists a routing $R$ of shortest paths such that for any vertex $u$, the edge-forwarding index is equal to

$$\pi(\Gamma) = \pi(\Gamma, R) = \lceil \frac{n}{|E|} \sum_{v \in V} d(u,v) \rceil.$$

**Conjecture 98** **([94])** In any distance-transitive graph there exists a routing of shortest paths for which the load is the same for any vertex, and almost the same for any edge.

No example is given in [94] of an arc-transitive graph which does not satisfy the conjectures. In view of the results obtained so far, the hypothesis of these conjectures can surely be reduced. One can also expect to prove the following conjecture.

**Conjecture 99** In any arc transitive Cayley graph $G = (V, E)$, there exists a routing $R$ of shortest paths such that the load is the same for any vertex, and almost the same for any edge.

## 5.6 Orbital regular graphs

P. Solé has introduced a new class of graphs which allows an easy calculation of the edge-forwarding index by averaging over the distances between pairs of distinct vertices [138]. We adopt here the presentation of [66] which is more precise than that of [138].

**Definition 100** Let $\Gamma = (V, E)$ be a simple connected graph and $H$ a subgroup of $Aut(\Gamma)$. $\Gamma$ is said to be *$H$-orbital regular* if:

(i) $H$ acts regularly on each of its orbits in $\{(x, y) : x, y \in V, x \neq y\}$.

(ii) One of these orbits, say $O$, is such that $E = \{xy : (x, y) \in O\}$.

A graph $\Gamma$ is said to be *orbital regular* if it is $H$-orbital regular for some subgroup $H$ of $Aut(\Gamma)$.

Note that an orbital regular graph is edge-transitive.

**Example 101** (*C. Delorme*) The Cayley graph defined on the additive group $\mathbb{Z}_{13}$ with generators $S = \{1, -1, 5, -5\}$ is orbital regular. The orbits are

$$\begin{aligned} O &= \{(x, y) : x - y = \pm 1, \pm 5\}, \\ \Omega_2 &= \{(x, y) : x - y = \pm 2, \pm 3\}, \\ \Omega_3 &= \{(x, y) : x - y = \pm 4, \pm 6\}. \end{aligned}$$

The automorphisms of $H$ are given by $f(x) = ax + b$, for $x \in \mathbb{Z}_{13}$, where $a \in S$ and $b \in \mathbb{Z}_{13}$.

In fact, almost all orbital regular graphs are Cayley graphs as follows from the characterization given in [66] using Frobenius groups. Let us recall that a *Frobenius group* is a group $G$ of permutations on a set $V$ such that the only element that fixes at least two elements of $V$ is the identity. For a Frobenius group $G$, a *G-Frobenius graph* is a graph $\Gamma$ with vertex set $V$ such that for some $G$-orbit $O$ in $\{(x, y) : x, y \in V, x \neq y\}$, the edge set of $\Gamma$ is $E = \{xy; (x, y) \in O\}$.

**Theorem 102 ([66])** *Let $\Gamma = (V, E)$ be a connected graph with $n$ vertices, and let $H$ be a subgroup of $Aut(\Gamma)$. Then $\Gamma$ is $H$-orbital regular if and only if the following holds :*

(i) *$\Gamma$ is an $H$-Frobenius graph, or*

(ii) *$\Gamma$ is a cycle of length $n$ and $H = \mathbb{Z}_n$ with $n \geq 3$, or*

(iii) *$\Gamma$ is the star $K_{1,n-1}$.*

Properties of Frobenius groups imply that any Frobenius graph is a Cayley graph [66] so that we get :

**Corollary 103 ([66])** *All orbital regular graphs except stars $K_{1,n-1}$ are Cayley graphs.*

We now return to the property which motivates the study of orbital regular graphs.

**Proposition 104 ([138])** *In an orbital regular graph with $n$ vertices the edge-forwarding index is*

$$\pi(\Gamma) = \frac{1}{|E|} \sum_{(u,v) \in V \times V} d(u, v).$$

**Proof** Let $\Gamma = (V, E)$ be an orbital regular graph. In each orbit $\Omega$ an ordered pair of vertices $(u, v)$ is selected and a shortest path $R(u, v)$ is chosen. For every other pair of vertices $(u', v')$ in the same orbit $\Omega$, put $R(u', v') = f(R(u, v))$, where $f$ is the unique automorphism of $H$ such that $f(u) = u'$ and $f(v) = v'$. Let $e$ and $e'$ be two different edges of $\Gamma$. Then by definition there exists a unique automorphism $h \in H$ such that $h(e) = e'$, so that the paths through $e'$ are the images by $h$ of the paths through $e$. □

Other results on the edge-forwarding index of orbital regular graphs can be found in [66].

## 5.7 Multicommodity flow problem

It is proved in [46] that the problem of determining the value of the vertex/edge-forwarding indices (respectively of the vertex/edge-forwarding indices in case of shortest paths) is an instance of the integral multicommodity flow problem (respectively integral flow with multipliers problem) for which good heuristics or approximation algorithms exist.

Another way to study forwarding indices is to enlarge the problem by considering it as an integer valued version of a real valued one. We adopt here the presentation of [136]. We first recall some definitions about multicommodity flows, and then present how multicommodity flows and forwarding indices are related.

Let $\Gamma$ be a connected graph considered as a symmetric digraph $\Gamma = (V, E)$ and $x, y$ two vertices of $\Gamma$. We denote by $\mathcal{P}_{x,y}$ the dipaths from $x$ to $y$ in $\Gamma$ and by $\mathcal{P}$ the set of all dipaths in $\Gamma$.

**Definition 105** A *uniform concurrent multicommodity flow* $f$ on a graph $\Gamma$ is a function $f : \mathcal{P} \to \mathbb{R}^+$ such that, for any two vertices $x$ and $y$ of $\Gamma$,

$$\sum_{p \in \mathcal{P}_{x,y}} f(p) = 1$$

This definition uses an "edge-path" formulation which is equivalent to the more usual node-arc formulation (see for example [27]). This means that the flow $f$ associates to any dipath a real value in such a manner that, for any vertices $x, y$, the sum taken over all the dipaths from $x$ to $y$ is equal to 1. If we restrict the values of $f$, and take integral values ($f(p) = 0$ or $f(p) = 1$), the choice of $f$ is equivalent to the choice of a directed path between any pair of vertices.

The flow in an arc $e$ of $\Gamma$ is defined by

$$f(e) = \sum_{\{p : e \in p\}} f(p),$$

and the *edge congestion* of the flow $f$ is

$$\mu_f = \max_{e \in E} f(e).$$

Similarly, the flow in a vertex $x$ of $\Gamma$ is defined by

$$f(x) = \sum_{\{p : x \in p\}} f(p),$$

and the *vertex congestion* of the flow $f$ is

$$m_f = \max_{x \in X} f(x).$$

A flow that minimizes $\mu_f$ (resp. $m_f$) is said to be *edge* (resp. *vertex*) *optimal*.

Let $f$ be an integral flow, and $R$ the associated routing defined by $R(x, y) = p, p \in \mathcal{P}_{x,y}, f(p) = 1$. Then the following equalities link the forwarding indices of $R$ to the congestions of the flow $f$ :

$$m_f = \xi(G, R) + 2(n - 1), \qquad \mu_f = \pi(G, R).$$

(Notice that the added value $2(n-1)$ corresponding to the fact that a vertex is the extremity of $2(n-1)$ dipaths.)

Therefore the following problems, denoted by UCFP in [136], are generalizations of forwarding index problems.

**Problem 106 (Uniform concurrent multicommodity flow problem on the edges)** *Find a flow $f$ which minimizes the edge congestion $\mu_f$.*

**Problem 107 (Uniform concurrent multicommodity flow problem on the vertices)** *Find a flow $f$ which minimizes the vertex congestion $m_f$.*

The UCFP can be solved in polynomial time using polynomial time linear programming algorithms (see [133]). But the resulting algorithms are not considered efficient enough for practical applications. On the other hand, efficient approximations exist [134], [133].

The *integral uniform concurrent flow problem* (IUCFP) is to compute a flow $f$ such that $f(p) \in \{0, 1\}$, for any $p \in \mathcal{P}$, which is either edge or vertex optimal.

Let $G$ be a subgroup of $Aut(\Gamma)$. A uniform concurrent multicommodity flow $f$ on $\Gamma$ is said to be *$G$-invariant* if for any $g \in G$ and any $p \in \mathcal{P}$, we have $f(p) = f(g(p))$.

A dipath $p$ such that $f(p) = 1$ is said to be an *active path*. The following result gives bounds on the edge and vertex congestions of $G$-invariant flows.

**Theorem 108 ([136])** *Let $\Gamma$ be a graph, $G$ a subgroup of $Aut(\Gamma)$ and $f$ a uniform concurrent multicommodity $G$-invariant flow on $\Gamma$ such that any active path is a path of shortest length. Then*

$$\mu_f \leq \frac{1}{|O_E|} \sum_{(u,v) \in V \times V} d(u,v), \qquad m_f \leq \frac{1}{|O_V|}(|V|(|V-1|) + \sum_{(u,v) \in V \times V} d(u,v)),$$

*where $|O_E|$ and $|O_V|$ are the smallest edge and vertex orbits of $\Gamma$ under $G$.*

Translated in terms of forwarding indices, this theorem gives upper bounds for $\xi_m(\Gamma)$ and $\pi_m(\Gamma)$ which are to be compared to the lower bounds of Propositions 96 and 66. Applied to a Cayley graph and an optimal (for the vertex load) routing constructed by automorphisms in it, Theorem 108 proves that such a routing has an optimal edge-forwarding index up to a factor $d$.

**Theorem 109 ([136])** *Let $\Gamma = Cay(G, S)$ be a Cayley graph of degree $d$. There exists an integral uniform flow $f$ in $\Gamma$ such that*

(i) *$f$ is $G$-invariant*[4]*;*

(ii) *each active path is a path of shortest length;*

(iii) $\mu_f \leq \frac{d}{|E|} \sum_{(u,v) \in V \times V} d(u,v)$.

[4]The group $G$ is identified with the isomorphic subgroup of $Aut(\Gamma)$ which is composed of the translations on $\Gamma$.

**Proof** The proof of the first two assertions (i) and (ii) is similar to the proof of Theorem 67, since the construction of $f$ uses paths constructed as in this proof. Part (iii) is a corollary of Theorem 108. Notice that in a vertex transitive graph, the equalities $d|V|/2 = |E|$ and $d_1|V|/2 = |O_E|$ (where $d_1$ is the degree of the subgraph induced by $O_E$) imply $|E| \leq d|O_E|$. Thus, the inequality of Theorem 108 gives

$$\mu_f \leq \frac{d}{|E|} \sum_{(u,v) \in V\times V} d(u,v).$$

Using the equality $\pi(\Gamma) = \mu_f$ and the inequality of Proposition 96, we get :

**Corollary 110** *If $\Gamma$ is a Cayley graph, then*

$$\frac{1}{|E|} \sum_{(u,v) \in V\times V} d(u,v) \leq \pi(\Gamma) \leq \frac{d}{|E|} \sum_{(u,v) \in V\times V} d(u,v).$$

By using the inequality of Proposition 96, Theorem 108 shows also that, if there exists a subgroup $H$ of $Aut(\Gamma)$ which acts transitively on the edges of $\Gamma$ (in which case $|O_E| = |E|$) and such that there exists an $H$-invariant routing of shortest paths $R$ of $\Gamma$, then the edge-forwarding index satisfies

$$\pi(G) = 1/|E| \sum_{(u,v) \in V\times V} d(u,v),$$

and therefore is optimal. Orbital regular graphs are graphs for which this is true.

In [137], F. Shahrokhi and L. Szekely construct integral flows corresponding to symmetric routings (they consider only one path instead of two dipaths between any pair of vertices) which are asymptotically vertex (resp. edge) optimal in vertex (resp. edge) transitive graphs. For this they construct first a *canonical* $Aut(\Gamma)$-invariant flow $\hat{f}$ in the graph as follows. For any pair of vertices $x$, $y$, and any (undirected) path $p$ between $x$ and $y$ they define $\hat{f}(p) = 1/2M_{x,y}$, where $M_{x,y}$ is the number of paths between $x$ and $y$. Then they apply to $\hat{f}$ a method called randomized rounding to obtain an integral flow $f$ such that the congestions of $f$ are close to the congestions of $\hat{f}$. In particular, if $|E| = o(n^2/\log n)$, where $n = |V|$, then putting $L = \frac{1}{2}\sum_{(u,v) \in V\times V} d(u,v)$, this flow satisfies:

- if the graph is edge transitive and $e$ is any edge,

$$\frac{1}{|E|}L(1-o(1)) \leq f(e) \leq \frac{1}{|E|}L(1+o(1));$$

- if the graph is vertex transitive and $x$ is any vertex of $V$;

$$(\frac{L}{|V|}+n-1)(1-o(1)) \leq f(x) \leq (\frac{L}{|V|}+n-1)(1+o(1)).$$

This proves Conjectures 97 and 70 asymptotically and for symmetric routings in graphs such that $|E| = o(n^2/\log n)$. The same authors prove also in [135] that the cube-connected cycles graph $CCC(n)$ and the butterfly graph $BF(n)$ have an integral flow which is asymptotically both edge and vertex optimal (recall that these Cayley graphs are not edge transitive). They introduce a class of graphs called *G-orbit proportional* which contains the class of the edge transitive graphs and for which they obtain some similar results.

Notice that [137] and some references cited therein contain applications of flows to parameters which are important for interconnection networks, such as the crossing number and the bisection width. Similarly, forwarding indices are related to expanding in [139].

Finally, the reader should be aware that some results exist for partial routings instead of routings. For example, flows are studied in [137] in a more general context of partial routings than what is presented here. A similar notion of *generalized forwarding index* is defined in [93] which considers the load of a partial routing on a given subset of vertices.

# 6 Conclusion

In this course we have presented some Cayley digraphs which are studied as models of interconnection networks. We have surveyed neither all the possible graphs that have been studied as good architectures for interconnection networks, nor all aspects of those graphs that we *have* discussed. We have confined our attention to problems of structure, diameter, routing and loads of routings. We have paid much attention to Cayley graphs generated by transpositions, rotational Cayley graphs, hypercubic graphs and to new families of vertex transitive graphs which have good properties as far as routing is concerned, such as quasi-Cayley graphs and regular orbital graphs.

To conclude, we want to underline some trends of research in this rich field.

- The problem of finding new (di)graphs with good parameters improving the already known graphs is still being studied. Numerous articles have been devoted to the study of generalizations of the most famous graphs for interconnection networks, for example : the supertoroidal networks which are 4-regular Cayley graphs [59], the multiply-twisted hypercube (a Cayley graph with the same vertex set , degree and connectivity as the hypercube, but a diameter equal to half the diameter of the hypercube) [102], Cayley graphs with $n2^n$ vertices, degree 3 and 4, logarithmic diameter and constructed on signed permutations, [141], [143],[142]. For all these graphs, the authors justify their relevance for interconnection networks by studying their parameters but also algorithms on them (routing, embedding, communication).

- General theorems are missing which would extend to *classes* of Cayley graphs properties that have been proved for particular graphs (for example, the calculus of the wide-diameter in Section 4.5).

- Very few good algorithms are known to be efficient for all Cayley digraphs (as is the case in [55]). One usually starts with a given algorithm, and asks for which Cayley digraphs it is good, as in [82].

- More group theory has to be exploited to find properties of Cayley graphs. For example, in [5] the subgroup structure of a group reveals ways to decompose some Cayley graphs on that group.

- Bus interconnection networks generalize the notion of point to point interconnection networks. In a bus network, some sets of processors share a communication medium called a bus. In a natural way, they are modeled by hypergraphs in which the vertices represent the processors, and the hyperedges represent the buses that group the processors sharing the same communication link. An increasing number of articles consider problems related to bus interconnection networks (see for example [61]). Different ways of extension of Cayley graphs related to hypergraphs have already been considered. One is to consider the dual hypergraph of a Cayley graph (see for example [23]); a second one is to consider hypergraphs the line graphs of which are Cayley graphs (for example, [12]). A definition of Cayley hypergraphs is given in [49].

• We only gave an overview of some static problems. Some dynamic problems interesting for distributed computing are considered in [105]. Parallel models of computation include automata. Cayley automata are defined in [75]. Studies of cellular automata working on Cayley graphs can be found in [109], [118] and [119].

Are Cayley graphs the most appropriate models for interconnection networks ? We have seen that quotients of Cayley digraphs can also be very convenient, and some people have confidence in non-symmetric graphs. What is certain is that the relation between Cayley graphs and interconnection networks is fruitful for the research in both fields.

# 7 Acknowledgements

Particular thanks are due to Bertrand Ducourthial who drew all the figures and took care of the readability of this article, and to Charles Delorme and Gert Sabidussi who carefully read this course and criticized to improve the mathematical substance and the references.

The author is also greatly indebted to a number of researchers for their active interest in this paper, in particular Denise Amar, Ginette Gauyacq, Geña Hahn, Patrick Solé, Dominique Sotteau, ... and to all the people who were patient during the writing of this article.

The author wishes to express her thanks for the helpful suggestions and references given in particular by Dominique Barth, Francesc Comellas, Evelyne Flandrin, Anna Lladó, Margarida Mitjana, Cheryl Praeger, Joel Puech, Michael Sampels, Raffaele Scapellato, Oriol Serra, Polina Strogova, Hal Sudborough, and Mark Watkins.

# 8 Appendix: some families of Cayley digraphs

This section contains definitions and drawings of some Cayley graphs studied or used as examples in this course. More information on them can be found for example in [3], [103], [106], and [47].

## 8.1 Cycle

The cycle $C_n$ is the connected graph with $n$ vertices and degree 2. It is the Cayley graph of the additive group $\mathbb{Z}_n$ generated by $\{-1,+1\}$. A complete rotation on $(\mathbb{Z}_n, \{-1,+1\})$ is defined by $(i \mapsto n-i)$.

$C_n$ is also the Cayley graph on $\mathfrak{S}_n$ and the subset consisting of the two cycles $\langle 1,2,\ldots,n\rangle$ and $\langle n,1,\ldots,n-1\rangle$. In this case a complete rotation $\omega$ is defined by $\omega(x) = \sigma x \sigma^{-1}$, where $\sigma$ is the permutation given by $\langle n, n-1, \ldots, 2, 1\rangle$.

$Diam(C_n) = \lfloor n/2 \rfloor$. $C_n$ is a distance transitive graph.

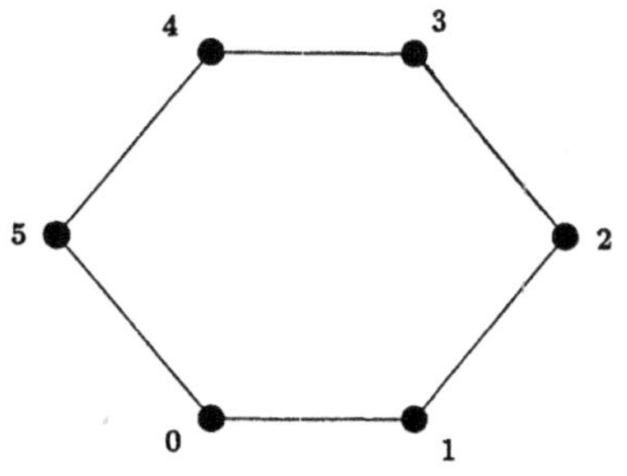


Figure 5: Cycle $C_6$

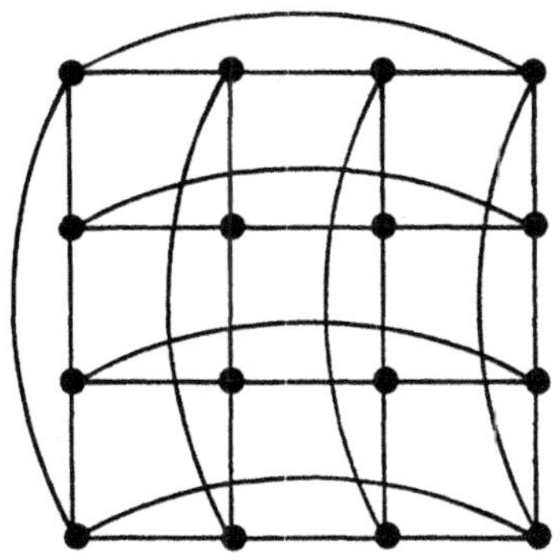

Figure 6: Torus $MT_{2,4}$

## 8.2 Multidimensional torus

The multidimensional torus $MT_{n,k}$ is the cartesian product of $n$ cycles of length $k$. $MT_{n,k}$ has $k^n$ vertices and degree $2n$. It is the Cayley graph of the group $\mathbb{Z}_k^n$ generated by the subset of $2n$ elements

$$S = \{(\underbrace{0,\ldots,0}_{i},1,\underbrace{0,\ldots,0}_{n-i-1}),(\underbrace{0,\ldots,0}_{i},-1,\underbrace{0,\ldots,0}_{n-i-1}) : 0 \leq i \leq n-1\}.$$

$MT_{n,k}$ is rotational (see Proposition 22). $Diam(MT_{n,k}) = n\lceil k/2 \rceil$.

## 8.3 Hypercube

The hypercube $H(n)$ is the graph with vertex set $\{x_1x_2\ldots x_n : x_i \in \{0,1\}\}$, two vertices $x_1x_2\ldots x_n$ and $y_1y_2\ldots y_n$ being adjacent if and only $x_i = y_i$ for all but one $i$ (see Figure 7). $H(n)$ can be considered as a particular case of torus, namely $MT_{n,2}$, since it is the cartesian product of $n$ complete graphs $K_2$. $H(n)$ is the Cayley graph of the group $(\mathbb{Z}_2^n,+)$ generated by the $n$ generators $\underbrace{0\ldots0}_{i}1\underbrace{0\ldots0}_{n-i-1}$, $0 \leq i \leq n-1$.

$H(n)$ is also the Cayley graph of the permutation group $G$ generated by the $n$ transpositions $\langle 2i-1, 2i\rangle$, $1 \le i \le n$, defined on the set of $2n$ elements $X = \{1, \ldots, 2n\}$ ($H(4)$ is shown in Figure 8 and the associated transposition graph in Figure 9). Indeed, each vertex $x_1x_2\ldots x_n$, where $x_i \in \{0,1\}$, can be renamed as the permutation $(a_1, a_2, \ldots, a_{2n})$, where $(a_{2i-1}, a_{2i}) = (2i-1, 2i)$ if $x_i = 0$ and $(a_{2i-1}, a_{2i}) = (2i, 2i-1)$ if $x_i = 1$.

$H(n)$ is rotational. A complete rotation $\omega$ is defined on $(G, S)$ by $\omega(x) = \sigma x \sigma^{-1}$, where $\sigma$ is the permutation given by $\sigma = (3, 4, \ldots, 2n-1,\ 2n, 1, 2) = \langle 1, 3, \ldots, 2n-1\rangle\langle 2, 4, \ldots, 2n\rangle$. Thus, $\sigma^{-1} = (2n-1, 2n, 1, 2, \ldots, 2n-3, 2n-2)$ and $\sigma^i\langle 1, 2\rangle\sigma^{-i} = \langle 2i+1, 2i+2\rangle$.

$Diam(H(n)) = n$. $H(n)$ is a distance transitive graph.

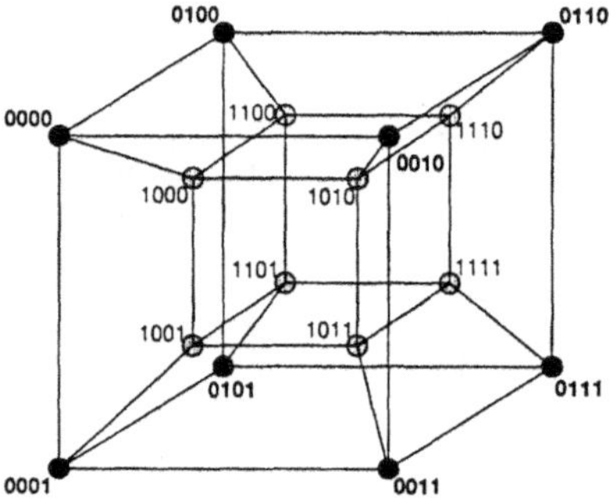


Figure 7: Hypercube graph $H(4)$

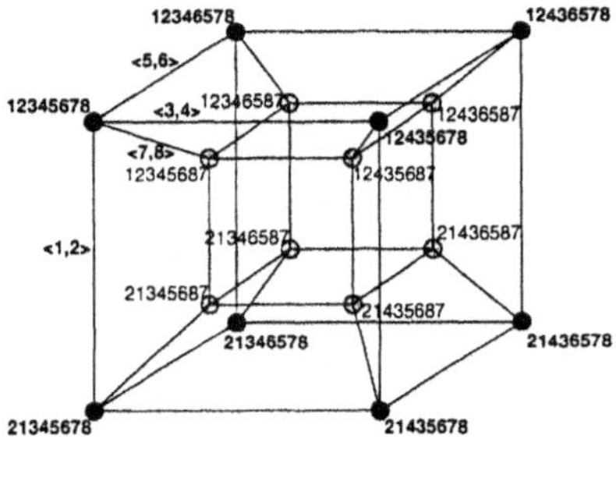


Figure 8: $H(4)$ is generated by transpositions

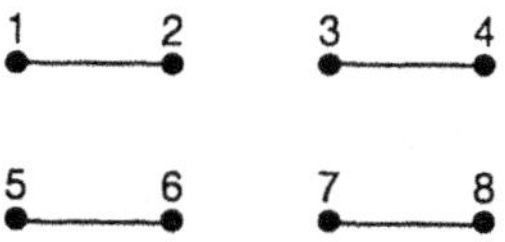


Figure 9: Transposition graph for $H(4)$

## 8.4 Star graph

The star graph $ST(n)$ is defined as the Cayley graph of the group $\mathfrak{S}_n$ generated by the $n-1$ transpositions $S = \{\langle 1, i\rangle, 1 < i \leq n\}$. The associated transposition graph is the star $K_{1,n-1}$ (see $ST(4)$ depicted in Figure 10 and the associated transposition graph in Figure 11). A complete rotation $\omega$ is defined on $(\mathfrak{S}_n, S)$ by $\omega(x) = \sigma x \sigma^{-1}$, where the permutation $\sigma$ given by $\sigma = (1, 3, 4, \ldots, n, 2) = \langle 2, 3, \ldots, n\rangle$.

$Diam(ST(n)) = \lfloor 3(n-1)/2 \rfloor$ (see [3]).

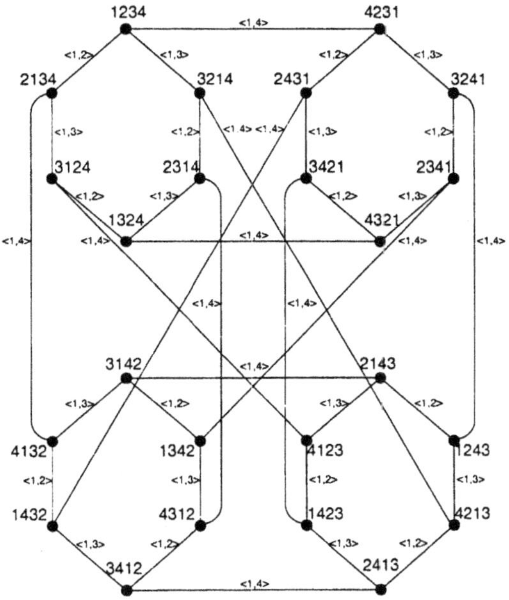


Figure 10: Star graph $ST(4)$

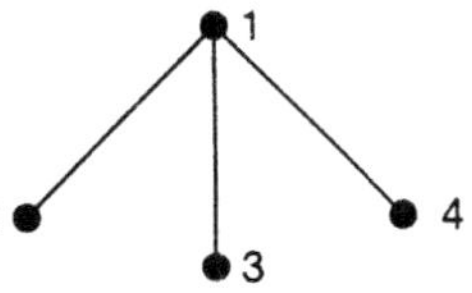


Figure 11: Transposition graph for $ST(4)$

## 8.5 Complete transposition graph

The complete transposition graph of dimension $n$, $CT_n$, is the Cayley graph of the group $\mathfrak{S}_n$ generated by all the transpositions $\{\langle i, j\rangle, 1 \leq i \leq n, 1 \leq j \leq n, i \neq j\}$. The associated transposition graph is the complete graph.

*Diam*$(CT_n) = n - 1$. $CT_n$ is arc transitive but not distance transitive [103].

## 8.6 Generalized star graph

The generalized star graph $GST(n,k)$ is defined in Section 3.2. $GST(n,k)$ is the Cayley graph of the group $\mathfrak{S}_n$ generated by the $k(n-k)$ transpositions $S = \{\langle i,j\rangle, 1 \le i \le k, k+1 \le j \le n\}$. The associated transposition graph is the bipartite complete graph $K_{k,n-k}$. $GST(4,2)$ is shown in Figure 12 and the associated transposition graph in Figure 13. The graph $GST(n,k)$ is rotational if $n = 4$ and $k = 2$ or if $k$ and $n - k$ are relatively prime (see Proposition 17).

*Diam*$(GST(k,n)) = n - 1 + \max(\lfloor \frac{k}{2} \rfloor, \lfloor \frac{n-k}{2} \rfloor)$ [128].

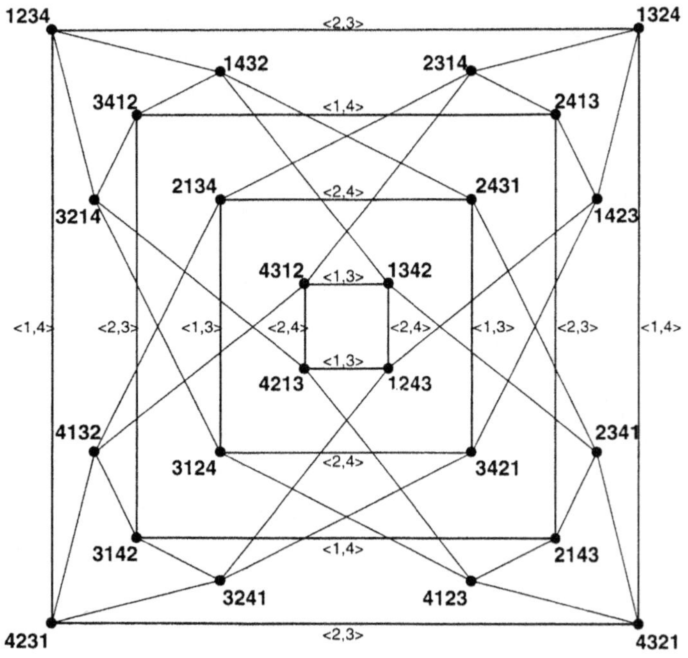


Figure 12: Generalized Star Graph $GST(4,2)$

## 8.7 Bubble sort graph

The bubble sort graph of dimension $n$ (or permutohedron), $BS(n)$, is the Cayley graph of $\mathfrak{S}_n$ generated by the $n-1$ transpositions $\{\langle i, i+1\rangle, 1 \le i < n\}$[5](see Figure 14). The associated transposition graph is a path of length $n - 1$.

[5]This Cayley graph is constructed on a Coxeter group [41].

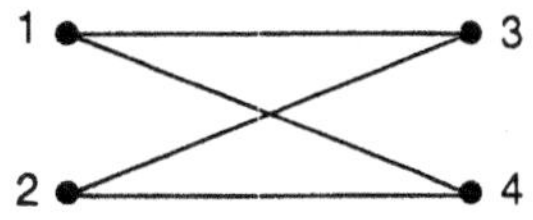


Figure 13: Transposition graph for $GST(4,2)$

$Diam(BS(n)) = n(n-1)/2$ [103].

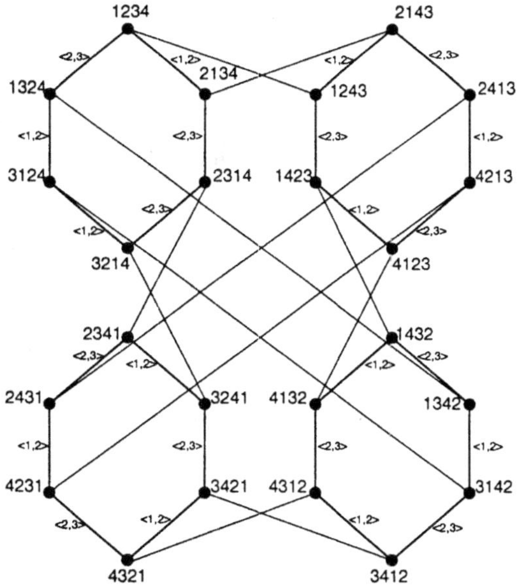


Figure 14: Bubble sort graph $BS(4)$

## 8.8 Modified bubble sort graph

The modified bubble sort graph of dimension $n$, $MBS(n)$, is defined as the bubble sort graph, adding a $n^{th}$ transposition: $\langle n, 1\rangle$. Thus the associated transposition graph is the cycle $C_n$. $MBS(n)$ has a complete rotation $\omega$ defined by $\omega(x) = \sigma x \sigma^{-1}$, where $\sigma$ is the cyclic permutation given by $\langle 1, 2, \ldots, n\rangle$.

$Diam(MBS(n)) = \lfloor n^2/4 \rfloor$ [128].

## 8.9 Pancake

The pancake graph, $PR(n)$, also called *prefix reversal graph*, is the Cayley graph of the group $\mathfrak{S}_n$ generated by the $n-1$ permutations $(\underbrace{i, i-1, \ldots, 1}_{i}, i+1, \ldots, n)$, $i = 2, \ldots, n$, of order 2.

$PR(2) = ST(2) = K_2$ and $PR(3) = ST(3) = C_6$. But $PR(n)$ is not edge transitive for $n \geq 4$ and therefore not rotational. $Diam(PR(n))$ is unknown for $n > 13$ (see Section 4).

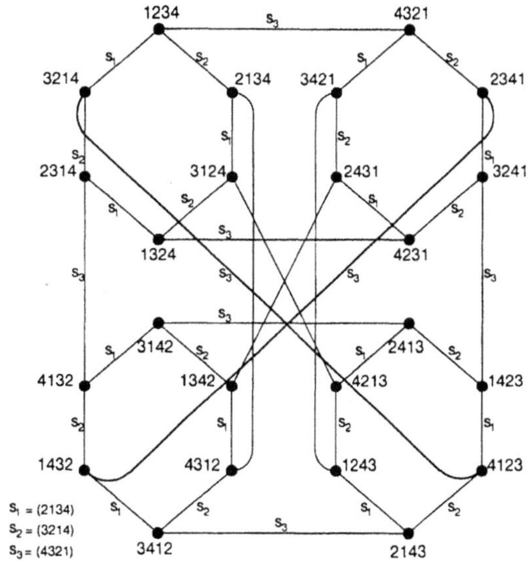


Figure 15: Pancake graph $PR(4)$

## 8.10 Cube Connected Cycle

The cube-connected cycles graph $CCC(n)$ is defined in Section 3.4.2. $CCC(1) = K_2$, $CCC(2) = C_8$. For $n \geq 3$, $CCC(n)$ is obtained from $H(n)$ by replacing each vertex by a cycle of length $n$ such that the degree is 3 (see Figure 16). The number of vertices is $n \times 2^n$. For $n \geq 3$,

$$Diam(CCC(n)) = 2n - 1 + \max(1, \lfloor \frac{n-2}{2} \rfloor).$$

This graph is not edge transitive and therefore not rotational for $n \geq 3$.

## 8.11 Butterfly

The butterfly graph $BF(n)$ is defined in Section 3.4.2. Its number of vertices is $n \times 2^n$. $BF(1) = K_2$, $BF(2) = H(3)$. For $n \geq 3$, the degree is 4 and $Diam(BF(n)) = n + \lfloor n/2 \rfloor$. This graph is not edge transitive and therefore not rotational for $n \geq 3$.

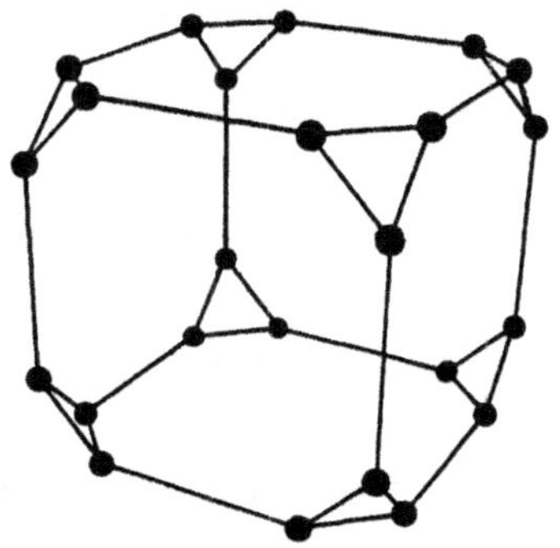

Figure 16: $CCC(3)$

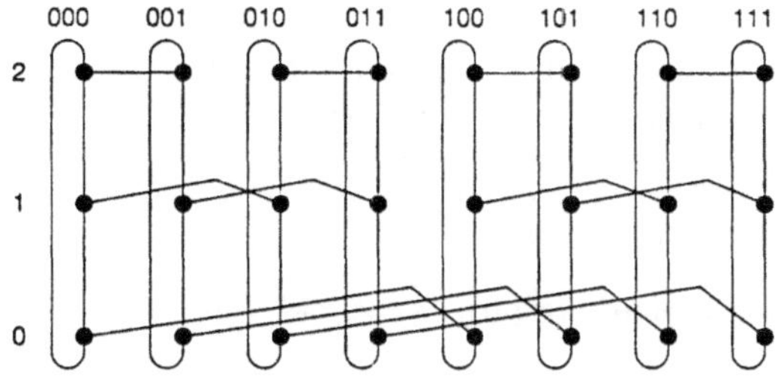


Figure 17: $CCC(3)$

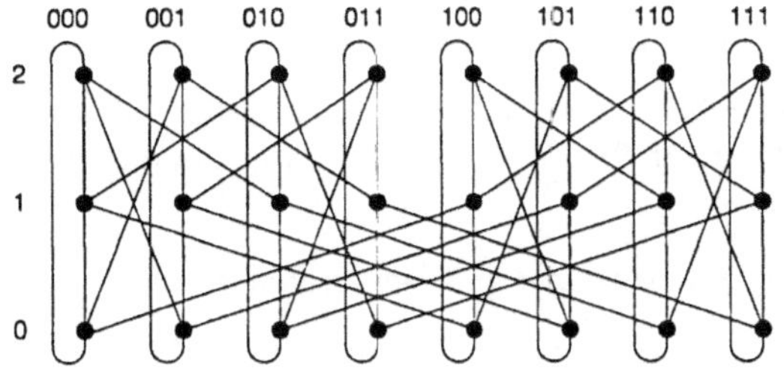


Figure 18: Butterfly graph $BF(3)$

## References


[1] S. Akers and B. Krishnamurthy, Group graphs as interconnection networks, *Proc. 14th Internat. Conf. Fault Tolerant Comput.*, 1984, 422–427.

[2] S. Akers and B. Krishnamurthy, On group graphs and their fault tolerance, *IEEE Trans. Comput.* **36** (1987), 885–888.

[3] S. Akers and B. Krishnamurthy, A group theoretic model for symmetric interconnection networks, *IEEE Trans. Comput.* **38** (1989), 555–566.

[4] B. Alspach, Cayley graphs with optimal fault tolerance, *IEEE Trans. Comput.* **41** (1992), 1337–1339.

[5] F. Annexstein and M. Baumslag, On the diameter and bisection size of Cayley graphs, *Math. Systems Theory* **26** (1993), 271–291.

[6] F. Annexstein, M. Baumslag, and A. L. Rosenberg, Group action graphs and parallel architectures, *SIAM J. Comput.* **19** (1990), 544–569.

[7] J.-M. Arnaudies and J. Bertin, *Groupes, algèbres et géométrie*, vol. 1, Ellipses, Paris, 1993.

[8] L. Babai, W. M. Kantor, and A. Lubotzky, Small diameter Cayley graphs for finite simple groups, *European J. Combin.* **10** (1989), 507–522.

[9] L. Babai and A. Seress, On the diameter of Cayley graphs of the symmetric group, *J. Combin. Theory Ser. A* **49** (1988), 175–179.

[10] E. Bannai, T. Ito, On Moore graphs, *J. Fac. Sci. Univ. Tokyo Sect. IA Math.* **20** (1973), 191–208.

[11] D. Barth, Private communication, 1996.

[12] D. Barth, A. Germa, M.-C. Heydemann, and D. Sotteau, Emulating networks by bus networks: Application to trees and hypermeshes, Technical Report 1073, L.R.I., Orsay, 1996.

[13] M. Baumslag, *Cayley networks : a group-theoretic approach to the design and analysis of parallel networks*, PhD thesis, The City University of New York, 1991.

[14] M. Baumslag, On the fault tolerance of quasi-minimal Cayley networks, in: *Advances in Comput. and Inf., ICCI '91*, Lecture Notes in Comput. Sci. **497**, Springer-Verlag, Berlin, 1992, 431–442.

[15] C. Berge, *Principes de combinatoire*, Dunod, Paris, 1968.

[16] C. Berge, *Graphes et hypergraphes*, Bordas, Paris, 1973.

[17] J.-C. Bermond, ed., *DAMIN*, *Discrete Applied Math.* **37-38** (1992).

[18] J.-C. Bermond, F. Comellas, and D. F. Hsu, Distributed loop computer networks: a survey, *J. Parallel Distrib. Comput.* **24** (1995), 2–10.

[19] J.-C. Bermond, C. Delorme, and J. Quisquater, Strategies for interconnection networks: some methods from graph theory, *J. Parallel Distrib. Comput.* **3** (1986), 433–449.

[20] J.-C. Bermond, C. Delorme, and J.-J. Quisquater, Table of large $(\Delta, D)$ graphs, *Discrete Appl. Math.* **37-38** (1992), 575–577.

[21] J.-C. Bermond, T. Kodate, and S. Perennes, Gossiping in Cayley graphs by packets, in: *Conf. CCS95 (8th Franco-Japanese and 4th Franco-Chinese Conf. Combin. Comput. Sci. (Brest, July 1995))*, Lecture Notes in Comput. Sci. **1120** , Springer-Verlag, Berlin, 1996, 301–305.

[22] P. Berthomé, A. Ferreira, and S. Perennes, Optimal information dissemination in star and pancake networks, in: *Proc. 5th IEEE Symp. Parallel Distribut. Process.*, IEEE Press, 1993, 720–724.

[23] L. Bhuyan and D. Agraval, Generalized hypercubes and hyperbus structures for a computer network, *IEEE Trans. Comput.* **33** (1984), 323–333.

[24] N. Biggs, *Algebraic Graph Theory*, Cambridge University Press, 1974.

[25] N. Biggs and A. White, *Permutation Groups and Combinatorial Structures*, London Math. Soc. Lecture Note Ser. **33**, Cambridge University Press, 1979.

[26] F. Boesch and R. Tindell, Circulants and their connectivities, *J. Graph Theory* **8** (1984), 487–499.

[27] B. Bollobás, *Graph Theory. An Introductory Course*, Graduate Texts in Math. **63**, Springer-Verlag, New York, 1979.

[28] J.A. Bondy and U.S.R. Murty, *Graph Theory and Applications*, Macmillan, London, 1976.

[29] A. Bouabdallah, M.-C. Heydemann, J. Opatrny, and D. Sotteau, Embeddings of complete binary trees into star graphs, in: *19th Conf. Math. Foundations of Comput. Sci.*, Lecture Notes in Comput. Sci. **841**, Springer-Verlag, Berlin, 1994, 266–275.

[30] J. Brunat, M. Espona, M. Fiol, and O. Serra, On Cayley line digraphs, *Discrete Math.* **138** (1995), 147–159.

[31] G. Butler, *Fundamental Algorithms for Permutation Groups*, Lecture Notes in Comput. Sci. **559**, Springer-Verlag, Berlin, 1991.

[32] J. Calais, *Eléments de la théorie des groupes*, 2nd ed., Presses universitaires de France, Paris, 1996.

[33] L. Campbell, G. Carlsson, M. Dinneen, V. Faber, M. Fellows, M. Langston, J. Moore, A. Mullhaupt, and H. Sexton, Small diameter symmetric networks from linear groups, *IEEE Trans. Comput.* **41** (1992), 218–220.

[34] J. J. Cannon, Construction of defining relators for finite groups, *Discrete Math.* **5** (1973), 105–129.

[35] F. Chung, E. Coffman, M. Reiman, and B. Simon, On the capacity and vertex-forwarding index of communication networks, *IEEE Trans. Inform. Theory* **33** (1987), 224–232.

[36] F. Comellas and M. Fiol, Vertex-symmetric digraphs with small diameter, *Discrete Appl. Math.* **58** (1995), 1–11.

[37] F. Comellas and M. Mitjana, Private communication, 1996.

[38] F. Comellas and M. Mitjana, Broadcasting in cycle prefix digraphs, submitted to *Discrete Appl. Math.*; DMAT Technical Report 05-9515, 1995.

[39] G. Cooperman and L. Finkelstein, New methods for using Cayley graphs in interconnection networks, *Discrete Appl. Math.* **37-38** (1992), 95–118.

[40] G. Cooperman, L. Finkelstein, and N. Sarawagi, Applications of Cayley graphs, in: *Applied Algebra, Algebraic Algorithms and Error-Correcting Codes (AAECC-8, 1990)*, Lecture Notes in Comput. Sci. **508**, Springer-Verlag, Berlin, 1991, 367–378.

[41] H.S.M. Coxeter and W. Moser, *Generators and Relations for Discrete Groups*, Springer-Verlag, New York, 1972.

[42] S. Curran and J. Gallian, Hamiltonian cycles and paths in Cayley graphs and digraphs - a survey, preprint, 1996.

[43] R.M. Damerell, On Moore graphs, *Proc. Cambridge Philos. Soc.* **74** (1973), 227–236.

[44] K. Day and A. Tripathi, Arrangements graphs: a class of generalized star graphs, *Inform. Process. Lett.* **42** (1992), 235–241.

[45] K. Day and A. Tripathi, A comparative study of topological properties of hypercubes and star graphs, *IEEE Trans. Parallel Distrib. Systems* **5** (1994), 31–38.

[46] W. F. de la Vega and Y. Manoussakis, Computation of the forwarding index via flows: a note, *Networks* **24** (1994), 273–276.

[47] J. de Rumeur, *Communications dans les réseaux de processeurs*, Études Rech. Inform., Masson, Paris, 1994.

[48] O. Delmas, Communications par commutation de circuits dans les réseaux d'interconnexion, PhD thesis, Université de Nice - Sophia Antipolis, Laboratoire I3S-CNRS URA 1376, 1997.

[49] C. Delorme, Graphes et hypergraphes sommet-transitifs, Technical Report 383, L.R.I., Orsay, 1987.

[50] C. Delorme, Construction de graphes possédant des routages de plus court chemin chargeant également les arêtes, Technical Report 482, L.R.I., Orsay, 1989.

[51] C. Delorme, Example of products giving large graphs with given degree and diameter, *Discrete Appl. Math.* **37-38** (1992), 157–167.

[52] C. Delorme, Isomorphism of transposition graphs, Technical Report, L.R.I., Orsay, 1997.

[53] D. Désérable, A family of Cayley graphs on the hexavalent grid, to appear in *Discrete Appl. Math.*, Special issue on Network Communication, 1997.

[54] W. Dörfler, Every regular graph is a quasigroup graph, *Discrete Math.* **10** (1974), 181–183.

[55] V. Dimakopoulos and N. Dimopoulos, Total exchange in Cayley networks, in: *Euro-Par96*, Lecture Notes in Comput. Sci. **1123**, Springer-Verlag, Berlin, 1996, 341–346.

[56] M. J. Dinneen and P. R. Hafner, New results on the degree/diameter problem, *Networks* **24** (1994), 359–367.

[57] D. Dolev, J. Halpern, B. Simons, and H. Strong, A new look at fault tolerant network routing, *Inform. and Comput.* **72** (1987), 180–196.

[58] L. Doty, R. Goldstone, and C. Suffel, Cayley graphs with neighbor connectivity one, *SIAM J. Discrete Math.* **9** (1997), 625–642.

[59] R. N. Draper, An overview of supertoroidal networks, in: *3rd ACM Symp. Parallel Algorithms and Architectures*, 1991, 95–102.

[60] D.-R. Duh, G.-H. Chen, and D. F. Hsu, Combinatorial properties of generalized hypercube graphs, *Inform. Process. Lett.* **57** (1996), 41–45.

[61] F. Ergincan and J.-C. Bermond, Bus interconnection networks, to appear in *Discrete Appl. Math.*

[62] M. Escudero, J. Fabrega, and P. Morillo, Fault tolerance routings in double-loop networks, *Ars Combin. A* **25** (1988), 187–198.

[63] B. Elspas, Topological constrainst on interconnection limited logic, in: *Proc. 5th Sympos. Switching Theory and Logical Design*, IEEE, 1964, 133–147.

[64] S. Even and O. Goldreich, The minimal-length generating sequence problem is NP-hard, *J. Algorithms* **2** (1981), 311–313.

[65] V. Faber, J. Moore, and W. Y. C. Chen, Cycle prefix digraphs for symmetric interconnection networks, *Networks* **23** (1993), 641–649.

[66] X. G. Fang, C. H. Li, and C. Praeger, On orbital regular graphs and Frobenius graphs, preprint, Univ. of Western Australia, 1996.

[67] R. J. Faudree, Some strong variations of connectivity, in: *Combinatorics, Paul Erdős is Eighty*, Vol. 1 (D. Miklós, V.T. Sós, T. Szőnyi, eds.), Coll. Math. Soc. János Bolyai **64**, Budapest, 1993, 125–144.

[68] R. Feldmann and W. Unger, The cube-connected cycles network is a subgraph of the butterfly network, *Parallel Process. Lett.* **2** (1992), 13–19.

[69] D. Fortin, C. Kirchner, and P. Strogova, Routing in regular networks using rewriting, in: *Proc. CADE Internat. Workshop on Automated Reasoning in Algebra (ARIA)*, (J. Slaney, ed.), 1994, 5–8.

[70] J. Fournier, Le groupe d'automorphismes des graphes de Cayley engendrés par des transpositions, Mémoire de maîtrise, Université de Montréal, 1997.

[71] P. Fragopoulou, Communication and fault tolerance algorithms on a class of interconnection networks, PhD thesis, Queen's University, Kingston, Ont., 1995.

[72] P. Fragopoulou and S. G. Akl, Optimal communication algorithms on star graphs using spanning tree constructions, *J. Parallel Distrib. Comput.* **24** (1995), 55–71.

[73] P. Fragopoulou and S. G. Akl, Optimal communication primitives on the generalized hypercube network, *J. Parallel Distrib. Comput.* **32** (1996), 173–187.

[74] P. Fragopoulou and S. G. Akl, Spanning graphs with applications to communication on a subclass of the Cayley graph based networks, submitted to *Discrete Appl. Math.*

[75] M. Garzon, Cayley automata, *Theoret. Comput. Sci.* **108** (1993), 83–102.

[76] W.H Gates and C.H. Papadimitriou, Bounds for sorting by prefix reversal, *Discrete Math.* **27** (1979), 47–57.

[77] G. Gauyacq, Edge-forwarding index of star graphs and other Cayley graphs, to appear in *Discrete Appl. Math.*, Research report LABRI, Bordeaux.

[78] G. Gauyacq, *Routages uniformes dans les graphes sommet-transitifs*, PhD thesis, LABRI, Bordeaux, 1995.

[79] G. Gauyacq, On quasi-Cayley graphs, to appear in *Discrete Appl. Math.*

[80] A. Germa, M.-C. Heydemann, and D. Sotteau, Cycles in cube-connected cycles, to appear in *Discrete Appl. Math.* Technical Report 1074, L.R.I., Orsay, 1996.

[81] C. D. Godsil, Connectivity of minimal Cayley graphs, *Arch. Math.* **37** (1981), 473–476.

[82] D. M. Gordon, Parallel sorting on Cayley graphs, *Algorithmica* **6** (1991), 554–564.

[83] C. GowriSankaran, Broadcasting on recursively decomposable Cayley graphs, *Discrete Appl. Math.* **53** (1994), 171–182.

[84] Q.-P. Gu and S. Peng, Fault tolerant routing in hypercubes and star graphs, *Parallel Process. Lett.* **6** (1996), 127–136.

[85] Y. O. Hamidoune, Sur quelques problèmes de connexité dans les graphes orientés, *J. Combin. Theory Ser. B* **30** (1981), 1–10.

[86] Y. O. Hamidoune, A. S. Lladó, and O. Serra, Vosperian and superconnected Abelian Cayley digraphs, *Graphs Combin.* **7** (1991), 143–152.

[87] Y. O. Hamidoune, A. S. Lladó, and O. Serra, The connectivity of hierarchical digraphs, *Discrete Appl. Math.* **37-38** (1992), 275–280.

[88] Y. O. Hamidoune, A. S. Lladó, and O. Serra, Small cuts in quasiminimal Cayley graphs, *Discrete Math.* **159** (1996), 131–142.

[89] Y. O. Hamidoune, A. S. Lladó, and O. Serra, An isoperimetric problem in Cayley graphs, preprint, 1997.

[90] Y. O. Hamidoune, A. S. Lladó, and O. Serra, On isoperimetric connectivity in vertex-transitive graphs, preprint, 1997.

[91] M. Heydari and H. Sudborough, On sorting by prefix reversals and the diameter of pancake networks, Lecture Notes in Comput. Sci. **678** (1993), 218–227.

[92] M.-C. Heydemann, N. Marlin, and S. Perennes, Cayley graphs with rotations, preprint, L.R.I., Orsay, 1997.

[93] M.-C. Heydemann, J.-C. Meyer, J. Opatrny, and D. Sotteau, Forwarding indices of consistent routings and their complexity, *Networks* **24** (1994), 75–82.

[94] M.-C. Heydemann, J.-C. Meyer, and D. Sotteau, On forwarding indices of networks, *Discrete Appl. Math.* **23** (1989), 103–123.

[95] C.-T. Ho and M.-Y. Kao, Optimal broadcast in all-port wormhole-routed hypercubes, *IEEE Trans. Parallel Distrib. Systems* **6** (1995), 200–204.

[96] A. Hoffman and R.R. Singleton, On Moore graphs with diameters 2 and 3, *IBM J. Res. Develop.* **4** (1960), 497–504.

[97] D. F. Hsu, Introduction to a special issue on interconnection networks, *Networks* **23** (1993), 211–213.

[98] D. F. Hsu, On container width and length in graphs, groups and networks, *IEICE Trans. Fundamentals E* **77**(A) (1994), 668–680.

[99] M. Jerrum, The complexity of finding minimal-length generating sequence, *Theoret. Comput. Sci.* **36** (1985), 265–289.

[100] W. H. Kautz, Bounds on directed $(d, k)$ graphs. Theory of cellular logic networks and machines, Final report, AFCRL-68-0668, 1968.

[101] M. S. Krishnamoorthy and B. Krishnamurthy, Fault diameter of interconnection networks, *Comput. Math. Appl.* **13** (1987), 577–1582.

[102] P. Kulasinghe and S. Bettayeb, On the multiply-twisted hypercube, in: *Parallel and Distributed Computing*, Lecture Notes in Comput. Sci. **805**, Springer-Verlag, Berlin, 1994, 267–278.

[103] S. Lakshmivarahan, J. Jwo, and S. K. Dhall, Symmetry in interconnection networks based on Cayley graphs of permutation groups: a survey, *Parallel Comput.* (1993), **19** 361–407.

[104] S. Latifi, On the fault diameter of the star graph, *Inform. Process. Lett.* **46** (1993), 143–150

[105] C. Lavault, Interconnection networks: graph and group-theoretic modelling, preprint 94-04, LIPN, Villetaneuse, France, 1994.

[106] F. T. Leighton, *Introduction to Parallel Algorithms and Architectures*, M. Kaufmann, San Mateo, CA, 1991.

[107] A. L. Liestman and D. Richards, eds., *Proc. Int. Workshop on Broadcasting and Gossiping, 1990; Discrete Applied Math.* **53** (1994).

[108] J. Liu, Hamiltonian decomposition of Cayley graphs on abelian groups of odd order, *J. Combin. Theory Ser. B* **66** (1996), 75–76.

[109] A. Machi and F. Mignosi, Garden of Eden configurations for cellular automata on Cayley graphs of groups, *SIAM J. Discrete Math.* **6** (1993), 44–56.

[110] W. Mader, Minimale n-fach kantenzusammenhängende Graphen, *Math. Ann.* **191** (1971), 21–28.

[111] Y. Manoussakis and Z. Tuza, The forwarding index of directed networks, *Discrete Appl. Math.* **68** (1996), 279–291.

[112] N. Marlin, Rotations complètes dans les graphes de Cayley, DEA, Université de Nice Sophia-Antipolis, 1996.

[113] E.F. Moore, Personal communication referred to in [96], 1958.

[114] J. Morris, Connectivity of Cayley graphs: A special family, *J. Combin. Math. Combin. Comput.* **20** (1996), 111–120.

[115] C. Padro and P. Morillo, Diameter vulnerability of iterated line digraphs, *Discrete Math.* **149** (1996), 189–204.

[116] F. P. Preparata and J. Vuillemin, The cube-connected cycles : a versatile network for parallel computation, *Comm. ACM* **24** (1981), 300–309.

[117] D. J. S. Robinson, *A Course in the Theory of Groups*, 2nd ed., Springer-Verlag, 1996.

[118] Z. Róka, Automates cellulaires sur graphes de Cayley, PhD thesis, École Normale Supérieure de Lyon, Lyon, 1994.

[119] Z. Róka, One-way cellular automata on Cayley graphs, *Theoret. Comput. Sci.* **132** (1994), 259–290.

[120] Y. Rouskov, S. Latifi, and P. K. Srimani, Conditional fault diameter of star graph networks, *J. Parallel Distrib. Comput.* **33** (1996), 91–97.

[121] Y. Rouskov and P. K. Srimani, Fault diameter of star graphs, *Inform. Process. Lett.* **48** (1993), 243–251.

[122] R. Saad, Complexity of the forwarding index problem, *SIAM J. Discrete Math.* **6** (1993), 418–427.

[123] Y. Saad and M. H. Schultz, Topological properties of hypercubes, *IEEE Trans. Comput.* **37** (1988), 867–872.

[124] G. Sabidussi, On a class of fixed-point-free graphs, *Proc. Amer. Math. Soc.* **9** (1958), 800–804.

[125] G. Sabidussi, Graph multiplication, *Math. Z.* **72** (1960), 446–457.

[126] G. Sabidussi, The lexicographic product of graphs, *Duke Math. J.* **28** (1961), 573–578.

[127] G. Sabidussi, Vertex transitive graphs, *Monatsh. Math.* **68** (1964), 426–438.

[128] J.-F. Saclé, Diameter of some Cayley graphs, private communication, 1997.

[129] M. Sampels, Algebraic constructions of efficient systolic architectures, in: *Proc. 2nd Internat. Conf. Massively Parallel Computing Systems (MPCS '96)*, IEEE, 1996.

[130] M. Sampels, Private communication, 1997.

[131] M. Sampels and S. Schöf, Massively parallel architectures for parallel discrete event simulation, in: *Proc. 8th European Simulation Symp. (ESS '96)*, vol. 2, SCS, 1996, 374–378.

[132] S. Schibell and R. Stafford, Processor interconnection networks from Cayley graphs, *Discrete Appl. Math.* **40** (1992), 333–357.

[133] F. Shahrokhi, Approximation algorithms for the maximum concurrent flow problem, *ORSA J. Comput.* **1** (1989), 62–69.

[134] F. Shahrokhi and D. Matula, The maximum concurrent flow problem, *J. Assoc. Comput. Mach.* **37** (1990), 318–334.

[135] F. Shahrokhi and L. Szekely, Formula for the optimal congestion of the uniform concurrent multicommodity flow, Research Report CRPDC-93-6, University of North Texas, 1993.

[136] F. Shahrokhi and L. Szekely, Integral uniform concurrent flows and all pairs packet routing, *Congr. Numer.* **97** (1993), 3–16.

[137] F. Shahrokhi and L. Szekely, On canonical concurrent flows, crossing numbers and graph expansion, *Combin. Probab. Comput.* **3** (1994), 523–543.

[138] P. Solé, The forwarding index of orbital regular graphs, *Discrete Math.* **130** (1994), 171–176.

[139] P. Solé, Expanding and forwarding, *Discrete Appl. Math.* **58** (1995), 67–78.

[140] P. Strogova, Finding a finite group presentation using rewriting, in: *Proc. Workshop on Symbolic Rewriting Techniques* (M. Bronstein, ed.), 1995, 35–40.

[141] P. Vadapelli and P. Srimani, Trivalent graphs for interconnection networks, *Inform. Process. Lett.* **54** (1995), 329–335.

[142] P. Vadapelli and P. Srimani, A new family of Cayley graph interconnection networks of constant degree four, *IEEE Trans. Parallel Distrib. Systems* **7** (1996), 26–32.

[143] P. Vadapelli and P. Srimani, Shortest routings in trivalent Cayley graphs networks, *Inform. Process. Lett.* **57** (1996), 183–188.

[144] E. VanDoorn, Connectivities of circulant digraphs, *J. Graph Theory* **10** (1986), 9–14.

[145] M. E. Watkins, Connectivity of transitive graphs, *J. Combin. Theory* **8** (1970), 23–29.

[146] M. E. Watkins, On the action of non-abelian groups on graphs, *J. Combin. Theory* **11** (1971), 95–104.

[147] A. White, *Graphs, Groups and Surfaces*, North-Holland, Amsterdam, 1984.

# Some applications of Laplace eigenvalues of graphs

Bojan MOHAR

*Department of Mathematics*
*University of Ljubljana*
*Jadranska 19*
*1111 Ljubljana*
*Slovenia*

Notes taken by Martin JUVAN

**Abstract**

In the last decade important relations between Laplace eigenvalues and eigenvectors of graphs and several other graph parameters were discovered. In these notes we present some of these results and discuss their consequences. Attention is given to the partition and the isoperimetric properties of graphs, the max-cut problem and its relation to semidefinite programming, rapid mixing of Markov chains, and to extensions of the results to infinite graphs.

## 1 Introduction

Applications of eigenvalue methods in combinatorics, graph theory and in combinatorial optimization have already a long history. For example, eigenvalue bounds on the chromatic number were formulated by Wilf [Wi] and Hoffman [Ho] already at the end of the sixties. Historically, the next applications related to combinatorial optimization, due to Fiedler [Fi1] and Donath and Hoffman [DH] in 1973, concerned the area of graph partition. A very important use of eigenvalues is Lovász' notion of the theta-function from 1979 [Lo]. Using it, he solved the long-standing Shannon capacity problem for the 5-cycle. The theta-function provides the only known way to compute the chromatic number of perfect graphs in polynomial time.

The next important result was the use of eigenvalues in the construction of superconcentrators and expanders by Alon and Milman [AM] in 1985. Their work motivated the study of eigenvalues of random regular graphs. Eigenvalues of random 01-matrices were studied already earlier by F. Juhász, who also analyzed the behavior of the theta-function on random graphs, and introduced eigenvalues in clustering [Ju]. Isoperimetric properties of graphs and their eigenvalues play a crucial role in the design of several randomized algorithms. These applications are based on the so-called rapidly mixing Markov chains. The most important discoveries in this area include random polynomial time algorithms for approximating the volume of a convex body (cf., e.g., [DFK, LS1, LS2]), polynomial time algorithms for approximate counting (e.g., approximating the permanent or counting the number of perfect matchings, see [Sin] for additional information), etc. Isoperimetric properties and related



*G. Hahn and G. Sabidussi (eds.), Graph Symmetry, 225–275.*

expansion properties of graphs are the basis for various other applications, ranging from the fast convergence of Markov chains, efficient approximation algorithms, randomized or derandomized algorithms, complexity lower bounds, and building efficient communication networks and networks for parallel computation.

There is an increasing interest in the application of eigenvalues in combinatorial optimization problems. To mention only some of them, Burkard, Finke, Rendl, and Wolkowicz [FBR, RW] used the eigenvalue approach in the study of the quadratic assignment problem and general graph partition problems, Delorme and Poljak [DP1, DP2] in the max-cut problem, and Juvan and Mohar [JM1, JM2] in the labeling problems. Spectral partitioning which is based on eigenvectors of Laplace eigenvalues of graphs has proved to be one of the most successful heuristic approaches in the design of partition algorithms [BS, HL, HMPR], in parallel computation [ST, Sim, Wil], in solving sparse linear systems [PSW], clustering [HK, CSZ] and in ranking [JM2, HMPR]. A similar kind of applications is based on the properties of the Perron-Frobenius eigenvector of a nonnegative matrix. This technique is suitable for the ranking problems. We refer to [MP2] for additional applications.

There are several ways of using eigenvalues in combinatorial optimization. The first possibility consists in formulating concrete bounds which involve eigenvalues of some related matrices. Examples of such bounds are given by the bounds on the separation properties in Section 3. Another way is to use the eigenvalues as a tool of transformation of combinatorial optimization problems to continuous optimization problems. Examples of this kind are provided by the bisection problem, max-cut problem, generalized partition problem, and the theta-function. The common point of these applications is the possibility of the change to a "continuous optimization", in particular to semidefinite programming. For example, in some cases there is a possibility of introducing a parameter $u \in \mathbf{R}^n$ and optimizing when $u$ is restricted to be an element of a convex set $K \subseteq \mathbf{R}^n$. This way we get improved bounds or methods for the problems in question. A classical example is the Lovász' theta-function [Lo, GLS, Kn]. Its use gives rise to polynomial time algorithms for determining the stability number, or the chromatic number of perfect graphs. A similar approach appears in relation to the following problems: bipartition width, graph partition, and the max-cut problem (cf. Section 3), the bandwidth [JM1, JM2, HMPR], etc. The max-cut problem and its relation to semidefinite programming is discussed in more detail in Section 4.

In these notes we first discuss some results on Laplace matrices and Laplace eigenvalues of weighted graphs in the general framework. After introducing the Laplace matrix of a graph and presenting its basic properties in Section 2, relations between Laplace eigenvalues and separation properties of graphs are considered in Section 3. Section 4 discusses semidefinite programming and its relations to graph partitions. Random walks and associated matrices are treated in Section 5. The paper concludes by a brief introduction to Laplacians of infinite graphs in Section 6.

There are several existing books and survey papers concerning graph eigenvalues, e.g., [Bi, CDSS , CDGT, Fi3, Go, Mo3, MP2, MW]. We do not intend to overlap our presentation with their contents except partially with [MP2]. Therefore we restrict ourselves to some problems which can be classified as applications of eigenvalue methods in the design of algorithms and in combinatorial optimization. In particular, we do not include discussion on expander graphs (whose constructions are accessible only via eigenvalue methods) and their applications, although they are quite an important tool in the design of algorithms and are extensively used in several other areas of theoretical computer science. Instead, we refer to the book [Lu1] by Lubotzky.

# 2 The Laplacian of a graph

## 2.1 Basic notation

Let $G$ be an undirected finite graph. By $V = V(G)$ and $E = E(G)$ we denote its vertex and edge set, respectively. Throughout these notes, the number of vertices and the number of edges of $G$ are denoted by $n$ and $m$, respectively. We also consider *weighted graphs*, i.e., graphs together with a *weight function*

$$w : V(G) \times V(G) \to \mathbf{R}^+$$

which assigns a nonnegative real weight $w(u, v)$ to each pair $u, v$ of vertices. It is required that $w$ satisfies the following properties:

(i) $w(u, v) > 0$ if $uv \in E(G)$, and $w(u, v) = 0$ if $uv \notin E(G)$.

(ii) $w(u, v) = w(v, u)$.

If $e = uv$ is an edge of $G$, property (ii) allows us to write $w(e)$ instead of $w(u, v)$. The weight $w(u, v)$ will also be denoted by $a_{uv}$ (or $a_{vu}$). Unweighted graphs can be viewed as weighted graphs where $w(u, v)$ is equal to the number of edges between $u$ and $v$.

In the unweighted case the degree $\mathrm{d}_u$ of the vertex $u \in V(G)$ is defined as the number of edges of $G$ incident with $u$. In the weighted case we set

$$\mathrm{d}_u = \sum_{v \in V} w(u, v) = \sum_{v \in V} a_{uv} \,.$$

The maximum and the minimum vertex degree in $G$ are denoted by $\Delta(G)$ and $\delta(G)$, respectively.

We will assume that the reader is familiar with basic notions of graph theory as found, e.g., in [BM].

## 2.2 The Laplacian

Given a graph $G$, its (*weighted*) *adjacency matrix* $A(G) = [a_{uv}]$ is an $n \times n$ matrix with rows and columns indexed by $V(G)$ whose entries $a_{uv}$ $(u, v \in V(G))$ are defined as above. Let $D(G) = \mathrm{diag}(\mathrm{d}_u; u \in V(G))$ be the diagonal matrix indexed by $V(G)$ and with vertex degrees on the diagonal. The difference

$$L(G) = D(G) - A(G) \tag{1}$$

is called the *Laplace matrix* of $G$ (or the (*difference*) *Laplacian* of $G$). Historically, one of the first applications of Laplace matrices of graphs is in the proof of the well-known Matrix-Tree Theorem of Kirchhoff [Ki] (see also [Bi, Chapter 6]) which states that the number of spanning trees in a graph $G$ is equal to the cofactor of any element of the matrix $L(G)$. In particular, this gives a simple algebraic proof of Cayley's formula [Ca] that there are $n^{n-2}$ distinct labeled trees on $n$ vertices.

Denote by $Q$ an *oriented incidence matrix* of $G$ which is defined as follows. Orient arbitrarily each edge of $G$. The matrix $Q$ is an $n \times m$ matrix with rows and columns indexed by $V(G)$ and $E(G)$, respectively, and entries $(Q)_{ve}$ equal to $-\sqrt{w(e)}$ if $v$ is the initial vertex of $e$, $\sqrt{w(e)}$ if $v$ is the terminal vertex of $e$, and 0 otherwise. The last possibility also includes loops, irrespective of their weights.

**Proposition 2.1** *Let $Q$ be an oriented incidence matrix with respect to some orientation of the edges of the graph $G$. Then*

$$L(G) = QQ^T. \tag{2}$$

**Proof** Given $u, v \in V(G)$ and $e \in E(G)$, it is easy to check that

$$(Q)_{ue}(Q^T)_{ev} = \begin{cases} -a_{uv}, & u \neq v \text{ and } u, v \text{ are the endvertices of } e \\ a_{uw}, & u = v \text{ and } e = uw, \text{ where } w \neq u \\ 0, & \text{otherwise.} \end{cases} \tag{3}$$

By summing (3) over all edges of $G$, the proposition follows. □

Note that Proposition 2.1 also shows that the product $QQ^T$ is independent of the chosen orientation of the edges of $G$.

Let $\mathbf{R}^V$ be the set of functions from $V$ to $\mathbf{R}$,

$$\mathbf{R}^V = \{f : V \to \mathbf{R}\}.$$

If $f \in \mathbf{R}^V$ and $u \in V$, we shall use the notation $f_u$ instead of the more customary $f(u)$ to denote the value of $f$ at the vertex $u$. With the usual operations of the sum $((f+g)_u = f_u + g_u)$ and the multiplication by real numbers $((\lambda f)_u = \lambda f_u)$, $\mathbf{R}^V$ becomes a real vector space of dimension $n$ endowed with the inner product

$$\langle f, g \rangle = \sum_{v \in V} f_v g_v.$$

The corresponding norm in $\mathbf{R}^V$ is

$$\|f\| = \langle f, f \rangle^{1/2} = \Big(\sum_{v \in V} f_v^2\Big)^{1/2}.$$

The matrix $L = L(G)$ (and other matrices indexed by the vertices of $G$) acts on $\mathbf{R}^V$ as a linear operator. Its action is determined by the rule of matrix-vector multiplication, i.e., $g = Lf$ is the function defined by the formula

$$g_u = (Lf)_u = \sum_{v \in V} (L)_{uv} f_v, \qquad u \in V.$$

There is a natural quadratic form associated with $L = L(G)$:

**Proposition 2.2**

$$\langle f, Lf \rangle = \sum_{uv \in E} a_{uv}(f_u - f_v)^2. \tag{4}$$

**Proof** Since $L = QQ^T$, the definition of $Q$ implies

$$\langle f, Lf\rangle = \langle f, QQ^T f\rangle = \langle Q^T f, Q^T f\rangle = \sum_{e\in E}(Q^T f)_e^2 = \sum_{uv\in E} a_{uv}(f_u - f_v)^2 . \qquad \square$$

Let us remark that in the case of an unweighted graph $G$, expression (4) reduces to

$$\langle f, Lf\rangle = \sum_{uv\in E}(f_u - f_v)^2 . \tag{5}$$

## 2.3 Laplacians of graphs and Laplace operators on manifolds

Let $M$ be a (compact) Riemannian manifold. The *Laplacian* $\Delta$ on $M$ is the second order differential operator acting on $\mathcal{C}^\infty(M)$,

$$\Delta(\varphi) = -\mathrm{div}(\mathrm{grad}(\varphi)).$$

Let $G$ be the graph of a triangulation of $M$. When looking for solutions of the Laplace equation

$$\Delta(\varphi) = \lambda\varphi \qquad (\varphi \in \mathcal{C}^\infty(M))$$

we may try to discretize the problem by looking at the restriction $f \in \mathbf{R}^V$ of $\varphi$ to the vertex set $V$ of $G$. For $uv \in E(G)$, define its weight as $a_{uv} = 1/(\rho(u,v))^2$ where $\rho(u,v)$ is the length of the edge $uv$ in $M$. Let $Q$ be the corresponding oriented incidence matrix of $G$. Then

$$(Q^T f)_{uv} = \frac{f_u - f_v}{\rho(u,v)}$$

is an approximation for the gradient of $\varphi$ at the edge $uv$. Similarly, the corresponding Laplace matrix of $G$ captures the properties of the Laplacian $\Delta$ on $M$: If the triangulation is "sufficiently fine", then the restriction of $\Delta\varphi$ to $V$ and $L(G)f$ are close to each other.

There is another relationship between graphs and Riemannian manifolds that carries over to their Laplacians. We will briefly describe this relationship only in case of dimension 2 although it has obvious analogues for higher dimensions as well.

Let $S$ be a surface of constant curvature $-1$ with three congruent boundary components such that $S$ is homeomorphic to the "pair of pants" surface (the disk with two holes). If $G$ is an arbitrary cubic graph, then we define a Riemannian surface $S(G)$ by taking a copy $S_v$ of $S$ for each vertex of $G$, and identifying the boundary circles of $S_v$ in pairs with boundary circles in copies $S_u$ of adjacent vertices $u$. This construction gives a surface of constant curvature $-1$ whose genus is equal to the cyclomatic number $g = |E(G)| - |V(G)| + 1$ of the graph $G$.

There is a close relationship between the Laplacian on $S(G)$ and the properties of the Laplace matrix of $G$. Moreover, some geometric properties of $S(G)$ are closely related to combinatorial properties of the graph. This relation gives an important bilateral link between spectral geometry of Riemannian manifolds and graph theory. It makes possible to use results about graphs in the study of Laplacians on manifolds and, conversely, gives a possibility of transferring results about Laplacians on manifolds to graphs. Each of these directions has proved to be useful in research. We refer to the book by Buser [Bu] and to the paper [Br] for more details.

## 2.4 Laplace eigenvalues

By definition, $L(G)$ is a real symmetric matrix, and (4) (together with the nonnegativity of the edge-weights) implies that it is positive semidefinite. Therefore it has $n$ nonnegative real eigenvalues $0 \le \lambda_1 = \lambda_1(G) \le \lambda_2 = \lambda_2(G) \le \cdots \le \lambda_n = \lambda_n(G)$ (repeated according to their multiplicities). These eigenvalues are called *Laplace eigenvalues* of the graph $G$. It is easy to see that 0 is always an eigenvalue of $L(G)$, and that $\mathbf{1} = (1, 1, \ldots, 1)^T$ is the corresponding eigenvector. More precisely, we have the following description of the multiplicity of 0 as an eigenvalue of $L(G)$.

**Proposition 2.3** *The multiplicity of the value 0 as an eigenvalue of $L(G)$ is equal to the number of connected components of $G$.*

**Proof** Let $H$ be a connected component of $G$. Denote by $f^H \in \mathbf{R}^V$ the characteristic function of $V(H)$, i.e., the 0-1 function whose value $f_v^H$ is equal to 1 if and only if $v \in V(H)$. Obviously, $L(G)f^H = 0$. Since the characteristic functions of different connected components are linearly independent, the multiplicity of the eigenvalue 0 is at least the number of connected components of $G$.

Conversely, let $f \in \mathbf{R}^V$ be a function from the kernel of $L(G)$. Then $\langle f, Lf \rangle = 0$, and (4) implies that $f$ is constant on each connected component of $G$. Therefore $f$ is a linear combination of the characteristic functions of connected components of $G$. □

In particular, Proposition 2.3 implies that $\lambda_1(G) = 0$ is a simple eigenvalue of $L(G)$ if and only if the graph $G$ is connected.

If $M$ is a matrix with real eigenvalues, we use the notation $\lambda_i(M)$ to denote the $i$-th smallest eigenvalue of $M$ (respecting the multiplicities). To denote the maximal eigenvalue of $M$ we will occasionally use the symbol $\lambda_{\max}(M)$. Consistently with this notation we may also write $\lambda_{\min}(M)$ instead of $\lambda_1(M)$.

Let $G$ be a $k$-regular (weighted) graph, i.e., $d_v = k$ for each $v \in V(G)$. Then (1) implies that $\lambda$ is an eigenvalue of $L(G)$ if and only if $k - \lambda$ is an eigenvalue of the weighted adjacency matrix $A(G)$. More precisely,

$$\lambda_i(L(G)) = k - \lambda_{n+1-i}(A(G)), \qquad i = 1, \ldots, n. \tag{6}$$

This result enables us to use known results about the eigenvalues of the adjacency matrix of a regular graph in the study of its Laplace eigenvalues. We refer to [CDSS ] and to [Go] for numerous results on the adjacency matrix eigenvalues of graphs. For example, the eigenvalues of the adjacency matrix of the complete graph $K_n$ are $n-1$ and $-1$ (the latter with multiplicity $n-1$). Therefore

$$\lambda_1(K_n) = 0, \text{ and } \lambda_k(K_n) = n \text{ for } 2 \le k \le n.$$

Similarly, the Laplace eigenvalues of the $n$-cycle $C_n$ are precisely the numbers

$$\nu_k = 2 - 2\cos\left(\frac{2k\pi}{n}\right), \qquad k = 1, \ldots, n.$$

If $G$ is a simple and unweighted graph and $\overline{G}$ is its complement, then

$$L(G) + L(\overline{G}) = nI - J \tag{7}$$

where $J$ is the matrix with all entries equal to 1. Let $f^1, f^2, \ldots, f^n$ be an orthogonal system of eigenvectors of $L(G)$ such that $f^1 = \mathbf{1}$ and $L(G)f^i = \lambda_i f^i$ for $i = 1, \ldots, n$. By (7) we get $L(\overline{G})f^1 = 0$ and for $2 \le i \le n$,

$$L(\overline{G})f^i = (n - \lambda_i)f^i.$$

This shows that $\lambda_1(\overline{G}) = 0$ and that for $2 \le i \le n$ we have

$$\lambda_i(\overline{G}) = n - \lambda_{n-i+2}(G).$$

In particular, if $G = K_{n,m}$, then $\overline{G} = K_m \cup K_n$. So, $\overline{G}$ has eigenvalues $0^{(2)}, m^{(m-1)}, n^{(n-1)}$ (with the numbers in the superscripts indicating multiplicities). Hence the Laplace eigenvalues of $K_{n,m}$ are $0^{(1)}, n^{(m-1)}, m^{(n-1)}$, and $(n+m)^{(1)}$.

Let us recall that the *Cartesian product* $G \,\Box\, H$ of graphs $G$ and $H$ has vertex set $V(G \,\Box\, H) = V(G) \times V(H)$, where $(u_1, u_2)$ is adjacent to $(v_1, v_2)$ if and only if $u_1 = v_1$ and $u_2v_2 \in E(H)$, or $u_2 = v_2$ and $u_1v_1 \in E(G)$. There is a simple description of the Laplace eigenvalues of $G \,\Box\, H$ in terms of the Laplace eigenvalues of $G$ and $H$.

**Proposition 2.4** *The Laplace eigenvalues of the Cartesian product $G \,\Box\, H$ are precisely the numbers*

$$\lambda_i(G) + \lambda_j(H) \quad (1 \le i \le |V(G)|, 1 \le j \le |V(H)|)$$

*where each number is obtained as many times as is its multiplicity as an eigenvalue of $L(G \,\Box\, H)$.*

The proof of Proposition 2.4 relies on the fact that the Laplace matrix of $G \,\Box\, H$ is equal to $L(G \,\Box\, H) = L(G) \otimes I_{V(H)} + I_{V(G)} \otimes L(H)$, where $\otimes$ denotes the Kronecker product of matrices. The details are left to the reader.

Proposition 2.4 holds also for weighted graphs $G$ and $H$ if the weights of edges in $G \,\Box\, H$ are determined by

$$w((u_1, u_2), (v_1, v_2)) = \begin{cases} w(u_1, v_1), & \text{if } u_2 = v_2 \\ w(u_2, v_2), & \text{if } u_1 = v_1 \\ 0, & \text{otherwise.} \end{cases}$$

As a consequence of Proposition 2.4 we obtain

$$\lambda_2(G \,\Box\, H) = \min\{\lambda_2(G), \lambda_2(H)\}, \tag{8}$$

and

$$\lambda_{\max}(G \,\Box\, H) = \lambda_{\max}(G) + \lambda_{\max}(H). \tag{9}$$

Proposition 2.4 can be used to determine the Laplace spectrum of several well-known families of graphs.

**Example 2.5** The *d-dimensional hypercube* $Q_d$ is the graph isomorphic to the Cartesian product of $d$ copies of $K_2$. Since the Laplace eigenvalues of $K_2$ are 0 and 2, Proposition 2.4 asserts that the Laplace spectrum of $Q_d$ consists of the numbers $2k$, $k = 0, 1, \ldots, d$. The multiplicity of $2k$ in the spectrum of $Q_d$ is equal to $\binom{d}{k}$.

## 2.5 Bounding the Laplace eigenvalues

There are various useful min-max formulae for the expression of eigenvalues of a symmetric matrix. If $M$ is a real symmetric matrix indexed by $V$, then

$$\begin{aligned}\lambda_1(M) &= \min\left\{\frac{\langle Mf,f\rangle}{\langle f,f\rangle} \mid 0 \neq f \in \mathbf{R}^V\right\} \\ &= \min\{\langle Mf,f\rangle \mid f \in \mathbf{R}^V,\ \|f\| = 1\}\end{aligned} \tag{10}$$

and similarly

$$\lambda_{\max}(M) = \max\{\langle Mf,f\rangle \mid f \in \mathbf{R}^V,\ \|f\| = 1\}. \tag{11}$$

The Rayleigh characterization (10) has a generalization, the min-max characterization of the $k$th smallest eigenvalue $\lambda_k(M)$, known also as the Courant-Fisher formula:

$$\lambda_k(M) = \min_U \max_f \{\langle Mf,f\rangle \mid \|f\| = 1,\ f \in U\}, \tag{12}$$

where the minimum is taken over all $k$-dimensional subspaces $U$ of $\mathbf{R}^V$. Another way of expressing (12) is

$$\lambda_k(M) = \min\{\langle Mf,f\rangle \mid \|f\| = 1, f \perp f_i, 1 \le i < k\}, \tag{13}$$

where $f_1,\ldots,f_{k-1}$ are pairwise orthogonal eigenvectors of $\lambda_1,\ldots,\lambda_{k-1}$, respectively.

Among the Laplace eigenvalues of $G$, the most important ones are the extreme nontrivial eigenvalues: the second smallest eigenvalue $\lambda_2(G)$ and the largest eigenvalue $\lambda_{\max}(G)$. (Let us observe that the relation (7) shows that $\lambda_2(G) = n - \lambda_{\max}(\overline{G})$. So, it is not surprising at all that the importance of one of these eigenvalues implies the importance of the other.) For a (weighted) graph $G$ with Laplace matrix $L = L(G)$, (13) reads

$$\lambda_2(G) = \min\{\langle Lf,f\rangle \mid \|f\| = 1,\ f \perp \mathbf{1}\} \tag{14}$$

since $\mathbf{1}$ is an eigenvector corresponding to $\lambda_1(G)$. Let us observe that $f$ is orthogonal to $\mathbf{1}$ if and only if the sum of the values of $f$ is 0, $\sum_{v\in V} f_v = 0$. Expression (14) can be used to get combinatorial upper bounds on $\lambda_2(G)$. For example:

**Lemma 2.6** *Let $s,t \in V(G)$ be nonadjacent vertices of a graph $G$. Then*

$$\lambda_2(G) \le \frac{1}{2}(\mathrm{d}_s + \mathrm{d}_t).$$

**Proof** Let $f \in \mathbf{R}^V$ be defined by

$$f_v = \begin{cases} 1, & v = s \\ -1, & v = t \\ 0, & \text{otherwise.}\end{cases}$$

Since $f \perp \mathbf{1}$, (14) yields

$$\lambda_2(G) \le \frac{\langle L(G)f,f\rangle}{\langle f,f\rangle} = \frac{\sum_{uv\in E} a_{uv}(f_u - f_v)^2}{\sum_{v\in V} f_v^2} = \frac{\mathrm{d}_s + \mathrm{d}_t}{2}. \qquad \square$$

The proof of Lemma 2.6 presents a basic technique to exploit expressions like (14). By inserting an appropriate function in the expression, the right-hand side gets a meaningful combinatorial interpretation, and this enables us to relate the combinatorial properties in question with the Laplace eigenvalues. To overcome the orthogonality restriction on $f$, Fiedler [Fi2] transformed (14) to a more suitable expression which does not require $f$ to be orthogonal to $\mathbf{1}$ in order to be applied.

**Proposition 2.7**

$$\lambda_2(G) = 2n \cdot \min \left\{ \frac{\sum\limits_{uv \in E} a_{uv}(f_u - f_v)^2}{\sum\limits_{u \in V} \sum\limits_{v \in V} (f_u - f_v)^2} \;\middle|\; f \neq c\mathbf{1} \text{ for } c \in \mathbf{R} \right\}. \qquad (15)$$

**Proof** Observe first that both the numerator and the denominator of the fraction in the minimum are invariant under the addition of a constant function. This observation together with (4) and (14) implies that it suffices to show that $2n\langle f, f\rangle = \sum_{u \in V} \sum_{v \in V} (f_u - f_v)^2$ whenever $f$ is orthogonal to $\mathbf{1}$. So, assume that $f \perp \mathbf{1}$. Then

$$\begin{aligned} \sum_{u \in V} \sum_{v \in V} (f_u - f_v)^2 &= \sum_{u \in V} \sum_{v \in V} f_u^2 - 2 \sum_{u \in V} \sum_{v \in V} f_u f_v + \sum_{u \in V} \sum_{v \in V} f_v^2 \\ &= 2n\langle f, f\rangle - 2\langle f, \mathbf{1}\rangle^2 = 2n\langle f, f\rangle . \end{aligned}$$

□

An almost identical proof yields a similar expression for $\lambda_{\max}(G)$:

$$\lambda_{\max}(G) = 2n \cdot \max \left\{ \frac{\sum\limits_{uv \in E} a_{uv}(f_u - f_v)^2}{\sum\limits_{u \in V} \sum\limits_{v \in V} (f_u - f_v)^2} \;\middle|\; f \neq c\mathbf{1} \text{ for } c \in \mathbf{R} \right\}. \qquad (16)$$

Let us conclude this section by presenting bounds on $\lambda_2(G)$ and $\lambda_{\max}(G)$ in terms of the maximum and minimum degrees of $G$.

**Proposition 2.8**

$$\lambda_2(G) \leq \frac{n}{n-1}\delta(G) \leq \frac{n}{n-1}\Delta(G) \leq \lambda_{\max}(G) \leq 2\Delta(G). \qquad (17)$$

**Proof** Choose a vertex $v \in V(G)$ and let $e^{(v)} \in \mathbf{R}^V$ be the function of the standard base corresponding to $v$, i.e., $e_v^{(v)} = 1$ and $e_u^{(v)} = 0$ for each $u \neq v$. By putting $e^{(v)}$ in (15), we get

$$\lambda_2(G) \leq 2n \frac{\mathrm{d}_v}{2(n-1)} = \frac{n}{n-1}\mathrm{d}_v .$$

Since $v$ was chosen arbitrarily, the first inequality follows. The lower bound on $\lambda_{\max}(G)$ is proved analogously using (16).

It remains to show that $\lambda_{\max}(G) \leq 2\Delta(G)$. Since the matrix $\Delta(G)I_{V(G)} - L(G)$ is (elementwise) nonnegative, the Perron-Frobenius theorem implies that its largest eigenvalue $\Delta(G)$ is greater than or equal to the absolute value of its smallest eigenvalue, i.e., $\Delta(G) \geq |\Delta(G) - \lambda_{\max}(G)|$. In particular, $\Delta(G) \geq \lambda_{\max}(G) - \Delta(G)$ as claimed. □

An alternative proof of the first part of Proposition 2.8 can be obtained by showing that the matrices $L(G)-\lambda_2(G)(I-\frac{1}{n}\mathbf{1}\,\mathbf{1}^T)$ and $\lambda_{\max}(G)(I-\frac{1}{n}\mathbf{1}\,\mathbf{1}^T)-L(G)$ are positive semidefinite, and by considering their diagonal entries.

The application of the Perron-Frobenius theorem in the proof of the last inequality in Proposition 2.8 can be strengthened to show that for a connected graph $G$, $\lambda_{\max}(G) = 2\Delta(G)$ if and only if $G$ is a bipartite $\Delta(G)$-regular graph. See, for example, [Go].

# 3 Partitions of graphs

## 3.1 Edge cuts and eigenvalues

For a subset $S \subseteq V(G)$, let $\bar{S} = V(G)\backslash S$ denote the complement of $S$ in $V(G)$. Given sets of vertices $A, B \subseteq V(G)$, let $E(A, B)$ be the set of those edges of $G$ that have one endvertex in $A$ and the other in $B$. We also set

$$e(A,B) = \sum_{u\in A}\sum_{v\in B} w(u,v) = \sum_{e\in E(A,B)} w(e)$$

to be the sum of the weights of the edges in $E(A, B)$. Note that for an unweighted graph $G$, $e(S, \bar{S})$ counts the number of edges in the cut $E(S, \bar{S})$.

The partition problems discussed in this section are mainly concerned with finding an appropriate subset $S \subseteq V(G)$ such that the edge cut $E(S, \bar{S})$ satisfies some specific extremal property. In particular, we focus our attention on the following quantities: the weight of a maximum cut, the bipartition width, and the isoperimetric number. The computational complexity of these problems is such that there is no hope of finding an optimal solution to a moderately sized problem in a reasonable amount of time. (More precisely, they are NP-hard.) Therefore any nontrivial bounds are desirable and potentially very important.

Let us start by a lemma that relates the weight of an edge cut $E(S, \bar{S})$ to the eigenvalues $\lambda_2(G)$ and $\lambda_{\max}(G)$ of the Laplacian of the underlying graph. Although simple to prove, the lemma is a fundamental tool for obtaining bounds on various graph properties related to partitions.

**Lemma 3.1** *Let $G$ be a (weighted) graph of order $n$, and $S \subseteq V(G)$. Then*

$$\lambda_2(G)\frac{|S|(n-|S|)}{n} \le e(S,\bar{S}) \le \lambda_{\max}(G)\frac{|S|(n-|S|)}{n}\,.$$

**Proof** Let $f \in \mathbf{R}^V$ be the characteristic function of $S$, i.e., $f_v = 1$ if $v \in S$ and $f_v = 0$ otherwise. Then

$$\sum_{u\in V}\sum_{v\in V}(f_u - f_v)^2 = 2|S|(n-|S|)$$

and

$$\sum_{uv\in E} a_{uv}(f_u - f_v)^2 = e(S,\bar{S})\,.$$

If $S \neq \emptyset$ and $S \neq V(G)$, (15) implies

$$\lambda_2(G) \le 2n\cdot\frac{e(S,\bar{S})}{2|S|(n-|S|)}\,.$$

After rearranging, this gives the lower bound. Obviously, the lower bound also holds for $S = \emptyset$ and $S = V(G)$. The upper bound is proved analogously using (16). □

It is an immediate, and also an important consequence of Lemma 3.1 that in a graph with all non-trivial Laplace eigenvalues being close together, i.e., when $\lambda_{\max} - \lambda_2$ is small, the weights of all edge cuts $E(S, \bar{S})$ corresponding to vertex sets $S$ of the same cardinality are approximately the same. In particular, this property holds for the cardinalities of edge cuts in random graphs. Therefore it is not surprising that many algorithms dealing with edge cuts perform very well on randomly chosen graphs.

## 3.2 The bipartition width

Let $G$ be a graph of order $n$. The *bipartition width* $\mathrm{bw}(G)$ of the graph $G$ is defined as

$$\mathrm{bw}(G) = \min\{e(S, \bar{S}) \mid S \subseteq V(G), |S| = \lfloor n/2 \rfloor\}.$$

In the unweighted case, the bipartition width of the graph $G$ is equal to the minimum number of edges whose deletion disconnects $G$ into two parts of the same size (up to a vertex). It is known that even for unweighted simple graphs the problem of determining $\mathrm{bw}(G)$ is NP-hard (see, e.g., [GJ, p. 210]).

Let us remark that the "reversed" problem of determining

$$\overline{\mathrm{bw}}(G) = \max\{e(S, \bar{S}) \mid S \subseteq V(G), |S| = \lfloor n/2 \rfloor\}$$

is closely related to the original one. Namely, given a weighted graph $G$ with weight function $w$, let $W = \max\{a_{uv} \mid u, v \in V(G)\}$. Define the "weighted complement" of $G$ as the weighted graph $\bar{G}$ with $V(\bar{G}) = V(G)$ and the weight function $\bar{w}(u,v) = W - w(u,v)$. Then

$$\mathrm{bw}(G) + \overline{\mathrm{bw}}(\bar{G}) = W \cdot \left\lfloor \frac{n}{2} \right\rfloor \left\lceil \frac{n}{2} \right\rceil$$

which implies that the problems of determining bw and $\overline{\mathrm{bw}}$ are actually equivalent.

Since all the sets, over which the minimum in the definition of $\mathrm{bw}(G)$ is taken, have size $\lfloor n/2 \rfloor$, Lemma 3.1 gives the following lower bound on $\mathrm{bw}(G)$.

**Corollary 3.2** *Let $G$ be a weighted graph of order $n$. If $n$ is even, then*

$$\mathrm{bw}(G) \geq \frac{n}{4} \cdot \lambda_2(G). \tag{18}$$

*If $n$ is odd, then*

$$\mathrm{bw}(G) \geq \frac{n^2 - 1}{4n} \cdot \lambda_2(G). \tag{19}$$

The bounds of Corollary 3.2 can be further improved by introducing a correction function. A function $c \in \mathbf{R}^V$ is called a *correction function* if $c \perp \mathbf{1}$. The following bound was proved by Boppana [Bo].

**Proposition 3.3** *Let $G$ be a weighted graph of even order $n$. Then*

$$\mathrm{bw}(G) \geq \frac{n}{4} \cdot \max_{c} \min_{f} \frac{\langle (L(G) + \mathrm{diag}(c)) f, f \rangle}{\langle f, f \rangle}, \tag{20}$$

*where the maximum is taken over all correction functions $c \in \mathbf{R}^V$ and the minimum is taken over all nonzero functions $f \in \mathbf{R}^V$ with $f \perp \mathbf{1}$.*

**Proof** Let $S \subseteq V(G)$ be a set of cardinality $\frac{n}{2}$ with $e(S, \bar{S}) = \mathrm{bw}(G)$ and let $g \in \mathbf{R}^V$ be its signed characteristic function, i.e., the function defined by $g_v = 1$ if $v \in S$ and $g_v = -1$ if $v \in \bar{S}$. Since $|S| = |\bar{S}|$, $g \perp \mathbf{1}$. Take an arbitrary correction function $c \in \mathbf{R}^V$. Since $c \perp \mathbf{1}$, we have

$$\langle \mathrm{diag}(c)g, g\rangle = \sum_{v\in V} c_v g_v^2 = \sum_{v \in V} c_v = 0\,. \tag{21}$$

Using (21) and applying (4) we get

$$\begin{aligned} \frac{\langle (L(G) + \mathrm{diag}(c))g, g\rangle}{\langle g, g\rangle} &= \frac{\langle L(G)g, g\rangle}{\langle g, g\rangle} = \frac{\sum_{uv\in E} a_{uv}(g_u - g_v)^2}{\sum_{v\in V} g_v^2} \\ &= \frac{4e(S, \bar{S})}{n} = \frac{4}{n} \cdot \mathrm{bw}(G). \end{aligned}$$

Since $c$ was arbitrary, the bound follows. □

For computational purposes, it is convenient to express the bound of Proposition 3.3 as a maximization of the smallest eigenvalue of an appropriate symmetric matrix. This can be done as follows. Let $Q = (q_1, \ldots, q_{n-1})$ be an $n \times (n-1)$ matrix such that the columns $q_i$ are pairwise orthogonal unit vectors with the property that $q_i \perp \mathbf{1}$ $(1 \le i < n)$. It is easy to see that for every $x \in \mathbf{R}^{n-1}$ we have $\langle Qx, Qx\rangle = \langle x, x\rangle$ and $Qx \perp \mathbf{1}$. This implies:

**Corollary 3.4** *We have*

$$\mathrm{bw}(G) \ge \frac{n}{4} \cdot \max_c \lambda_{\min}(Q^T(L(G) + \mathrm{diag}(c))Q),$$

*where the maximum is taken over all correction functions* $c \in \mathbf{R}^V$.

We shall see in Section 4 that the bound of Corollary 3.4 can be formulated as a semidefinite program, and can therefore be computed to an arbitrary precision in polynomial time using known polynomial time methods for solving such programs.

## 3.3 The max-cut problem

The *maximum cut* (or *max-cut*) *problem* is similar to the "reversed" bipartition width problem except that the restrictions on the sizes of the subsets over which the maximum is taken are omitted. More precisely, let $\mathrm{mc}(G)$ denote the maximum weight of an edge cut in $G$,

$$\mathrm{mc}(G) = \max\{e(S, \bar{S}) \mid \emptyset \ne S \subset V(G)\}\,.$$

As far as the computational complexity is concerned, the problem of determining $\mathrm{mc}(G)$ is NP-hard [Karp] (see also [GJ, p. 210]). Even more, it is known that there exists a constant $\varepsilon > 0$ such that there is no polynomial time $(1-\varepsilon)$-approximation algorithm for the max-cut problem unless $P = NP$ [PY, A-S]. On the other hand, we will see in Section 4 that it is possible to find a 0.878-approximation to $\mathrm{mc}(G)$ in polynomial time. Unlike the bipartition width problem, the "reversed" problem of max-cut, the *minimum cut problem*, which requires to determine

$$\min\{e(S, \bar{S}) \mid \emptyset \ne S \subset V(G)\}\,,$$

is easy to solve in polynomial time using standard flow techniques.

Lemma 3.1 implies the following upper bound on $\mathrm{mc}(G)$ which was first observed by Mohar and Poljak [MP1].

**Corollary 3.5** *Let $G$ be a weighted graph of order $n$. Then*

$$\mathrm{mc}(G) \le \frac{n}{4} \cdot \lambda_{\max}(G)\,. \tag{22}$$

Similarly to the bipartition width problem, the bound (22) can be further improved using correction functions. The following eigenvalue bound was introduced by Delorme and Poljak [DP1].

**Proposition 3.6** *Let $G$ be a weighted graph of order $n$. Then*

$$\mathrm{mc}(G) \le \frac{n}{4} \cdot \min_c \lambda_{\max}(L(G) + \mathrm{diag}(c)), \tag{23}$$

*where the minimum is taken over all correction functions $c \in \mathbf{R}^V$.*

**Proof** The proof consists of similar steps as in the proof of Proposition 3.3. Choose $S \subset V(G)$ such that $e(S, \bar{S}) = \mathrm{mc}(G)$ and let $g \in \mathbf{R}^V$ be the signed characteristic function of $S$, i.e., $g_v = 1$ if $v \in S$ and $g_v = -1$ if $v \in \bar{S}$. Take an arbitrary correction function $c \in \mathbf{R}^V$. As in the proof of Proposition 3.3, we have

$$\langle \mathrm{diag}(c) g, g \rangle = 0$$

and

$$\frac{\langle (L(G) + \mathrm{diag}(c)) g, g \rangle}{\langle g, g \rangle} = \frac{4 e(S, \bar{S})}{n} = \frac{4}{n} \cdot \mathrm{mc}(G)\,.$$

Since

$$\frac{\langle (L(G) + \mathrm{diag}(c)) g, g \rangle}{\langle g, g \rangle} \le \lambda_{\max}(L(G) + \mathrm{diag}(c))$$

and $c$ was chosen arbitrarily, the bound follows. □

The performance of (23) was investigated in [DP2]. It will be further discussed in Section 4 where it will be formulated as a semidefinite program.

## 3.4 Edge-forwarding index

An important notion used in the theory of communication networks is the forwarding index of a graph. A relation between the forwarding index and the Laplace eigenvalues was observed by Solé [So].

Let $G$ be a connected graph and suppose that for each (ordered) pair $u, v$ of distinct vertices of $G$ we have a path $R_{u,v}$ from $u$ to $v$. The collection of paths

$$\mathcal{R} = \{R_{u,v} \mid u, v \in V(G), u \ne v\}$$

is called a *routing* in $G$. For each $e \in E(G)$, denote by $\xi(G, \mathcal{R}, e)$ the number of paths $R_{u,v}$ which contain the edge $e$. Let

$$\xi(G, \mathcal{R}) = \max\{\xi(G, \mathcal{R}, e) \mid e \in E(G)\}.$$

The minimum

$$\xi(G) = \min \xi(G, \mathcal{R})$$

taken over all routings $\mathcal{R}$ of $G$ is called the *edge-forwarding index* of $G$. The edge-forwarding index is an important parameter in the study of communication networks (see, e.g., [CCRS, HMS] or the notes of Heydemann [He] in this volume).

**Proposition 3.7** *Let $G$ be unweighted connected graph of order $n$ with edge-forwarding index $\xi(G)$, and let $S \subseteq V(G)$. Then*

$$\xi(G) \geq \frac{2|S|(n-|S|)}{e(S,\bar{S})} \geq \frac{2n}{\lambda_{\max}(G)}.$$

**Proof** Let $\mathcal{R}$ be a routing. Each path in $\mathcal{R}$ joining a vertex of $S$ with a vertex in $\bar{S}$ contains at least one edge in $E(S,\bar{S})$. Since there are $2|S|(n-|S|)$ such paths, this proves the first inequality. The second inequality follows by Lemma 3.1. □

## 3.5 Isoperimetric inequalities

Isoperimetric problems are related to questions in which one considers the ratio between the surface area and the volume of $d$-dimensional bodies or more general (Riemannian) manifolds. In graph theory, a natural analogue to the volume is the number of vertices or the sum of vertex degrees in a subset $S$ of vertices of the graph, while the counterpart for the measure of the surface area is the number $e(S,\bar{S})$ of edges with one end in $S$ and the other end outside the set $S$. Problems in which one considers ratios of the form $e(S,\bar{S})/|S|$ are therefore called isoperimetric problems for graphs. In other words, isoperimetric properties concern the size of the neighborhood of a set of vertices. The related term "expansion" usually means that the size of the neighborhood can be bounded from below by some function of the size of the subset. Such isoperimetric properties provide the foundation for many recent developments in applications of graph theory in theoretical computer science which are mentioned in the introduction.

The *isoperimetric number* $i(G)$ of a graph $G$ of order $n \geq 2$ is defined as

$$i(G) = \min\left\{\frac{e(S,\bar{S})}{|S|} \;\middle|\; S \subset V(G), 0 < |S| \leq \frac{n}{2}\right\}.$$

Computationally the problem of determining $i(G)$ is NP-hard [Mo2]. Let us remark that a seemingly related quantity

$$\max\left\{\frac{e(S,\bar{S})}{|S|} \;\middle|\; S \subset V(G), 0 < |S| \leq \frac{n}{2}\right\}$$

is not very interesting since it is always equal to $\Delta(G)$.

A straightforward application of Lemma 3.1 yields the following eigenvalue lower bound on $i(G)$.

**Corollary 3.8** *Let $G$ be a weighted graph of order $n$. Then*

$$i(G) \geq \frac{\lambda_2(G)}{2}.$$

**Proof** Let $S \neq \emptyset$ be a set of vertices. By Lemma 3.1, we have

$$\frac{e(S,\bar{S})}{|S|} \geq \lambda_2(G)\frac{n-|S|}{n}.$$

If $|S| \leq \frac{n}{2}$, then $(n - |S|)/n \geq 1/2$. This implies the bound on $i(G)$. □

It is difficult to obtain useful lower bounds on $i(G)$ by combinatorial means. On the other hand, Corollary 3.8, although easy to prove, gives a nontrivial lower bound on $i(G)$. Let us present an example where the bound of Corollary 3.8 is, in fact, tight.

**Example 3.9** Take the $d$-dimensional cube $Q_d = K_2 \square Q_{d-1}$ and let $S = \{0\} \times Q_{d-1}$. Since $|S| = e(S, \bar{S}) = 2^{d-1}$, $i(Q_d) \leq 1$. On the other hand, we know from Example 2.5 that $\lambda_2(Q_d) = 2$. Using Corollary 3.8 we conclude that $i(Q_d) = 1$.

The following quantity is sometimes easier to deal with than $i(G)$:

$$i^*(G) = \min\Big\{ \frac{e(S,\bar{S})}{|S|\,|\bar{S}|} \cdot \frac{|V|}{2} \;\Big|\; \emptyset \neq S \subset V \Big\}.$$

Clearly, $i(G)/2 \leq i^*(G) \leq i(G)$. It turns out that $i^*(G)$ satisfies a relation similar to (8):

$$i^*(G \square H) = \min\{i^*(G), i^*(H)\}.$$

A proof of this can be found in [HT]; see also [CT].

The above inequality implies

$$\frac{1}{2}\min\{i(G), i(H)\} \leq i(G \square H) \leq \min\{i(G), i(H)\}.$$

While the inequality on the right can be strict (see [Mo2]), it is not known whether the factor $\frac{1}{2}$ on the left is the best possible constant.

Corollary 3.8 may be used to get bounds on $i(G)$ since $\lambda_2$ is usually easier to compute (or to estimate) than $i(G)$. Sometimes, however, one can estimate $i(G)$, while a bound on $\lambda_2$ is desired. (Such examples appear, for instance, in the study of the rate of convergence of Markov chains; cf. Section 5.) In such a case, one would like to use a converse inequality, obtaining an upper bound on $i(G)$ in terms of $\lambda_2(G)$. Such inequalities are known as *Cheeger inequalities* since they are discrete analogues of their continuous counterpart [Che] (see also [Br, Bu, Cha]) arising in the study of Laplace operators on Riemannian manifolds.

The classical form of Cheeger's bound adapted to graphs reads as

$$i(G) \leq \sqrt{2\Delta(G)\lambda_2(G)}\,. \tag{24}$$

In the sequel we shall present an improved eigenvalue upper bound of Cheeger's type from [Mo2].

**Theorem 3.10** *Let $G$ be a weighted graph of order $n$, $G \neq K_n$, and let $\Delta = \Delta(G)$ and $\lambda_2 = \lambda_2(G)$. Then*

$$i(G) \leq \sqrt{(2\Delta - \lambda_2)\lambda_2}\,.$$

**Proof** Since for a disconnected graph $G$ we have $\lambda_2 = i(G) = 0$, we may assume that $G$ is connected. Let $f$ be an eigenfunction of $L(G)$ corresponding to $\lambda_2$. Set $W = \{v \in V(G) \mid$

$f_v > 0\}$. After possible replacement of $f$ by $-f$ we may assume that $|W| \le \frac{n}{2}$. Let $g \in \mathbf{R}^V$ be defined by $g_v = f_v$ if $v \in W$ and $g_v = 0$ otherwise. Then

$$\begin{aligned}
\lambda_2 \sum_{u\in W} f_u^2 &= \sum_{u\in W} (Lf)_u f_u = \sum_{u\in W} \Big(\mathrm{d}_u f_u - \sum_{v\in V} a_{uv} f_v\Big) f_u \\
&= \sum_{u\in W}\sum_{v\in V} a_{uv}(f_u - f_v) f_u \\
&= \sum_{u\in W}\sum_{v\in W} a_{uv}(f_u - f_v) f_u + \sum_{u\in W}\sum_{v\notin W} a_{uv}(f_u - f_v) f_u \\
&\ge \sum_{u\in W}\sum_{v\in W} a_{uv}(f_u - f_v) f_u + \sum_{u\in W}\sum_{v\notin W} a_{uv} f_u^2 \\
&= \sum_{uv\in E(W,W)} a_{uv}(g_u - g_v)^2 + \sum_{uv\in E(W,\bar W)} a_{uv}(g_u - g_v)^2 \\
&= \sum_{uv\in E} a_{uv}(g_u - g_v)^2 = \langle Lg, g\rangle .
\end{aligned}$$

Since $\sum_{v\in W} f_v^2 = \sum_{v\in V} g_v^2 = \langle g, g\rangle$, we have

$$\lambda_2 \ge \frac{\langle Lg, g\rangle}{\langle g, g\rangle} =: K .$$

Then

$$\begin{aligned}
\sum_{uv\in E} a_{uv}(g_u + g_v)^2 &= 2\sum_{uv\in E} a_{uv}(g_u^2 + g_v^2) - \sum_{uv\in E} a_{uv}(g_u - g_v)^2 \\
&= 2\sum_{v\in V} \mathrm{d}_v g_v^2 - \langle Lg, g\rangle \le 2\Delta \sum_{v\in V} g_v^2 - \langle Lg, g\rangle \\
&= (2\Delta - K)\langle g, g\rangle.
\end{aligned}$$

This inequality together with the Cauchy-Schwarz inequality implies

$$\begin{aligned}
K &= \frac{\sum_{uv\in E} a_{uv}(g_u - g_v)^2 \sum_{uv\in E} a_{uv}(g_u + g_v)^2}{\langle g, g\rangle \sum_{uv\in E} a_{uv}(g_u + g_v)^2} \\
&\ge \frac{\Big(\sum_{uv\in E} a_{uv}|g_u^2 - g_v^2|\Big)^2}{(2\Delta - K)\langle g, g\rangle^2} . \qquad (25)
\end{aligned}$$

The quantity

$$M := \sum_{uv\in E} a_{uv}|g_u^2 - g_v^2|$$

can be interpreted as a measure of optimality of the partitions of $V$ determined by $\lambda_2$ and $f$. A lower bound on $M$ in terms of the isoperimetric number can be obtained in a similar manner as in the setting of Riemannian manifolds. The following calculation is a discrete analogue of the integration by parts. Let $0 = t_0 < t_1 < \cdots < t_m$ be all distinct values of $g_v$

$(v \in V)$. For $k = 0, 1, \ldots, m$, let $V_k := \{v \in V \mid g_v \geq t_k\}$. Note that $|V_k| \leq |W| \leq \frac{n}{2}$ if $k > 0$. Then

$$\begin{aligned}
M &= \sum_{uv \in E} a_{uv} |g_u^2 - g_v^2| = \sum_{k=1}^{m} \sum_{\substack{uv \in E \\ g_v < g_u = t_k}} a_{uv}(g_u^2 - g_v^2) \\
&= \sum_{k=1}^{m} \sum_{\substack{g_u = t_k \\ g_v = t_l, l < k}} a_{uv}(t_k^2 - t_{k-1}^2 + t_{k-1}^2 - \cdots - t_{l+1}^2 + t_{l+1}^2 - t_l^2) \\
&= \sum_{k=1}^{m} \sum_{u \in V_k} \sum_{v \notin V_k} a_{uv}(t_k^2 - t_{k-1}^2) \\
&= \sum_{k=1}^{m} e(V_k, \bar{V}_k) \cdot (t_k^2 - t_{k-1}^2) \\
&\geq i(G) \sum_{k=1}^{m} |V_k|(t_k^2 - t_{k-1}^2) = i(G) \sum_{k=1}^{m} t_k^2(|V_k| - |V_{k+1}|) \\
&= i(G) \sum_{v \in W} g_v^2 .
\end{aligned} \tag{26}$$

In the last line we used the fact that $|V_{m+1}| = 0$.

Combining (25) and (26), we get

$$K \geq \frac{i(G)^2}{2\Delta - K} . \tag{27}$$

Since $\lambda_2 \geq K$, (27) implies

$$\lambda_2 \geq \Delta - \sqrt{\Delta^2 - i(G)^2} . \tag{28}$$

Since $G \neq K_n$, Lemma 2.6 shows that $\lambda_2 \leq \Delta$. Rearranging (28), this inequality finally yields $i^2(G) \leq (2\Delta - \lambda_2)\lambda_2$. □

Let us note that the bound of Theorem 3.10 also holds for $G = K_n$ if $n \geq 3$.

Other forms of discrete Cheeger inequalities have been proved by Alon [Alo], Mohar [Mo1], Sinclair and Jerrum [SJ], Diaconis and Stroock [DSt], Chung [Chu], and others. See also Section 6.

By using (14), inequality (24) can be expressed as follows:

$$\frac{i^2(G)}{2\Delta(G)} \leq \min\Big\{ \frac{\sum_{uv \in E} a_{uv}(f_u - f_v)^2}{\sum_{v \in V} f_v^2} \;\Big|\; f \in \mathbf{R}^V, f \perp \mathbf{1} \Big\}.$$

Recent results of [HT] provide a generalization of the above inequality by replacing the right hand side by the $p$th powers of $\ell_p$-norms (and more generally to Orlicz norms), and replacing the constant 2 on the left by a constant $c_p$ which depends only on the norm. The main tool in this approach is the notion of a discrete gradient.

Although difficult to prove, the Cheeger bound seems to be less important than the bound of Corollary 3.8. However, the combination of both bounds shows that for every sequence of graphs of bounded degree, their isoperimetric numbers tend to 0 if and only if their second smallest Laplace eigenvalues tend to 0. This observation has an important meaning for the design of "expanding graphs" since it shows that families of "expanders" correspond to families of graphs whose second Laplace eigenvalue is uniformly bounded away from 0.

Another important observation following from the proof of Theorem 3.10 is that the partition based on the eigenfunction $f$ of $\lambda_2$ is not too far from the optimal one since in the proof of the lower bound on $M$, only the sets $V_k$ $(1 \le k \le m)$ have been considered. Constructing partitions based on the eigenfunctions of $\lambda_2$ has proved to be one of the most successful heuristics in parallel computation and for divide-and-conquer approach of processor distribution. Several references with evidence about this approach are given in the introduction.

# 4 Semidefinite programming

## 4.1 Introduction

At the end of the seventies Khachiyan [Kh] showed how to use the ellipsoid method to derive a polynomial time algorithm for solving linear programs (see also [GLS]). After the algorithm of Karmarkar [Karm] appeared in 1984, the field of mathematical programming faced a rapid development. Many new algorithms were introduced and several of them were applied to more general classes of problems.

Semidefinite programs constitute such a class of problems. A *semidefinite program* is the problem of minimizing a linear function of a variable $x = (x_1, \ldots, x_n)^T \in \mathbf{R}^n$ subject to positive semidefiniteness of a certain matrix $F(x)$. The problem input consists of $c \in \mathbf{R}^n$ and $n+1$ symmetric matrices $F_0, \ldots, F_n \in \mathbf{R}^{m \times m}$ and asks to

$$\begin{array}{ll} \text{minimize} & \langle c, x \rangle \\ \text{subject to} & F(x) \succeq 0 \end{array} \tag{SP}$$

where

$$F(x) = F_0 + \sum_{i=1}^{n} x_i F_i \,.$$

The inequality sign in $F(x) \succeq 0$ means that $F(x)$ is a positive semidefinite matrix, i.e., $\langle F(x)z, z\rangle \ge 0$ for all $z \in \mathbf{R}^m$. Let us remark at this point that we will use the fact that for every positive semidefinite matrix $F$ there exists a unique positive semidefinite matrix $S$ such that $F = S^2$. For the obvious reason, the matrix $S$ is denoted by the symbol $F^{1/2}$.

A semidefinite program is a convex optimization problem since its objective function $x \mapsto \langle c, x\rangle$ is a convex function, and the set of feasible solutions is a convex set. This is shown by the next claim.

**Claim 4.1** *The feasible region*

$$\mathcal{F} = \{x \in \mathbf{R}^n \mid F(x) \succeq 0\}$$

*of the problem* (SP) *is a convex set.*

**Proof** Suppose that $x, y \in \mathcal{F}$ and take $\lambda \in [0,1]$. Then for every $z \in \mathbf{R}^m$, we have

$$\langle(\lambda F(x) + (1-\lambda)F(y))z, z\rangle = \lambda\langle F(x)z, z\rangle + (1-\lambda)\langle F(y)z, z\rangle \ge 0 \,.$$

Therefore $\lambda F(x) + (1-\lambda)F(y) = F(\lambda x + (1-\lambda)y) \succeq 0$, and hence $\lambda x + (1-\lambda)y \in \mathcal{F}$. □

Let us now show that semidefinite programming is indeed a generalization of linear programming.

**Example 4.2** Consider a linear program

$$\begin{array}{ll} \text{minimize} & \langle c, x \rangle \\ \text{subject to} & Ax + b \geq 0 \end{array} \tag{LP}$$

where $c \in \mathbf{R}^n$, $A \in \mathbf{R}^{m\times n}$, and $b \in \mathbf{R}^m$. Since $Ax + b \in \mathbf{R}^m$, the inequality in (LP) denotes componentwise inequality. To formulate (LP) as a semidefinite program, observe first that a vector $x \geq 0$ (componentwise) if and only if $\operatorname{diag}(x) \succeq 0$ (i.e., the diagonal matrix containing components of $x$ on the diagonal is positive semidefinite). Denote by $a_1, \ldots, a_n$ the columns of $A$. Since

$$\operatorname{diag}(Ax + b) = \sum_{i=1}^{n} x_i \operatorname{diag}(a_i) + \operatorname{diag}(b)\,,$$

the problem (LP) is equivalent to the semidefinite program with $F(x) = F_0 + \sum_{i=1}^{n} x_i F_i$, where $F_0 = \operatorname{diag}(b)$ and $F_i = \operatorname{diag}(a_i)$, $i = 1, \ldots, n$.

Another important class of semidefinite programs are (convex) quadratically constrained quadratic programs.

**Example 4.3** Every convex quadratic function $f : \mathbf{R}^n \to \mathbf{R}$ can be expressed as $f(x) = \langle Ax, x\rangle - \langle b, x\rangle - c$, where $A \in \mathbf{R}^{n\times n}$ is a positive semidefinite matrix, $b \in \mathbf{R}^n$, and $c \in \mathbf{R}$. The eigenvalues of the matrix

$$M = \begin{bmatrix} I & A^{1/2}x \\ (A^{1/2}x)^T & \langle b, x\rangle + c \end{bmatrix}$$

are $\det(M) = -f(x)$ with multiplicity 1, and 1 with multiplicity $n$. Therefore the constraint $f(x) \leq 0$ is equivalent to $M \succeq 0$. Observe also that $M$ depends affinely on the vector $x$ and can therefore be expressed as

$$M = F_0 + x_1 F_1 + \cdots + x_n F_n$$

with

$$F_0 = \begin{bmatrix} I & 0 \\ 0 & c \end{bmatrix} \quad \text{and} \quad F_i = \begin{bmatrix} 0 & a_i \\ a_i^T & b_i \end{bmatrix}, \quad i = 1, \ldots, n\,,$$

where $a_1, \ldots, a_n$ are the columns of $A^{1/2}$.

A general (convex) quadratically constrained quadratic program (QP) is the following problem:

$$\begin{array}{ll} \text{minimize} & f_0(x) \\ \text{subject to} & f_i(x) \leq 0, \quad i = 1, \ldots, m, \end{array} \tag{QP}$$

where each $f_i : \mathbf{R}^n \to \mathbf{R}$ $(0 \leq i \leq m)$ is a convex quadratic function given by $f_i(x) = \langle A_i x, x\rangle - \langle b_i, x\rangle - c_i$, $A_i \succeq 0$, $b_i \in \mathbf{R}^n$, $c_i \in \mathbf{R}$. By the above, the problem (QP) can be restated as a semidefinite program

$$\begin{array}{ll} \text{minimize} & t \\ \text{subject to} & \begin{bmatrix} I & A_0^{1/2}x \\ (A_0^{1/2}x)^T & \langle b_0, x\rangle + c_0 + t \end{bmatrix} \succeq 0 \end{array}$$

$$\begin{bmatrix} I & A_i^{1/2}x \\ (A_i^{1/2}x)^T & \langle b_i, x\rangle + c_i \end{bmatrix} \succeq 0, \quad i = 1, \dots, m.$$

Since a block-diagonal matrix is positive semidefinite if and only if each of its blocks is positive semidefinite, the above $m+1$ constraints can also be viewed as a single constraint on a large block-diagonal matrix.

Let us remark that it is a standard procedure to introduce an additional variable, in our case $t$, when transforming a problem into a semidefinite form in order to linearize the objective function and to move its nonlinearity into the constraints.

For our applications, the most important semidefinite programs are those for minimizing the largest eigenvalue of an affine combination of symmetric matrices.

**Example 4.4** Let $A_0, \dots, A_n \in \mathbf{R}^{m\times m}$ be symmetric matrices, and for $x \in \mathbf{R}^n$, set

$$A(x) = A_0 + \sum_{i=1}^{n} x_i A_i .$$

We would like to

$$\begin{array}{ll} \text{minimize} & \lambda_{\max}(A(x)) \\ \text{subject to} & x \in U \end{array}$$

where $U$ is a (linear) subspace of $\mathbf{R}^n$. Similarly as in the previous example, the above problem can be translated into a semidefinite program by introducing an auxiliary variable $t \in \mathbf{R}$ :

$$\begin{array}{ll} \text{minimize} & t \\ \text{subject to} & tI - A(x) \succeq 0, \\ & x \in U. \end{array}$$

The first constraint is equivalent to $t \geq \lambda_{\max}(A(x))$. We leave it to the reader to see how both conditions can be combined into a single condition on the positive semidefiniteness of a matrix as required in the standard form of semidefinite programs.

In particular, the bound on the max-cut of a graph stated in Proposition 3.6 is an instance of the problem from Example 4.4, where we take $U$ to be the orthogonal complement of the vector $\mathbf{1}$. Also, the bound on the bipartition width from Corollary 3.4 can be viewed as a semidefinite program of the same form since

$$\max_{x\in U} \lambda_{\min}(A(x)) = \max_{x\in U}\{-\lambda_{\max}(-A(x))\} = -\min_{x\in U} \lambda_{\max}(-A(x)).$$

There exist efficient and practically useful algorithms for solving semidefinite programs. Given any $\varepsilon > 0$, the given semidefinite program can be solved within an additive error of $\varepsilon$ in polynomial time (where $\varepsilon$ is part of the input, its size being proportional to $\log(1/\varepsilon)$). For this purpose one can use the ellipsoid algorithm (cf. [GLS]), other polynomial time algorithms for convex programming [Vai], or the interior-point methods [Ali, NN1, NN2]. Since the work of Nesterov and Nemirovskii [NN1], and Alizadeh [Ali] there has been much development in the design and analysis of interior-point methods for semidefinite programming, yielding practically useful algorithms whose performance is comparable, and even superior, to the performance of the simplex method.

## 4.2 Max-cut and semidefinite programming

Given a (weighted) graph $G$, the max-cut problem for $G$ can be formulated as a quadratic integer program with variables $y_v$, $v \in V(G)$:

$$\begin{array}{ll} \text{maximize} & \frac{1}{2} \sum_{uv \in E(G)} a_{uv}(1 - y_u y_v) \\ \text{subject to} & y_v \in \{-1, 1\} \text{ for every } v \in V(G). \end{array} \tag{MCQI}$$

If $y$ is a feasible solution, then $1 - y_v y_u$ is equal to either 0 or 2. Given a solution $y$ of (MCQI), the set $S$ with the property $\mathrm{mc}(G) = e(S, \bar{S})$ is determined by $S = \{v \in V(G) \mid y_v = 1\}$.

Goemans and Williamson [GW] considered the following relaxation of (MCQI):

$$\begin{array}{ll} \text{maximize} & \frac{1}{2} \sum_{uv \in E(G)} a_{uv}(1 - \langle z_v, z_u \rangle) \\ \text{subject to} & z_v \in S^{n-1} \text{ for every } v \in V(G). \end{array} \tag{MCQR}$$

Here, $S^{n-1}$ denotes the $(n-1)$-dimensional sphere in $\mathbf{R}^n$,

$$S^{n-1} = \{x \in \mathbf{R}^n \mid \|x\| = 1\}.$$

It is important to observe that the problem (MCQR) can be reformulated as a semidefinite program:

$$\begin{array}{ll} \text{maximize} & \frac{1}{2} \sum_{uv \in E(G)} a_{uv}(1 - y_{uv}) \\ \text{subject to} & y_{vv} = 1 \text{ for every } v \in V(G) \text{ and} \\ & Y = [y_{uv}]_{u,v \in V(G)} \succeq 0. \end{array} \tag{MCSP}$$

**Claim 4.5** *Problems* (MCQR) *and* (MCSP) *are equivalent.*

**Proof** Denote by $\Phi_1$ and $\Phi_2$ the objective functions of (MCQR) and (MCSP), respectively, and let $\Phi_1^*$ and $\Phi_2^*$ be the corresponding optimum values. Suppose that $z = \{z_v \in S^{n-1} \mid v \in V(G)\}$ is an optimum solution to (MCQR). Form the matrix $Y = [y_{uv}]_{u,v \in V(G)}$, where $y_{uv} = \langle z_u, z_v \rangle$. This matrix, known as the *Gram matrix* of vectors $z_v$ ($v \in V(G)$), is easily seen to be positive semidefinite. Clearly, $y_{vv} = 1$ for each $v \in V(G)$. Since also $\Phi_1^* = \Phi_1(z) = \Phi_2(Y)$, we have $\Phi_1^* \leq \Phi_2^*$.

Conversely, suppose that $Y$ is an optimum solution to (MCSP). Let $z = \{z_v \mid v \in V(G)\}$ be the columns of $Y^{1/2}$. Since $y_{vv} = 1$, we have $\langle z_v, z_v \rangle = 1$. Therefore $z$ is a feasible solution to (MCQR). Clearly, $\Phi_2(Y) = \Phi_1(z)$, and hence $\Phi_2^* \leq \Phi_1^*$. □

Poljak and Rendl [PR] were the first to realize that the bound of the relaxation (MCSP) is in fact equivalent to the eigenvalue upper bound of Proposition 3.6.

**Proposition 4.6** *Let $G$ be a weighted graph of order $n$. Then the value of an optimum solution to* (MCSP) *is equal to*

$$\frac{n}{4} \cdot \min_c \lambda_{\max}(L(G) + \mathrm{diag}(c)),$$

*where the minimum is taken over all vectors $c \in \mathbf{R}^V$ with $\sum_{v \in V(G)} c_v = 0$.*

For a proof (which uses elegant duality theory of semidefinite programs) we refer to [PR].

Goemans and Williamson suggested the following randomized algorithm based on the relaxation (MCsp) that finds an approximation to the max-cut of $G$ (and also returns the corresponding set $S \subset V(G)$):

**Algorithm 4.7 (Randomized max-cut approximation algorithm)**

*Input:* A weighted graph $G$ of order $n$.
*Output:* A set $S \subseteq V(G)$ and $e(S,\bar{S})$.
1. *Solve the program* (MCsp), *and let* $Y$ *be the obtained solution.*
2. *Compute the Cholesky decomposition of* $Y = Z^T Z$, *and let* $z_v$, $v \in V(G)$, *be the columns of* $Z$.
3. *Choose a random vector* $c \in S^{n-1}$ *(according to the uniform distribution), and set* $S := \{v \in V(G) \mid \langle c, z_v \rangle \geq 0\}$.
4. *Compute* $e(S,\bar{S})$.
5. *Return* $S$ *and* $e(S,\bar{S})$.

The algorithm can be implemented to run in polynomial time in the size of the input. (Recall that the size of the input depends on $n$ and $\log w(E)$, where $w(E) = \sum_{e \in E(G)} w(e)$.) See [GW] for more details. In the next section we will learn how one can upgrade Algorithm 4.7 to a deterministic 0.878-approximation algorithm for the max-cut problem.

## 4.3 Approximating the max-cut

For a weighted graph $G$ with the weight function $w$, denote by $w(E)$ the sum of the weights $w(e)$ over all edges $e$ of $G$. Let us first compute the expected value of the weight $e(S,\bar{S})$.

**Claim 4.8** *Let $S$ be a randomly chosen subset of $V(G)$, where each vertex $v \in V(G)$ belongs to $S$ with probability $\frac{1}{2}$. Then*

$$\mathrm{Exp}[\, e(S,\bar{S})\,] = \frac{1}{2}\, w(E).$$

**Proof** For each $e \in E(G)$, there are exactly $2^{n-1}$ subsets $S \subseteq V(G)$ such that $e \in E(S,\bar{S})$. Therefore

$$\begin{aligned}\mathrm{Exp}[\, e(S,\bar{S})\,] &= \sum_{S \subseteq V(G)} \frac{1}{2^n} \cdot e(S,\bar{S}) = \frac{1}{2^n} \sum_{S \subseteq V(G)} \sum_{e \in E(S,\bar{S})} w(e) \\ &= \frac{1}{2^n} \sum_{e \in E(G)} 2^{n-1} \cdot w(e) = \frac{1}{2} w(E)\,. \qquad \square\end{aligned}$$

It is also easy to describe a deterministic procedure that finds a $\frac{1}{2}$-approximation to $\mathrm{mc}(G)$, i.e., returns a subset $S \subset V(G)$ such that $e(S,\bar{S}) \geq \frac{1}{2}\mathrm{mc}(G)$.

**Algorithm 4.9 (Deterministic $\frac{1}{2}$-approximation algorithm for the max-cut)**

*Input:* A weighted graph $G$.
*Output:* A set $S \subset V(G)$ and $e(S,\bar{S})$ such that $e(S,\bar{S}) \geq \frac{1}{2}\mathrm{mc}(G)$.
1. *Set* $S_1 := \emptyset$ *and* $S_2 := \emptyset$.

2. *Take $v \in V(G)\backslash(S_1 \cup S_2)$ and compute $e(\{v\}, S_1)$ and $e(\{v\}, S_2)$.*
3. *If $e(\{v\}, S_1) \geq e(\{v\}, S_2)$, let $S_2 := S_2 \cup \{v\}$.*
   *Otherwise, let $S_1 := S_1 \cup \{v\}$.*
4. *If $S_1 \cup S_2 \neq V(G)$, goto Step 2.*
5. *Set $S := S_1$ and compute $e(S, \bar{S})$.*
6. *Return $S$ and $e(S, \bar{S})$.*

Note that throughout the algorithm, we have

$$e(S_1, S_2) \geq \frac{1}{2} \sum_{u,v \in S_1 \cup S_2} w(uv)\,,$$

which proves that at the end

$$e(S, \bar{S}) \geq \frac{1}{2} w(E) \geq \frac{1}{2} \operatorname{mc}(G)$$

holds. There exist examples where the value $e(S, \bar{S})$ returned by Algorithm 4.9 is arbitrarily close to $\frac{1}{2}\operatorname{mc}(G)$.

It is much more difficult to find an $\alpha$-approximation to $\operatorname{mc}(G)$ for a constant $\alpha > \frac{1}{2}$ in polynomial time. Let us also recall that there exists a constant $\varepsilon > 0$ such that there is no polynomial time $(1 - \varepsilon)$-approximation algorithm for the max-cut problem unless $P = NP$ [PY, A-S].

Following [GW], we will now show that Algorithm 4.7 can be used to find such a 0.878-approximation in polynomial time.

**Lemma 4.10** *The expected weight of the cut $E(S, \bar{S})$ produced by Algorithm 4.7 is equal to*

$$\operatorname{Exp}[\, e(S, \bar{S})\,] = \frac{1}{\pi} \sum_{uv \in E(G)} a_{uv} \arccos\langle z_u, z_v\rangle\,,$$

*where $z_v$, $v \in V(G)$, is an optimum solution of* (MCQR).

**Proof** Observe first that

$$\Pr[\, u \in S, v \notin S\,] = \Pr[\, \langle c, z_u\rangle \geq 0, \langle c, z_v\rangle < 0\,] = \frac{\arccos\langle z_u, z_v\rangle}{2\pi}\,.$$

Therefore

$$\begin{aligned}\operatorname{Exp}[\, e(S, \bar{S})\,] &= \sum_{uv \in E(G)} w(uv) \Pr[\, uv \in E(S, \bar{S})\,] = 2 \sum_{uv \in E(G)} w(uv) \Pr[\, u \in S, v \notin S\,] \\ &= 2 \sum_{uv \in E(G)} w(uv) \frac{\arccos\langle z_u, z_v\rangle}{2\pi} = \frac{1}{\pi} \sum_{uv \in E(G)} w(uv) \arccos\langle z_u, z_v\rangle\,. \qquad \square\end{aligned}$$

It is an easy exercise in calculus to show that for $x \in [0, 1]$, we have

$$\frac{1}{\pi} \arccos(x) \geq \beta \cdot \frac{1}{2}(1 - x)\,, \tag{29}$$

where $\beta = 0.87856$. Therefore Lemma 4.10 implies:

**Corollary 4.11** *For a weighted graph $G$, let $E(S, \bar{S})$ be the cut obtained by Algorithm 4.7. Then*

$$\mathrm{Exp}[\, e(S, \bar{S})\,] \geq \beta \cdot \mathrm{mc}(G)\,.$$

**Proof** Lemma 4.10 and inequality (29) imply that

$$\mathrm{Exp}[\, e(S, \bar{S})\,] = \frac{1}{\pi} \sum_{uv \in E(G)} w(uv) \arccos\langle z_u, z_v\rangle \geq \beta \cdot \frac{1}{2} \sum_{uv \in E(G)} w(uv)(1 - \langle z_u, z_v\rangle)\,.$$

Since $z_v$, $v \in V(G)$, is an optimum solution to (MCQR), and since (MCQR) gives an upper bound on $\mathrm{mc}(G)$, we finally get

$$\mathrm{Exp}[\, e(S, \bar{S})\,] \geq \beta \cdot \mathrm{mc}(G)\,.$$ □

It remains to describe how to convert the randomized Algorithm 4.7 into a deterministic one (without substantially increasing its time complexity and without worsening its approximation performance).

Following [GW], the idea of derandomization is to maintain a set of unit vectors $y_v \in S^m$, $v \in V(G)$, with the property

$$\mathrm{Exp}[\, e(S, \bar{S})\,] \geq \beta \cdot \mathrm{mc}(G)\,, \tag{30}$$

where the partitioning is performed with respect to a random hyperplane in $\mathbf{R}^{m+1}$, and to gradually decrease the dimension $m$ to 0 by transforming $y_v$, $v \in V(G)$, to a new set of unit vectors $y'_v \in S^{m-1}$, $v \in V(G)$, still having the property (30) (with respect to the partition determined by a random hyperplane in $\mathbf{R}^m$). For $m = 0$, $S^0 = \{-1, 1\}$, and there is only one "random" hyperplane in $\mathbf{R}$. Therefore the final vectors $y_v$, $v \in V(G)$, determine a set $S := \{v \in V(G) \mid y_v = 1\}$ with the property $e(S, \bar{S}) \geq \beta \cdot \mathrm{mc}(G)$.

Initially, $y_v := z_v \in S^{n-1}$, where $z_v$, $v \in V(G)$, is an optimum solution to (MCQR). Let us now describe how to decrease the dimension $m$ by 1. Suppose that we have vectors $y_v \in S^m$, $v \in V(G)$, satisfying (30). Decompose $S^m$ into a disjoint union

$$S^m = \bigcup_{\theta \in (-\pi, \pi]} S^m_\theta\,,$$

where

$$S^m_\theta = \{x = (x_1, \ldots, x_{m+1})^T \in S^m \mid x_m = s\cos\theta,\ x_{m+1} = s\sin\theta \text{ for some } s \in \mathbf{R}\}\,.$$

Then there exists an angle $\theta$ such that choosing the normal $c$ only from $S^m_\theta$ still gives (possibly even stronger) inequality (30). Moreover, such an angle $\theta$ can be efficiently computed. For every $v \in V(G)$, let $t_v$ and $\gamma_v$ be the absolute value and the polar angle of the complex number $x_m + i \cdot x_{m+1}$, respectively. Define

$$y^\circ_v = (y_{v,1}, \ldots, y_{v,m-1}, t_v \cos(\gamma_v - \theta))^T \in \mathbf{R}^m$$

and set

$$y'_v = \frac{y^\circ_v}{\|y^\circ_v\|} \in S^{m-1}\,.$$

It remains to prove that the transition from $y_v$ to $y'_v$ does not violate (30). The key observation that guarantees this is the following:

**Claim 4.12** *Suppose that* $x = (x_1, \ldots, x_{m-1}, s\cos\theta, s\sin\theta)^T \in S_\theta^m$. *Define* $x' := (x_1, \ldots, x_{m-1}, s)^T \in S^{m-1}$. *Then*

$$\langle x, y_v \rangle = \langle x', y_v^\circ \rangle .$$

**Proof** Standard properties of the cosine give

$$\begin{aligned}\langle x, y_v \rangle - \langle x', y_v^\circ \rangle &= s\cos\theta \cdot t_v \cos\gamma_v + s\sin\theta \cdot t_v \sin\gamma_v - s \cdot t_v \cos(\gamma_v - \theta) \\ &= s \cdot t_v (\cos\theta\cos\gamma_v + \sin\theta\sin\gamma_v - \cos(\gamma_v - \theta)) = 0.\end{aligned}$$ □

Consequently, the sign of $\langle x, y_v \rangle$ is the same as the sign of $\langle x', y'_v \rangle$, and therefore the expected value $\mathrm{Exp}[\, e(S, \bar{S})\,]$ of $e(S, \bar{S})$ on $S^{m-1}$ with respect to $y'_v$, $v \in V(G)$, is the same as the expectation $\mathrm{Exp}[\, e(S, \bar{S})\,]$ of $e(S, \bar{S})$ on $S_\theta^m$ with respect to $y_v$, $v \in V(G)$, which is in turn at least $\beta$ mc$(G)$. This proves that the derandomization works as claimed above.

After the announcement of the approximation algorithm of Goemans and Williamson, semidefinite programming has been successfully applied in the design of a number of other approximation algorithms, for example the max-$k$-cut (Frieze and Jerrum [FJ]), a 0.931-approximation algorithm for MAX 2-SAT (Feige and Goemans [FG]), a 0.859-approximation algorithm for MAX DICUT [FG], and approximate coloring (Karger, Motwani, and Sudan [KMS]).

# 5 Random walks on graphs

Isoperimetric properties and eigenvalues treated in the previous sections are closely related to the convergence rates of Markov chains. Several important randomized algorithms discovered in the last decade increased applicability of random walks and Markov chains in solving previously untractable problems. In this section we briefly describe relations between simple random walks in graphs and parameters similar to those treated in previous sections. At the end we mention how these results are used in the design of efficient randomized approximation algorithms. Additional reading on the results related to the presentation in this section can be found, e.g., in [MR, Kan, Ros, Sin].

## 5.1 Markov chains

A *Markov chain* consists of a measurable *state space* $\mathcal{V}$, an initial probability distribution (i.e., a probability measure) $\mu_0$ on $\mathcal{V}$, and *transition probabilities* $P(u, \cdot)$ which give, for each state $u \in \mathcal{V}$, a distribution $P(u, \cdot)$ on $\mathcal{V}$ (which represents the probabilities of where the Markov chain will go one step after being in the state $u$). It is assumed that $u \mapsto P(u, A)$ is a measurable function on $\mathcal{V}$ for each fixed measurable subset $A \subseteq \mathcal{V}$. If $\mathcal{V}$ is a discrete space (in particular, finite), then the initial distribution $\mu_0$ can be specified by the function $x^{(0)} : \mathcal{V} \to \mathbf{R}^+$ such that $\sum_{u \in \mathcal{V}} x_u^{(0)} = 1$. Similarly, the transition probabilities $P(u, v)$ can be described by the *transition matrix* $P = (p_{uv})_{u,v \in \mathcal{V}}$, indexed by $\mathcal{V}$, which has nonnegative real entries $p_{uv} = P(u, \{v\})$ such that for each $u \in \mathcal{V}$, we have $\sum_{v \in \mathcal{V}} p_{uv} = 1$.

Given the initial distribution $\mu_0$ and transition probabilities $P(u, \cdot)$, we can inductively construct distributions $\mu_t$ on $\mathcal{V}$, representing the probabilities of where the Markov chain will be after $t$ steps:

$$\mu_t(A) = \int_{\mathcal{V}} P(u, A) \mu_{t-1}(du) , \quad t = 1, 2, \ldots . \tag{31}$$

We will be interested only in Markov chains with discrete state space $\mathcal{V}$, which we assume in what follows. In such a case, (31) can be presented more directly as

$$x_v^{(t)} = \sum_{u \in \mathcal{V}} p_{uv} x_u^{(t-1)} \tag{32}$$

or in matrix form:

$$x^{(t)} = P^T x^{(t-1)} = \cdots = (P^T)^t x^{(0)}, \tag{33}$$

where the action of the transpose matrix $P^T$ on the elements of $\mathbf{R}^{\mathcal{V}}$ is determined by the rule of matrix-vector multiplication. The formula (33) is easy to explain informally. It states that to be in the state $v$ after $t$ steps, we must have been at some state $u$ at step $t-1$ (with probability $x_u^{(t-1)}$), and then jumped from $u$ to $v$ in the next step (with probability $p_{uv}$).

If the state space $\mathcal{V}$ contains a nonempty proper subset $\mathcal{V}' \subset \mathcal{V}$ such that $P(v, \mathcal{V}') = 1$ for all $v \in \mathcal{V}'$, then the Markov chain is said to be *reducible*. If there is no such subset $\mathcal{V}'$, the Markov chain is *irreducible*. The Markov chain is *periodic*, if the state space $\mathcal{V}$ contains disjoint nonempty subsets $\mathcal{V}_0, \ldots, \mathcal{V}_{d-1}$ ($d \geq 2$) such that $P(u, \mathcal{V}_i) = 1$ for all $u \in \mathcal{V}_{i-1}$ (where the indices are taken modulo $d$). If there is no such decomposition, the Markov chain is *aperiodic*.

For states $u, v \in \mathcal{V}$ and $t > 0$, let $r_{uv}^{(t)}$ denote the probability that the Markov chain with the initial state $u$ visits the state $v$ for the first time in step $t$. The probability that the Markov chain with the initial state $u$ visits the state $v$ after some number of steps is denoted by $r_{uv}^{(\infty)}$, i.e.,

$$r_{uv}^{(\infty)} = \sum_{t=1}^{\infty} r_{uv}^{(t)} .$$

A state $v \in \mathcal{V}$ is *transient* if $r_{vv}^{(\infty)} < 1$; otherwise it is *recurrent* (or *persistent*).

The expected number of steps to reach the state $v$ starting from the state $u$ is denoted by $h_{uv}$. Then

$$h_{uv} = \sum_{t=1}^{\infty} t\, r_{uv}^{(t)} .$$

A *stationary distribution* for the Markov chain with transition matrix $P$ is a probability distribution $x$ such that $x = P^T x$.

The following theorem summarizes the basic properties of finite Markov chains (i.e., Markov chains whose state space $\mathcal{V}$ is finite).

**Theorem 5.1** *Any finite, irreducible, aperiodic Markov chain with the state space $\mathcal{V}$ has the following properties:*

(i) *All states $v \in \mathcal{V}$ are aperiodic and recurrent and $h_{vv} < \infty$.*

(ii) *There is a unique stationary distribution $x^{(\infty)}$, and $x_v^{(\infty)} > 0$ for every $v \in \mathcal{V}$.*

(iii) *For every $v \in \mathcal{V}$, $r_{vv}^{(\infty)} = 1$ and $h_{vv} = 1/x_v^{(\infty)}$.*

(iv) *Choose an initial distribution $x^{(0)}$, and let $N(v,t)$ denote the expected number of times the Markov chain visits state $v$ in $t$ steps, i.e., $N(v,t) = \sum_{i=1}^{t} x_v^{(i)}$. Then*

$$\lim_{t \to \infty} \frac{N(v,t)}{t} = x_v^{(\infty)} .$$

The proof of Theorem 5.1 is elementary and can be found in most textbooks treating Markov chains. We refer, for example, to [Fe]. The reader can also consult a similar proof of Theorem 5.7.

## 5.2 Random walks on graphs

Given a (finite) weighted graph $G$ (possibly containing loops), a *random walk* on $G$ is a Markov chain with state space $\mathcal{V} = V(G)$ and with transition probabilities $p_{uv}$ such that $p_{uv} > 0$ if and only if $uv \in E(G)$. The most common rule to assign values to the transition probabilities $p_{uv}$ to step from $u$ to $v$, which we assume in what follows, is to take

$$p_{uv} = \frac{a_{uv}}{\mathrm{d}_u} .$$

This random walk is called the *simple random walk* on $G$. For the simple random walk on $G$, the transition matrix $P(G)$ is equal to

$$P(G) = D^{-1}A(G) ,$$

where $D = \mathrm{diag}(\mathrm{d}_v; v \in V(G))$ is the diagonal matrix with vertex degrees on the diagonal. Although $P$ is not necessarily symmetric, it has only real eigenvalues. This can be seen as follows. Let $L'(G)$ be the *transition Laplacian matrix* of $G$ defined by

$$L'(G) = D^{-1/2}L(G)D^{-1/2} .$$

In particular, we have

$$(L'(G))_{uv} = \begin{cases} 1 - \frac{a_{uu}}{\mathrm{d}_u}, & u = v \\ -\frac{a_{uv}}{\sqrt{\mathrm{d}_u \mathrm{d}_v}}, & uv \in E(G) \\ 0 , & \text{otherwise.} \end{cases} \tag{34}$$

Observe that $L'(G)$ is a symmetric matrix and that for a $d$-regular graph $G$, we have

$$L(G) = d \cdot L'(G) .$$

Notice also that the difference Laplacian matrix $L(G)$ of $G$ does not change if we add loops to $G$, while the transition Laplacian matrix $L'(G)$ is changed under addition of loops. We will see that this property can be used to construct graphs with simple random walks that converge rapidly to the stationary distribution.

The transition matrix $P(G)$ of the simple random walk on $G$ can be expressed as follows:

**Claim 5.2** $P(G) = I - D^{-1/2}L'(G)D^{1/2}$.

**Proof** Since $I - D^{-1/2}L'(G)D^{1/2} = I - D^{-1}L(G)$, the diagonal elements of the matrix are $1 - \mathrm{d}_u^{-1}(L(G))_{uu} = 1 - (\mathrm{d}_u - a_{uu})/\mathrm{d}_u = a_{uu}/\mathrm{d}_u$ $(u \in V(G))$, and the off-diagonal entries are equal to $-\mathrm{d}_u^{-1}(L(G))_{uv} = a_{uv}/\mathrm{d}_u = p_{uv}$ $(u, v \in V(G), u \neq v)$. □

In particular, $I - P(G)$ and $L'(G)$ have the same eigenvalues. Moreover, Claim 5.2 implies that $P(G)$ (and also $P(G)^T$) can be diagonalized, i.e., it has $n$ pairwise orthogonal eigenvectors.

A proof similar to the proof of Proposition 2.8 gives the following result.

**Claim 5.3** *Let $G$ be a weighted graph with at least two vertices. Then*

$$-1 \le \lambda_{\min}(P(G)) < \lambda_{\max}(P(G)) = 1 .$$

*If $G$ is connected, then $\lambda_{\max}(P(G)) = 1$ is a simple eigenvalue, and we have $\lambda_{\min}(P(G)) = -1$ if and only if $G$ is bipartite.*

It is easy to determine, just by looking at the structure of the graph $G$, whether a random walk on $G$ is aperiodic or irreducible.

**Claim 5.4** *Let $G$ be a weighted graph. Then:*

(i) *The simple (and hence every) random walk on $G$ is periodic if and only if $G$ is bipartite.*

(ii) *The simple (and hence every) random walk on $G$ is irreducible if and only if $G$ is connected.*

The easy proof of Claim 5.4 is left to the reader.

Theorem 5.1 states that for every finite, irreducible, aperiodic Markov chain, there exists a unique stationary distribution $x^{(\infty)}$. In case of simple random walks on connected nonbipartite weighted graphs, the distribution $x^{(\infty)}$ can be easily found.

**Proposition 5.5** *Let $G$ be a connected, nonbipartite weighted graph. Then the stationary distribution of the simple random walk on $G$ is given by*

$$x_v^{(\infty)} = \frac{\mathrm{d}_v}{2m}, \qquad v \in V(G),$$

*where $2m = \sum_{v \in V(G)} \mathrm{d}_v$. In particular, if $G$ is d-regular, then the stationary distribution is uniform,*

$$x_v^{(\infty)} = \frac{1}{n}, \qquad v \in V(G).$$

**Proof** It suffices to check that $x^{(\infty)}$ is an eigenfunction of $P(G)^T$ corresponding to the eigenvalue 1:

$$\begin{aligned}
(P(G)^T x^{(\infty)})_v &= \sum_{u \in V(G)} P(G)^T_{vu} x_u^{(\infty)} \\
&= \sum_{u \in V(G)} \frac{a_{uv}}{\mathrm{d}_u} \cdot \frac{\mathrm{d}_u}{2m} = \frac{\mathrm{d}_v}{2m} = x_v^{(\infty)} .
\end{aligned}$$

□

Let us conclude this subsection by an example.

**Example 5.6 Card shuffling** Suppose that a deck of $n$ cards $c_1, \ldots, c_n$ is lying on the table. We would like to shuffle the cards according to the following rule: select randomly $i, j$ $(1 \le i, j \le n)$ and exchange the cards $c_i$ and $c_j$. To simulate this shuffling process as a simple random walk on a graph, consider the Cayley graph $\mathrm{Cay}(S_n, T)$ of the symmetric group $S_n$, where $T$ is the set of all transpositions in $S_n$. Since this graph is bipartite, it is a standard trick to add a loop to each vertex to avoid the periodicity of the random walk. Denote the obtained graph by $G$. To assure that our card-shuffling process is faithfully modelled by the simple random walk on $G$, we have to assign weights to the edges of $G$ appropriately. There are $n^2$ distinct possibilities to choose the pair $i, j$: two for each of the $\binom{n}{2}$ transpositions, and $n$ to choose $i = j$ in which case the deck remains unchanged. Therefore we set

$$a_{uv} = \begin{cases} \frac{1}{n}, & u = v \\ \frac{2}{n^2}, & u \neq v. \end{cases} \tag{35}$$

Well, shuffling cards is fun, but we would like to know how long do we have to shuffle so that the cards are properly shuffled. Or, more formally, how fast does the simple random walk on $G$ converge to the stationary distribution? We address this type of questions in the next subsection.

Example 5.5 belongs to a broader class of random walks on groups. Suppose that we have a (finite) group $\Gamma$ and a symmetric probability distribution $p$ on $\Gamma$, i.e., a probability distribution with the property $p(g) = p(g^{-1})$ for every $g \in \Gamma$. Define $S = \{g \in \Gamma \mid p(g) > 0\}$. Then a random walk on $\Gamma$ is actually a random walk on the graph $G = \mathrm{Cay}(\Gamma, S)$ where the weight of the edge $gh \in E(G)$ is equal to $p(g^{-1}h)$. In particular, since Cayley graphs are regular, the uniform distribution is stationary for any random walk on any (finite) group.

## 5.3 Rate of convergence of a random walk

As suggested by the results of the previous subsections, there is a close relationship between the eigenvalues of $P(G)$ and the rate of convergence of the simple random walk on $G$ to the stationary distribution.

**Theorem 5.7** *Let $G$ be a connected, nonbipartite weighted graph and $x^{(0)}$ the initial probability distribution on $V(G)$. Set*

$$\lambda = \max\{|\lambda_i(P(G))| \mid \lambda_i(P(G)) \neq 1\}.$$

*For the simple random walk on $G$ we have*

$$\|x^{(t)} - x^{(\infty)}\| < \lambda^t.$$

**Proof** For $i = 1, \ldots, n$, let $y^{(i)}$ be the eigenfunction of $Q = P(G)^T$ corresponding to the eigenvalue $\lambda_i = \lambda_i(Q) = \lambda_i(P(G))$. We may assume that $\|y^{(i)}\| = 1$ and that $y^{(i)}$, $i = 1, \ldots, n$, are pairwise orthogonal. Recall that $\lambda_n(Q) = 1$ and that the assumptions in the theorem imply $\lambda < 1$. Moreover, the Perron-Frobenius theorem shows that all components of $y^{(n)}$ are positive.

Let us write

$$x^{(0)} = \sum_{i=1}^{n} \alpha_i y^{(i)}$$

and observe that $\alpha_n = \langle x^{(0)}, y^{(n)} \rangle \neq 0$. Then

$$x^{(t)} = Q^t x^{(0)} = \sum_{i=1}^{n} \alpha_i \lambda_i^t y^{(i)} = \sum_{i=1}^{n-1} \alpha_i \lambda_i^t y^{(i)} + \alpha_n y^{(n)}.$$

Since $\lambda < 1$, $x^{(t)}$ converges to $\alpha_n y^{(n)} = x^{(\infty)}$. Therefore,

$$\begin{aligned} \|x^{(t)} - x^{(\infty)}\|^2 &= \Big\| \sum_{i=1}^{n-1} \alpha_i \lambda_i^t y^{(i)} \Big\|^2 = \sum_{i=1}^{n-1} \|\alpha_i \lambda_i^t y^{(i)}\|^2 \\ &= \sum_{i=1}^{n-1} \alpha_i^2 \lambda_i^{2t} \|y^{(i)}\|^2 \le \lambda^{2t} \sum_{i=1}^{n-1} \alpha_i^2 . \end{aligned}$$

Since also

$$\sum_{i=1}^{n-1} \alpha_i^2 < \sum_{i=1}^{n} \alpha_i^2 = \|x^{(0)}\|^2 \le 1 ,$$

we finally get $\|x^{(t)} - x^{(\infty)}\| < \lambda^t$. □

**Example 5.8** Let us consider the random walk on the $d$-dimensional hypercube $Q_d$. Note that $Q_d$ is a Cayley graph of the group $\mathbf{Z}_2^d$. Since $Q_d$ is bipartite, we add a loop to each vertex, and denote the obtained graph by $G$. We would like to assign the weights to the loops and proper edges of $G$ in such a way that $\lambda$ becomes as small as possible. Let $\alpha$ be the weight assigned to the loops and $\beta$ the weight of the proper edges of $G$. We also require $\alpha + d\beta = 1$. Then

$$P(G) = \alpha I + \beta A(Q_d),$$

where $A(Q_d)$ is the adjacency matrix of $Q_d$. Example 2.5 shows that the eigenvalues of $P(G)$ are $\alpha + \beta(d - 2i)$, $i = 0, \ldots, d$, where the multiplicity of $\lambda_i = \alpha + \beta(d - 2i)$ is equal to $\binom{d}{i}$. Therefore

$$\lambda = \max\{\alpha + \beta(d-1), |\alpha - \beta d|\} .$$

By balancing both terms, we get equality when $\alpha = \beta = \frac{1}{d+1}$. Hence

$$\lambda = \frac{d-1}{d+1} = 1 - \frac{2}{d+1} .$$

By Theorem 5.7, the rate of convergence can be estimated by

$$\lambda^t = \left(\frac{d-1}{d+1}\right)^t = \left(1 - \frac{2}{d+1}\right)^t \le \exp\left(-\frac{2t}{d+1}\right) .$$

Let us remark that similar estimates can be established for other Abelian groups (cf., e.g., [Ros]).

**Example 5.9 Card shuffling** In this case, the general analysis is more complicated than for the $d$-cube since it involves group representations of $S_n$. In the special case of Example 5.6 we obtain

$$\lambda = 1 - \frac{2}{n} \le \exp(-2/n) ,$$

and consequently

$$\|x^{(t)} - x^{(\infty)}\| \le \exp(-\frac{2t}{n}) .$$

See [DSh] for more details.

**Example 5.10 Usual card shuffling** This problem was considered by Bayer and Diaconis [BD]. They showed that $\frac{3}{2}\log_2 n$ iterations are enough (and also necessary) for good card shuffling of the deck of $n$ cards. An interesting result of their analysis of the rate of convergence of random walks on $S_n$ corresponding to card shuffling is the cut-off phenomenon which states that $n$ cards become "well shuffled" (with high probability) all of a sudden in a very short period at the time $\frac{3}{2}\log_2 n + o(\log_2 n)$. In particular, 7 ordinary "riffle" shuffles are (required and) enough to properly mix the ordinary deck of 52 cards. For additional background see [Di].

Perhaps the most useful distance between probability distributions $x$, $y$ of a Markov chain is the so-called "chi-squared" distance which is defined as follows:

$$\chi^2(x,y) = \sum_{v\in V(G)} \frac{(x_v - y_v)^2}{y_v}.$$

Obviously, the definition makes sense only when $y_v > 0$ for every $v \in V(G)$. A proof similar to the proof of Theorem 5.7 gives:

**Theorem 5.11** *For the simple random walk on a non-bipartite weighted graph we have*

$$\chi^2(x^{(t)}, x^{(\infty)}) \le \lambda^t \cdot \chi^2(x^{(0)}, x^{(\infty)}).$$

The quantity

$$\frac{1}{1-\lambda}$$

is called the *mixing time* of the random walk and is obviously of interest since the above theorem implies that in this many steps, the chi-squared distance to the stationary distribution $x^{(\infty)}$ is cut by a constant factor.

The quantity $1-\lambda$ is related to the second smallest eigenvalue $\lambda_2(I-P)$ which is the same as the second smallest eigenvalue of the transition Laplace matrix $L'(G)$.

The isoperimetric constant related to the transition Laplacian is also known as *conductance* of the corresponding Markov chain (cf., e.g., [Sin]). In the case of the simple random walk on the graph $G$, the conductance is defined as

$$i_P(G) = \min\left\{ \frac{\sum_{u\in S}\sum_{v\notin S} p_{uv}\mathrm{d}_u}{\mathrm{d}(S)} \;\middle|\; S\subset V(G), 0<\mathrm{d}(S)\le m \right\}$$

where

$$\mathrm{d}(S) = \sum_{v\in S}\mathrm{d}_v \quad \text{and} \quad m = \frac{1}{2}\sum_{v\in V}\mathrm{d}_v = \frac{1}{2}\mathrm{d}(V).$$

If the graph $G$ is regular, then the conductance coincides with the isoperimetric number of $G$ (divided by the degree).

Similar proofs as given in Section 3 (for Corollary 3.8 and Theorem 3.10) give the following bounds:

**Theorem 5.12** *For the simple random walk on the graph $G$ we have*

$$\frac{1}{2}\, i_P^2(G) \le \lambda_2(I-P) = \lambda_2(L'(G)) \le 2\, i_P(G).$$

For proofs we refer, for example, to [Mo1, Mo4].

Since $\lambda_2(I-P) \ge 1-\lambda$, Theorem 5.12 gives a bound on the mixing time as well.

## 5.4 Some applications

Rapidly mixing Markov chains proved to be an important tool in the design of polynomial time randomized algorithms. In what follows, we will describe two such applications, and we refer the reader to [MR, Kan, Ros] for additional applications and for more detailed presentation.

First we describe an application which concerns computing an approximation of the volume of a convex body. Suppose that the convex body $K \subseteq \mathbf{R}^n$ is given by a *membership oracle*, i.e., for each point $x \in \mathbf{R}^n$, one can decide in constant (or polynomial) time if $x \in K$ or not. Bárány and Füredi [BF] proved that an approximation $V$ for the volume of $K$ such that

$$\frac{1}{2}\mathrm{Vol}(K) < V < 2\mathrm{Vol}(K)$$

(or even with 2 being replaced by a polynomial function in $n$) cannot be computed in polynomial time. See also [LS1, LS2]. However, Dyer, Frieze, and Kannan [DFK] found a polynomial time randomized approximation algorithm for computing the volume of the convex set $K$ based on sampling random points in $K$. Lovász and Simonovits [LS1, LS2] improved their algorithm by providing random walk sampling with better mixing and by proving better isoperimetric estimates. Further improvements have been obtained since; cf. [Kan].

The basic idea of the randomized algorithms for approximating the volume of the convex body $K$ is as follows. Let $B$ be a ball contained in $K$. (Usually, such a ball is part of the input.) The algorithm constructs a sequence of convex bodies $K_0 = B \subset K_1 \subset \cdots \subset K_t = K$ such that:

(i) There is a constant $c > 0$ such that for $i = 1, \ldots, t$, we have

$$\mathrm{Vol}(K_i) \le c \cdot \mathrm{Vol}(K_{i-1}).$$

(ii) The length $t$ of the sequence is polynomial in $n$.

In such a situation, it suffices to construct a fully polynomial randomized approximation scheme that estimates the ratio

$$\frac{\mathrm{Vol}(K_i)}{\mathrm{Vol}(K_{i-1})} \qquad (i = 1, \ldots t).$$

The idea is to choose a random point in $K_i$ and check whether it also belongs to $K_{i-1}$. Property (i) guarantees that such an approach works. The nontrivial part of this procedure is to choose a random point in $K_i$. This is done by imposing a suitably fine grid over $K_i$ and performing a random walk on the vertices of the grid. The crucial step in the method is to prove that the random walk is rapidly mixing: in a number of steps that is polynomial in $n$, the probability distribution of the walk becomes almost uniform on all vertices of the grid. The reader is invited to consult [DFK, LS1, LS2] for details.

Another application, in a way similar to the first one, deals with approximate counting of perfect matchings in a (bipartite) graph (and therefore also with the approximation of the permanent of a 01-matrix). Let us recall that computing the permanent of a 01-matrix is $\#P$-hard [Val]. Each 01-matrix $A$ of size $n \times n$ determines a bipartite graph $G$ on $n + n$ vertices in which the $i$th vertex in the first bipartition class is adjacent to the $j$th vertex in the second class if and only if $a_{ij} = 1$. Then per$(A)$ is equal to the number of perfect matchings of $G$. Broder [Bro] proposed the following Markov chain approach to approximating the number of perfect matchings of $G$.

One starts with a bipartite graph $G$ on $n+n$ vertices having at least one perfect matching. The state space $\mathcal{V}$ of the Markov chain is the set of all perfect and all near-perfect matchings. (The matching is said to be *near-perfect* if it covers all vertices but one in each bipartition class of the graph.) We start with an arbitrary matching $M \in \mathcal{V}$. The transitions in the Markov chain are as follows. Let $M$ be the current state. We stay with probability 1/2 in the same state, and with probability 1/2 we do the following. Pick an edge $e = uv$ of $G$ uniformly at random and:

(a) If $M$ is a perfect matching and $e \in M$, then move to $M \backslash \{e\}$.

(b) If $M$ is a near-perfect matching and no edge of $M$ contains $u$ or $v$, then move to the matching $M \cup \{e\}$.

(c) If $M$ is a near-perfect matching, an edge $uw$ incident with $u$ is in $M$, and no edge of $M$ contains $v$, then move to $(M \cup \{e\}) \setminus \{uw\}$.

(d) In all other cases stay at $M$.

Jerrum and Sinclair [JS, SJ] proved (by estimating the isoperimetric number) that this random walk mixes rapidly if the graph is "dense enough" (e.g., if each row and column of $A$ contains at least $n/2$ ones). This yields a polynomial time randomized approximation algorithm for computing the number of perfect matchings of "dense" bipartite graphs. For details, the reader is invited to consult [JS] or [Sin].

## 6 Infinite graphs

In this section we show how to extend the results of previous sections to infinite locally finite graphs. Most of the results can be derived for weighted graphs where the local finiteness condition is replaced by the requirement that $\mathrm{d}_v < \infty$ for each $v \in V(G)$. However, some results, for instance those about the growth, do not extend to the weighted case.

In considering the appropriate notions of infinite graphs there is a possibility of introducing the *essential graph invariants*, examples of which are presented in Subsection 6.1. In Subsection 6.2 we introduce the Laplace matrix and the Laplace spectrum of infinite graphs, and prove some counterparts to the results of Section 3. In the last subsection we touch the concepts of amenability and transience of graphs. A thorough treatment of these results can be found in [Mo4] whose presentation is partially followed also in this section.

### 6.1 Essential invariants of infinite graphs

Graphs $G$ and $H$ are said to be (*cofinitely*) *equivalent* or *almost isomorphic* if there is a bijection $\psi : V(G) \to V(H)$ which maps adjacent, or non-adjacent pairs of vertices of $G$ to adjacent and non-adjacent pairs in $H$, respectively, with finitely many exceptions. Equivalently, $H$ can be obtained from $G$, up to an isomorphism, by first deleting and then adding finitely many edges.

We shall denote by $\Delta(G)$ and $\delta(G)$ the maximum and minimum vertex degree in $G$, respectively. It is possible that $\Delta(G) = \infty$. These numbers have their "essential" counterparts, the *essential maximum degree* $\Delta'(G)$ and the *essential minimum degree* $\delta'(G)$, defined by

$$\Delta'(G) := \inf \{\Delta(H) \mid H \text{ equivalent to } G\}$$

and

$$\delta'(G) := \sup \{\delta(H) \mid H \text{ equivalent to } G\} .$$

It may happen that $\Delta'(G) = \infty$ and $\delta'(G) = \infty$. Clearly, $\Delta'(G)$ is equal to the minimal value $d$ for which there are only finitely many vertices of $G$ of degree $> d$. Similarly, $\delta'(G)$ is the minimal $d$ with infinitely many vertices of degree $d$. It is obvious that $\delta'(G) \leq \Delta'(G)$. These two quantities are defined in such a way that no finite perturbation of $G$ changes their values, i.e., if $G$ and $H$ are equivalent then $\Delta'(G) = \Delta'(H)$ and $\delta'(G) = \delta'(H)$. Graph invariants with this property are said to be *essential.* So, $\Delta'(G)$ and $\delta'(G)$ are the first examples of essential graph invariants we met.

Let $v \in V(G)$. Denote by $B_n(v)$ the set of all vertices of $G$ at distance at most $n$ from $v$ (the *ball* of radius $n$ centred at $v$), and let $b_n(v) := |B_n(v)|$. If it is necessary to mention the graph explicitly, we write $b_n(G, v)$ and $B_n(G, v)$. The graph $G$ has *exponential growth* (from the vertex $v$) if $b_n(v) \geq Cq^n$ for some constants $C > 0$, $q > 1$, and each $n \geq 0$. It has *polynomial growth* if $b_n(v) \leq p(n)$ where $p(.)$ is a polynomial. $G$ has *subexponential growth* if it does not grow exponentially. Note that the type of the growth is independent of $v$ if $G$ is connected. Let

$$\varepsilon(G, v) := \limsup_{n\to\infty} (b_n(G, v))^{1/n}$$

and

$$\tau(G, v) := \liminf_{n\to\infty} (b_n(G, v))^{1/n} .$$

If $v$ and $u$ are vertices in the same component of $G$, then $b_n(u) \leq b_{n+d}(v) \leq b_{n+2d}(u)$, where $d = \text{dist}(u, v)$. It follows that $\varepsilon(G, v)$ and $\tau(G, v)$ are constant on each component of the graph. Define

$$\varepsilon(G) := \sup \ \{\varepsilon(G, v) \mid v \in V(G)\}$$

and

$$\tau(G) := \sup \ \{\tau(G, v) \mid v \in V(G)\} .$$

By the above conclusion, the suprema may be taken on representatives of each of the connected components only.

**Lemma 6.1** *If $e$ is an edge of the graph $G$, then $\varepsilon(G - e) = \varepsilon(G)$ and $\tau(G - e) \leq \tau(G)$. If $e$ is not a cut-edge then $\tau(G - e) = \tau(G)$.*

**Proof** The proof of $\varepsilon(G - e) = \varepsilon(G)$ can be found in [Mo1]. To verify the relations for $\tau$, let $e = uv$. Clearly, $\tau(G, v) \geq \tau(G - e, v)$ and $\tau(G, v) = \tau(G, u) \geq \tau(G - e, u)$. Consequently, $\tau(G) \geq \tau(G - e)$. Assume now that $e$ is not a cut-edge. Then $u$ and $v$ belong to the same component of $G - e$. Let $d$ be the distance between $u$ and $v$ in $G - e$. Clearly, $b_n(G, v) \leq b_{n+d}(G - e, v)$. Therefore $\tau(G, v) \leq \tau(G - e, v)$. The same holds for $u$ instead of $v$. This implies that $\tau(G) \leq \tau(G - e)$, so by the converse inequality proved above, $\tau(G) = \tau(G - e)$. □

It may happen that $\tau(G-e) < \tau(G)$. There are also cases where $\tau(G) > 1$ and $\tau(G-e) = 1$. Examples are not hard to construct.

**Corollary 6.2** *The numbers $\varepsilon(G)$ and $\tau'(G) := \inf\{\tau(H) \mid H$ equivalent to $G\}$ are essential graph invariants.*

**Proof** $\varepsilon$ is essential by Lemma 6.1, since addition or deletion of edges does not affect its value. On the other hand, $\tau'$ is essential since for $G$ and $G'$ equivalent, the set of graphs which are equivalent to $G$ is equal to the equivalence class of $G'$. □

**Proposition 6.3** *Let $G$ be a graph and $v \in V(G)$. Then $\tau(G,v) > 1$ if and only if $G$ has exponential growth at $v$.*

**Proof** First, if $b_n(v) \geq Cq^n, q > 1, C > 0$, then $\tau(G,v) \geq q > 1$. Conversely, if $\tau(G,v) > 1$, then for an arbitrarily small $\alpha > 0$,

$$b_n(v)^{1/n} \geq \tau(G,v) - \alpha \tag{36}$$

for all but finitely many values of $n$. Take $\alpha := \frac{1}{2}(\tau(G,v) - 1)$ and $q := \tau(G,v) - \alpha > 1$. By (36), $b_n(v) \geq q^n$ for all but finitely many $n$, thus $b_n(v) \geq Cq^n$ for some $C > 0$ and for each $n$. □

The *isoperimetric number* $i(G)$ of $G$ is the number

$$i(G) := \inf\left\{\frac{e(S,\bar{S})}{|S|} \mid S \subset V(G),\, 0 < |S| < \infty\right\}.$$

The *essential isoperimetric number* of $G$ is

$$i'(G) := \sup\ \{i(G') \mid G' \text{ equivalent to } G\}\ .$$

It is clear that $i'(G) \geq i(G)$. It is also immediate that $i(G) \leq \delta(G)$ and $i'(G) \leq \delta'(G)$. However, these inequalities need not be very tight since there are examples where $i'(G) = 0$ and $\delta'(G) = \infty$. On the other hand,

$$i'(G) \leq \Delta'(G) - 2. \tag{37}$$

This can be shown as follows. Let $G'$ be equivalent to $G$. Then it has only finitely many vertices of degree greater than $\Delta'(G)$. Therefore one can find an arbitrarily large finite $S \subset V(G')$ such that all vertices in $S$ have degree at most $\Delta'(G)$ in $G'$, and the subgraph of $G'$ induced on $S$ is connected. Then, in $G'$,

$$e(S,\bar{S}) \leq \Delta'(G) \cdot |S| - 2e(S,S) \leq \Delta'(G)|S| - 2(|S| - 1)\ ,$$

which implies (37). This bound is best possible since for the $\Delta$-regular tree $T_\Delta$, $i(T_\Delta) = i'(T_\Delta) = \Delta - 2$.

**Theorem 6.4** *Let $G$ be a connected locally finite graph with $\Delta'(G) < \infty$. Then*

$$i(G) \leq i'(G) \leq \Delta'(G)\frac{\tau(G) - 1}{\tau(G) + 1}\ . \tag{38}$$

**Proof** The first inequality is obvious. By Lemma 6.1, $\tau(G)$ does not change after adding finitely many edges to $G$ since $G$ is connected. Thus we may assume that $i(G)$ is arbitrarily close to $i'(G)$, and so it is sufficient to show that

$$i(G) \leq \Delta'(G) \cdot \frac{\tau(G) - 1}{\tau(G) + 1} \,. \tag{39}$$

Choose a vertex $v \in V(G)$ and let $s_n := b_n(v) - b_{n-1}(v)$. For $n$ large enough, all vertices at distance $n$ from $v$ have degree at most $\Delta'(G)$. Hence

$$\begin{aligned}\Delta'(G)s_{n+1} &\geq e(B_{n+1}, V\backslash B_{n+1}) + e(B_n, V\backslash B_n)\\ &\geq i(G)(b_{n+1} + b_n) = i(G)(2b_n + s_{n+1}).\end{aligned}$$

It follows that $(\Delta'(G) - i(G))s_{n+1} \geq 2i(G)b_n$ and hence

$$b_{n+1} = b_n + s_{n+1} \geq b_n \frac{\Delta'(G) + i(G)}{\Delta'(G) - i(G)}$$

for $n$ large enough. Therefore

$$b_n \geq C \cdot \left(\frac{\Delta'(G) + i(G)}{\Delta'(G) - i(G)}\right)^n$$

for some $C > 0$ and each $n \geq 1$. So $\tau(G) \geq (\Delta'(G)+i(G))/(\Delta'(G)-i(G))$. This is equivalent to (39). □

**Corollary 6.5** *If $\Delta'(G) < \infty$ and $i'(G) > 0$, then $\tau(G) > 1$, and so $G$ has exponential growth in some of its components.*

There are graphs with exponential growth and with $i'(G) = 0$. For example, Cayley graphs of some soluble groups which are not nilpotent by finite are known to grow exponentially but having $i'(G) = 0$, see [Ro]. On the other hand, there are graphs with $\Delta'(G) = \infty$ and $i'(G) > 0$ and having polynomial growth. Take, for example, the graph whose vertices are all integers, and between $i$ and $i+1$ there are $|i|+1$ parallel edges, $-\infty < i < +\infty$. This graph is easily seen to have $i(G) = i'(G) = 1$, but it grows linearly. If we want simple graphs, we may add a vertex of degree two on each edge, or replace each vertex $i$ by a clique of order $|i|+1$, and between any two consecutive cliques put a complete join. This shows that the assumptions in Corollary 6.5 cannot be omitted.

There are other possibilities to define isoperimetric constants of graphs, by taking other distances in graphs. The *transition isoperimetric number* $i_P(G)$ is defined as follows. If $G$ is a graph with no edges, then we set $i_P(G) = 0$. Otherwise

$$i_P(G) := \inf \left\{\frac{e(S, \bar{S})}{\mathrm{d}(S)} \;\middle|\; S \subset V(G), \mathrm{d}(S) \neq 0, |S| < \infty\right\}$$

where $\mathrm{d}(S) := \sum_{v \in S} \mathrm{d}_v$ is the sum of the degrees of the vertices in $S$. Clearly, $0 \leq i_P(G) < 1$. In contrast to $i(G)$, the transition isoperimetric number does not properly measure the connectivity of the graph since it may happen that adding an edge to a graph decreases its value. It is easily seen that $\delta(G)i_P(G) \leq i(G) \leq \Delta(G)i_P(G)$. Therefore $i_P(G)$ has advantage over $i(G)$ in case $\Delta(G) = \infty$ only. The *essential transition isoperimetric number* $i'_P(G)$ is equal to the supremum of all $i_P(G')$, where $G'$ is cofinitely equivalent to $G$. A similar result as Theorem 6.4 holds for $i_P(G)$.

**Theorem 6.6** *Let $G$ be a connected locally finite graph. Then:*

(a) *Regardless of the degrees in $G$,*

$$\tau(G) \geq \sqrt{\frac{1+i'_P(G)}{1-i'_P(G)}}\,, \quad i.e., \quad i_P(G) \leq i'_P(G) \leq \frac{\tau^2(G)-1}{\tau^2(G)+1}\,. \tag{40}$$

(b) *For a vertex $v \in V(G)$, let $D_n$ be the maximum degree of vertices at distance $n$ from $v$. If $\limsup_{n\to\infty} D_n^{1/n} = 1$, then*

$$\tau(G) \geq \frac{1+i'_P(G)}{1-i'_P(G)}\,, \quad i.e., \quad i_P(G) \leq i'_P(G) \leq \frac{\tau(G)-1}{\tau(G)+1}\,. \tag{41}$$

(c) *If there is a constant $M < \infty$ such that each vertex of $G$ is contained in at most $M$ edge-disjoint cycles, then (41) holds.*

For a proof we refer to [Mo4].

**Corollary 6.7** *If $i'_P(G) > 0$, then $\tau(G) > 1$, and so $G$ has exponential growth in some of its components.*

## 6.2 The Laplace spectrum of infinite graphs

We define the Laplace matrix $L(G)$ of an infinite locally finite graph $G$ in the same way as for finite graphs.

Let $\ell^2(V)$ denote the Hilbert space of all functions $f : V \to \mathbf{R}$ such that

$$\|f\|^2 = \sum_{v\in V} |f_v|^2 < \infty\,.$$

The inner product in $\ell^2(V)$ is defined as $\langle x, y\rangle = \sum_{v\in V} f_v g_v$. The Laplace matrix $L(G)$ acts naturally on $\ell^2(V)$ as a linear operator. In the following it will be assumed that $\Delta(G) < \infty$. This condition implies that $L(G)$ determines a bounded self-adjoint linear operator on $\ell^2(V)$. Its spectrum $\sigma_L(G)$ is real and (cf., e.g., [MW])

$$\sigma_L(G) \subseteq [0, 2\Delta(G)]\,.$$

The reader is referred to [MW] for more details about the spectra of infinite graphs.

There is an alternative way of defining $\sigma_L(G)$ since the spectrum of $L(G)$ is the *approximate point spectrum* in $\ell^2(V)$. This means that $\lambda \in \sigma_L(G)$ if and only if there is a sequence of (not necessarily distinct) unit[1] functions $f^{(n)}$ $(n = 1, 2, 3, \ldots)$ such that

$$\|L(G)f^{(n)} - \lambda f^{(n)}\| \to 0\,, \qquad \text{as} \quad n \to \infty\,.$$

The number $\lambda$ is an *eigenvalue* of $L$ if there exists a function $f \in \ell^2(V)$ such that $Lf = \lambda f$. The dimension of the *eigenspace* $\{f \in \ell^2(V) \mid Lf = \lambda f\} \subseteq \ell^2(V)$ is called the *multiplicity* of $\lambda$.

[1] A function $f \in \ell^2(V)$ is called a *unit* function if $\|f\| = 1$.

The value related to the second smallest eigenvalue of finite graphs is the infimum of $\sigma_L(G)$:

$$\lambda_1(G) := \inf \sigma_L(G).$$

It can be expressed as

$$\lambda_1(G) = \inf \left\{ \frac{\langle L(G)f, f\rangle}{\langle f, f\rangle} \;\middle|\; f \in \ell^2(V), f \neq 0 \right\}. \tag{42}$$

The "essential" invariant corresponding to the spectrum is the *essential spectrum* of $L(G)$. The essential spectrum of a self-adjoint linear operator $B$ on a Hilbert space consists of all those elements in the spectrum of $B$ which are not isolated eigenvalues of finite multiplicity. It is well known (cf., e.g., [Kat, §X.2]) that these are exactly those $\lambda \in \sigma(B)$ which remain in the spectrum whenever $B$ is changed by a compact perturbation, i.e., $\lambda \in \sigma(B + K)$ for every compact linear operator $K$. It follows that the essential spectrum is an essential graph invariant since cofinite equivalence perturbs the graph matrices in finitely many entries only and thus presents only compact perturbations. The elements of the essential spectrum are also characterized as the approximate eigenvalues of infinite multiplicity [FSW] (cf. also [Kat]). This means that $\lambda$ is in the essential spectrum of $B$ if and only if there are *pairwise orthogonal* unit functions $f^{(n)}$, $n = 1, 2, 3, \ldots$, such that

$$\|Bf^{(n)} - \lambda f^{(n)}\| \to 0\,, \qquad \text{as} \quad n \to \infty\,.$$

We shall denote by $\lambda_1'(G)$ the infimum of the essential spectrum of $L(G)$. Clearly, $\lambda_1(G) \leq \lambda_1'(G)$. Later we shall need the following inequality:

$$\lambda_1'(G) \leq \frac{\Delta'(G) + \delta'(G)}{2} - 1 < \Delta'(G) \tag{43}$$

if $\delta'(G) \geq 1$. Its proof goes as follows. There are infinitely many pairs $u_i, v_i$ of adjacent vertices, such that $\deg(u_i) \leq \delta'(G)$ and $\deg(v_i) \leq \Delta'(G)$. Let $g^{(1)}, g^{(2)}, \ldots$ be their normalized characteristic functions (i.e., $g^{(i)}_{v_i} = g^{(i)}_{u_i} = \frac{1}{\sqrt{2}}$, and $g^{(i)}_v = 0$ at all other vertices $v$). Then

$$\langle L(G)g^{(i)}, g^{(i)}\rangle \leq \frac{1}{2}(\mathrm{d}_{u_i} + \mathrm{d}_{v_i} - 2) \leq \frac{\Delta'(G) + \delta'(G)}{2} - 1.$$

Since this happens for infinitely many pairwise orthogonal unit functions $g^{(i)}$, an element of the essential spectrum must be $\leq \frac{1}{2}(\Delta'(G) + \delta'(G)) - 1$. The last conclusion is proved as follows. We claim that if $\langle Lf^{(i)}, f^{(i)}\rangle \leq t$ for infinitely many pairwise orthogonal unit functions $f^{(i)}$, then there is some $\lambda \leq t$ in the essential spectrum of $L$. It is known [St] that for some compact symmetric operator $K$, the spectrum of $L + K$ is equal to the essential spectrum of $L$. Since $K$ is compact, $\langle Kf^{(i)}, f^{(i)}\rangle \to 0$ as $i \to \infty$. Consequently

$$\begin{aligned} \lambda_1'(L) &= \lambda_1(L + K) = \inf\{\langle (L + K)f, f\rangle \mid \|f\| = 1\} \\ &\leq \inf\{\langle Lf^{(i)}, f^{(i)}\rangle + \langle Kf^{(i)}, f^{(i)}\rangle\} \leq t. \end{aligned}$$

**Theorem 6.8** *Let $G$ be a locally finite graph, let $\Delta = \Delta(G) < \infty$, and let $\Delta' = \Delta'(G)$. Then*

$$i(G) \geq \lambda_1(G) \frac{\Delta^2}{\Delta^2 - \Delta - \lambda_1(G)} \geq \lambda_1(G) \tag{44}$$

*and*

$$i'(G) \geq \lambda_1'(G)\frac{\Delta'^2}{\Delta'^2 - \Delta' - \lambda_1'(G)} \geq \lambda_1'(G) . \tag{45}$$

**Proof** Let $U$ be an arbitrary finite subset of $V = V(G)$. Define $f \in \ell^2(V)$ by setting $f_v = 1$ if $v \in U$, and otherwise $f_v = n_v/\Delta$, where $n_v$ is the number of edges joining $v$ to vertices in $U$. Denote by $W$ the set of those vertices of $V \setminus U$ which have a neighbor in $U$. By (42),

$$\lambda_1(G)\|f\|^2 \leq \langle L(G)f, f\rangle . \tag{46}$$

Similarly as in the finite case we have

$$\langle L(G)f, f\rangle = \sum_{uv\in E(G)} (f_u - f_v)^2 . \tag{47}$$

Since $f_u - f_v$ is non-zero only if $v$ or $u$ lies in $W$, we get from (46) and (47):

$$\begin{aligned} \lambda_1\|f\|^2 &= \lambda_1\Big(|U| + \sum_{v\in W}\frac{n_v^2}{\Delta^2}\Big) \leq \sum_{vu\in E(G)} (f_u - f_v)^2 \\ &\leq \sum_{v\in W}\Big(n_v\Big(1 - \frac{n_v}{\Delta}\Big)^2 + (\Delta - n_v)\Big(\frac{n_v}{\Delta}\Big)^2\Big) \\ &= \sum_{v\in W} n_v - \sum_{v\in W}\frac{n_v^2}{\Delta} = e(U,\bar{U}) - \frac{1}{\Delta}\sum_{v\in W} n_v^2 . \end{aligned} \tag{48}$$

By rearranging this inequality and using the fact $n_v^2 \geq n_v$, one immediately gets

$$e(U,\bar{U}) - \lambda_1|U| \geq \frac{\Delta+\lambda_1}{\Delta^2}\sum_{v\in W} n_v \geq \frac{\Delta+\lambda_1}{\Delta^2}e(U,\bar{U}).$$

It follows that

$$\frac{e(U,\bar{U})}{|U|} \geq \lambda_1\frac{\Delta^2}{\Delta^2 - \Delta - \lambda_1}$$

which implies (44) since $U$ was arbitrary.

The proof of (45) needs more care. Fix an arbitrarily small $\varepsilon > 0$. First we shall prove that there exists a graph $G_1$, which can be obtained from $G$ by adding finitely many edges, such that $\lambda_1(G_1) \geq \lambda_1'(G) - \varepsilon$.

The only elements in the Laplacian spectrum of $G$ which are smaller than $l = \lambda_1'(G)$ are isolated eigenvalues of finite multiplicity. Let $N$ be the span of the eigenspaces of all those eigenvalues of $L(G)$ which are smaller than $l - \frac{\varepsilon}{2}$, and let $n := \dim N < \infty$. Fix an orthonormal basis $f^{(1)}, f^{(2)}, \ldots, f^{(n)}$ of $N$. It should be mentioned at once that for any $f \in \ell^2(V)$ which is orthogonal to $N$

$$\langle L(G)f, f\rangle \geq (l - \frac{\varepsilon}{2})\|f\|^2 . \tag{49}$$

Let $U$ be a finite subset of $V(G)$ such that for each $i = 1, 2, \ldots, n$ ,

$$\sum_{u\in U}\Big(f_u^{(i)}\Big)^2 > 1 - \delta , \quad \text{i.e.,} \quad \sum_{u\notin U}\Big(f_u^{(i)}\Big)^2 < \delta , \tag{50}$$

where $\delta > 0$ is a small number to be fixed later (it will depend on $\varepsilon, n, G$).

Let $G_1$ be the graph which is obtained from $G$ by adding, for each vertex $u \in U$, $m$ edges joining $u$ to $m$ new distinct neighbors which lie outside $U$. Denote by $L_1 := L(G_1)$ its Laplacian matrix. We shall prove that $\lambda_1(G_1) \geq l - \varepsilon$, assuming $m$ is large enough. Note that $m$ may depend on $\varepsilon, n$, and $G$.

Take an arbitrary $f \in \ell^2(V)$, $\|f\| = 1$. We shall prove that $\langle L_1 f, f\rangle \geq l - \varepsilon$ which will imply, by (42), that $\lambda_1(G_1) \geq l - \varepsilon$. It is easy to see that we may assume that $f_v \geq 0$ for each $v \in V(G)$ (apply, for example, (47)). Write now $f = g + h$, where $g$ agrees with $f$ on $U$, and $h$ agrees with $f$ at other coordinates.

If $\|g\|^2 \leq \delta$, then we do the following calculation. Let $f = p + q$ where $p \in N$ and $q \perp N$. By (49), $\langle Lq, q\rangle \geq (l - \frac{\varepsilon}{2})\|q\|^2$, while $\langle Lp, p\rangle$ can be estimated as follows. If $p = \sum_{i=1}^n \alpha_i f^{(i)}$, then $\|p\|^2 = \sum_{i=1}^n |\alpha_i|^2$ and

$$|\alpha_i| = |\langle f, f^{(i)}\rangle| = \sum_{u\in U} f_u f_u^{(i)} + \sum_{v\notin U} f_v f_v^{(i)} < \|g\| + \|f\|\sqrt{\delta} \leq 2\sqrt{\delta}\ . \tag{51}$$

In the last inequality we applied the Cauchy-Schwarz inequality and (50). Consequently, $\|p\|^2 < 4n\delta = \varepsilon/2\Delta(G)$ if we choose $\delta := \varepsilon/8n\Delta(G)$. Therefore

$$\|q\|^2 = \|f\|^2 - \|p\|^2 > 1 - \frac{\varepsilon}{2\Delta(G)}.$$

Using this estimate, the fact that $\langle Lp, q\rangle = \langle Lq, p\rangle = 0$, and (43) we get

$$\begin{aligned} \langle L_1 f, f\rangle &\geq \langle Lf, f\rangle = \langle Lp, p\rangle + \langle Lq, q\rangle \\ &\geq (l - \frac{\varepsilon}{2})\|q\|^2 > (l - \frac{\varepsilon}{2})(1 - \frac{\varepsilon}{2\Delta(G)}) \geq l - \varepsilon\ . \end{aligned}$$

The other possibility to consider is the case when $\|g\|^2 > \delta = \varepsilon/8n\Delta(G)$ . In this case let $B := L_1 - L(G)$ and let $F := E(G_1) \setminus E(G)$. If $h = p + q$ where $p = \sum_{i=1}^n \alpha_i f^{(i)} \in N$, $q \perp N$, then we see as in (51) that

$$|\alpha_i| = |\langle h, f^{(i)}\rangle| < \sqrt{\delta}\|h\|$$

and hence $\|p\|^2 < n\delta\|h\|^2$, so $\|q\|^2 > (1 - n\delta)\|h\|^2$. Therefore

$$\langle L_1 h, h\rangle \geq \langle Lh, h\rangle = \langle Lp, p\rangle + \langle Lq, q\rangle \geq (l - \frac{\varepsilon}{2})\|q\|^2 > (l - \varepsilon)\|h\|^2\ . \tag{52}$$

Finally:

$$\begin{aligned} \langle L_1 f, f\rangle &= \langle L_1 g, g\rangle + 2\langle L_1 g, h\rangle + \langle L_1 h, h\rangle \\ &\geq (l - \varepsilon)\|h\|^2 + \langle Bg, g\rangle + 2\langle L_1 g, h\rangle \\ &= (l - \varepsilon)\|h\|^2 + \sum_{uv\in F} (g_u - g_v)^2 + 2 \sum_{uv\in E(G_1)} (g_u - g_v)(h_u - h_v) \\ &\geq (l - \varepsilon)\|h\|^2 + m\sum_{u\in U} g_u^2 - 2 \sum_{uv\in E(G_1)} g_u h_v \\ &\geq (l - \varepsilon)\|f\|^2 + (m - l)\|g\|^2 - 2\sum_{uv\in F_1} g_u h_v \end{aligned}$$

where $F_1 = F \cup E(U, \bar{U}) \subseteq E(G_1)$, and it is assumed that $u \in U, v \notin U$. It remains to show that

$$(m - l)\|g\|^2 - 2 \sum_{uv \in F_1} g_u h_v \geq 0 . \tag{53}$$

By the Cauchy-Schwarz inequality and assuming that each edge $uv \in F_1$, $u \in U$, has a different end-vertex $v \notin U$, we estimate:

$$\Big(\sum_{uv \in F_1} g_u h_v\Big)^2 \leq \sum_{uv \in F_1} |g_u|^2 \sum_{uv \in F_1} |h_v|^2 \leq (m + \Delta(G))\|g\|^2 \cdot (\Delta(G) + 1)\|h\|^2 \quad .$$

This inequality and the assumption $\|g\|^2 > \delta$ imply that

$$(m - l)\|g\|^2 - 2 \sum_{uv \in F_1} g_u h_v \geq (m - l)\sqrt{\delta} - 2\sqrt{m + \Delta(G)}\sqrt{\Delta(G) + 1}\sqrt{1 - \delta} .$$

Obviously, the last expression is $\geq 0$ if $m$ is large enough. It should be added that $m$ depends on $l$, $\Delta(G)$ and $\delta = \delta(\varepsilon, n, \Delta(G))$ but not on $f$. This finally establishes (53). Thus we succeeded to show $\langle L_1 f, f\rangle \geq l - \varepsilon$, and consequently that $\lambda_1(G_1) \geq l - \varepsilon$.

It is clear that adding edges to a graph cannot decrease the isoperimetric number. Therefore there exists a graph $G_2$ which can be obtained from $G_1$ by adding finitely many edges so that $i(G_2) \geq i'(G) - \varepsilon$. Clearly, also $\lambda_1(G_2) \geq \lambda_1(G_1) \geq l - \varepsilon$.

Let $S \subset V(G_2)$ be the set containing all vertices of degree greater than $\Delta'(G)$ and all their neighbors. Let $G_3$ be the graph obtained from $G_2$ by adding $m$ parallel edges between any two vertices of $S$, where $m = \Delta'(G)|S|$.

Choose a finite $U_1 \subset V(G_3)$ such that $e_{G_3}(U_1, \bar{U}_1)/|U_1| \leq i(G_3) + \varepsilon$ (the index $G_3$ in $e_{G_3}(U_1, \bar{U}_1)$ indicates that the edges are in the graph $G_3$). If $U_1 \cap S = \emptyset$, then let $U := U_1$, otherwise let $U := U_1 \cup S$. So $U$ either contains $S$, or it is disjoint from $S$. Therefore no vertex adjacent to $U$ in $G_2$ has degree greater than $\Delta'(G)$. It is easily seen that in each case

$$\frac{e_{G_2}(U, \bar{U})}{|U|} = \frac{e_{G_3}(U, \bar{U})}{|U|} \leq \frac{e_{G_3}(U_1, \bar{U}_1)}{|U_1|} \leq i(G_3) + \varepsilon \leq i'(G) + \varepsilon \tag{54}$$

since $m = \Delta'(G)|S|$.

Now we may repeat the first part of the proof for the graph $G_3$, using $\Delta'(G)$ instead of $\Delta$. Note that in (48) we need the fact that $n_v \leq \Delta'(G)$ for each vertex $v$ adjacent to $U$. This completes the proof since $i(G_3)$ and $\lambda(G_3)$ are arbitrarily close to their essential values. □

The Cheeger inequality of Theorem 3.10 extends to infinite graphs as shown by the next theorem [Mo4]

**Theorem 6.9** *For an arbitrary locally finite graph $G$ with bounded degrees*

$$i(G) \leq \sqrt{\lambda_1(G)(2\Delta(G) - \lambda_1(G))} \tag{55}$$

*and*

$$i'(G) \leq \sqrt{\lambda'_1(G)(2\Delta'(G) - \lambda'_1(G))} \ . \tag{56}$$

**Proof** The spectrum $\sigma_L(G)$ is closed, so $\lambda_1 = \lambda_1(G) \in \sigma_L(G)$. For each small enough $\varepsilon > 0$ there is a function $f \in \ell^2(V(G))$ with finite support (only finitely many $f_v$ are nonzero) and with $\|f\| = 1$ such that

$$\langle Lf, f\rangle \le \lambda_1 + \varepsilon \le \Delta(G)\ . \tag{57}$$

If

$$\eta := \sum_{uv\in E(G)} |f_u^2 - f_v^2|$$

then it can be shown using summation by parts that

$$\eta \ge i(G) \tag{58}$$

(see the proof of Theorem 3.10). On the other hand,

$$\begin{aligned}
\eta^2 &= \Big(\sum_{uv\in E} |f_u^2 - f_v^2|\Big)^2 = \Big(\sum_{uv\in E} |f_u + f_v||f_u - f_v|\Big)^2 \\
&\le \sum_{uv\in E} (f_u + f_v)^2 \cdot \sum_{uv\in E} (f_u - f_v)^2 \\
&= \sum_{uv\in E} (2f_u^2 + 2f_v^2 - (f_u - f_v)^2) \cdot \langle Lf, f\rangle \\
&= \Big(2\sum_{v\in V} \mathrm{d}_v f_v^2 - \langle Lf, f\rangle\Big) \cdot \langle Lf, f\rangle \\
&\le (2\Delta(G)\|f\|^2 - \langle Lf, f\rangle)\langle Lf, f\rangle \\
&\le 2\Delta(G)(\lambda_1 + \varepsilon) - (\lambda_1 + \varepsilon)^2\ .
\end{aligned} \tag{59}$$

In the first inequality we used the Cauchy-Schwarz inequality, while in the last one we needed (57). Since $\varepsilon$ was arbitrarily small, it follows that $i^2(G) \le \eta^2 \le 2\Delta(G)\lambda_1 - \lambda_1^2$, and the proof of (55) is done.

To prove (56), let $G_1$ be a graph obtained from $G$ by adding finitely many edges and such that $i(G_1) \ge i'(G) - \varepsilon$, where $\varepsilon > 0$ is arbitrarily small again. As we know from the proof of Theorem 6.8, there is a graph $G_2$, obtainable from $G_1$ by the addition of finitely many edges, such that $\lambda_1(G_2) \ge \lambda_1'(G) - \varepsilon$. Since $i(G_2) \ge i(G_1)$, we have $i(G_2) \ge i'(G) - \varepsilon$.

Let $S$ be the set of vertices of $G_2$ having degree $> \Delta'(G)$. The set $S$ is clearly finite. Since $\lambda_1' = \lambda_1'(G) = \lambda_1'(G_2)$ is in the essential spectrum, there are arbitrarily many unit functions $f^{(i)}$, $i = 1, 2, \ldots$, such that $\langle L(G_2)f^{(i)}, f^{(i)}\rangle \le \lambda_1' + \varepsilon$, and such that $\langle f^{(i)}, f^{(j)}\rangle = 0$ for $i \ne j$. Let us take $N = |S| + 1$ such functions $f^{(i)}$. They are linearly independent, so any nontrivial linear combination is nonzero. On the other hand, there exist $\alpha_1, \alpha_2, \ldots, \alpha_N$ with some $\alpha_i \ne 0$ such that $\sum_{i=1}^N \alpha_i f^{(i)} =: f$ has the values $f_v = 0$ for each $v \in S$. If we require that $\sum \alpha_i^2 = 1$, then

$$\|f\|^2 = \langle \sum_{i=1}^N \alpha_i f^{(i)}, \sum_{i=1}^N \alpha_i f^{(i)}\rangle = \sum_{i=1}^N \alpha_i^2 = 1\ .$$

The functions $f^{(i)}$ may also be assumed to satisfy $\|L(G_2)f^{(i)} - \lambda_1' f^{(i)}\| \le \varepsilon/N$. Then:

$$\langle L(G_2)f, f\rangle = \sum_{i=1}^N \sum_{j=1}^N \alpha_i\alpha_j \langle L(G_2)f^{(i)}, f^{(j)}\rangle$$

$$
\begin{aligned}
&= \sum_{i=1}^{N} \alpha_i^2 \langle \lambda_1' f^{(i)}, f^{(i)} \rangle + \sum_{i=1}^{N}\sum_{j=1}^{N} \alpha_i\alpha_j \langle L(G_2) f^{(i)} - \lambda_1' f^{(i)}, f^{(j)} \rangle \\
&\le \lambda_1' + \sum_{i=1}^{N}\sum_{j=1}^{N} |\alpha_i\alpha_j| \| L(G_2) f^{(i)} - \lambda_1' f^{(i)} \| \cdot \| f^{(j)} \| \\
&\le \lambda_1' + \frac{\varepsilon}{N} \sum_{i=1}^{N}\sum_{j=1}^{N} |\alpha_i\alpha_j| \le \lambda_1' + \varepsilon.
\end{aligned}
$$

Now we may carry out the same calculation as used to obtain (58) and (59), this time for $G_2$. The inequality $d_v f_v^2 \le \Delta'(G) f_v^2$ is clearly satisfied, and the only remaining fact to verify is $\lambda_1' + \varepsilon \le \Delta'(G)$. But we have shown this by (43) for any graph with $\delta'(G) \ge 1$. The remaining case $\delta'(G) = 0$ is unimportant since in this case $i'(G) = 0$ and $l = 0$, so (56) is trivial. □

## 6.3 Amenability

In this section we shall shortly exhibit the notion of amenability of infinite graphs. This concept was originally introduced as a property of locally compact groups. We refer to [Pa, Pi] for more information.

Let $\Gamma$ be a locally compact group. By $L^\infty(\Gamma)$ we denote the set of all functions $\Gamma \to \mathbf{R}$ which are bounded a.e. with respect to the Haar measure $\lambda(.)$ on $\Gamma$. The group $\Gamma$ is *amenable* if $L^\infty(\Gamma)$ has an *invariant mean*, i.e., a linear functional $m : L^\infty(\Gamma) \to \mathbf{R}$ which satisfies the following conditions:

(i) For $f \in L^\infty(\Gamma)$, if $f \ge 0$ (a.e.) then $m(f) \ge 0$.

(ii) $m(\chi_\Gamma) = 1$, where $\chi_\Gamma$ is the constant function 1 on $\Gamma$.

(iii) It is $\Gamma$-invariant, i.e., $m(g \cdot f) = m(f)$ for every $g \in \Gamma$. (Note that $(g \cdot f)(x) = f(g^{-1}x)$.)

By a theorem of Følner [Fo], the amenability is equivalent to the following condition:

(F) Given $\varepsilon > 0$ and a compact set $K \subseteq \Gamma$ there is a Borel set $U \subseteq \Gamma$ of positive finite measure $\lambda(U) < \infty$ such that

$$\lambda(kU \Delta U) < \varepsilon \lambda(U)$$

for all $k \in K$, where $\Delta$ denotes the symmetric difference of sets.

For a discrete group, the Haar measure counts the number of elements of the set. Then it is easy to see that (F) is equivalent to:

(F′) For every $\varepsilon > 0$ and finite $K \subseteq \Gamma$ one can find a finite set $U \subseteq \Gamma$ such that

$$|KU \Delta U| < \varepsilon |U| \ .$$

**Lemma 6.10** *Let $\Gamma$ be a group generated by a finite set of generators $S = \{g_1, \ldots, g_k\}$. Then $\Gamma$ is amenable if and only if for every $\varepsilon > 0$ there is a finite set $U \subseteq \Gamma$ such that $|SU \setminus U| < \varepsilon |U|$.*

**Proof** ($\Rightarrow$) Let $K := S \cup \{e\}$, where $e$ is the identity in $\Gamma$. Then $KU \Delta U = SU \setminus U$. We are done by the amenability of $\Gamma$.

($\Leftarrow$) Choose $\varepsilon > 0$ and a finite $K \subseteq \Gamma$, $K = \{k_1, k_2, \dots, k_t\}$. It may be assumed that $e \notin K$. If $k_i$ can be written as a word of length $l(i)$ in terms of $S \cup S^{-1}$, let

$$\varepsilon_i := \frac{\varepsilon}{2t \cdot l(i)}$$

and let $U \subseteq \Gamma$ be a set for which $|SU \setminus U| < \varepsilon' |U|$, where $\varepsilon' = \min \varepsilon_i$. For each generator $g_j \in S$,

$$|g_j U \setminus U| \le |SU \setminus U| < \varepsilon' |U|.$$

From this we also see that $|g_j U \cap U| > (1 - \varepsilon')|U|$, therefore also $|g_j^{-1} U \cap U| > (1 - \varepsilon')|U|$, so $|g_j^{-1} U \setminus U| < \varepsilon' |U|$. It follows that

$$|k_i U \setminus U| < l(i) \varepsilon' |U| \le \frac{\varepsilon}{2t} |U|.$$

This implies that $|KU \setminus U| < \frac{\varepsilon}{2}|U|$. Since $|KU| \ge |U|$, we get $|KU \,\Delta\, U| < \varepsilon |U|$. □

**Corollary 6.11** *A finitely generated infinite group* $\Gamma$ *is amenable if and only if the Cayley graph* $\mathrm{Cay}(\Gamma, S)$ *with respect to some (and hence every) finite generating set* $S$ *has the isoperimetric number* $i(\mathrm{Cay}(\Gamma, S)) = 0$.

By Corollary 6.11 it follows trivially that every finitely generated group with polynomial or subexponential growth is amenable. The converse is not true. For example, all soluble groups are amenable but some of them have exponential growth [Ro].

The importance of Corollary 6.11 lies in the fact that it gives a possibility to extend the notion of amenability to graphs in such a way that a finitely generated group $\Gamma$ is amenable if and only if it has an amenable Cayley graph. So, let us call a graph $G$ *amenable* if $i(G) = 0$.

The case of graphs with bounded degrees was investigated in detail by Peter Gerl and we refer to his works [Ge1, Ge2], where several conditions equivalent to amenability are derived. For a possibility of extending the notion to graphs with unbounded degrees see [Mo4]. We refer to [SW] for additional results on amenability of graphs.

## 6.4 Random walks on infinite graphs

The *transition matrix* $P(G) = [p_{uv}]$ of a locally finite infinite graph $G$ has entries $p_{uv} = a_{uv}/\mathrm{d}_u$. This matrix is usually assumed to act on the Hilbert space $\ell^2_d(V)$ endowed with the inner product given by

$$\langle f, g \rangle_d = \sum_{v \in V} f_v g_v \mathrm{d}_v,$$

and whose elements $f$ satisfy $\langle f, g \rangle_d < \infty$. Then $P(G)$ defines a bounded self-adjoint linear operator on $\ell^2_d(V)$ (even though the matrix $P(G)$ need not be symmetric). It is important that $G$ is locally finite but the requirement $\Delta(G) < \infty$ is not needed for $P(G)$ to be everywhere defined and bounded. The spectrum $\sigma_P(G)$ of $P(G)$ is real and always contained in the interval $[-1, 1]$ (cf., e.g., [MW]). We define

$$\rho_1(G) := \sup \sigma_P(G).$$

The Rayleigh characterization of $\rho_1$, similar to (42), is

$$\rho_1(G) = \sup \left\{ \frac{\langle P(G)f, f\rangle_d}{\langle f, f\rangle_d} \;\Big|\; f \in \ell_d^2(V), f \neq 0 \right\}. \tag{60}$$

We can also define the essential spectrum and set $\rho_1'(G)$ to be the supremum of the essential spectrum of $P(G)$.

Relations between $\rho_1(G), \rho_1'(G)$ and isoperimetric invariants are proved in [Mo4]. They seem to be more natural than the corresponding relations with the Laplacians.

The transition matrix is related to simple random walks on the graph $G$. The most important problem is the *type problem* for the simple random walks on graphs, which asks if the random walk is recurrent or transient. It is well known that the random walk on $\mathbf{Z}^2$ (the 2-dimensional integer lattice) is recurrent, while the random walk on $\mathbf{Z}^3$ is transient. For this type of graphs, the threshold between recurrence and transience lies somewhere between the quadratic and cubic growth (cf. [DS]). More generally, Varopoulos [Var] proved:

**Theorem 6.12** *The simple random walk on a Cayley graph of a finitely generated group is recurrent if and only if the graph has polynomial growth of degree at most two.*

This result was extended to general (locally finite) infinite graphs by Thomassen [Th] as follows. We say that $G$ satisfies the *connected $\phi$-isoperimetric inequality* at the vertex $v_0 \in V(G)$, where $\phi : \mathbf{N} \to \mathbf{N}$, if there is a constant $c > 0$ such that every finite connected subgraph $H$ of $G$ with $v_0 \in V(H)$ satisfies

$$e(V(H), V(G) \backslash V(H)) \geq c\,\phi(|V(H)|).$$

Let us observe that $G$ satisfies the connected isoperimetric inequality for the identity function if and only if $i(G) > 0$.

**Theorem 6.13** *If $G$ satisfies the connected $\phi$-isoperimetric inequality at some vertex $v_0$ and if $\sum_{k=1}^{\infty} \phi(k)^{-2} < \infty$, then the simple random walk on $G$ is transient.*

For further reading we refer to [Woe].

**Acknowledgement** I am greatly indebted to my notetaker Dr. Martin Juvan who wrote the first draft of these notes.

## References

[Ali] F. Alizadeh, Interior point methods in semidefinite programming with applications to combinatorial optimization, *SIAM J. Optim.* **5** (1995), 13–51.

[Alo] N. Alon, Eigenvalues and expanders, *Combinatorica* **6** (1986), 83–96.

[AM] N. Alon, V. D. Milman, $\lambda_1$, isoperimetric inequalities for graphs and superconcentrators, *J. Combin. Theory Ser. B* **38** (1985), 73–88.

[A-S] S. Aurora, C. Lund, R. Motwani, M. Sudan, M. Szegedy, Proof verification and hardness of approximation problems, *Proc. 33rd Ann. Sympos. FOCS*, 1992, 14–23.

[BF] I. Bárány, Z. Füredi, Computing the volume is difficult, *Discrete Comput. Geom.*, **2** (1987), 319–326.

[BS] S. T. Barnard, H. D. Simon, A fast multilevel implementation of recursive spectral bisection for partitioning unstructured problems, Technical Report RNR-92-033, NASA Ames Res. Center, 1992.

[BD] D. Bayer, P. Diaconis, Trailing the dovetail shuffle to its lair, *Ann. Appl. Probab.* **2** (1992), 294–313.

[Bi] N. L. Biggs, *Algebraic Graph Theory* (2nd edition), Cambridge University Press, Cambridge, 1993.

[BM] J. A. Bondy, U. S. R. Murty, *Graph Theory with Applications*, North Holland/Elsevier, Amsterdam, 1976.

[Bo] R. B. Boppana, Eigenvalues and graph bisection: An average case analysis, *28th Ann. Sympos. Found. Comp. Sci.*, IEEE, 1987, 280–285.

[Bro] A. Z. Broder, How hard is to marry at random? (On the approximation of the permanent), *Proc. 18th ACM STOC*, 1986, 50–58. Erratum in *Proc. 20th ACM STOC,* 1988, 551.

[Br] R. Brooks, Spectral geometry and the Cheeger constant, in: *Expanding Graphs,* (J. Friedman, ed.), DIMACS Ser. Discrete Math. Theoret. Comput. Sci. **10**, Amer. Math. Soc., Providence, RI, 1993, 5–19.

[Bu] P. Buser, *Geometry and Spectra of Compact Riemann Surfaces,* Birkhäuser, Boston, 1992.

[Ca] A. Cayley, A theorem on trees, *Quart. J. Math.* **23** (1889), 376–378.

[CSZ] P. K. Chan, M. Schlag, J. Zien, Spectral $k$-way ratio cut partitioning and clustering, *Proc. Symp. Integrated Systems*, 1993.

[Cha] I. Chavel, *Eigenvalues in Riemannian Geometry,* Academic Press, San Diego, 1984.

[Che] J. Cheeger, A lower bound for the smallest eigenvalue of the Laplacian, in: *Problems in Analysis* (R.C. Gunnig, ed.), Princeton University Press, 1970, 195–199.

[Chu] F. R. K. Chung, Laplacians of graphs and Cheeger's inequalities, in: *Combinatorics, Paul Erdős is Eighty,* Vol.2, János Bolyai Math. Soc., Budapest, 1996, 157–172.

[CCRS] F. R. K. Chung, E. G. Coffman, M. I. Reiman, B. E. Simon, On the capacity and vertex-forwarding index of communication networks, *IEEE Trans. Inform. Theory* **33** (1987), 224–232.

[CT] F. R. K. Chung, P. Tetali, Isoperimetric inequalities for Cartesian products of graphs, submitted to *Combin. Probab. Comput.,* 1996.

[CDGT] D. M. Cvetković, M. Doob, I. Gutman, A. Torgašev, *Recent Results in the Theory of Graph Spectra, Ann. Discrete Math.* **36**, North-Holland, 1988.

[CDSS ] D. M. Cvetković, M. Doob, H. Sachs, *Spectra of Graphs,* Academic Press, New York, 1979.

[DP1] C. Delorme, S. Poljak, Laplacian eigenvalues and the maximum cut problem, *Math. Programming* **62** (1993), 557–574.

[DP2] C. Delorme, S. Poljak, Combinatorial properties and the complexity of a max-cut approximation, *European J. Combin.* **14** (1993), 313–333.

[Di] P. Diaconis, Group Representations in Probability and Statistics, IMS Lecture Notes-Monograph Ser. **11**, Hayward, CA, 1988.

[DSh] P. Diaconis, M. Shahshahani, Generating a random permutation with random transpositions, *Z. Wahrsch. Verw. Gebiete* **57** (1981), 159–179.

[DSt] P. Diaconis, D. Stroock, Geometric bounds for eigenvalues of Markov chains, *Ann. Appl. Probab.* **1** (1991), 36–61.

[DH] W. E. Donath, A. J. Hoffman, Lower bounds for the partitioning of graphs, *IBM J. Res. Develop.* **17** (1973), 420–425.

[DS] P. G. Doyle, J. L. Snell, *Random Walks and Electric Networks*, Math. Assoc. of America, Washington, DC, 1984.

[DFK] M. E. Dyer, A. Frieze, R. Kannan, A random polynomial-time algorithm for approximating the volume of convex bodies, *J. Assoc. Comput. Mach.* **38** (1991), 1–17.

[FG] U. Feige, M. X. Goemans, Approximating the value of two prover proof systems, with applications to MAX 2SAT and MAX DICUT, *Proc. Third Israel Sympos. Theory of Computing and Systems,* 1995, 182–189.

[Fe] W. Feller, *An Introduction to Probability Theory and Its Applications,* Wiley, New York, 1968.

[Fi1] M. Fiedler, Algebraic connectivity of graphs, *Czechoslovak Math. J.* **23** (98) (1973), 298–305.

[Fi2] M. Fiedler, A property of eigenvectors of nonnegative symmetric matrices and its application to graph theory, *Czechoslovak Math. J.* **25** (100) (1975), 619–633.

[Fi3] M. Fiedler, Laplacian of graphs and algebraic connectivity, in: *Combinatorics and Graph Theory* (Z. Skupień and M. Borowiecki, eds.), Banach Center Publ. **25**, PWN, Warsaw, 1989, 57–70.

[FSW] P. A. Fillmore, J. G. Stampfli, J. P. Williams, On the essential numerical range, the essential spectrum, and a problem of Halmos, *Acta Sci. Math. (Szeged)* **33** (1972), 179–192.

[FBR] G. Finke, R. E. Burkard and F. Rendl, Quadratic assignment problems, in: *Surveys in Combinatorial Optimization* (S. Martello et al., eds.), *Ann. Discrete Math.* **31** (1987), 61-82.

[Fo] E. Følner, On groups with full Banach mean values, *Math. Scand.* **3** (1955), 243-254.

[FJ] A. Frieze, M. Jerrum, Improved approximation algorithms for MAX $k$-CUT and MAX BISECTION, *Proc. 4th IPCO*, Springer-Verlag, Berlin, 1995.

[GJ] M. R. Garey, D. S. Johnson, *Computers and Intractability: A Guide to the Theory of NP-Completeness,* Freeman, San Francisco, 1979.

[GJS] M. R. Garey, D. S. Johnson, R. L. Stockmeyer, Some simplified NP-complete problems, in: *Proc. 6th ACM Sympos. Theory of Computing,* 1974, 47–63.

[Ge1] P. Gerl, Random walks on graphs with a strong isoperimetric property, *J. Theoret. Probab.* **1** (1988), 171–188.

[Ge2] P. Gerl, Amenable groups and amenable graphs, in: *Harmonic Analysis* (P. Eymard and J.-P. Pier, eds.), Lecture Notes in Math. **1359**, Springer-Verlag, 1988, 181–190.

[GW] M. X. Goemans, D. P. Williamson, Improved approximation algorithms for maximum cut and satisfiability problems using semidefinite programming, *J. Assoc. Comput. Mach.* **42** (1995), 1115–1145.

[Go] C. D. Godsil, *Algebraic Combinatorics,* Chapman and Hall, New York, 1993.

[GLS] M. Grötschel, L. Lovász, A. Schrijver, *Geometric Algorithms and Combinatorial Optimization,* Springer-Verlag, Berlin, 1988.

[HK] L. Hagen, A. B. Kahng, New spectral methods for ratio cut partitioning and clustering, *IEEE Trans. Comput.-Aided Design* **11** (1992), 1074–1085.

[HMPR] C. Helmberg, B. Mohar, S. Poljak, F. Rendl, A spectral approach to bandwidth and separator problems in graphs, *Linear and Multilinear Algebra* **39** (1995), 73–90.

[HT] C. Hendré, P. Tetali, Isoperimetric invariants for product Markov chains and graph products, submitted to *Ann. Appl. Probab.,* 1996.

[HL] B. Hendrickson, R. Leland, An improved spectral graph partitioning algorithm for mapping parallel computations, *SIAM J. Sci. Comput.* **16** (1995), 452–469.

[He] M.-C. Heydemann, Cayley graphs and interconnection networks, this volume,

[HMS] M.-C. Heydemann, J.-C. Meyer, D. Sotteau, On forwarding indices of networks, *Discrete Appl. Math.* **23** (1989), 103–123.

[Ho] A. J. Hoffman, On eigenvalues and colorings of graphs, in: *Graph Theory and Its Applications* (B. Harris, ed.), Academic Press, New York, 1970, 79–91.

[JS] M. Jerrum, A. Sinclair, Approximating the permanent, *SIAM J. Comput.* **18** (1989), 1149–1178.

[Ju] F. Juhász, On a method of cluster analysis, *Z. Angew. Math. Mech.* **64** (1984), 335–336.

[JM1] M. Juvan, B. Mohar, Optimal linear labelings and eigenvalues of graphs, *Discrete Appl. Math.* **36** (1992), 153–168.

[JM2] M. Juvan, B. Mohar, Laplace eigenvalues and bandwidth-type invariants of graphs, *J. Graph Theory* **17** (1993), 393-407.

[Kan] R. Kannan, Markov chains and polynomial time algorithms, *Proc. 35th Ann. Sympos. FOCS*, 1994, 656–671.

[KMS] D. Karger, R. Motwani, M. Sudan, Approximate graph coloring by semidefinite programming, *Proc. 35th Ann. Sympos. FOCS*, 1994, 2–13.

[Karm] N. Karmarkar, A new polynomial-time algorithm for linear programming, *Combinatorica* **4** (1984), 373–395.

[Karp] R. M. Karp, Reducibility among combinatorial problems, in: *Complexity of Computer Computation* (R. E. Miller and J. W. Thather, eds.), Plenum Press, New York, 1972, 85–103.

[Kat] T. Kato, *Perturbation Theory for Linear Operators*, Springer-Verlag, Berlin, 1966.

[Kh] L. G. Khachiyan, A polynomial algorithm in linear programming, *Soviet. Math. Dokl.* **20** (1979), 191–194.

[Ki] G. Kirchhoff, Über die Auflösung der Gleichungen, auf welche man bei der Untersuchung der linearen Verteilung galvanischer Ströme geführt wird, *Ann. Phys. Chem.* **72** (1847), 497–508. Translated by J. B. O'Toole in *I.R.E. Trans. Circuit Theory* **5** (1958), 4.

[Kn] D. E. Knuth, The sandwich theorem, *Electron. J. Combin.* **1** (1994), A#1.

[Lo] L. Lovász, On the Shannon capacity of a graph, *IEEE Trans. Inform. Theory* **25** (1979) 1–7.

[LS1] L. Lovász, M. Simonovits, The mixing rate of Markov chains, an isoperimetric inequality, and computing the volume, *31st Ann. Sympos. FOCS*, 1990, 346–354.

[LS2] L. Lovász, M. Simonovits, Random walks in a convex body and an improved volume algorithm, *Random Structures Algorithms* **4** (1993), 359–412.

[Lu1] A. Lubotzky, *Discrete Groups, Expanding Graphs and Invariant Measures,* Birkhäuser, Basel, 1994.

[Mo1] B. Mohar, Isoperimetric inequalities, growth, and the spectrum of graphs, *Linear Algebra Appl.* **103** (1988), 119–131.

[Mo2] B. Mohar, Isoperimetric numbers of graphs, *J. Combin. Theory Ser. B* **47** (1989), 274–291.

[Mo3] B. Mohar, The Laplacian spectrum of graphs, in: *Graph Theory, Combinatorics, and Applications,* (Y. Alavi et al., eds.), Wiley, New York, 1991, 871–898.

[Mo4] B. Mohar, Some relations between analytic and geometric properties of infinite graphs, *Discrete Math.* **95** (1991), 193–219.

[MP1] B. Mohar, S. Poljak, Eigenvalues and the max-cut problem, *Czechoslovak Math. J.* **40** (115) (1990), 343–352.

[MP2] B. Mohar, S. Poljak, Eigenvalues in combinatorial optimization, in: *Combinatorial and Graph-Theoretical Problems in Linear Algebra* (R. A. Brualdi, S. Friedland, V. Klee, eds.), IMA Vol. Math. Appl. **50**, Springer-Verlag, 1993, 107–151.

[MW] B. Mohar, W. Woess, A survey on spectra of infinite graphs, *Bull. London Math. Soc.* **21** (1989), 209–234.

[MR] R. Motwani, P. Raghavan, *Randomized Algorithms,* Cambridge University Press, New York, 1995

[NN1] Yu. Nesterov, A. Nemirovskii, *Self-Concordant Functions and Polynomial Time Methods in Convex Programming,* Central Economic and Mathematical Institute, USSR Academy of Science, Moscow, 1989.

[NN2] Yu. Nesterov, A. Nemirovskii, *Interior Point Polynomial Methods in Convex Programming,* SIAM, Philadelphia, PA, 1994.

[PY] C. H. Papadimitriou, M. Yannakakis, Optimization, approximation, and complexity classes, *J. Comput. System. Sci.* **43** (1991), 425–440.

[Pa] A. L. T. Paterson, *Amenability*, Math. Surveys Monographs **29**, Amer. Math. Soc., Providence, RI, 1988.

[Pi] J.-P. Pier, *Amenable Locally Compact Groups,* Wiley, New York, 1984.

[PR] S. Poljak, F. Rendl, Nonpolyhedral relaxations of graph-bisection problems, *SIAM J. Optim.* **5** (1995), 467–487.

[PSW] A. Pothen, H. D. Simon, L. Wang, Spectral nested dissection, preprint, 1992.

[RW] F. Rendl, H. Wolkowicz, Applications of parametric programming and eigenvalue maximization to the quadratic assignment problem, *Math. Programming* **53** (1992), 63-78.

[Ro] J. M. Rosenblatt, Invariant measures and growth conditions, *Trans. Amer. Math. Soc.* **193** (1974), 33–53.

[Ros] J. S. Rosenthal, Convergence rates for Markov chains, *SIAM Rev.* **37** (1995), 387–405.

[Sim] H. D. Simon, Partitioning of unstructured problems for parallel processing, *Comput. Syst. in Engin.* **2** (1991), 135–148.

[Sin] A. Sinclair, *Algorithms for Random Generation and Counting: A Markov Chain Approach,* Birkhäuser, Boston, 1993.

[SJ] A. J. Sinclair, M. R. Jerrum, Approximate counting, uniform generation and rapidly mixing Markov chains, *Inform. and Comput.* **82** (1989), 93–133.

[SW] P. M. Soardi, W. Woess, Amenability, unimodularity, and the spectral radius of random walks on infinite graphs, *Math. Z.* **205** (1990), 471–486.

[So] P. Solé, Expanding and forwarding, *Discrete Appl. Math.* **58** (1995), 67–78.

[ST] D. A. Spielman, S.-H. Teng, Spectral partitioning works: Planar graphs and finite element meshes, preprint, 1996.

[Th] C. Thomassen, Trees, ends, and transience, in: *Harmonic Analysis and Discrete Potential Theory (Frascati 1991)* (M. A. Picardello, ed.) Plenum, New York, 1992, 259–266.

[St] J. G. Stampfli, Compact perturbations, normal eigenvalues, and a problem of Salinas, *J. London Math. Soc.* **9** (1974), 165–175.

[Vai] P. M. Vaidya, A new algorithm for minimizing convex functions over convex sets, *Proc. 30th Ann. Sympos. FOCS*, 1989, 338–343.

[Val] L. G. Valiant, The complexity of computing the permanent, *Theoret. Comput. Sci.* **8** (1979), 189–201.

[Var] N. T. Varopoulos, Isoperimetric inequalities and Markov chains, *J. Funct. Anal.* **63** (1985), 215–239.

[Wi] H. S. Wilf, The eigenvalues of a graph and its chromatic number, *J. London Math. Soc.* **42** (1967), 330–332.

[Wil] R. D. Williams, Performance of dynamic load balancing algorithms for unstructured mesh calculations, Technical Report, California Inst. of Technology, 1990.

[Woe] W. Woess, Random walks on infinite graphs and groups – A survey on selected topics, *Bull. London Math. Soc.* **26** (1994), 1–60.

# Finite transitive permutation groups and finite vertex-transitive graphs

Cheryl E. PRAEGER

with the assistance of
Cai Heng LI and Alice C. NIEMEYER

*Department of Mathematics*
*The University of Western Australia*
*Nedlands, WA 6907*
*Australia*

**Abstract**

The theory of vertex-transitive graphs has developed in parallel with the theory of transitive permutation groups. The chapter explores the way the two theories have influenced each other. Examples are drawn from the enumeration of vertex-transitive graphs of small order, the classification problem for finite distance transitive graphs, and the investigations of finite 2-arc transitive graphs, finite primitive and quasiprimitive permutation groups, and finite locally primitive graphs. The nature of the group theoretic techniques used range from elementary ones to some involving the finite simple group classification. In particular the theorem of O'Nan and Scott for finite primitive permutation groups, and a generalisation of it for finite quasiprimitive permutation groups is discussed.

## 1 Introduction

The theory of vertex-transitive graphs has developed in parallel with the theory of transitive permutation groups. In this chapter we explore some of the ways the two theories have influenced each other. On the one hand each finite transitive permutation group corresponds to several vertex-transitive graphs, namely the generalised orbital graphs which we shall discuss below. On the other hand, each finite vertex-transitive graph gives rise to (usually) several transitive permutation groups, namely the vertex-transitive subgroups of the full automorphism group of the graph. We shall study pairs $(\Gamma, G)$ where $\Gamma$ is a finite graph and $G$ is a vertex-transitive subgroup of its automorphism group $\mathrm{Aut}\Gamma$.

In doing so we shall be bringing together, and learning from, two mathematical cultures: group theory and graph theory. We shall see the interchange of techniques and ideas between the theory of transitive permutation groups and the theory of vertex-transitive graphs. More specifically, we will look at various ways in which permutation group theory has been used to solve problems about finite vertex-transitive graphs. Sometimes only elementary group theoretic techniques were required, while in other cases quite sophisticated group theory was necessary, occasionally involving the finite simple group classification. In one case, the



*G. Hahn and G. Sabidussi (eds.), Graph Symmetry,* 277–318.

necessary group theory was not available, but the desire to solve the graph theoretic problem stimulated its development.

The problems we shall examine relate to the following areas of finite graph theory:

(a) the enumeration of vertex-transitive graphs of small order;

(b) distance transitive graphs;

(c) $s$-arc transitive graphs;

(d) locally-primitive graphs.

In addition (and motivated by the lectures of Brian Alspach at this NATO Advanced Study Institute) a problem about isomorphisms of finite Cayley graphs, and its solution, will be described. The discussion involves an overview of several parts of permutation group theory, including the O'Nan-Scott Theorem on finite primitive permutation groups, and the structure theory of finite quasiprimitive permutation groups.

## 2 Transitive groups and generalised orbital graphs

We begin with a short review of some of the basic concepts of permutation groups, including in particular a description of the construction of the family of generalised orbital graphs associated with a transitive permutation group.

For a set $\Omega$, the *symmetric group* $\mathrm{Sym}(\Omega)$ on $\Omega$ is the group of all permutations of $\Omega$, and a *permutation group on* $\Omega$ is simply a subgroup of $\mathrm{Sym}(\Omega)$. If $|\Omega| = n$ and $G \leq \mathrm{Sym}(\Omega)$, then $G$ is said to be a permutation group of *degree* $n$. For $\alpha \in \Omega$, the image of $\alpha$ under a permutation $g$ of $\Omega$ is denoted by $\alpha^g$. A permutation group $G$ on $\Omega$ induces an equivalence relation $\sim_G$ on $\Omega$ given by

$$\alpha \sim_G \beta \text{ if and only if } \alpha^g = \beta \text{ for some } g \in G.$$

The equivalence classes of $\sim_G$ are called *G-orbits* in $\Omega$. So the $G$-orbit containing $\alpha$ is $\alpha^G := \{\alpha^g \mid g \in G\}$; $G$ is said to be *transitive* on $\Omega$ if there is only one $G$-orbit, that is, if $\Omega = \alpha^G$ for some and hence every point $\alpha \in \Omega$.

When studying automorphism groups of graphs we wish to consider their actions on several different sets: the vertices, edges, arcs etc. Thus it will be helpful to introduce the concept of a permutation representation. A *permutation representation* of a group $G$ on a set $\Omega$ is a group homomorphism $\varphi : G \to \mathrm{Sym}(\Omega)$. The image $(G)\varphi$ of $G$ under $\varphi$ is the permutation group on $\Omega$ induced by $G$ under the representation $\varphi$ and is often denoted by $G^\Omega$. The representation $\varphi$ is said to be *faithful* if $\mathrm{Ker}(\varphi) = 1$. When there is no likelihood of confusion we often suppress $\varphi$ and write the permutation $(g)\varphi$ of $\Omega$ simply as $g$, or sometimes as $g^\Omega$, for $g \in G$.

In fact it is useful to consider more than one permutation representation when analysing any permutation group. For example, any transitive permutation group $G$ on $\Omega$ has a faithful permutation representation on $\Omega \times \Omega$ given by $(\alpha, \beta)^g := (\alpha^g, \beta^g)$ for $g \in G$ and $\alpha, \beta \in \Omega$. The $G$-orbits in $\Omega \times \Omega$ are called the *G-orbitals* on $\Omega$ (or simply orbitals), and, for each point $\alpha \in \Omega$, there is a 1-1 correspondence between the $G$-orbitals and the orbits in $\Omega$ of the stabilizer $G_\alpha$ of $\alpha$ (where $G_\alpha := \{g \in \Gamma \mid \alpha^g = \alpha\}$). If $\Delta$ is a $G$-orbital then the corresponding $G_\alpha$-orbit in $\Omega$ is $\Delta(\alpha) := \{\beta \mid (\alpha,\beta) \in \Delta\}$. Also we define a *generalised orbital* to be a non-empty $G$-invariant subset of $\Omega \times \Omega$, that is, a non-empty union of $G$-orbitals. For each (generalised) orbital $\Delta$, there is a *paired (generalised) orbital* $\Delta^*$, namely $\Delta^* := \{(\beta,\alpha) \mid (\alpha,\beta) \in \Delta\}$. If $\Delta = \Delta^*$ we say that the (generalised) orbital $\Delta$ is *self-paired.* Also, for each transitive group $G$ on $\Omega$, the set $\Delta_0 := \{(\alpha,\alpha) \mid \alpha \in \Omega\}$ is a $G$-orbital and is called the *trivial* $G$-orbital. The number of $G$-orbitals, or equivalently the number of $G_\alpha$-orbits in $\Omega$, is called the *rank* of $G$.

With each generalised $G$-orbital we will associate a di-graph, and with each self-paired generalised $G$-orbital not containing $\Delta_0$ we will associate a graph; each such di-graph and graph will have vertex set $\Omega$ and will admit the group $G$ as a vertex-transitive subgroup of automorphisms.

First we explain our notation. By a *di-graph* (or *directed graph*) we mean a pair $\Gamma = (\Omega, E)$ where $\Omega$ is a set whose elements are called the *vertices* of $\Gamma$, and $E$ is a subset of $\Omega \times \Omega$ whose elements are called the *edges* or *arcs* of $\Gamma$. Similarly a *graph* is a pair $\Gamma = (\Omega, E)$ where $\Omega$ is the set of *vertices* of $\Gamma$, and $E$ is a subset of unordered pairs of distinct vertices which are called the *edges* of $\Gamma$; the ordered pairs $(\alpha,\beta)$ of adjacent vertices (that is those for which $\{\alpha,\beta\} \in E$) are called the *arcs* of $\Gamma$. Thus, for a di-graph, edges and arcs are the same, but this is not the case for a graph. Note that, according to our definition, a di-graph may or may not have some *loops* (edges of the form $(\alpha,\alpha)$), but a graph does not have any loops. An *automorphism* of a graph or di-graph $\Gamma = (\Omega, E)$ is an element $g$ of $\mathrm{Sym}(\Omega)$ such that $e^g \in E$ if and only if $e \in E$; and the set of all automorphisms forms a subgroup $\mathrm{Aut}\Gamma$ of $\mathrm{Sym}(\Omega)$ called the *automorphism group* of $\Gamma$. If a subgroup $G$ of $\mathrm{Aut}\Gamma$ is transitive on the vertex set $\Omega$, the edge set $E$, or the set of arcs of $\Gamma$, we say that $\Gamma$ is *G-vertex-transitive, G-edge-transitive*, or *G-arc-transitive* respectively. Often we omit the prefix "$G$'.

Now let $\Delta$ be a (generalised) $G$-orbital. Then the associated *(generalised) orbital di-graph* is the di-graph $(\Omega, \Delta)$. As it will seldom cause confusion, we shall denote this di-graph also by $\Delta$. It follows from this definition that $G \leq \mathrm{Aut}\Delta$, and $G$ is transitive on the vertices of $\Delta$. Also $G$ is transitive on the arcs of $\Delta$ if and only if $\Delta$ is a $G$-orbital. Now a (generalised) orbital $\Delta$ has the property

$$(\alpha,\beta) \in \Delta \text{ if and only if } (\beta,\alpha) \in \Delta$$

if and only if $\Delta$ is self-paired. Thus, if $\Delta$ is self-paired and $\Delta_0 \not\subseteq \Delta$, then the di-graph $\Delta$ contains no loops and the edges occur in pairs of the form $\{(\alpha,\beta),(\beta,\alpha)\}$. If this is the case then we may replace each such pair by the unordered pair $\{\alpha,\beta\}$ and thereby obtain a graph, called the *(generalised) orbital graph* associated with $\Delta$. Again if no confusion arises we denote the (generalised) orbital graph by $\Delta$. As for the (generalised) orbital di-graph, this (generalised) orbital graph also admits $G$ as a subgroup of its automorphism group; $G$ is transitive on vertices, and $G$ is transitive on arcs if and only if $\Delta$ is a self-paired nontrivial $G$-orbital. As examples of generalised orbital graphs $\Delta$ for a transitive permutation group $G \leq \mathrm{Sym}(\Omega)$, we always have the *empty graph* $(\Omega, \emptyset)$ on $\Omega$, by taking $\Delta = \Delta_0$, and the *complete graph* on $\Omega$, that is the graph in which every pair of distinct vertices is an edge, by taking $\Delta = (\Omega \times \Omega) \backslash \Delta_0$. If $|\Omega| = n$ then the complete graph on $\Omega$ is denoted by $K_n$.

The most important (but elementary) observations to be made about generalised orbital graphs of transitive permutation groups on a set $\Omega$ are the following.

**Theorem 2.1**

(a) *$G$ is a vertex-transitive subgroup of automorphisms of a graph $\Gamma = (\Omega, E)$ if and only if $\Gamma$ is a generalised orbital graph for $G$, namely for the self-paired generalised orbital $\Delta := \{(\alpha, \beta) \mid \{\alpha, \beta\} \in E\}$.*

(b) *$G$ is an arc-transitive group of automorphisms of a graph $\Gamma = (\Omega, E)$ if and only if $\Gamma$ is an orbital graph for $G$, namely for the self-paired orbital $\Delta := \{(\alpha, \beta) \mid \{\alpha, \beta\} \in E\}$.*

We have already seen that generalised orbital graphs have the asserted properties. For the converse observe that, if $G \leq \mathrm{Aut}\Gamma$ and $G$ is transitive on the vertex set $\Omega$ of $\Gamma$, then the set of arcs of $\Gamma$ is a $G$-invariant subset of $\Omega \times \Omega$ and hence is a generalised orbital for the transitive permutation group $G$ on $\Omega$.

## 3 Partitions, blocks and primitivity

One of the traditional ways of studying a transitive permutation group $G \leq \mathrm{Sym}(\Omega)$ is to examine the $G$-invariant partitions of $\Omega$. A partition $\mathcal{P}$ of $\Omega$ is said to be *$G$-invariant* if the elements of $G$ *permute the parts of $\mathcal{P}$ block-wise*, that is, $P^g \in \mathcal{P}$ for all $P \in \mathcal{P}$ and $g \in G$ (where $P^g := \{\alpha^g \mid \alpha \in P\}$). The *trivial partitions* $\{\Omega\}$ and $\{\{\omega\} \mid \omega \in \Omega\}$ are $G$-invariant for all transitive groups $G$, and a transitive permutation group $G$ on $\Omega$ is said to be *primitive* on $\Omega$ if these are the only $G$-invariant partitions of $\Omega$. If $G$ is transitive, but not primitive, on $\Omega$, then $G$ is said to be *imprimitive* on $\Omega$.

Let $G$ be a transitive permutation group $G$ on $\Omega$. Then for every non-empty subset $B$ of $\Omega$, the collection of subsets $\mathcal{B} := \{B^g \mid g \in G\}$ is $G$-invariant in the sense that $D^g \in \mathcal{B}$ for all $g \in G$ and $D \in \mathcal{B}$. Thus $\mathcal{B}$ is a $G$-invariant partition of $\Omega$ if and only if it is in fact a partition. It turns out that this is the case if and only if the subset $B$ is a block of imprimitivity which is defined as follows. A non-empty subset $B$ of $\Omega$ is a *block of imprimitivity for* $G$ in $\Omega$ if, for all $g \in G$, either $B^g = B$ or $B^g \cap B = \emptyset$. It is not difficult to verify that a partition of $\Omega$ is $G$-invariant if and only if all of its parts are blocks of imprimitivity, and that the latter is true if and only if at least one of the parts is a block of imprimitivity. In line with our definitions for $G$-invariant partitions, a block of imprimitivity is said to be *trivial* if either it contains only one element or it is equal to $\Omega$. Thus $G$ is primitive on $\Omega$ if and only if the only blocks of imprimitivity for $G$ in $\Omega$ are the trivial ones.

For any subset $P$ of $\Omega$ the *setwise stabilizer* $G_P$ of $P$ in $G$ is defined as $G_P = \{g \in G \mid P^g = P\}$; clearly $G_P$ is a subgroup of $G$. For a $G$-invariant partition $\mathcal{P}$ of $\Omega$, if $P$ is the part of $\mathcal{P}$ containing a point $\alpha$, then $G_P$ contains the point stabilizer $G_\alpha$. Conversely, if $G_\alpha \leq H \leq G$ then the $H$-orbit $P := \alpha^H$ containing $\alpha$ determines a $G$-invariant partition of $\Omega$, namely $\mathcal{P} := \{P^g \mid g \in G\}$. In fact the lattice of all $G$-invariant partitions of $\Omega$ is isomorphic to the lattice of subgroups of $G$ containing $G_\alpha$, and hence $G$ is primitive on $\Omega$ if and only if $G_\alpha$ is a maximal subgroup of $G$.

For a $G$-invariant partition $\mathcal{P}$ of $\Omega$, there correspond two (usually smaller) transitive permutation groups, namely the group $G^{\mathcal{P}}$ of permutations of $\mathcal{P}$ induced by $G$, and the permutation group $G_P^P$ induced on $P$ by $G_P$, where $P \in \mathcal{P}$. We describe these groups below.

The map $g \mapsto g^P$ $(g \in G)$ defines a permutation representation of $G_P$ on $P$ with image the permutation group $G_P^P$. Of course, choosing different parts $P, P' \in \mathcal{P}$ gives us different transitive permutation groups $G_P^P$ and $G_{P'}^{P'}$, but they are essentially the same in that they are permutationally equivalent. As in [71, p.32] by a *permutational equivalence* between two permutation groups $H, K$ on $\Delta$, $\Sigma$ respectively we mean a pair $(f, \varphi)$ such that $f : H \to K$ is a group isomorphism, $\varphi : \Delta \to \Sigma$ is a bijection, and for all $h \in H$, $\delta \in \Delta$, we have $(\delta^h)\varphi = (\delta\varphi)^{(h)f}$. (Also, if $H = K$ so that the isomorphism $f$ is an automorphism, then we call $(f, \varphi)$ a *permutational isomorphism*.) It is easy to verify that each element $x \in G$ such that $P^x = P'$ induces a bijection $\bar{x}$ from $P$ to $P'$ (by restriction) and an isomorphism $\hat{x}$ from $G_P^P$ to $G_{P'}^{P'}$ (by conjugation), and the pair $(\hat{x}, \bar{x})$ is a permutational equivalence.

For $g \in G$, we let $g^{\mathcal{P}}$ denote the permutation of $\mathcal{P}$ induced by $g$. Then the map $g \mapsto g^{\mathcal{P}}$ defines a permutation representation of $G$ on $\mathcal{P}$. Thus, as an abstract group, the image $G^{\mathcal{P}}$ of this permutation representation is a quotient group of $G$, and we view the permutation group $G^{\mathcal{P}}$ on $\mathcal{P}$ as a quotient of the permutation group $G$ on $\Omega$. If we choose $\mathcal{P}$ such that $G_P$ is maximal in $G$ for some, and hence all, $P \in \mathcal{P}$, we see that $G^{\mathcal{P}}$ must be primitive on $\mathcal{P}$. Thus every transitive permutation group has at least one primitive quotient. We may view the transitive group $G$ as being "built up" of the two permutation groups $G^{\mathcal{P}}$ and $G_P^P$. If $\Omega$ is finite then, by repeating this procedure for the transitive group $G_P^P$ (and so on), we will eventually obtain a finite sequence of primitive permutation groups of which, in some sense, $G$ is "composed".

The traditional approach to the study of finite transitive permutation groups is to assert, at this point, that the primitive permutation groups are the building blocks for transitive permutation groups, and that the central problem for finite permutation groups is to understand the structure of finite primitive permutation groups, and, incidentally, how they can be fitted back together to form the transitive groups.

In our exploration of vertex-transitive graphs in this chapter we shall examine the extent to which this traditional approach has yielded an appropriate group theoretical framework for applications to problems about finite vertex-transitive graphs. For the problem of classifying distance-transitive graphs we shall see that this approach was ideal; for the problem of enumeration of vertex-transitive graphs of small order it yielded helpful but insufficient information; while for the problem of describing the structure of 2-arc transitive graphs we shall see that it was regrettably inappropriate, but that a generalisation of the structure theory of finite primitive groups to quasiprimitive groups was both achievable and useful.

## 4 Quotients of vertex-transitive graphs

Here we look briefly at the exact analogue of the discussion in Section 3 for finite vertex-transitive graphs. First we define quotient graphs relative to arbitrary partitions of the vertex set. Let $\mathcal{P}$ be a partition of the vertex set $\Omega$ of a graph $\Gamma = (\Omega, E)$. Then the *quotient graph* $\Gamma_{\mathcal{P}}$ of $\Gamma$ relative to $\mathcal{P}$ is defined to be the graph $\Gamma_{\mathcal{P}} = (\mathcal{P}, E_{\mathcal{P}})$, where $\{P, P'\} \in E_{\mathcal{P}}$ if and only if there exist $\alpha \in P$ and $\alpha' \in P'$ such that $\{\alpha, \alpha'\} \in E$. Quotient graphs inherit some of the properties of the original graph. For example quotient graphs of connected graphs are connected. Moreover, if $\mathcal{P}$ is invariant under some subgroup $G$ of Aut$\Gamma$, then $\Gamma_{\mathcal{P}}$ admits $G^{\mathcal{P}}$ as a subgroup of automorphisms. (A graph $\Gamma = (\Omega, E)$ is *connected* if, for all $\alpha, \beta \in \Omega$, there is a finite sequence $\gamma_0, \ldots, \gamma_n$ of vertices such that $\gamma_0 = \alpha$, $\gamma_n = \beta$, and $\{\gamma_i, \gamma_{i+1}\} \in E$ for all $0 \leq i < n$. If $\Gamma$ is not connected then we say that $\Gamma$ is *disconnected*.)

Also, for $P \subseteq \Omega$, let $[P]$ denote the *subgraph of* $\Gamma$ *induced on* $P$, that is, the graph with vertex set $P$ such that a pair of vertices of $P$ is an edge of $[P]$ if and only if it is an edge of $\Gamma$. Again, if $P$ is invariant under some subgroup $G$ of Aut$\Gamma$ (that is, if $P$ is fixed setwise by

$G$), then the permutation group $G^P$ induced on $P$ by $G$ is a subgroup of Aut$[P]$.

Now suppose that $\Gamma$ is *$G$-vertex-transitive.* If $G$ is imprimitive on $\Omega$ (that is, $\Gamma$ is *$G$-vertex-imprimitive*) and $\mathcal{P}$ is a nontrivial $G$-invariant partition of $\Omega$, then $\Gamma_{\mathcal{P}}$ is $G^{\mathcal{P}}$-vertex-transitive, and $[P]$ is $G_P^P$-vertex-transitive. Furthermore the graphs $[P]$, for $P \in \mathcal{P}$, are all isomorphic, and the transitive permutation groups $G_P^P$, for $P \in \mathcal{P}$, are permutationally equivalent. Thus in some sense the vertex-transitive graph $\Gamma$ is "composed of" the two smaller vertex-transitive graphs $\Gamma_{\mathcal{P}}$ and $[P]$, and we could refine this decomposition (just as for transitive permutation groups in the previous section) to give a sequence of vertex-primitive graphs. However for several reasons this may not always be the most helpful path to an understanding of the structure of a finite vertex-transitive graph $\Gamma$.

For example, it follows from the discussion in Section 3 that if $G$ is arc-transitive on a connected graph $\Gamma$ then, for a nontrivial $G$-invariant partition $\mathcal{P}$, the group $G^{\mathcal{P}}$ is arc-transitive on $\Gamma_{\mathcal{P}}$. Thus if $\Gamma$, and hence also $\Gamma_{\mathcal{P}}$, are connected, then all edges of $\Gamma$ join vertices in different parts of $\mathcal{P}$; so in particular the induced subgraphs $[P]$, for $P \in \mathcal{P}$, are all *empty graphs* (that is they have no edges). Thus the subgraphs $[P]$ may not give much information about $\Gamma$ (although a more imaginative approach to the induced structure on $P$ is possible and is the subject of Tony Gardiner's lectures for this NATO Advanced Study Institute, see also [33]). From the point of view described here, most of the information will be contained in the quotient graph $\Gamma_{\mathcal{P}}$, and by choosing the nontrivial $G$-invariant partition $\mathcal{P}$ to have $|\mathcal{P}|$ as small as possible, we may ensure that the group $G^{\mathcal{P}}$ is primitive. This suggests that one might study vertex-primitive graphs to gain insights into the nature of quotients of finite connected vertex-transitive graphs.

However, if we are interested in a family of vertex-transitive graphs with some extra defining property, there are some additional difficulties. We will have in general no guarantee that this extra property will be inherited by $\Gamma_{\mathcal{P}}$ and/or by $[P]$. Sometimes just one of $\Gamma_{\mathcal{P}}$ and $[P]$ will have this extra property (this is the case for example with the property of being connected and arc-transitive), and sometimes neither will inherit it. The examples discussed in this chapter will demonstrate a range of possibilities.

## 5 Enumeration of vertex-transitive graphs of small order

The *order* of a finite graph $\Gamma = (\Omega, E)$ is the number $|\Omega|$ of vertices. In this section we discuss briefly some of the ideas and techniques used to enumerate all vertex-transitive graphs of certain (small) orders. It turned out that a straight-forward application of the approach outlined in the previous section for analysing the structure of vertex-transitive graphs was not the most efficient way of proceeding. In fact, this example was chosen to demonstrate that, even though a theoretical framework may give a good description of the structure of certain finite vertex-transitive graphs, it may not provide a computationally efficient approach to their enumeration. First we make a few general elementary observations about finite vertex-transitive graphs.

1. If a vertex-transitive graph $\Gamma$ of order $n$ is disconnected, then all of its connected components are isomorphic, and are themselves vertex-transitive. Thus, if all vertex-transitive graphs of order a proper divisor of $n$ have been enumerated, then we may read off from these lists, up to isomorphism, all the disconnected vertex-transitive graphs of order $n$.

2. If $\Gamma = (\Omega, E)$ is a vertex-transitive graph of order $n$, then also its *complementary graph* $\overline{\Gamma} := (\Omega, \overline{E})$ is vertex-transitive of order $n$ (where, for distinct points $\alpha$ and $\beta$ of $\Omega$, $\{\alpha, \beta\} \in \overline{E}$ if and only if $\{\alpha, \beta\} \notin E$). Also $\mathrm{Aut}\Gamma = \mathrm{Aut}\overline{\Gamma}$.

3. Every finite vertex-transitive graph $\Gamma$ is *regular*, that is, there is a constant $k$ such that every vertex $\alpha$ is adjacent to exactly $k$ other vertices. The constant $k$ is called the *valency* of $\Gamma$, and is denoted by $\mathrm{val}(\Gamma)$.

Thus to enumerate all vertex-transitive graphs $\Gamma$ of a given order $n$, we may assume that both $\Gamma$ and $\overline{\Gamma}$ are connected, and that $\mathrm{val}(\Gamma) < n/2$. Moreover, we might expect (naively perhaps) that those of valency close to $n/2$ might be more difficult to enumerate than those of small valency.

Up to 1985 the only catalogues of vertex-transitive graphs available in the literature were the list of vertex-transitive graphs of order at most 13 by H. P. Yap [94] (excluding those of valency 5 on 12 vertices), and the list of all vertex-transitive graphs of order at most 19 by B. D. McKay [63, 64]. In the latter case, McKay had used complicated matrix methods to construct his list. The vertex-transitive graphs of orders $n = 20, 21, 22$ and 23 were constructed by McKay and Royle [67]. These constructions were based on an elementary group-theoretic lemma exploiting the fact that in all cases the prime decomposition of $n$ contained a relatively large prime factor $p$. The lemma showed that every vertex-transitive subgroup of the automorphism group of such a graph contained a permutation of order $p$ which had relatively few cycles of length $p$ and no fixed vertices.

However, this method did not easily carry over to the case $n = 24$ as the existence of such automorphisms could not be guaranteed. The enumeration of the vertex-transitive graphs of order 24 was completed by G. F. Royle and the author [81] making use of the theory outlined in the previous section. Further, it required the classification of the transitive permutation groups of degree 12 (see [80]), it involved heavy use of the computer sytem CAYLEY (which has now evolved into the new system MAGMA [17]) and a suite of programs to compute second cohomology groups developed by D. F. Holt [39]. We outline some of the main ideas of this enumeration.

All primitive groups of degree 24 are 2-transitive. (A permutation group $G$ on $\Omega$ is said to be *2-transitive* on $\Omega$ if $G$ is transitive on the ordered pairs of distinct points of $\Omega$.) Hence the only vertex-primitive graphs of order 24 are $24K_1$ and $K_{24}$. So we may assume that the automorphism group of a vertex-transitive graph $\Gamma = (\Omega, E)$ of order 24 is imprimitive, and by our remarks above we may also assume that $\Gamma$ is connected and has valency less than 12. Let $G$ be a vertex-transitive subgroup of $\mathrm{Aut}\Gamma$. Then certainly $G$ is imprimitive on $\Omega$, and for each nontrival $G$-invariant partition $\mathcal{P}$ of $\Omega$, the two graphs $\Gamma_{\mathcal{P}}$ and $[P]$ (where $P \in \mathcal{P}$) are vertex-transitive with orders being proper divisors of 24. Since all vertex-transitive graphs of order a proper divisor of 24 are known, the search procedure might be subdivided as follows. For each divisor $d$ of 24 with $1 < d < 24$, find all pairs $(\Gamma, G)$ with $\Gamma$ a graph of order 24, $G$ a vertex-transitive subgroup of $\mathrm{Aut}\Gamma$, and such that there is a $G$-invariant partition $\mathcal{P}$ of vertices for which $|\mathcal{P}| = d$ and $|\mathcal{P}|$ is maximal over all nontrivial $G$-invariant partitions $\mathcal{P}$.

Although this was the basic underlying strategy used for the enumeration, the large number of graphs to be identified presented serious problems of storage and isomorphism testing, besides the other group theoretic problems associated with the search. (It turned out that there are, up to isomorphism, 7753 vertex-transitive graphs of order 24.) From their experience with the classification of the smaller vertex-transitive graphs, McKay and Royle guessed that the majority of the examples would be Cayley graphs, and suggested that these could be found separately. Then, when following the search strategy above, whenever a graph was found we could first test to see if it was a Cayley graph and, if so, discard it.

We now explain what Cayley graphs are.

**Definition 5.1** Let $G$ be a finite group, and let $S \subseteq G$ such that $1_G \notin S$. The *Cayley digraph* $\mathrm{Cay}(G, S)$ of $G$ relative to $S$ is the di-graph with vertex set $G$ such that, for $x, y \in G$, the pair $(x, y)$ is an arc if and only if $xy^{-1} \in S$. Further, if $S = S^{-1} := \{s^{-1} \mid s \in S\}$, then $(x, y)$ is an arc if and only if $(y, x)$ is an arc, and we define the *Cayley graph* of $G$ relative to $S$ as the graph with vertex set $G$ such that $\{x, y\}$ is an edge if and only if $xy^{-1} \in S$. We abuse notation somewhat and denote this graph also by $\mathrm{Cay}(G, S)$, as its meaning will be clear from the context. A subset $S$ of $G$ such that $1 \notin S$ and $S = S^{-1}$ will be called a *Cayley subset* of $G$.

The group $G$ acting on itself in its right regular representation (that is, $g: x \mapsto xg$) is a subgroup of the automorphism group of $\mathrm{Cay}(G, S)$ and acts transitively on $G$. Thus all Cayley graphs are vertex-transitive. Further, this action of $G$ on itself is regular, where we say that a permutation group $H$ on a set $\Delta$ is *regular* on $\Delta$ if $H$ is transitive on $\Delta$ and the only element of $H$ which fixes a point of $\Delta$ is the identity $1_H$. The existence of a subgroup of $\mathrm{Aut}\Gamma$ acting regularly on vertices characterises Cayley graphs.

**Lemma 5.2 ([13, Lemma 16.3])** *A graph $\Gamma = (\Omega, E)$ is isomorphic to a Calyey graph for a group $G$ if and only if* $\mathrm{Aut}\Gamma$ *has a subgroup $\overline{G}$ isomorphic to $G$ such that $\overline{G}$ is regular on $\Omega$.*

Thus the search for incomplete vertex-transitive graphs of order 24 had two phases, finding the Cayley graphs of order 24, and finding the remaining examples, the so-called *non-Cayley graphs* of order 24. We require a list containing exactly one copy of a graph from each isomorphism class of vertex-transitive graphs of order 24. Note that, if $\Gamma = \mathrm{Cay}(G, S)$, and if $\sigma \in \mathrm{Aut}(G)$, then $\sigma$ induces an isomorphism from $\Gamma$ to $\mathrm{Cay}(G, S^\sigma)$. There may be other isomorphisms between Cayley graphs for $G$, and it is possible for a graph to be a Cayley graph for more than one group.

**Cayley graphs of order** 24: All groups of order 24 are known, up to isomorphism. For each such group $G$, the following procedure, written schematically below, was used to find, up to isomorphism, all Cayley graphs for $G$. Then graphs in the resulting collection were tested for isomorphism.

*for each group $G$ of order* 24
 *for each orbit of* $\mathrm{Aut}(G)$ *on Cayley subsets*
  *choose one representative $S$ and test isomorphism*
  *of the resulting list of Cayley graphs* $\mathrm{Cay}(G, S)$.

The isomorphism test was performed using McKay's program nauty for computing graph automorphisms and isomorphisms (see [65]). In this way 7697 pairwise non-isomorphic Cayley graphs of order 24 were found.

**Non-Cayley graphs of order** 24: For each vertex- transitive non-Cayley graph $\Gamma = (\Omega, E)$ of order 24, the automorphism group Aut$\Gamma$ has at least one *minimal transitive subgroup* $G$, that is, a subgroup $G$ which is transitive on $\Omega$ but has the property that every proper subgroup of $G$ is intransitive. Since $\Gamma$ is a non-Cayley graph, $G$ is not regular on $\Omega$. Moreover $G$ is imprimitive on $\Omega$. Let $d$ be maximal such that $d < 24$ and there exists a $G$-invariant partition of $\Omega$ having $d$ parts, and let $\mathcal{P} = \{P_1 \mid \ldots \mid P_d\}$ be such a partition.

This phase of the search found, for all divisors $d$ of 24 with $1 < d < 24$, all pairs $(\Gamma, G)$ with $\Gamma, G$ as above. Since the choice of the minimal transitive subgroup $G$ was not necessarily unique, isomorphism testing was necessary. We sketch some of the details. It should be emphasised that, even if $\Gamma$ admits a minimal transitive subgroup $G$ which is not regular on vertices, $\Gamma$ may still be a Cayley graph for some other subgroup of Aut$\Gamma$. Thus, our analysis may produce Cayley graphs, and this must be checked for.

Let $\Gamma, G, d$ be as above. Let $K$ be the kernel $G_{(\mathcal{P})}$ of the action of $G$ on $\mathcal{P}$, so that $G^{\mathcal{P}} = G/K$. By the maximality of $d$, the group $G_{P_1}^{P_1}$ is primitive. If $K = 1$, then $G^{\mathcal{P}} \cong G$ is a transitive, but not necessarily minimal transitive permutation group of degree $d$, and 24 divides $|G^{\mathcal{P}}|$. If $K \neq 1$, then $G^{\mathcal{P}}$ is a minimal transitive permutation group of degree $d$. A careful theoretical analysis showed, using the fact that $\Gamma$ is a non-Cayley graph, that $d \geq 6$ and $d \neq 8$. Also, in the case $d = 6$ it turned out that all the possibilities for $\Gamma$ were Cayley graphs, but discovering this required a combination of theoretical analysis and some computation.

The heart of the problem was the case $d = 12$. Here, if $\alpha \in P_1$ then $G_\alpha$ fixes some union of parts of $\mathcal{P}$ pointwise. A combination of computation and theory showed that the number $f$ of fixed points of $G_\alpha$ was not 2, 4, or 6. So $f = 8, 12$ or 24. Thus $|K| \leq 2^3$. Suppose first that $K \neq 1$. Then $G^{\mathcal{P}} = G/K$ is one of the 17 minimal transitive groups of degree 12. (Classifying these formed part of the project!) Thus $G$ is an extension of the elementary abelian 2-group $K$ by a known minimal transitive group $G^{\mathcal{P}}$ of degree 12. The programs of Derek Holt (see [39]) were used to construct the second cohomology group for each pair $(K, G^{\mathcal{P}})$ for each possible action of $G^{\mathcal{P}}$ on $K$. This yielded all possibilities for the group $G$ together with the setwise stabilizer $G_{P_1}$. Next all possibilities for the point stabilizer $G_\alpha$ (an index 2 subgroup of $G_{P_1}$) had to be identified, and then all generalised orbital graphs for these permutation groups had to be examined to determine which of them were non-Cayley graphs. In this way a list of 56 pairwise non-isomorphic non-Cayley graphs of order 24 were constructed.

If $K = 1$ the group $G^{\mathcal{P}} \cong G$ is transitive of degree 12, but is not necessarily a minimal transitive group. Each group on Royle's list [80] of transitive groups of degree 12 was therefore a candidate for $G^{\mathcal{P}}$. Each of these was examined as a possibility for $G$, with $G_{P_1}$ a point stabilizer in the action of degree 12. The candidates for $G_\alpha$ were the subgroups of $G_{P_1}$ of index 2. These were all located and the corresponding transitive permutation groups of degree 24 were computed and their generalised orbital graphs constructed. The only non-Cayley graphs found were among the list of 56 graphs already constructed.

The fact that most of the vertex-transitive graphs of order at most 24 are Cayley graphs led McKay and Praeger [48] to make the following conjecture.

**Conjecture 5.3** *For* $n \geq 1$, *let* $\mathrm{vtr}(n)$ *and* $\mathrm{cay}(n)$ *denote the numbers of isomorphism types of vertex-transitive graphs and Cayley graphs, respectively, with order at most* $n$. *Then*

$$\lim_{n\to\infty} \frac{\mathrm{cay}(n)}{\mathrm{vtr}(n)} = 1.$$

# 6 Primitive groups and the O'Nan-Scott Theorem

For a proper discussion of our next problem, that of classifying finite distance transitive graphs, we need to understand the possible structures of finite primitive permutation groups. The type of information needed is contained in the so-called O'Nan-Scott Theorem (see [61] or [84]) which was proved independently by Michael O'Nan and Leonard Scott in 1980. It has become one of the most important and useful theorems for studying finite primitive permutation groups and their applications.

The O'Nan-Scott Theorem provides essentially an identification of several types of finite primitive permutation groups such that for each type we have additional information about either the abstract group theoretical structure, or the nature of the action, or both. The ordering and the presentation of the various types, and the amount of subdivision of the types preferred by those who wish to use this theorem varies according to the requirements of the different applications. For some requirements the nature of the socle or the minimal normal subgroups is most important, while (for example) for others it is the nature of the suborbits, or the existence of regular subgroups that is needed. (The *socle* $\mathrm{soc}(G)$ of a group $G$ is the product of the minimal normal subgroups of $G$.) For this exposition I have chosen a subdivision into types which was suggested by Laci Kovacs in 1985. It is a little finer than the subdivision used in [61] or [84] (but it is the subdivision used in [74]).

We shall define eight types of finite primitive permutation groups. The O'Nan-Scott Theorem states that every finite primitive permutation group belongs to exactly one of these types. Let $G$ be a finite primitive permutation group on $\Omega$ and let $\alpha \in \Omega$. Then $G$ has at most two minimal normal subgroups (see the first part of the proof in [61, Section 2]), and if there are two minimal normal subgroups then they are isomorphic and each is equal to the centraliser of the other. In particular, if $G$ has an abelian minimal normal subgroup $N$ then $N$ is the unique minimal normal subgroup of $G$, that is, $N$ is the socle of $G$. Further $N$ is elementary abelian and is regular on $\Omega$, and $G$ is the semidirect product $G = NG_\alpha$, for $\alpha \in \Omega$. Thus we may identify $\Omega$ with $N = \mathbb{Z}_p^d$, which may be viewed as a $d$-dimensional vector space over a field of prime order $p$, and choosing $\alpha$ as the zero vector we then have that $G_\alpha$ is an irreducible subgroup of nonsingular linear transformations of $N$. This is our first type, and is named HA since $G$ is contained in the Holomorph of the Abelian group $N$. The *holomorph* $\mathrm{Hol}(N)$ of a group $N$ is the semidirect product $N \cdot \mathrm{Aut}(N)$, where $\mathrm{Aut}(N)$ acts naturally on the normal subgroup $N$. In the case where $N$ is an elementary abelian group $\mathbb{Z}_p^d$, $\mathrm{Hol}(N)$ is the affine group $\mathrm{AGL}(d,p)$ of $N$ regarded as a finite affine space.

HA (holomorph of an abelian group): $\Omega = \mathbb{Z}_p^d$ for a prime $p$ and positive integer $d$ and $G$ is the semidirect product $G = N \cdot G_o$, a subgroup of the affine group $\mathrm{AGL}(d,p)$ on $\Omega$, where $N$ is the group of translations and $G_o$ is an irreducible subgroup of $\mathrm{GL}(d,p)$.

For all other types each minimal normal subgroup $N$ of $G$ is nonabelian, and hence $N = T_1 \times \ldots \times T_k$ for some positive integer $k$ where each $T_i \cong T$, a nonabelian simple group. If there is a second minimal normal subgroup $M \neq N$, then $M \cong N$, both $M$ and $N$ are regular on $\Omega$, and $\mathrm{soc}(G) = N \times M$. In this case $G$ is contained in the holomorph of $N$, and the second minimal normal subgroup $M = \{x\varphi_x^{-1} \mid x \in N\}$, where $\varphi_x$ is the *inner automorphism* of $N$ induced by $x$, that is, $\varphi_x : g \mapsto g^x$. This case is subdivided into two primitive types, namely the cases where $k = 1$ and $k > 1$. In the former case the type is named HS since $G$ is contained in the Holomorph of the Simple group $T$. The case where $k > 1$ is named HC since

$G$ is contained in the Holomorph of a "Compound" group $N$. The word compound is used here because $G$ is also a subgroup of a group of compound diagonal type, a type which will be described below. For a group $N$, and for $x \in N$ and $\sigma \in \mathrm{Aut}(N)$, we denote the image of $x$ under $\sigma$ by $x^\sigma$. Also the set $\{\varphi_x \mid x \in N\}$ of all inner automorphisms of $N$ forms the *inner automorphism group* $\mathrm{Inn}(N)$ of $N$.

HS (holomorph of a simple group) $\Omega = T$, a finite nonabelian simple group, and we have $T \cdot \mathrm{Inn}(T) \leq G \leq \mathrm{Hol}(T) = T \cdot \mathrm{Aut}(T)$, where for $x \in \Omega$, $t \in T$ and $\sigma \in \mathrm{Aut}(T)$, $t\sigma : x \mapsto x^\sigma t^\sigma$. If $\alpha = 1_T \in \Omega$, then $\mathrm{Inn}(T) \leq G_\alpha \leq \mathrm{Aut}(T)$.

HC (holomorph of a compound group) $\Omega = T^k$, where $k > 1$ and $T$ is a finite nonabelian simple group, and $N \cdot \mathrm{Inn}(N) \leq G \leq \mathrm{Hol}(N) = N \cdot \mathrm{Aut}(N)$. As for type HS, for $x, n \in N$ and $\sigma \in \mathrm{Aut}(N)$, $n\sigma : x \mapsto x^\sigma n^\sigma$. If $\alpha = 1_N \in \Omega$, then $\mathrm{Inn}(N) \leq G_\alpha \leq \mathrm{Aut}(N)$ and $G_\alpha$ acts transitively by conjugation on the simple direct factors of $N$.

There are five further primitive types and for each of these types, $N = \mathrm{soc}(G)$ is the unique minimal normal subgroup of $G$. These five types are the type AS, where $G$ is Almost Simple; type SD, where $G$ has a Simple Diagonal action on $\Omega$; type CD, where $G$ has a Compound Diagonal action on $\Omega$; type TW, where $G$ is a Twisted Wreath product with regular socle; and type PA, where $T$ induces a Product Action on a cartesian product structure on $\Omega$. We begin with the description of the type AS. A finite group $G$ is said to be *almost simple* if $T \leq G \leq \mathrm{Aut}(T)$ for some finite nonabelian simple group $T$.

AS (almost simple) $T \leq G \leq \mathrm{Aut}(T)$, where $T$ is a finite nonabelian simple group, and $G = TG_\alpha$ with $G_\alpha \not\geq T$; the stabilizer $G_\alpha$ must be maximal in $G$, and in particular $G_\alpha \neq \{1_G\}$.

The next type is the simple diagonal type SD. A primitive permutation group of type SD is a subgroup of the group

$$W = \{(a_1, \ldots, a_k) \cdot \pi \mid a_i \in \mathrm{Aut}(T), \pi \in S_k, a_i \equiv a_j \ (\mathrm{mod}\ \mathrm{Inn}(T)) \text{ for all } i, j\} \qquad (1)$$

where $\pi^{-1}(a_1, \ldots, a_k)\pi = (a_{1\pi^{-1}}, \ldots, a_{k\pi^{-1}})$. Further, $\mathrm{soc}(W)$ is $\{(t_1, \ldots, t_k) \mid t_i \in \mathrm{Inn}(T)\}$, and a primitive action of $W$ on $T^{k-1}$ (which we identify with $\mathrm{Inn}(T)^{k-1}$) is defined by

$$(a_1, \ldots, a_k) : (t_1, \ldots, t_{k-1}) \mapsto (a_k^{-1} t_1 a_1, \ldots, a_k^{-1} t_{k-1} a_{k-1}),$$

and

$$\pi : (t_1, \ldots, t_{k-1}) \mapsto (t_{k\pi^{-1}}^{-1} t_{1\pi^{-1}}, \ldots, t_{k\pi^{-1}}^{-1} t_{(k-1)\pi^{-1}}),$$

for $(a_1, \ldots, a_k)\pi \in W$, where $t_k = 1_T$. Thus for $\alpha = (1_T, \ldots, 1_T) \in T^{k-1}$, $W_\alpha = A \times S_k$ where $A = \{(a, \ldots, a) \mid a \in \mathrm{Aut}(T)\}$. The condition for a subgroup $G$ of $W$ which contains $\mathrm{soc}(W)$ to be primitive in this action is that $G$ permutes the simple direct factors of $N$ primitively.

SD (simple diagonal) $N = \mathrm{soc}(W) \leq G \leq W$, $\Omega = T^{k-1}$ with the action defined above, and $G$ acts primitively by conjugation on the $k$ simple direct factors of $N$, $(k > 1)$.

The name simple diagonal comes from the fact that $N_\alpha$ is the full diagonal subgroup $\{(t,\dots,t)\,|\,t\in \mathrm{Inn}(T)\}$ of $N$. For the compound diagonal type CD the group $G$ preserves a *product structure* on $\Omega$, that is $\Omega=\Delta^l$ for some $l\geq 2$ and $G\leq \mathrm{Sym}(\Delta)\wr S_l$ where, for $\delta=(\delta_1,\dots,\delta_l)\in\Omega$ and $(a_1,\dots,a_l)\pi\in \mathrm{Sym}(\Delta)\wr S_l$,

$$(a_1,\dots,a_l):\delta\mapsto(\delta_{1\pi^{-1}},\dots,\delta_{l\pi^{-1}})\text{ and }\pi:\delta\mapsto(\delta_1^{a_1},\dots,\delta_l^{a_l}).$$

Moreover, the subgroup of Sym($\Delta$) involved is a primitive group of type SD, hence the name Compound Diagonal.

CD (compound diagonal) $\Omega=\Delta^l$, and $N=T^k\leq G\leq H\wr S_l\leq \mathrm{Sym}(\Delta)\wr S_l$, for some divisor $l$ of $k$, where $l\geq 2$ and $k/l\geq 2$, and $H\leq \mathrm{Sym}(\Delta), \mathrm{soc}(H)=T^{k/l}$ and $H$ is primitive of type SD; $G$ acts transitively by conjugation on the simple direct factors of $N$.

The next type is the twisted wreath type TW. The original definition of a twisted wreath product was given by B. H. Neumann in [69]. The exposition here follows that in Suzuki's book [86, p. 269] and is the same as in [61]. Its use was suggested by Laci Kovacs. The *core* of a subgroup $H$ of a group $G$ is $\mathrm{core}_G(H):=\bigcap_{g\in G}H^g$. The *twisted wreath product* $T\mathrm{twr}_\varphi P$ of groups $T$ and $P$ relative to $\varphi$ may be defined as follows. Let $P$ have a transitive action on $\{1,\dots,k\}$ and let $Q$ be the stabilizer of the point 1 in this action. Suppose that there is a homomorphism $\varphi:Q\to \mathrm{Aut}(T)$ such that $\mathrm{core}_P(\varphi^{-1}(\mathrm{Inn}(T))=\{1_P\}$. Define

$$N:=\{f:P\to T\,|\,f(pq)=f(p)^{\varphi(q)}\text{ for all }p\in P,q\in Q\}.$$

Then $N$ is a group under pointwise multiplication and $N\cong T^k$. Let $P$ act on $N$ by

$$f^p(x):=f(px)\text{ for all }p,x\in P.$$

We define $T\mathrm{twr}_\varphi P$ to be the semidirect product of $N$ by $P$ relative to this conjugation action of $P$. Such a twisted wreath product $T\mathrm{twr}_\varphi P$ has a transitive action on $N$ such that $N$ acts regularly by right multiplication and for $f\in N$ and $p\in P$, $p:f\mapsto f^p$. Thus the stabilizer of the point $1_N$ in $T\mathrm{twr}_\varphi P$ is the subgroup $P$. The conditions on $N$ and $P$, for $P$ to be maximal in $T\mathrm{twr}_\varphi P$, are quite subtle (see [4, 50]) and imply that the image of $\varphi$ contains Inn($T$), and that $\mathrm{core}_P(Q)=1$.

TW (twisted wreath product) $G$ is the twisted wreath product $T\mathrm{twr}_\varphi P$ and $\Omega=N$ with the action as defined above; moreover, $\mathrm{core}_P(Q)=1$, $\varphi(Q)$ contains Inn($T$), and $\varphi$ cannot be extended to any larger subgroup of $P$.

For the final primitive type PA, $G$ preserves a product structure $\Delta^k$ on $\Omega$ and the subgroup of Sym($\Delta$) involved is primitive of type AS with socle $T$. Note that, for $\alpha=(\delta,\dots,\delta)\in\Omega$ where $\delta\in\Delta$, we have $N_\alpha=T_\delta^k\neq 1$.

PA (product action) $\Omega=\Delta^k$, $N=T^k\leq G\leq H\wr S_k\leq \mathrm{Sym}(\Delta)\wr S_k$, where $H$ is a primitive permutation group on $\Delta$ of type AS with socle $T$, and $G$ acts transitively by conjugation on the $k$ simple direct factors of $N$ $(k>1)$.

It is clear from the descriptions of these eight types that a finite primitive permutation group belongs to at most one of these types. The O'Nan-Scott Theorem asserts that the converse is also true.

**Theorem 6.1 ([61, 84])** *Each finite primitive permutation group is permutationally equivalent to a primitive group in exactly one of the primitive types* HA, HS, HC, AS, SD, CD, TW, PA.

# 7 Distance transitive graphs

The problem of classifying the finite distance transitive graphs is one for which the theoretical developments in group theory in the 1980's were almost perfectly suited to the graph theoretic application.

## 7.1 Definitions and examples

First we explain what these graphs are and give some examples. For vertices $\alpha$, $\beta$ of a graph $\Gamma = (\Omega, E)$, a *path* from $\alpha$ to $\beta$ is a finite sequence of vertices $\gamma_0, \ldots, \gamma_n$ such that $\gamma_0 = \alpha$, $\gamma_n = \beta$, and $\{\gamma_i, \gamma_{i+1}\} \in E$ for all $i$ with $0 \leq i < n$; and $n$ is called the *length* of the path. The smallest value for $n$ such that there is a path of length $n$ from $\alpha$ to $\beta$ is called the *distance* from $\alpha$ to $\beta$ and is denoted $d(\alpha, \beta)$. The *diameter* $d(\Gamma)$ of a connected graph $\Gamma$ is the maximum of $d(\alpha, \beta)$ over all $\alpha, \beta \in \Omega$. If $\Gamma$ is connected then, for each $i \in \{0, 1, \ldots, d(\Gamma)\}$, we set $\Gamma_i := \{(\alpha, \beta) \mid d(\alpha, \beta) = i\}$, the set of all pairs of vertices at distance $i$. Then, for $G \leq \mathrm{Aut}\Gamma$, we say that $\Gamma$ is a *G-distance transitive graph* if $G$ is transitive on each of the sets $\Gamma_0, \ldots, \Gamma_d$ where $d = d(\Gamma)$; $\Gamma$ is *distance transitive* if it is $(\mathrm{Aut}\Gamma)$-distance transitive. Notice that, for a $G$-distance transitive graph $\Gamma = (\Omega, E)$, the requirements that $G$ be transitive on $\Gamma_0$, and $\Gamma_1$ are equivalent to the conditions that $G$ be transitive on vertices and arcs, respectively. Also, the sets $\Gamma_i$ are the $G$-orbitals in $\Omega$.

It is easy to see that, for each $n$, the complete graph $K_n$ and the cycle $C_n$ on $n$ vertices are distance transitive, as is the complete bipartite graph $K_{n,n}$ on $2n$ vertices. (For $n \geq 3$, the *cycle* $C_n$ is the graph with vertex set $\mathbb{Z}_n$ and edges $\{i, i+1\}$ for each $i \in \mathbb{Z}_n$; and for $n \geq 2$, the *complete bipartite graph* $K_{n,n}$ is the graph with vertex set $\mathbb{Z}_n \times \mathbb{Z}_2$ such that $\{(i,j), (i',j')\}$ is an edge if and only if $j \neq j'$.) There are other interesting infinite familes of distance transitive graphs, and we describe a few of them below.

**Example 7.1** (a) Two of the families of finite distance transitive graphs have, as vertex set, the set $\Omega$ of $k$-element subsets of a set of size $n$, for some $k$, $1 \leq k < n$. For the first family, the *Johnson graphs* $J(n,k)$, two $k$-element subsets $\alpha, \beta$ are adjacent if and only if the intersection $\alpha \cap \beta$ has size $k-1$. The valency of $J(n,k)$ is $k(n-k)$, and $\mathrm{Aut}\, J(n,k) = S_n$. For the second family, the *Odd graphs* $O_{k+1}$, we have $n = 2k+1$, and two $k$-element subsets $\alpha, \beta$ are adjacent if and only if they are disjoint. The valency of $O_{k+1}$ is $k+1$ and $\mathrm{Aut}\, O_{k+1} = S_n$. (See [18, Section 9.1].)

(b) The *Hamming graph* $H(d,n)$ (where $d \geq 2, n \geq 2$) has vertex set $\mathbb{Z}_n^d$ and the vertices $\alpha$ and $\beta$ are joined by an edge if and only if $\alpha - \beta$ has exactly one non-zero entry. The valency of $H(d,n)$ is $d(n-1)$, and the diameter is $d$; its automorphism group is $S_n \wr S_d$, and it is vertex-primitive if and only if $n \geq 3$. When $n = 2$, the Hamming graph $H(d,2) = H(d)$, is often called the *d-cube graph*. (See [18, Section 9.2].)

For a vertex $\alpha$, and $i \leq d(\Gamma)$, define $\Gamma_i(\alpha) := \{\beta \mid d(\alpha, \beta) = i\}$, the set of vertices at distance $i$ from $\alpha$. The following elementary lemma follows immediately from our assertions in Section 2, since a graph $\Gamma$ is $G$-distance transitive if and only if the sets $\Gamma_i$ are the $G$-orbitals. It shows that, for a vertex-transitive graph, distance transitivity may be determined by examining the actions of the point stabilizer $G_\alpha$ on the sets $\Gamma_i(\alpha)$.

**Lemma 7.2** *Let $\Gamma = (\Omega, E)$ be a connected graph, let $\alpha \in \Omega$, and let $G \leq \mathrm{Aut}\Gamma$. Then $\Gamma$ is $G$-distance transitive if and only if $G$ is transitive on $\Omega$ and $G_\alpha$ is transitive on $\Gamma_i(\alpha)$ for all $i \in \{0, \ldots, d(\Gamma)\}$.*

Thus the sets $\Gamma_i(\alpha)$ are the $G_\alpha$-orbits in $\Omega$.

## 7.2 Parameters of finite distance transitive graphs

Let $\Gamma = (\Omega, E)$ be a finite distance-transitive graph, and define the *parameters* $k_i$, $a_i$, $b_i$ and $c_i$ of $\Gamma$ as follows. Let $\alpha \in \Omega$. For each $i = 0, \ldots, d(\Gamma)$, define $k_i := |\Gamma_i(\alpha)|$. For each $i$, choose a vertex $\beta \in \Gamma_i(\alpha)$, and let $a_i, b_i, c_i$ be the number of vertices joined to $\beta$ and lying in $\Gamma_i(\alpha)$, $\Gamma_{i+1}(\alpha)$ (if $i < d$), and $\Gamma_{i-1}(\alpha)$ (if $i > 0$), respectively. By distance transitivity these numbers are independent of the choices of the points $\alpha$ and $\beta$.

The parameters are subject to many restrictions (see for example [18, Section 4.1]). We mention here only a few of them. If both the valency and the diameter of $\Gamma$ are at least 3, then the parameters $k_i$, $i = 0, \ldots, d(\Gamma)$, have a *unimodal property* [78, Lemma 1.1], namely, there exist $i$ and $j$ such that $i + j \leq d(\Gamma)$,

$$k_1 < k_2 < \cdots < k_i = \cdots = k_{i+j}, \quad \text{and if } i + j < d(\Gamma) \text{ then} \quad k_{i+j} > \cdots > k_d.$$

Thus each positive integer can occur as a value of $k_i$ for at most two subscripts $i$, unless it is equal to the maximum of the $k_i$. It is also true that the valency of $\Gamma$, which by definition is equal to $k_1$, must be the smallest or the second smallest of the $k_i$.

## 7.3 Imprimitive distance transitive graphs

Now suppose that $G$ acts imprimitively on the vertex set $\Omega$ of $\Gamma$, that is, there is a nontrivial $G$-invariant partition $\mathcal{P} = \{P_1 \mid \ldots \mid P_r\}$ of $\Omega$. Let $\alpha, \beta \in P_1$ and set $i := d(\alpha, \beta)$. As each $g \in G_\alpha$ fixes $\alpha$, it must fix $P_1$ setwise since $P_1$ is a block of imprimitivity for $G$ in $\Omega$. Hence $\Gamma_i(\alpha) \subseteq P_1$. This observation was strengthened by D. H. Smith [85] to show that there are just two different kinds of $G$-invariant partitions possible, corresponding to bipartite and antipodal graphs. A graph $\Gamma$ is called *bipartite* if its vertex set may be partitioned into two sets, say $P_1, P_2$, such that, for $i = 1, 2$, no two vertices of $P_i$ are joined by an edge of $\Gamma$. A graph $\Gamma$ is said to be *antipodal* if its diameter $d(\Gamma) = d$ is greater than 1, and (the generalised $\mathrm{Aut}\Gamma$-orbital graph) $\Gamma_d$ is a disjoint union of complete graphs.

**Theorem 7.3 ([85])** *Let $\Gamma = (\Omega, E)$ be a finite $G$-distance transitive graph of valency greater than 2. Then $G$ is imprimitive on $\Omega$ if and only if $\Gamma$ is either bipartite or antipodal.*

Thus if $P_1$ is a nontrivial block of imprimitivity containing $\alpha$, then either $\Gamma$ is bipartite and $P_1 = \{\alpha\} \cup \Gamma_2(\alpha) \cup \Gamma_4(\alpha) \cup \ldots$, or $\Gamma$ is antipodal of diameter $d$ and $P_1 = \{\alpha\} \cup \Gamma_d(\alpha)$. Note that an imprimitive distance-transitive graph may be both bipartite and antipodal (for example the $d$-cube graph for all $d \geq 2$). In each of these two imprimitive cases it is possible to produce smaller distance transitive graphs. If the $G$-distance transitive graph $\Gamma$ is bipartite, then we may represent $\Gamma$ by the following diagram.

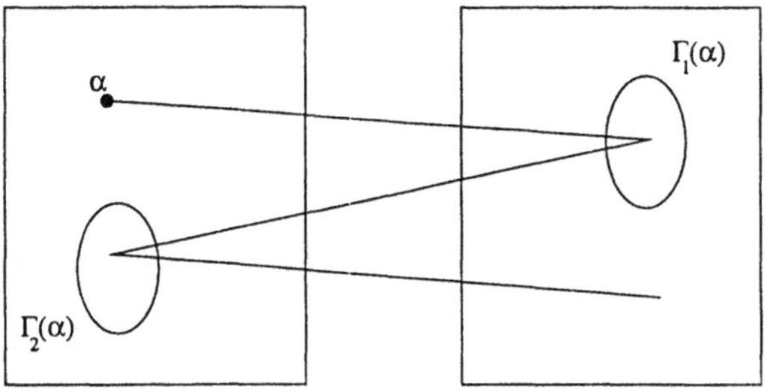


The quotient graph $\Gamma_{\mathcal{P}}$ of $\Gamma$ modulo the bipartition $\mathcal{P} = \{P_1, P_2\}$ is the complete graph $K_2$, and the induced subgraph $[P_1]$ is an empty graph. It turns out that it is helpful to consider the *distance-2 graph*, that is the orbital graph corresponding to the $G$-orbital $\Gamma_2$. If $\Gamma$ is connected and of diameter at least 2 then $\Gamma_2$ has two connected components, namely $P_1$ and $P_2$ and the graphs induced on these connected components are denoted by $\Gamma^+$ and $\Gamma^-$ and are called the *bipartite halves* of $\Gamma$. Smith [85] proved that $\Gamma^+$ and $\Gamma^-$ are isomorphic distance-transitive graphs. If $\Gamma$ is an antipodal $G$-distance transitive graph, then we can represent $\Gamma$ by the following diagram.

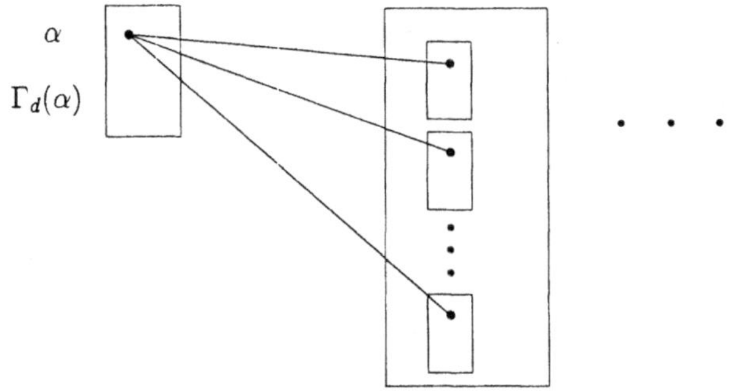


In this case Smith [85] proved that the *antipodal quotient* $\Gamma_{\mathcal{P}} = (\mathcal{P}, E_{\mathcal{P}})$ of $\Gamma$ relative to the $G$-invariant partition $\mathcal{P}$ generated by $\{\alpha\} \cup \Gamma_{d(\Gamma)}$ (called the *antipodal partition*) is a distance transitive graph. Moreover $\Gamma$ is a cover of $\Gamma_{\mathcal{P}}$. (For a graph $\Gamma = (\Omega, E)$ and a partition $\mathcal{P}$ of $\Omega$, we say that $\Gamma$ *covers* $\Gamma_{\mathcal{P}} = (\mathcal{P}, E_{\mathcal{P}})$, or $\Gamma$ is a *cover* of $\Gamma_{\mathcal{P}}$, if, whenever $\{P, P'\} \in E_{\mathcal{P}}$, each point of $P, P'$ is joined to exactly one point of $P', P$ respectively. In other words, the induced subgraph $[P \cup P']$ is a *perfect matching* between $P$ and $P'$.) The distance transitive graph $\Gamma$ is called an *antipodal cover* of $\Gamma_{\mathcal{P}}$.

Following Cameron's discussion in [21, p. 11], if $\Gamma$ is a finite bipartite distance transitive graph of valency greater than 2, then its bipartite half $\Gamma^+$ is not bipartite, and either $\Gamma^+$

is primitive, or $\Gamma^+$ is antipodal and the antipodal quotient of $\Gamma^+$ is primitive; on the other hand, if $\Gamma$ is a finite non-bipartite distance transitive graph, then its antipodal quotient $\Gamma_{\mathcal{P}}$ is neither antipodal nor bipartite, and hence is primitive. Thus by forming a sequence of antipodal quotients and/or bipartite halves of a given finite distance transitive graph $\Gamma$, we obtain a primitive distance transitive graph in at most three steps, and in at most two steps if we do the process carefully.

Thus the classification problem for finite distance transitive graphs may be divided into the problem of classifying the finite, vertex-primitive, distance transitive graphs, and the problem of reconstructing general finite distance transitive graphs from the primitive examples.

## 7.4 Primitive distance transitive graphs

The analysis of finite primitive distance transitive graphs using the O'Nan-Scott Theorem was begun by the author and Jan Saxl in 1983. It was this project which led to the statement and proof of the O'Nan-Scott Theorem given in [61]. The O'Nan-Scott Theorem identifies eight types of finite primitive permutation groups, and it turns out that only three of these types can occur as vertex-primitive groups $G$ acting distance transitively, namely the types HA, AS, and PA. Moreover all the primitive distance transitive graphs of type PA could be identified. In proving this result, the most unpleasant types to deal with were the types HS and SD, and the original rather ugly proof was superseded by a more pleasant combinatorial approach due to Kazuhiro Yokoyama.

It is possible to get a good understanding of a graph $\Gamma = (\Omega, E)$ admitting a vertex-primitive group $G$ of type HS or SD by considering the case where $G = T \times T$, for some non-abelian simple group $T$. We may identify the set $\Omega$ with $T$, and then the action of $G$ on $\Omega$ is given by $(x, y) : g \mapsto x^{-1}gy$. The stabilizer $G_\alpha$ of the point $\alpha = 1_T$ is $G_\alpha = \{(x, x) \mid x \in T\}$, and the $G_\alpha$-orbits in $\Omega$ are the conjugacy classes of $T$. Moreover each of the simple direct factors $T$ of $G$ acts regularly on $\Omega$, and hence by Lemma 5.2, $\Gamma$ is a Cayley graph $\mathrm{Cay}(T, S)$ for $T$, for some Cayley subset $S$ of $T$. Since $S$ is $G_\alpha$-invariant, it is a union of conjugacy classes of $T$. If now $\Gamma$ is $G$-distance transitive, the set $S$ must be a single conjugacy class of non-identity elements of $T$ by Lemma 7.2, and must be the smallest or second smallest of these (see Subsection 7.2). So the question was: Can such a Cayley subset give rise to a distance transitive Cayley graph? The negative answer involved a delicate interplay of group theory and combinatorics.

The result of this first analysis of finite primitive distance transitive graphs is the following theorem.

**Theorem 7.4 ([78])** *Let $\Gamma = (\Omega, E)$ be a graph and $G \le \mathrm{Aut}\Gamma$. If $\Gamma$ is $G$-distance transitive, and $G$ is primitive on $\Omega$, then one of the following holds:*

(i) *$\Gamma$ is a Hamming graph $H(d, n)$ or its complement $\overline{H(2, n)}$ (if $d = 2$), for some $n \ge 3$, and $G$ is of type* PA.

(ii) *$G$ is of type* AS.

(iii) *$G$ is of type* HA.

Thus to complete the classification of finite primitive distance transitive graphs, it is necessary to find all those admitting primitive groups of types HA and AS. Suppose that $\Gamma = (\Omega, E)$ is $G$-distance transitive with $G$ primitive on $\Omega$.

If $G$ is almost simple, then $T \leq G \leq \mathrm{Aut}(T)$, for some nonabelian simple group $T$. Consider the *permutation character* of $G$ on $\Omega$, that is the map $\pi : G \to \mathbf{C}$ defined by $\pi(g) = |\mathrm{fix}_\Omega g|$, the number of fixed points of $g$ in $\Omega$. It is a basic result from group representation theory (see, for example [42, Theorem 2.8]) that $\pi$ is the sum of so-called irreducible characters. Moreover, because $\Gamma$ is $G$-distance transitive, all of the $G$-orbitals are self-paired, and as a consequence (see [70, Theorem 8]) the irreducible characters occurring in $\pi$ are distinct. Since $\pi(1_G) = |\Omega|$, it follows that $|\Omega|$ is at most the sum of the degrees of all irreducible characters of $G$. (The *degree* of a character $\chi : G \to \mathbf{C}$ is the $\chi(1_G)$.) This places a strong restriction on the order of $\Gamma$. More importantly it means that $|G_\alpha| = |G|/|\Omega|$ has to be quite large relative to $|G|$, and since $G$ is primitive on $\Omega$, $G_\alpha$ is a maximal subgroup of $G$. In most cases, all maximal subgroups of $G$ of such a large size have been classified. Thus the possibilities for the permutation representation of $G$ on $\Omega$ are known, and $\Gamma$ must be an orbital graph for one of these permutation representations. In fact $\Gamma$ must correspond to one of the two smallest $G_\alpha$-orbits. Thus the strategy is, for each of these permutation representations, to look for the two smallest $G_\alpha$-orbits, which are the possibilities for $\Gamma_1(\alpha)$ and to examine these orbital graphs to decide whether they are distance transitive. By its very nature, this approach to a classification depends on the finite simple group classification. An excellent reference for a more detailed account of the current state of the finite distance transitive graph classification is the survey article of A. A. Ivanov [44]. We give a brief overview of the situation.

The strategy suggested above for classifying the primitive distance transitive graphs of type AS has been followed for some classes of almost simple groups $G$, and has led to a classification of $G$-vertex-primitive, $G$-distance transitive graphs for these groups. In the case where $\mathrm{soc}(G) = A_n$ with $n > 18$, the possibilities for the maximal subgroup $G_\alpha$ for which the associated permutation character is multiplicity free were determined by Saxl [83], and the corresponding distance transitive graphs, together with the examples in the small cases $5 \leq n \leq 18$ were classified by Liebeck, Saxl and the author in [60]. Independently the classification of distance transitive graphs with $\mathrm{soc}(G) = A_n$ (not just the primitive examples) was done by A. A. Ivanov [43]. Next, the primitive distance transitive graphs with $\mathrm{soc}(G) = \mathrm{PSL}(n, q)$ have been classified as a result of work by several groups, see [16, 15, 29, 41]. Those with $\mathrm{soc}(G)$ one of the other classical groups were studied by Inglis in his thesis [40] and a precise strategy for a classification for classical groups of dimension at least 13 was formulated (including a classification of the multiplicity free permutation representations for these groups). However the full details of such a classification for the classical groups has not yet been published, though a group including van Bon, Inglis and Saxl have made good progress on it. Similarly the classification for $\mathrm{soc}(G)$ an exceptional Lie type simple group is still an open problem. The examples for which $\mathrm{soc}(G)$ is one of the 26 sporadic simple groups have very recently been classified in a joint research effort involving heavy and creative use of computation by A. A. Ivanov, S. A. Linton, K. Lux, J. Saxl and L. Soicher [46].

**Problem 7.5** Complete the classification of the finite primitive distance transitive graphs of type AS.

The situation for the classification of the primitive distance transitive graphs of type HA is not so clear. We outline what seems to be the most helpful strategy, and refer the reader to [44] for more details. If $\Gamma = (\Omega, E)$ is $G$-distance transitive, with $G$ primitive on $\Omega$ of type

HA, then $G = NG_0$ with $N = \mathbb{Z}_p^n$ and $G_0$ an irreducible subgroup of $\mathrm{GL}(n,p)$, for some prime $p$ and integer $n \geq 1$. Further, $N$ is regular on $\Omega$ and so by Lemma 5.2, $\Gamma$ is isomorphic to the Cayley graph $\mathrm{Cay}(N, \Gamma_1(\alpha))$, for $\alpha$ the zero vector of the $n$-dimensional vector space $N$ over $\mathrm{GF}(p)$. Without loss of generality we may assume that $\Gamma$ is not a complete graph or a cycle, so $d(\Gamma) \geq 2$ and $k_1 = |\Gamma_1(\alpha)| \geq 3$. Moreover the classification of the examples having diameter 2 follows from Martin Liebeck's classification of the finite affine rank 3 permutation groups [59]. So we may assume that $d(\Gamma) \geq 3$. Choose $x \in \Gamma_1(\alpha)$, and suppose that the 1-dimensional space $\langle x \rangle$ over $\mathrm{GF}(p)$ generated by $x$ is not contained in $\Gamma_1(\alpha) \cup \{\alpha\}$. Then there exist $\lambda_1, \lambda_2 \in \mathrm{GF}(p)^{\#}$ such that $\lambda_1 x, \lambda_2 x \in \Gamma_1(\alpha)$, but $y := (\lambda_1 + \lambda_2)x \notin \Gamma_1(\alpha)$. It follows from the definition of a Cayley graph that $y \in \Gamma_2(\alpha)$, and so we find that $\Gamma_2(\alpha) = (\lambda_1 + \lambda_2)\Gamma_1(\alpha)$ and hence that $k_1 = |\Gamma_1(\alpha)| = |\Gamma_2(\alpha)| = k_2$. By an extension of the unimodality property discussed in Subsection 7.2, see [44, Lemma 2.3.2 (vi)], it follows that $\Gamma$ is an antipodal 2-fold covering of a complete graph and in particular is not vertex-primitive. This is a contradiction, and hence $\langle x \rangle \subseteq \Gamma_1(\alpha) \cup \{\alpha\}$. Since $\Gamma$ is connected, $\Gamma_1(\alpha)$ is a spanning set for $N$, and it follows that every element of $N$ can be expressed as a sum of at most $n$ elements of $\Gamma_1(\alpha)$. In other words, $\Gamma$ has diameter $d(\Gamma) \leq n$. Hence $G_0$ has at most $n$ orbits on the non-zero vectors of $N$, and it turns out that this is a strong restriction on $G_0$. In organising the possibilities for $G_0$, it is helpful also to use the Aschbacher classification of the subgroups of $\mathrm{GL}(n,p)$, see [3, 49]. Jon van Bon has been pursuing this approach to analysing the possibilities for $\Gamma$. However it is not clear how far he is towards a classification.

**Problem 7.6** Complete the classification of the finite primitive distance transitive graphs of type HA.

# 8 Classifying finite imprimitive distance transitive graphs

Several aspects of the problem of constructing all the finite distance transitive graphs from the primitive examples have received attention. Again we refer to [44] for an account of the status of this up to 1994, and make only a few remarks. J. Hemmeter, and A. Brouwer [19, 36, 37, 38] have classified the bipartite doublings of some of the infinite families of finite distance transitive graphs of large dimension, and Ivanov [44, Section 6.8] discusses this problem for the small diameter case. A general study of distance transitive antipodal covers $\Gamma$ of finite distance transitive graphs of valency $k$ was made by Tony Gardiner [30]: he showed that the antipodal blocks must have size at most $k$, and in the case of equality, either the graph $\Gamma$ is one of two well known examples, or its antipodal quotient is the complete bipartite graph $K_{k,k}$. In the latter case, he showed further in [31] that the examples were in 1–1 correspondence with a certain family of finite projective planes. This family of projective planes has been determined explicitly (see [22, 62]) and consists of the Desarguesian projective planes and a family of projective planes defined over the twisted fields of Albert. Van Bon and Brouwer [14] have classified the distance transitive antipodal covers of classical distance transitive graphs of sufficently large diameter (finding no previously unknown examples). On the other hand, the recent classification [45] of the distance transitive antipodal covers of the complete bipartite graphs $K_{k,k}$ produced several new examples. The distance transitive double covers of complete graphs are in 1–1 correspondence with the so-called doubly-transitive regular two-graphs, and these have recently been classified by Don Taylor [87]. A general investigation of distance transitive antipodal covers of complete graphs [34] has also produced some new families of examples.

# 9 $s$-arc transitive graphs

For a positive integer $s$, an $s$-*arc* in a graph $\Gamma = (\Omega, E)$ is an $(s+1)$-tuple $(\alpha_0, \dots, \alpha_s)$ of vertices with the property that $(\alpha_{i-1}, \alpha_i) \in E$ for $1 \leq i \leq s$ and $\alpha_{i-1} \neq \alpha_{i+1}$ for $1 \leq i \leq s-1$. Further, $\Gamma$ is said to be $(G,s)$-arc transitive if $G \leq \mathrm{Aut}\Gamma$ and $G$ is transitive on the $s$-arcs of $\Gamma$; also an $(\mathrm{Aut}\Gamma, s)$-arc transitive graph will be called simply $s$-*arc transitive.* A 1-arc is just an arc, and hence $\Gamma$ is $(G,1)$-arc transitive if and only if it is an orbital graph for $G$. Thus the property of being $s$-arc transitive is mainly of interest if $s \geq 2$. In this section we shall study the property of $s$-arc transitivity and the structure of finite $s$-arc transitive graphs, for $s \geq 2$.

## 9.1 Examples and Tutte's Theorem

Although $s$-arc transitivity ($s \geq 2$) seems a strong restriction on a finite graph, there are many families of examples known. We list a few of them.

**Example 9.1** (a) For $n \geq 3$, the cycle $C_n$ is $s$-arc transitive for all $s \geq 1$.

(b) For $n \geq 4$, the complete graph $\Gamma = K_n$ is 2-arc transitive, but not 3-arc transitive.

(c) For $n \geq 3$, the complete bipartite graph $\Gamma = K_{n,n}$ is 3-arc transitive, but not 4-arc transitive.

(d) Also, for $n \geq 3$, the Odd graph $\Gamma = O_n$ (defined in Example 7.1 (a)) is 3-arc transitive but not 4-arc transitive.

There are also examples known of 3-valent graphs which are 4-arc transitive but not 5-arc transitive, and 5-arc transitive but not 6-arc transitive, for example the Coxeter graph on 28 vertices (see [18, Section 12.3]) and the $(3,8)$-cage on 30 vertices (see [18, Section 6.9]). A remarkable theorem of Tutte (proved in 1947 and 1959) shows that for 3-valent graphs the parameter $s$ can be no larger than 5.

**Theorem 9.2 ([89, 90])** *If $\Gamma = (\Omega, E)$ is a $(G,s)$-arc transitive graph of valency 3, with $G$ transitive on $\Omega$, then $s \leq 5$.*

This surprising result raised the question of whether there might be an absolute bound on $s$ for finite vertex-transitive, $s$-arc transitive graphs of valencies greater than 3. This question received much attention in the literature and was ultimately answered in the affirmative by Richard Weiss in 1983. However, whereas Tutte's Theorem had a beautiful and elementary proof, the proof of Weiss's Theorem relies on the finite simple group classification. We continue our discussion of this below. An excellent survey of the status of the work in 1978, including a proof of Tutte's theorem, can be found in [91].

## 9.2 Vertex-transitivity, local 2-transitivity, and Weiss's Theorem

Although every 1-arc transitive graph is vertex-transitive this is not necessarily the case when $s \geq 2$. However every connected $s$-arc transitive graph which is not vertex-transitive, and which contains at least one $s$-arc, is a tree. (A *tree* is a graph which has no cycles of length greater than 2.)

**Theorem 9.3** *If $\Gamma = (\Omega, E)$ is a connected $(G,s)$-arc transitive graph containing at least one $s$-arc, for some positive integer $s$, then one of the following holds.*

(a) *$G$ is transitive on $\Omega$ and $\Gamma$ is $(G,i)$-arc transitive for each $i$ satisfying $1 \leq i \leq s$;*

(b) *$\Gamma$ is a tree of diameter $s$, and if $\alpha = (\alpha_0, \alpha_1, \ldots, \alpha_s)$ and $\beta = (\beta_0, \beta_1, \ldots, \beta_s)$ are $s$-arcs of $\Gamma$, then for some $\omega \in \{\beta_0, \beta_s\}$ and $\omega' \in \{\alpha_0, \alpha_s\}$, $d(\omega, \omega') = s$.*

**Proof** Set $v := |\Gamma_1(\alpha_0)|$. Suppose first that $v \geq 2$. We shall show that (a) holds. For each $\gamma \in \Gamma_1(\alpha_0) \setminus \{\alpha_1\}$, $(\gamma, \alpha_0, \alpha_1, \ldots, \alpha_{s-1})$ is an $s$-arc, and so there exists an element $g \in G$ which maps $\alpha$ to this $s$-arc, and hence $\alpha_0^g = \gamma$. Thus $|\Gamma_1(\gamma)| = |\Gamma_1(\alpha_0)| = v \geq 2$ for each such vertex $\gamma$. Repeating this argument, with $\gamma$ and $(\gamma, \alpha_0, \alpha_1, \ldots, \alpha_{s-1})$ replacing $\alpha_0$ and $\alpha$ respectively, we see that $|\Gamma_1(\omega)| = v$ for each vertex $\omega$ at distance at most 2 from $\alpha_0$. Continuing in this way, we deduce, since $\Gamma$ is connected, that $\Gamma$ is regular of valency $v$. Now choose $\gamma_0 \in \Omega$, $\gamma_0 \neq \alpha_0$. Since $\Gamma$ is connected there exists a path $\gamma = (\gamma_0, \gamma_1, \ldots, \gamma_t)$ from $\gamma_0$ to $\alpha_0$ (that is, $\gamma_t = \alpha_0$), and we choose such a path with $t$ as small as possible. Then $\gamma$ is a $t$-arc. If $t \geq s$ then $(\gamma_0, \gamma_1, \ldots, \gamma_s)$ is an $s$-arc, while if $t < s$ then, since $v \geq 2$, we may extend $\gamma$ to an $s$-arc. Since $\Gamma$ is $(G,s)$-arc transitive, some element of $G$ maps $\alpha$ to this $s$-arc, and hence maps $\alpha_0$ to $\gamma_0$. Thus $G$ is transitive on $\Omega$. Moreover, for each $i$ such that $1 \leq i < s$, each $i$-arc may be extended to an $s$-arc. Then a simlar argument shows that $\Gamma$ is $(G,i)$-arc transitive also.

Thus we may suppose that $v = 1$. Since $\Gamma$ is $(G,s)$-arc transitive, the first vertex $\beta_0$ of every $s$-arc satisfies $|\Gamma_1(\beta_0)| = 1$ also. We claim that $\Gamma$ is a tree. For if not, then some vertex $\omega$ lies in a cycle of $\Gamma$ of length at least 3, and hence occurs as the first vertex of an $s$-arc (winding around this cycle) while $|\Gamma_1(\omega)| \geq 2$. Thus $\Gamma$ is a tree. Since $\Gamma$ contains $s$-arcs, the diameter $d(\Gamma) \geq s$. However, if $d(\Gamma)$ were greater than $s$ then $\Gamma$ would contain a $d(\Gamma)$-arc $(\delta_0, \delta_1, \ldots, \delta_{d(\Gamma)})$ with $|\Gamma_1(\delta_0)| = 1$. Since $\Gamma$ is $(G,s)$-arc transitive, there exists $g \in G$ such that $(\delta_0, \delta_1, \ldots, \delta_s)^g = (\delta_1, \delta_2, \ldots, \delta_{s+1})$, which is a contradiction since $|\Gamma_1(\delta_1)| \geq 2$. Therefore, $d(\Gamma) = s$. Note that, since $\Gamma$ is a tree, for vertices $\omega, \omega'$, there is a unique path $p(\omega, \omega')$ with no repeated edges from $\omega$ to $\omega'$ (having length zero if $\omega = \omega'$).

Let $\beta = (\beta_0, \beta_1, \ldots, \beta_s)$ be an arbitrary $s$-arc. If, for each $\omega \in \{\beta_0, \beta_s\}$ and $\omega' \in \{\alpha_0, \alpha_s\}$, $p(\omega, \omega')$ had length strictly less than $s/2$, then we would have $d(\alpha_0, \alpha_s) \leq d(\alpha_0, \beta_0) + d(\beta_0, \alpha_s) < s$, which is not the case. Thus we may assume without loss of generality that $p(\beta_0, \alpha_0)$ has length $d_0 = d(\beta_0, \alpha_0) \geq s/2$. Of course $d_0 \leq d(\Gamma) = s$. Let $i$ be the greatest integer such that $0 \leq i \leq s$ and $\alpha_i$ belongs to $p(\beta_0, \alpha_0)$. Then $d(\beta_0, \alpha_i) = d_0 - i$. Also since $s = d(\beta_0, \beta_s) \leq d(\beta_0, \alpha_i) + d(\alpha_i, \beta_s)$, it follows that $d(\alpha_i, \beta_s) \geq s - d_0 + i$, with equality if and only if $\alpha_i$ belongs to the $s$-arc $\beta$. Suppose that $\alpha_i$ does not belong to $\beta$. Let $j$ be the greatest integer such that $\beta_j$ belongs to $p(\beta_0, \alpha_0)$. Then $\beta_j$ lies on the path $p(\beta_0, \alpha_i)$ and $j < d_0 - i$, see the diagram below.

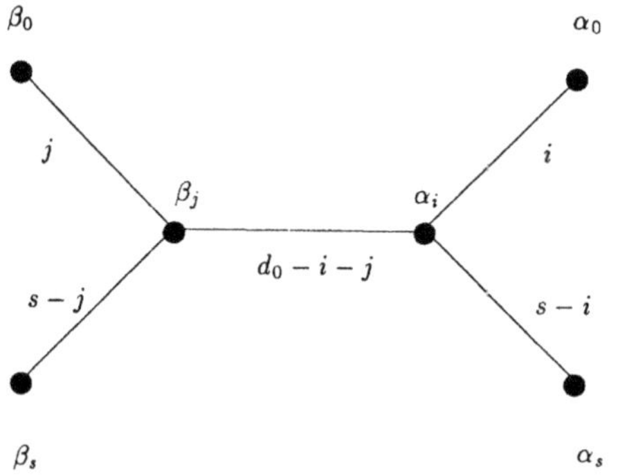


By the maximality of $i$ and $j$, the shortest path from $\beta_s$ to $\alpha_s$ runs first along part of the $(s-j)$-arc $\beta^- := (\beta_s, \ldots, \beta_0)$ from $\beta_s$ to $\beta_j$, then along part of $p(\beta_0, \alpha_0)$ from $\beta_j$ to $\alpha_i$, and finally along part of $\alpha$ from $\alpha_i$ to $\alpha_s$. Thus

$$s \geq d(\beta_s, \alpha_s) = (s-j) + (d_0 - i - j) + (s-i) = 2s + d_0 - 2(i+j),$$

whence $i + j \geq (s + d_0)/2$. However we have also shown that $i + j < d_0$, and it follows that $d_0 > s$ which is a contradiction. Hence $\alpha_i$ belongs to $\beta$ and $d(\beta_s, \alpha_i) = s - d_0 + i$. With the same meaning for $j$ as above, we see that the point $\beta_j$ must lie on the path $p(\alpha_i, \alpha_0)$, so $\beta_j = \alpha_t$ for some $t$ such that $0 \leq t \leq i$. So the shortest path from $\beta_s$ to $\alpha_s$ runs first along $\beta^-$ from $\beta_s$ to $\beta_j = \alpha_t$, then along $\alpha$ from $\alpha_t$ through $\alpha_i$ to $\alpha_s$. Thus $s \geq d(\beta_s, \alpha_s) = (s-j) + (s-t)$, whence $j + t \geq s$. However $d_0 = d(\beta_0, \alpha_0) = d(\beta_0, \beta_j) + d(\alpha_t, \alpha_0) = j + t$ (since $\beta_j = \alpha_t$), and $d_0 \leq s$. It follows that $d_0 = s$, as required. □

We shall be concerned with finite $G$-vertex-transitive, $(G, s)$-arc transitive graphs of valency at least 3, with $s \geq 2$. By Theorem 9.3, such graphs are $(G, 2)$-arc transitive, and this property for vertex-transitive graphs is equivalent to a condition on the vertex stabiliser $G_\alpha$, a so-called "local condition" on $\Gamma$. For a group-theoretic property $\mathcal{Q}$, we shall say that $\Gamma$ is *$G$-locally $\mathcal{Q}$* if the permutation group $G_\alpha^{\Gamma_1(\alpha)}$ has property $\mathcal{Q}$.

**Lemma 9.4** *Suppose that the graph $\Gamma = (\Omega, E)$ is $G$-vertex-transitive and let $\alpha \in \Omega$. Then $\Gamma$ is $(G,2)$-arc transitive if and only if $G_\alpha^{\Gamma_1(\alpha)}$ is 2-transitive on $\Gamma_1(\alpha)$.*

**Proof** Suppose that $\Gamma$ is $(G,2)$-arc transitive. Let $(\beta_1, \beta_2), (\gamma_1, \gamma_2)$ be pairs of distinct points of $\Gamma_1(\alpha)$. Then $(\beta_1, \alpha, \beta_2)$ and $(\gamma_1, \alpha, \gamma_2)$ are 2-arcs of $\Gamma$. Thus there exists $g \in G$ such that $(\beta_1, \alpha, \beta_2)^g = (\gamma_1, \alpha, \gamma_2)$. In particular, $g \in G_\alpha$ and $(\beta_1, \beta_2)^g = (\gamma_1, \gamma_2)$. So $G_\alpha^{\Gamma_1(\alpha)}$ is 2-transitive.

Conversely, suppose that for some, and hence for every, $\alpha \in \Omega$, the permutation group $G_\alpha^{\Gamma_1(\alpha)}$ is 2-transitive. Let $(\alpha_1, \alpha_2, \alpha_3)$ and $(\beta_1, \beta_2, \beta_3)$ be 2-arcs of $\Gamma$. Since $G$ is transitive on $\Omega$, there exists $g \in G$ such that $\alpha_2^g = \beta_2$. Then $\alpha_1^g, \alpha_3^g$ are distinct points of $\Gamma_1(\beta_2)$, as are $\beta_1, \beta_3$. Since $G_{\beta_2}$ is 2-transitive on $\Gamma_1(\beta_2)$, there exists $h \in G_{\beta_2}$ such that $(\alpha_1^g, \alpha_3^g)^h = (\beta_1, \beta_3)$. Then $gh \in G$ and $gh$ maps $(\alpha_1, \alpha_2, \alpha_3)$ to $(\beta_1, \beta_2, \beta_3)$, whence $\Gamma$ is $(G,2)$-arc transitive. □

It follows from the finite simple group classification that all finite 2-transitive permutation groups are known, see [20]. Thus all possibilities for the group $G_\alpha^{\Gamma_1(\alpha)}$ with $|\Gamma_1(\alpha)| \geq 3$ are known. It was proved in a series of papers that, for each of the possibilities for $G_\alpha^{\Gamma_1(\alpha)}$, the graph $\Gamma$ could not be as much as $(G, 8)$-arc transitive. The final steps of the proof were completed by Richard Weiss [92] in 1983 (and references to the other papers contributing to the proof can be found in [92]).

**Theorem 9.5 (Weiss)** *Suppose that* $\Gamma = (\Omega, E)$ *is a finite* $G$*-vertex-transitive,* $(G, s)$*-arc transitive graph of valency at least* 3*. Then* $s \leq 7$*.*

## 9.3 Quotients of finite 2-arc transitive graphs

In Section 4 we defined the quotient $\Gamma_{\mathcal{P}} = (\mathcal{P}, E_{\mathcal{P}})$ of a graph $\Gamma = (\Omega, E)$ relative to a partition $\mathcal{P}$ of its vertex set $\Omega$. In the special case where $\Gamma$ is a cover of $\Gamma_{\mathcal{P}}$ (as defined in Section 7.3), $\Gamma$ and $\Gamma_{\mathcal{P}}$ have very similar local properties. In particular, if $\Gamma$ is regular, then also $\Gamma_{\mathcal{P}}$ is regular, and $\mathrm{val}(\Gamma) = \mathrm{val}(\Gamma_{\mathcal{P}})$. In this subsection we begin a discussion of the role of covers in describing the structure of finite 2-arc transitive graphs.

Although 2-arc transitivity is a very strong group theoretic condition for the automorphism group of a graph to satisfy, the family of finite, connected, vertex-transitive, 2-arc transitive graphs is very large. Laci Babai showed in [6] that every finite regular graph has a finite 2-arc transitive cover. Thus the family of all finite regular graphs is contained in the class of all quotient graphs of finite 2-arc transitive graphs; and we note that for most regular graphs, their only automorphism is the identity map. This observation of Babai suggests that the family of finite 2-arc transitive graphs might be a rather uncontrollable class for which no sensible organisational principle could be expected.

However, in 1985, a preliminary study was conducted in [73] of certain families of arc-transitive graphs with extra group theoretic defining properties, one of these properties being 2-arc transitivity. The philosophy underlying [73] was that, in order to understand the structure of typical graphs in such a family $\mathcal{F}$, the only quotients of graphs from $\mathcal{F}$ which should be studied were those relative to a vertex-partition invariant under a group which witnessed membership of the graph in $\mathcal{F}$. For a graph $\Gamma = (\Omega, E)$ and a given subgroup $G$ of $\mathrm{Aut}\Gamma$, most partitions $\mathcal{P}$ of $\Omega$ will neither be invariant under $G$, nor have the property that $\Gamma$ is a cover of $\Gamma_{\mathcal{P}}$. Babai's approach was to consider partitions $\mathcal{P}$ with the latter property, while the approach in [73] was to consider partitions with the former property. However, although the quotient $\Gamma_{\mathcal{P}}$ of a $(G, 2)$-arc transitive, $G$-vertex-transitive graph $\Gamma$ relative to a $G$-invariant partition $\mathcal{P}$ is $(G^{\mathcal{P}}, 1)$-arc transitive, it is in general far from being $(G^{\mathcal{P}}, 2)$-arc transitive. It turned out that, by choosing a rather natural sub-class of $G$-invariant partitions, it could be guaranteed that for such partitions $\mathcal{P}$ the quotient $\Gamma_{\mathcal{P}}$ was $(G^{\mathcal{P}}, 2)$-arc transitive, and also that $\Gamma$ was a cover of $\Gamma_{\mathcal{P}}$. These partitions are called normal partitions and will be discussed in the next section.

The assertion that a quotient of a $(G, 2)$-arc transitive graph relative to a $G$-invariant partition might (by chance) be, but certainly need not be 2-arc transitive, can be illustrated effectively by the case where $G = \mathrm{Sz}(q)$, a Suzuki simple group. The $(\mathrm{Sz}(q), 2)$-arc transitive graphs were classified by Fang and the author in [26], and it turns out that there are essentially two kinds of such graphs.

**Theorem 9.6 ([26])** *Let $G = \mathrm{Sz}(q)$ where $q = 2^c$ for some odd integer $c \geq 3$, and suppose that $\Gamma = (\Omega, E)$ is a $(G,2)$-arc transitive graph. Then, for $\alpha \in \Omega$, one of the following holds, and in each case examples of such graphs exist.*

(a) *$\Gamma$ has valency $2^a$, for some divisor $a$ of $c$ with $a > 1$, and $G_\alpha \cong \mathbb{Z}_2^a \rtimes \mathbb{Z}_{2^a-1}$;*

(b) *$\Gamma$ has valency 5, and $G_\alpha \cong \mathbb{Z}_5 \rtimes \mathbb{Z}_4$.*

Each of the graphs $\Gamma(2^a) = (\Omega, E)$ occurring in part (a) has a quotient $\Gamma(2^a)_{\mathcal{P}}$ (relative to a certain $\mathrm{Sz}(q)$-invariant partition $\mathcal{P}$ of $\Omega$) which is a complete graph $K_{q^2+1}$ and on which $\mathrm{Sz}(q)$ acts 2-transitively on vertices: hence $\mathrm{Sz}(q)$ is primitive on vertices and transitive on 2-arcs of this quotient $\Gamma(2^a)_{\mathcal{P}}$. Moreover, if $a$ divides $a'$ which divides $c$, then $\Gamma(2^{a'})$ is a quotient of $\Gamma(2^a)$, and $\Gamma(2^c)$ is a quotient of both $\Gamma(2^a)$ and $\Gamma(2^{a'})$, all relative to $\mathrm{Sz}(q)$-invariant partitions. The situation for the graph $\Gamma(5)$ in part (b) is quite different. The graph $\Gamma(5) = (\Omega, E)$ has a unique vertex-primitive quotient $\Gamma(5)_{\mathcal{P}}$ relative to a $\mathrm{Sz}(q)$-invariant partition $\mathcal{P}$ of $\Omega$, and $\mathrm{Sz}(q)$ is far from transitive on its 2-arcs. In fact this quotient is not even $\mathrm{Sz}(q)$-locally primitive if $q > 8$. Also in none of these examples is the original graph a cover of the quotient.

## 10 Normal quotients of finite locally primitive graphs

For a transitive permutation group $G$ on a set $\Omega$, a partition $\mathcal{P}$ of $\Omega$ will be called *$G$-normal* if $\mathcal{P}$ is the set of $N$-orbits in $\Omega$ for some normal subgroup $N$ of $G$. It is not difficult to see that all $G$-normal partitions of $\Omega$ are $G$-invariant.

**Lemma 10.1** *Let $G$ be a permutation group on a set $\Omega$, and let $N$ be a normal subgroup of $G$. Then the set $\mathcal{P}_N$ of $N$-orbits in $\Omega$ is a $G$-invariant partition of $\Omega$.*

**Proof** By definition $\mathcal{P}_N$ is a partition on $\Omega$. We need to show that it is $G$-invariant. Let $P \in \mathcal{P}_N$ and $g \in G$. Suppose that $P = \alpha^N$, the $N$-orbit containing $\alpha$. Then $P = \{\alpha^n \,|\, n \in N\}$, and so

$$P^g = \{\alpha^{ng} \,|\, n \in N\} = \{(\alpha^g)^{g^{-1}ng} \,|\, n \in N\}.$$

Since $N$ is normal in $G$, each element $g^{-1}ng$ lies in $N$, and moreover $g^{-1}ng$ runs over all elements of $N$ as $n$ does. Thus $P^g = \{(\alpha^g)^n \,|\, n \in N\}$, that is $P^g$ is the $N$-orbit containing $\alpha^g$. □

The set $\{\{\alpha\} | \alpha \in \Omega\}$ of orbits of the identity subgroup, and the set $\{\Omega\}$ consisting of the single $G$-orbit $\Omega$, are both $G$-normal partitions (corresponding to the identity subgroup and to $G$), and they are called *trivial* $G$-normal partitions. All other $G$-normal partitions of $\Omega$ (if any) are said to be *nontrivial*. It may be the case that the only $G$-normal partitions are the trivial ones, and in this case $G$ is said to be *quasiprimitive* on $\Omega$. From this definition it is clear that every primitive permutation group is quasiprimitive, but there are many quasiprimitive permutation groups which are not primitive. For example, any transitive permutation representation of a nonabelian simple group for which a point stabiliser is a non-maximal subgroup determines a quasiprimitive permutation group which is not primitive.

Suppose now that $\Gamma = (\Omega, E)$ is a finite $G$-vertex-transitive graph, for some subgroup $G$ of $\mathrm{Aut}\Gamma$. Let $N$ be a normal subgroup of $G$ which is not transitive on $\Omega$, and let $\mathcal{P}_N$ denote the set of $N$-orbits in $\Omega$ and $\Gamma_N$ the corresponding quotient graph $\Gamma_{\mathcal{P}_N}$ relative to the $G$-normal partition $\mathcal{P}_N$. Then $\Gamma_N$ is called a *G-normal quotient graph* of $\Gamma$.

We shall consider normal quotients of graphs belonging to a larger family of finite vertex-transitive graphs than the family of finite vertex-transitive, 2-arc transitive graphs, namely we shall consider the family of finite vertex-transitive, locally primitive graphs. Recall from Section 9.2 that a $G$-vertex-transitive graph $\Gamma = (\Omega, E)$ is said to be *G-locally primitive* if, for $\alpha \in \Omega$, $G_\alpha$ is primitive on $\Gamma_1(\alpha)$. By Lemma 9.4, if $\Gamma$ is $(G, 2)$-arc transitive and $G$-vertex-transitive, then $\Gamma$ is $G$-locally 2-transitive. Since every 2-transitive permutation group is primitive, it follows that such a graph $\Gamma$ is $G$-locally primitive.

Now let $\Gamma = (\Omega, E)$ be a connected $G$-locally primitive graph, for some vertex-transitive subgroup $G$ of Aut$\Gamma$, and let $N$ be a normal subgroup of $G$ which is intransitive on $\Omega$. It was shown in [73] that, provided $|\mathcal{P}_N| > 2$, $\Gamma_N$ is $G^{\mathcal{P}_N}$-locally primitive and $G^{\mathcal{P}_N}$-vertex-transitive, $\Gamma$ is a cover of $\Gamma_N$, and $G^{\mathcal{P}_N} = G/N$ with $N$ semiregular on $\Omega$. (A permutation group $N$ on $\Omega$ is *semiregular* if the only element of $N$ which fixes a point of $\Omega$ is the identity element.)

**Theorem 10.2** *Let $\Gamma = (\Omega, E)$ be a connected $G$-vertex-transitive and $G$-locally primitive graph of valency $v$, and let $N$ be a normal subgroup of $G$. Then one of the following holds.*

(a) *$N$ is transitive on $\Omega$; or*

(b) *$\Gamma$ is bipartite, the $N$-orbits in $\Omega$ are the two parts of the bipartition, and the graph $\Gamma_N$ is the complete graph $K_2$; or*

(c) *$\Gamma_N = (\mathcal{P}_N, E_N)$ is a connected $G^{\mathcal{P}_N}$- vertex-transitive and $G^{\mathcal{P}_N}$-locally primitive graph of valency $v$, and $\Gamma$ is a cover of $\Gamma_N$. The subgroup $N$ is semiregular on $\Omega$, and $G^{\mathcal{P}_N} \cong G/N$. Moreover if $\Gamma$ is $(G, 2)$-arc transitive then $\Gamma_N$ is $(G^{\mathcal{P}_N}, 2)$-arc transitive.*

**Proof** Suppose that $|\mathcal{P}_N| \geq 2$. We have already observed that $\Gamma_N$ is connected and $G^{\mathcal{P}_N}$-vertex-transitive. Let $\alpha$ be a vertex in a part $P \in \mathcal{P}_N$ such that some vertex $\beta \in \Gamma_1(\alpha)$ belongs to a part $P'$ of $\mathcal{P}_N$ different from $P$. It is easy to check that $\Gamma_1(\alpha) \cap P'$ is a block of imprimitivity for the permutation group $G_\alpha^{\Gamma_1(\alpha)}$. Since this group is primitive, $\Gamma_1(\alpha) \cap P'$ is a trivial block of imprimitivity, and hence is equal either to $\{\beta\}$ or to the whole set $\Gamma_1(\alpha)$. Suppose the latter holds so that $\Gamma_1(\alpha) \subseteq P'$. Since $G$ is transitive on arcs, there is an element $g \in G$ which maps the arc $(\alpha, \beta)$ to $(\beta, \alpha)$, and hence interchanges $\alpha$ and $\beta$. This element $g$ must interchange the parts $P$ and $P'$ containing $\alpha$ and $\beta$, and also $g$ must interchange $\Gamma_1(\alpha)$ and $\Gamma_1(\beta)$. It follows that $\Gamma_1(\beta) \subseteq P$, and hence that $P$ contains all vertices which are at distance 2 from $\alpha$. Continuing this argument, and using the fact that $\Gamma$ is connected, we see that all vertices of $\Gamma$ must lie in either $P$ or $P'$, and hence that part (b) holds.

Thus we may assume that $\Gamma_1(\alpha) \cap P' = \{\beta\}$. This means that the $v$ vertices adjacent to $\alpha$ lie in exactly $v$ distinct parts of $\mathcal{P}_N$, say $P_1 = P', P_2, \ldots, P_v$. Since $P$ and each of the parts $P_i$ are $N$-orbits, it follows that each vertex of $P$ is adjacent to exactly one vertex of each of the $P_i$. Similarly, for each $i$, each vertex of $P_i$ is adjacent to exactly one vertex of $P$. Thus the subgraph induced on $P \cup P_i$ is a complete matching, and since $G$ is transitive on the arcs of $\Gamma$ it follows that this assertion holds for every pair of parts adjacent in $\Gamma_N$. Thus $\Gamma$ is a cover of $\Gamma_N$. Now the setwise stabiliser $G_P$ of $P$ is equal to $NG_\alpha$ (since $NG_\alpha$ clearly fixes $P$ setwise, contains $G_\alpha$ and is transitive on $P$), and it is not difficult to check that the permutation group $G_P^{\Gamma_N(P)}$ induced by $G_P$ on the $v$ parts of $\Gamma_N$ adjacent to $P$ is permutationally equivalent to the permutation group induced by $G_\alpha$ on $\Gamma_1(\alpha)$. In particular $G_P^{\Gamma_N(P)}$ is primitive, and is even 2-transitive if $G_\alpha^{\Gamma_1(\alpha)}$ is 2-transitive. Thus $\Gamma_N$ is $G^{\mathcal{P}_N}$-locally primitive and, if $\Gamma$ is $(G,2)$-arc transitive, then $\Gamma_N$ is $(G^{\mathcal{P}_N},2)$-arc transitive.

Finally we show that $N$ is semiregular and $G^{\mathcal{P}_N} = G/N$. Let $K$ be the kernel of the action of $G$ on $\mathcal{P}_N$. Then $N \leq K$. Suppose that $K_\gamma \neq 1$ for some vertex $\gamma$. Conjugating $K_\gamma$ by an element of $G$ which maps $\gamma$ to $\alpha$ we have that $K_\alpha \neq 1$. Consider $K_\alpha^{\Gamma_1(\alpha)}$. Since $K_\alpha$ fixes each part of $\mathcal{P}_N$ setwise, and since distinct vertices of $\Gamma_1(\alpha)$ lie in distinct parts of $\mathcal{P}_N$, it follows that $K_\alpha$ fixes $\Gamma_1(\alpha)$ pointwise. Similarly, for each $\beta \in \Gamma_1(\alpha)$, we have that $K_\beta$ fixes $\Gamma_1(\beta)$ pointwise, and it follows that $K_\alpha$ fixes all points at distance 2 from $\alpha$. Continuing this argument, we see, since $\Gamma$ is connected, that $K_\alpha$ fixes each vertex of $\Gamma$, contradicting the assumption that $K_\alpha \neq 1$. Thus $K$ is semiregular, and hence its subgroup $N$ is also semiregular. Moreover, for each $g \in K$, $\alpha^g \in \alpha^N$ (since $K$ fixes $\alpha^N$ setwise). Hence $\alpha^g = \alpha^n$ for some $n \in N$, and so $gn^{-1} \in K \cap G_\alpha = K_\alpha = 1$. Therefore $K = N$ and $G^{\mathcal{P}_N} = G/N$. □

**Remark 10.3** It was shown in the proof that, in Theorem 10.2 (c), the permutation groups $G_\alpha^{\Gamma(\alpha)}$ and $G_P^{\Gamma_N(P)}$ are permutationally equivalent, where $P = \alpha^N$. Further we did not assume, and the proof did not use, finiteness of $\Gamma$.

Suppose now that $\Gamma = (\Omega, E)$ is a finite graph as in Theorem 10.2, and that $\Gamma$ is not bipartite. Thus, choosing $N$ to be maximal by inclusion subject to being intransitive on $\Omega$, we find that $G^{\mathcal{P}_N} = G/N$ is quasiprimitive on $\mathcal{P}_N$ in addition to $\Gamma_N$ being $G^{\mathcal{P}_N}$-locally primitive. We say that a graph $\Gamma$ is a *quasiprimitive, locally primitive graph* if there is a subgroup $G \leq \mathrm{Aut}\Gamma$ such that $\Gamma$ is $G$-locally primitive, and also $G$ is quasiprimitive on vertices. Similarly, $\Gamma$ is a *quasiprimitive, 2-arc transitive graph* if there is a subgroup $G \leq \mathrm{Aut}\Gamma$ such that $\Gamma$ is $(G,2)$-arc transitive, and also $G$ is quasiprimitive on vertices. Thus we have proved the following result.

**Theorem 10.4** *Let $\Gamma = (\Omega, E)$ be a finite connected $G$-vertex-transitive and $G$-locally primitive graph which is not bipartite. Then there is a semiregular normal subgroup $N$ of $G$ which is intransitive on $\Omega$ (possibly $N = 1$), such that $\Gamma$ is a cover of $\Gamma_N = (\mathcal{P}_N, E_N)$, $G^{\mathcal{P}_N} \cong G/N$, $G^{\mathcal{P}_N}$ is quasiprimitive on $\mathcal{P}_N$, and $\Gamma_N$ is $G^{\mathcal{P}_N}$-locally primitive. Thus $\Gamma_N$ is a quasiprimitive, locally primitive normal quotient of $\Gamma$ which is covered by $\Gamma$. Moreover if $\Gamma$ is $(G,2)$-arc transitive then $\Gamma_N$ a quasiprimitive, 2-arc transitive normal quotient of $\Gamma$.*

**Remark 10.5** (a) Similar assertions may be made about the existence of special kinds of vertex-transitive, locally primitive quotients covered by the finite bipartite graphs satisfying the conditions of Theorem 10.2 (by choosing normal subgroups of the given group which are maximal with respect to having more than two orbits on vertices, see [76]). However, the structure for such graphs is not as easily described as that for the non-bipartite examples.

(b) Theorem 10.4 shows that each finite, non-bipartite, vertex-transitive, locally primitive graph $\Gamma$ is a cover of at least one of its quasiprimitive, locally primitive, normal quotients. If in addition $\Gamma$ is 2-arc transitive, then $\Gamma$ is a cover of at least one of its quasiprimitive, 2-arc transitive, normal quotients. Thus there is, after all, a strong organisational principle underlying the structure of finite, vertex-transitive, 2-arc transitive graphs. To understand the effectiveness of these observations, we need information about the structure of finite quasiprimitive permutation groups. At the time when Theorems 10.2 and 10.4 were proved, no information of this kind was available in the literature. However, the wish to explore the structure of finite 2-arc transitive graphs stimulated an investigation of finite quasiprimitive groups leading to a theorem akin to the O'Nan-Scott Theorem. We discuss this theorem in Section 12, and its application to 2-arc transitive graphs in Section 13.

(c) The quasiprimitive, locally primitive normal quotient $\Gamma_N$ obtained in Theorem 10.4 is not in general determined uniquely (even up to isomorphism) by $\Gamma$ and $G$. In fact it is possible for there to be an arbitrarily large number of pairwise non-isomorphic quasiprimitive 2-arc transitive normal quotients of a finite 2-arc transitive graph, as Example 11.2 in the next section demonstrates.

Before moving to the next section, we mention the following conjecture of Richard Weiss.

**Conjecture 10.6 (Weiss [91])**There exists a function $f$ on the natural numbers $\mathbf{N}$ such that if $\Gamma = (\Omega, E)$ is a finite, $G$-vertex-transitive, $G$-locally primitive graph of valency $v$ then, for $\alpha \in \Omega$, $|G_\alpha| \leq f(v)$.

For non-bipartite graphs $\Gamma$, it follows from Theorem 10.4 (see [73]) that Conjecture 10.6 is true if it is true for finite, quasiprimitive, locally primitive graphs.

## 11 Sabidussi's construction of arc-transitive graphs

Let $\Gamma = (\Omega, E)$ be a $G$-arc transitive graph. Then by Theorem 2.1, $\Gamma$ is an orbital graph for the transitive permutation group $G$ on $\Omega$. In [82] Gert Sabidussi introduced a way of identifying the self-paired orbital involved, and a group theoretic method of constructing an isomorphic copy of $\Gamma$ which has proved useful in classification problems.

Let $\alpha$ and $\beta$ be adjacent vertices of $\Gamma$, and let $H := G_\alpha$, the stabilizer of $\alpha$. Then the subgroup $H$ is *core-free* in $G$, that is $\bigcap_{x\in G} H^x = 1$, since $G$ acts faithfully on the vertices of $\Gamma$. Moreover since $G$ is transitive on the arcs of $\Gamma$, there is an element $g \in G$ which maps the arc $(\alpha, \beta)$ to $(\beta, \alpha)$, and hence interchanges $\alpha$ and $\beta$. Clearly we may require $g$ to be a 2-element (that is, to have order a power of 2). It is not difficult to see that $g \notin N_G(H)$ (since $H^g = G_\beta \neq G_\alpha$), and $g^2 \in H \cap H^g = G_{\alpha\beta}$, and a little more tricky to see that $\langle H, g\rangle = G$ if and only if $\Gamma$ is connected (since the elements of $\langle H, g\rangle$ map $\alpha$ to vertices in the connected component of $\Gamma$ containing $\alpha$). It was shown by Sabidussi [82] that the given $G$-arc transitive graph $\Gamma$ is isomorphic to the graph $\Gamma(G, H, HgH) = (\Omega^*, E^*)$ defined by

$$\Omega^* := [G : H], \text{ and } E^* := \{\{Hx, Hy\} | xy^{-1} \in HgH\},$$

where we write $[G : H] = \{Hx | x \in G\}$. Conversely, if a group $G$ is given with a core-free subgroup $H$ and a 2-element $g$ such that $g \notin N_G(H)$, $g^2 \in H \cap H^g$, and $\langle H, g\rangle = G$, then the graph $\Gamma(G, H, HgH)$ is connected and $G$-arc transitive, with $G$ acting on $\Omega^*$ by right multiplication.

Sabidussi's results were refined in [26, Theorem 2.1] yielding precise conditions under which $\Gamma(G, H, HgH)$ is $(G, 2)$-arc transitive.

**Theorem 11.1** *Let $G$ be a finite group with a core-free subgroup $H$ and a 2-element $g$. Then the graph $\Gamma(G,H,HgH)$ is a finite, connected, $(G,2)$-arc transitive graph with $G$ transitive on vertices (acting by right multiplication) if and only if*

$$g \notin N_G(H), g^2 \in H, \langle H, g\rangle = G,$$

*and the action of $H$ on $[H : H \cap H^g]$ by right multiplication is 2-transitive.*

Thus a classification of all connected $(G,2)$-arc transitive graphs is equivalent to a classification of all core-free subgroups $H$ and 2-elements $g$ of $G$ satisfying the conditions of Theorem 11.1. We illustrate this construction with a family of finite non-bipartite, vertex-transitive, 2-arc transitive graphs of valency 4. These examples also demonstrate that such a graph may have an arbitrarily large number of pairwise non-isomorphic quasiprimitive, 2-arc transitive quotients.

**Example 11.2 ([32, Example 2.1])** Let $p_i$, where $i = 1,\ldots,n$, be distinct primes such that $p_i \equiv \pm 3 \pmod 8$ for all $i$. For each $i$, let $G_i = \mathrm{PSL}(2,p_i)$ and let $H_i$ be a subgroup of $G_i$ with $H_i \cong A_4$. Then there exists a 2-element $g_i \in G_i$ such that $G_i = \langle H_i, g_i\rangle$ and $H_i \cap H_i^{g_i} \cong \mathbb{Z}_3$. Note that, for each $i$, the triple $G_i, H_i, g_i$ satisfies the conditions of Theorem 11.1, and hence the graph $\Gamma(i) := \Gamma(G_i, H_i, H_i g_i H_i)$ is connected, $G_i$-vertex-transitive, and $(G_i,2)$-arc transitive of valency 4. Moreover, $G_i$ is quasiprimitive on the vertices of $\Gamma(i)$. However, we wish to push the construction a little further.

Set $G := G_1 \times \ldots \times G_n$, and let $H$ be a diagonal subgroup of $H_1 \times \ldots \times H_n$, that is, for each $i$ choose an isomorphism $\varphi_i : A_4 \to H_i$ and set

$$H := \{(g^{\varphi_1},\ldots,g^{\varphi_n}) \,|\, g \in A_4\}.$$

Then $H \cong A_4$. Further let $g = g_1 \ldots g_n$. Then $H$ is core-free in $G$, $G = \langle H, g\rangle$ and $H \cap H^g \cong \mathbb{Z}_3$. It follows from Theorem 11.1 that the graph $\Gamma := \Gamma(G,H,HgH)$ is connected, $G$-vertex transitive, and $(G,2)$-arc transitive of valency 4. Since $G$ has no subgroups of index 2, $\Gamma$ is not bipartite. For any $i \leq n$, set $N_i := \prod_{j\neq i} G_j$. Then $N_i$ is normal in $G$ and $G/N_i \cong G_i$, and hence $N_i$ is a maximal normal subgroup of $G$. Moreover $N_i \cap H = 1$, and hence $N_i$ is semiregular on the vertex set of $\Gamma$. It follows from Theorem 10.4 that the quotient graph $\Gamma_{N_i}$ is a $(G_i,2)$-arc transitive graph with $G_i$ quasiprimitive on vertices. In fact it is not difficult to see that $\Gamma_{N_i} \cong \Gamma(i)$. Thus $\Gamma$ is a cover of each of its (pairwise non-isomorphic) quasiprimitive quotients $\Gamma(1),\ldots,\Gamma(n)$.

## 12 Finite quasiprimitive permutation groups

The class of finite quasiprimitive permutation groups may be described (see [75]) in a fashion very similar to the description given by the O'Nan-Scott Theorem 6.1 for finite primitive permutation groups. Eight types of finite quasiprimitive permutation groups can be defined, which in most cases parallel the primitive types from the O'Nan-Scott Theorem, and it is proved in [75] that every finite quasiprimitive permutation group belongs to exactly one of these types. The subdivision into types given here was used in [74] for primitive groups and in [11] for quasiprimitive groups; and moreover in the latter paper a refinement of it was needed.

Let $G$ be a finite quasiprimitive permutation group on $\Omega$ and let $\alpha \in \Omega$. Then, as in the case of finite primitive groups, $G$ has at most two minimal normal subgroups (see the first portion of [75, Section 3]), and if there are two minimal normal subgroups then they are isomorphic and each is equal to the centraliser of the other. The first three quasiprimitive types, namely types HA, HS, and HC, are exactly the same as the corresponding primitive types, that is, all quasiprimitive permutation groups of these types are primitive. We therefore suppose that $G$ is not of these types, and hence (see the discussion in Section 6) $G$ has a unique minimal normal subgroup $N = \mathrm{soc}(G) = T_1 \times \ldots \times T_k$, for some positive integer $k$, where each $T_i \cong T$, a nonabelian simple group.

Each of the five remaining quasiprimitive types corresponds to a primitive type. However, unlike the types HA, HS, and HC, each of these five types contains some imprimitive quasiprimitive permutation groups. For the first four of them the parallel with the corresponding primitive type is very close. These four types are the types AS, SD, CD, and TW. We describe these four types below, highlighting the differences from the corresponding primitive types.

AS (almost simple) $T \leq G \leq \mathrm{Aut}(T)$, where $T$ is a finite nonabelian simple group, and $G = TG_\alpha$. In this type we may have $G_\alpha = \{1_G\}$, that is $N = T$ may be regular on $\Omega$.

Recall that for $G$ to be primitive of type AS the stabilizer $G_\alpha$ must be maximal in $G$, and in particular $G_\alpha \neq \{1_G\}$. The next type is the simple diagonal type SD. A quasiprimitive permutation group of type SD is a subgroup of the group $W$ defined in Section 6. The difference here from the primitive type SD is that $G$ is only required to act transitively, not primitively, on the simple direct factors of $N$.

SD (simple diagonal) $N = \mathrm{soc}(W) \leq G \leq W$, $\Omega = T^{k-1}$, where $k > 1$, with the action defined in Section 6, and $G$ acts transitively by conjugation on the $k$ simple direct factors of $N$.

Similarly for the compound diagonal type CD the group $G$ preserves a product structure $\Omega = \Delta^l$, for some $l \geq 2$, and the subgroup of $\mathrm{Sym}(\Delta)$ involved is a quasiprimitive group, rather than a primitive group, of type SD.

CD (compound diagonal) $\Omega = \Delta^l$, and $N = T^k \leq G \leq H \wr S_l \leq \mathrm{Sym}(\Delta) \wr S_l$, for some divisor $l$ of $k$, where $l \geq 2$ and $k/l \geq 2$, and $H \leq \mathrm{Sym}(\Delta), \mathrm{soc}(H) = T^{k/l}$ and $H$ is quasiprimitive of type SD; $G$ acts transitively by conjugation on the simple direct factors of $N$.

For the quasiprimitive type TW, again $G = T\mathrm{twr}_\varphi P = N.P$, the subgroup $Q := N_P(T_1)$ has index $k$ in $P$ $(k > 1)$, $\varphi : Q \to \mathrm{Aut}(T)$, and $\mathrm{core}_P(\varphi^{-1}(\mathrm{Inn}(T)) = \{1_P\}$.

TW (twisted wreath product) $G$ is the twisted wreath product $T\mathrm{twr}_\varphi P$, and $\Omega = N$ with the action as defined in Section 6.

The differences between the conditions on $P, Q$ for primitivity and quasiprimitivity of $T\mathrm{twr}_\varphi P$ are rather subtle. A discussion can be found in [75, Remark 2.1]. We note in particular that for the quasiprimitive type TW we do not require that the image of $\varphi$ contains $\mathrm{Inn}(T)$.

For the final quasiprimitive type PA, $G$ preserves a product structure $\Delta^k$ on a $G$-invariant partition of $\Omega$ and the subgroup of $\mathrm{Sym}(\Delta)$ involved is quasiprimitive of type AS with socle $T$. Thus a quasiprimitive group $G$ of type PA induces a faithful product action on this partition of $\Omega$. For a group $R$ and positive integer $k \geq 2$, a *subdirect product* of $R^k$ is a subgroup of $R^k$ which projects onto each of the $k$ direct factors $R$.

PA (product action) $N = T^k \leq G \leq H \wr S_k \leq \mathrm{Sym}(\Delta) \wr S_k$, where $H$ is a quasiprimitive permutation group on $\Delta$ of type AS with non-regular socle $T$, and $G$ acts transitively by conjugation on the simple direct factors of $N$. Choose $\delta \in \Delta$ and set $R := T_\delta$. There is a (possibly trivial) $G$-invariant partition $\Omega'$ of $\Omega$ such that for some $\omega \in \Omega'$, $N_\omega = R^k$, and for $\alpha \in \omega$, $N_\alpha$ is a subdirect product of $R^k$.

This last quasiprimitive type is the furthest from the corresponding primitive type. It is clear from the descriptions of these eight types that a finite quasiprimitive permutation group belongs to at most one of these types. The main theorem in [75] shows that the converse is also true.

**Theorem 12.1 ([75, Theorem 1])** *Each finite quasiprimitive permutation group is permutationally equivalent to a quasiprimitive group in exactly one of the quasiprimitive types* HA, HS, HC, AS, SD, CD, TW, PA.

In the next section we describe how this structure theorem for finite quasiprimitive permutation groups can be used to study the structure of finite quasiprimitive 2-arc transitive graphs. We have already noted, in Remark 10.5 that the structure of the finite bipartite, vertex-transitive, 2-arc transitive graphs is more complicated than that of the non-bipartite examples. However, quasiprimitive groups certainly arise when describing their normal 2-arc transitive quotients (see [76]).

Finally we remark that this structure theorem has also been used very recently, by Conder, Li and the author, to study Weiss's Conjecture 10.6 and it seems that we can now show that Conjecture 10.6 is true provided it is true for almost simple, locally primitive graphs, that is, for graphs $\Gamma$ which admit a quasiprimitive subgroup $G$ of $\mathrm{Aut}\Gamma$ of type AS such that $\Gamma$ is $G$-locally primitive.

## 13 Finite quasiprimitive 2-arc transitive graphs

It is important to test the effectiveness or power of our observations about the role of quasiprimitive graphs in understanding typical finite, vertex-transitive, 2-arc transitive graphs. To this end, Theorem 12.1 was used to study the possible structures of the finite quasiprimitive 2-arc transitive graphs. It turned out that only half of the quasiprimitive types could occur as quasiprimitive 2-arc transitive groups of automorphisms.

**Theorem 13.1 ([75, Lemmas 5.2 and 5.3])** *If* $\Gamma = (\Omega, E)$ *is a finite* $(G,2)$*-arc transitive graph such that* $G$ *is quasiprimitive on* $\Omega$, *then* $G$ *has quasiprimitive type* HA, AS, TW, *or* PA.

Further, it was observed in [75, Section 6] and [77] that, for each of the types HA, AS, TW, and PA, there are examples of $(G,2)$-arc transitive graphs with $G$ quasiprimitive of the given type. The next step in the study of quasiprimitive 2-arc transitive graphs was to investigate further the nature of the graphs corresponding to each of these quasiprimitive types. It was in fact possible to complete the classification of those of type HA. This was done in [47] by Ivanov and the author. Recall that all quasiprimitive permutation groups of type HA are primitive.

**Theorem 13.2 ([47])** *If $\Gamma = (\Omega, E)$ is a finite $(G,2)$-arc transitive graph of valency $k$ such that $G$ is (quasi)primitive on $\Omega$ of type* HA, *then $|\Omega| = 2^n$, $G = \mathbb{Z}_2^n{\cdot}G_o \leq \mathbb{Z}_2^n{\cdot}(\mathrm{Aut}\Gamma \cap \mathrm{GL}(n,2))$ with $G_o$ irreducible, and $\Gamma, n, k, \mathrm{Aut}\Gamma \cap \mathrm{GL}(n,2)$ are as in one of the lines of Table* 1.

| $\Gamma$ | $d$ | $n$ | $\mathrm{Aut}(\Gamma) \cap \mathrm{GL}(d,2)$ | Comments |
|---|---|---|---|---|
| $K_{k+1}$ | $\log_2(k+1)$ | $k = 2^n - 1$ | $\mathrm{GL}(n,2)$ | – |
| $\Box_k$ | $k-1$ | $k$ odd | $S_k$ | – |
| $P_m(a)$ | $m^a$ | $(2^{am}-1)/(2^a-1)$ | $\mathrm{P\Gamma L}(m,2^a)$ | $m \geq 3$ |
| $U(q)$ | $\geq q^2-q+1$ | $q^3+1$ | $\mathrm{P\Gamma U}(3,q)$ | $q \equiv 3 \pmod 4$; not confirmed |
| $\Gamma(C_{23})$ | 11 | 23 | $M_{23}$ | – |
| $\Gamma(C_{22})$ | 10 | 22 | $M_{22} \cdot 2$ | – |

Table 1: Quasiprimitive 2-arc transitive graphs of type HA

If $\Gamma = (\Omega, E)$ is a finite primitive $(G,2)$-arc transitive graph of type HA having valency $k \geq 3$, then it was shown in [47] that $G$ must be of the form $G = N \cdot G_o$, where the normal subgroup $N = \mathbb{Z}_2^n$ acts regularly on vertices, and the subgroup $G_o$ (which is the stabilizer of the identity element of $N$) has a faithful 2-transitive action of degree $k$. Furthermore it was shown that $N$ may be identified with a quotient of the natural $\mathrm{GF}(2)G_o$-permutation module $\mathbb{Z}_2^k$ for the 2-transitive action of $G_o$ relative to some maximal submodule $W$ such that $G_o$ is faithful on $\mathbb{Z}_2^k/W$, and that $N$ and hence $W$ determine $\Gamma$. In fact we have $\Omega = N = \mathbb{Z}_2^k/W$ and $\{x+W, y+W\}$ is an edge if and only if $x + y \equiv e \pmod W$ for some weight 1 vector $e \in \mathbb{Z}_2^k$. Thus, for a 2-transitive permutation group $G_o$ of degree $k$, the finite primitive $(G,2)$-arc transitive graphs of type HA, with point stabilizer $G_o$, are in one-to-one correspondence with the maximal $\mathrm{GF}(2)G_o$-submodules $W$ of the $\mathrm{GF}(2)G_o$-permutation module $\mathbb{Z}_2^k$ such that $G_o$ is faithful on $\mathbb{Z}_2^k/W$; alternatively they are in one-to-one correspondence with the minimal faithful $\mathrm{GF}(2)G_o$-submodules of $\mathbb{Z}_2^k$. Most of the examples in Theorem 13.2 are therefore defined in terms of a maximal submodule of the $\mathrm{GF}(2)G_o$-permutation module for the 2-transitive group $G_o$. Occasionally the examples turn out to be isomorphic to some well-known graphs, and in such cases we use their more familiar descriptions. This is the case for the complete graphs in line 1 of the table, and is also the case for the graphs in line 2. The graph $\Box_k$ is the *folded cube*, the antipodal quotient of the $k$-cube graph; it may be defined as the graph with vertex set $\Omega = \mathbb{Z}_2^{k-1}$ with $\{x,y\}$ an edge if and only if $x - y$ has weight either 1 or $k-1$.

The graph $P_m(a)$ $(m \geq 3, a \geq 1)$ is defined in terms of the $\mathrm{GF}(2)G_o$-permutation module $\mathbb{Z}_2^k$ for $G_o$ where $\mathrm{PSL}(m,2^a) \leq G_o \leq \mathrm{P\Gamma L}(m,2^a)$ and $k = (2^{ma}-1)/(2^a-1)$. Let $W$ be the $\mathrm{GF}(2)G_o$-submodule generated by the characteristic functions of all the complements of hyperplanes of the projective geometry $\mathrm{PG}_{m-1}(2^a)$. Then the vertex set of $P_m(2^a)$ is the quotient module $\Omega := \mathbb{Z}_2^k/W^\perp$ with edges as described above.

The examples in line 4 probably do not exist (that is, they are probably not quasiprimitive examples). When $q \equiv 3 \pmod 4$, the submodule structure of the $GF(2)G_o$-permutation module $V = \mathbb{Z}_2^k$ for the 2-transitive unitary groups $G_o$, where $PSU(3,q) \leq G_o \leq P\Gamma U(3,q)$ and $k = q^3 + 1$, has not been completely determined. It is believed, but has not yet been proved, that $V$ has no faithful minimal $G$-submodule and therefore provides no quasiprimitive examples in Theorem 13.2. This has been proved by Jane McCorkindale [66] in her D. Phil. thesis in the case where $q \equiv 3 \pmod 8$, $q + 1 \not\equiv 0 \pmod 3$, and has also been verified computationally when $q = 3, 7, 11$ by Andries Brouwer (private communication).

The graphs $\Gamma(C_{23})$ and $\Gamma(C_{22})$ are constructed in a similar manner from the GF(2)-permutation modules for the 2-transitive groups $M_{23}$ and $M_{22} \cdot 2$ respectively. Descriptions of these graphs, including distance diagrams, may be found in [18, Sections 11.3.4 and 11.3.5] and in [47, 1.8].

We next turn to the classification of $(G,2)$-arc transitive graphs with $G$ quasiprimitive of type AS, say $T \leq G \leq \mathrm{Aut}(T)$ for a finite nonabelian simple group $T$. In this case the approach of Section 11 is the most useful way both of constructing examples, and of classifying all possible examples for certain classes of almost simple groups. For each almost simple group $G$, we need to locate first all subgroups $H$ not containing $\mathrm{soc}(G)$ which have a 2-transitive permutation representation of degree at least 3. Then we need to find the 2-elements $g \in G$ such that $H \cap H^g$ is a point stabiliser in this representation. For each such triple $G, H, g$, if the conditions of Theorem 11.1 hold, then the graph $\Gamma(G, H, HgH)$ is $G$-vertex-transitive and $(G,2)$-arc transitive. The remaining problem then is to determine, where possible, any isomorphisms between the examples constructed.

It seemed to the author that it should be possible to use this strategy to classify all $(G,2)$-arc transitive graphs for some classes of Lie type almost simple groups $G$ of low Lie rank. This has been achieved for the groups $PSL(2,q), Sz(q)$ and $Ree(q)$ in [25, 26, 27, 35]. Each of these classifications involved the constructions of several new infinite families of examples.

On the other hand it can be demonstrated that such an explicit classification is not possible for the family of finite alternating and symmetric groups. This was done in [79] by Jie Wang and the author, where a study was made of primitive permutation representations of $G = A_n$ or $S_n$ such that a point stabilizer $G_\alpha$ is 2-transitive on one of its orbits $\Gamma(\alpha)$. All examples were classified except those for which $G_\alpha$ was a primitive permutation group of degree $n$ of primitive type AS and had a faithful 2-transitive permutation representation of degree $|\Gamma(\alpha)|$. Although all possibilities for the 2-transitive group $G_\alpha$ are known (using the classification of the finite simple groups, see [20]) we still do not know all faithful primitive representations of all such groups. Thus it is not feasible at present to complete the classification in [79], and consequently it is not feasible to classify even the $(G,2)$-arc transitive graphs with $G$ primitive on vertices, and $G = A_n$ or $S_n$.

Robert Baddeley [9] has made a study of finite $(G,2)$-arc transitive graphs with $G$ quasiprimitive or primitive on vertices of type TW. There are many examples and his paper gives a general approach to their construction. The situation for finite $(G,2)$-arc transitive graphs with $G$ quasiprimitive of type PA is not quite so clear. Examples may be constructed as follows (again using the approach described in Section 11). However we do not know what "typical" quasiprimitive graphs of this type are like.

**Example 13.3** Let $\Delta = (V_\Delta, E_\Delta)$ be a connected $(H,2)$-arc transitive graph, where $H$ is a nonabelian simple group. Then, for $\delta \in V_\Delta$, there is a 2-element $h \in H$ such that $\delta^h$ is adjacent to $\delta$ in $\Delta$, $\delta^{h^2} = \delta$, and $H = \langle H_\delta, h \rangle$. Let $G := H \wr \mathbb{Z}_2 = (H \times H) \cdot \langle \pi \rangle$, where $\pi^2 = 1$ and $(h_1, h_2)^\pi = (h_2, h_1)$ for all $h_i \in H$, and let $K = \{(x,x) \mid x \in H_\delta\}$ and $g = (1_H, h)\pi$. Consider $\Gamma := \Gamma(G, K, KgK)$. Now $K \cap K^g = \{(x,x) \mid x \in H_\delta \cap H_\delta^h\}$ and $\langle K, g \rangle = G$, and it follows from Theorem 11.1 that $\Gamma$ is a connected $(G,2)$-arc transitive graph and $G$ is quasiprimitive of type PA on vertices.

Finally we make some remarks about the full automorphism groups of finite, quasiprimitive, 2-arc transitive graphs.

**Remark 13.4** Suppose that $G < H \leq \mathrm{Sym}(\Omega)$. It follows from the definition of a primitive permutation group that if $G$ is primitive on $\Omega$, then $H$ will also be primitive. However, if $G$ is an imprimitive, quasiprimitive permutation group on $\Omega$, then $H$ need not be quasiprimitive on $\Omega$. For example, if $G$ preserves a nontrivial partition $\mathcal{P}$ of $\Omega$, we could take $H$ to be the full stabiliser of $\mathcal{P}$ in $\mathrm{Sym}(\Omega)$: $H$ would be a wreath product of symmetric groups, and its base group would have the parts of $\mathcal{P}$ as orbits.

Now consider the situation for graphs. Suppose that $G < H \leq \mathrm{Aut}\Gamma$ and that $\Gamma = (\Omega, E)$ is $(G,2)$-arc transitive (or $G$-locally primitive) with $G$ quasiprimitive on $\Omega$. Certainly, for $\alpha \in \Omega$, $G_\alpha^{\Gamma_1(\alpha)} \leq H_\alpha^{\Gamma_1(\alpha)}$ so that $\Gamma$ is $(H,2)$-arc transitive (or $H$-locally primitive respectively) on $\Gamma$. However it is not clear that $H$ will be quasiprimitive on $\Omega$. Thus, for example, it may be the case that the full automorphism group of a quasiprimitive, 2-arc transitive graph is not quasiprimtive.

Robert Baddeley [9] gave the first example of a 2-arc transitive graph $\Gamma = (\Omega, E)$ and subgroup $G \leq \mathrm{Aut}\Gamma$, with $G$ quasiprimitive on $\Omega$ and transitive on 2-arcs of $\Gamma$, for which some over-group $H$ of $G$ in $\mathrm{Aut}\Gamma$ was not quasiprimitive. In his example, the group $G$ was quasiprimitive of type TW. Recently Cai Heng Li [53] has constructed an infinite family of such graphs $\Gamma$, where the group $G$ is quasiprimitive of type AS, and $\mathrm{Aut}\Gamma$ is not quasiprimitive on vertices. Further, Xin Gui Fang and the author [28] have made a study of over-groups in $\mathrm{Aut}\Gamma$ of a quasiprimitive, locally primitive subgroup $G$, where $G$ is of type AS.

These considerations suggest some fundamental questions about the nature of over-groups in $\mathrm{Sym}(\Omega)$ of a quasiprimitive permutation group $G$. For an imprimitive, quasiprimitive permutation group $G$ on $\Omega$, if $G < H < \mathrm{Sym}(\Omega)$, then one of the following is true.

(1) $H$ is not quasiprimitive on $\Omega$ (for example, $H$ might be the full stabiliser of a nontrivial $G$-invariant partition of $\Omega$);

(2) $H$ is quasiprimitive and imprimitive on $\Omega$;

(3) $H$ is primitive on $\Omega$.

Several studies have been made: in [74] in the case where $G$ is primitive, and in [11] in the case where $G$ is imprimitive and quasiprimitive, and the over-group $H$ is primitive. A more detailed discussion of the group theoretic problems which have arisen through attempts to give a precise description of the possibilities can be found in [77, Section 7].

## 14 Isomorphisms of Cayley graphs

This section was included because of the discussion in the lectures of Brian Alspach at this Advanced Study Institute of certain problems about finite Cayley graphs. We shall consider the nature of isomorphisms between Cayley graphs for a given finite group $G$. Let $\Gamma := \mathrm{Cay}(G, S)$, for a Cayley subset $S$ of $G$. Recall that the group $G$, acting by right multiplication on the vertex set $G$ of $\Gamma$, induces a subgroup, which we shall denote by $\hat{G}$, of $\mathrm{Aut}\Gamma$. Thus we have $\hat{G} \cong G$ and $\hat{G} \leq \mathrm{Aut}\Gamma \leq \mathrm{Sym}(G)$. We also have, as a subgroup of $\mathrm{Sym}(G)$, the group $\mathrm{Aut}(G)$ acting in the natural way: $\sigma \in \mathrm{Aut}(G)$ maps $g \in G$ to its image $g^\sigma$ under $\sigma$. This subgroup $\mathrm{Aut}(G)$ of $\mathrm{Sym}(G)$ intersects the subgroup $\hat{G}$ in the identity subgroup. Observe that each element $\sigma \in \mathrm{Aut}(G)$ induces a graph isomorphism from $\Gamma = \mathrm{Cay}(G, S)$ to $\mathrm{Cay}(G, S^\sigma)$. The graph $\mathrm{Cay}(G, S)$ is called a CI-*graph* of $G$ if, whenever $\mathrm{Cay}(G, S) \cong \mathrm{Cay}(G, T)$ for some Cayley subset $T$ of $G$, there is an element $\sigma \in \mathrm{Aut}(G)$ such that $S^\sigma = T$. (CI stands for *Cayley Invariant.*)

The general question arising here is: *which Cayley graphs for a group $G$ are* CI-*graphs*? In the special case where all Cayley graphs for $G$ are CI-graphs, the group $G$ is said to be a CI-*group*. Interest in this question stems from a conjecture in 1967 of Ádám [1] that all finite cyclic groups are CI-groups. Ádám's conjecture was disproved by Elspas and Turner [24] in 1970, and has received considerable attention in the literature (see for example [68, 72] and the references cited in these papers, and also the chapter by Alspach in the present volume). In 1977, Babai [5] began a study of this question for arbitrary groups. In particular he observed in [5, Lemma 3.1], as did Alspach and Parsons in [2, Theorem 1], that there is a simple group theoretic criterion which determines whether or not a Cayley graph is a CI-graph.

**Theorem 14.1** *Let $\Gamma$ be a Cayley graph for a group $G$, and let $\hat{G}$ denote the subgroup of* $\mathrm{Sym}(G)$ *consisting of the permutations of $G$ induced by the elements of $G$ acting by right multiplication. Then $\Gamma$ is a CI-graph if and only if, whenever $G^\tau \leq \mathrm{Aut}\Gamma$, for $\tau \in \mathrm{Sym}(G)$, there exists $\sigma \in \mathrm{Aut}\Gamma$ such that $G^\sigma = G^\tau$.*

It has very recently been proved by Cai Heng Li [51, 52] that all finite CI-groups are soluble. His result improves and completes results of Babai and Frankl [7] dating from 1978: they proved that, if $G$ is a CI-group of odd order, then either $G$ is abelian, or $G$ has an abelian normal subgroup of index 3 and its Sylow 3-subgroup is either elementary abelian, or cyclic of order 9 or 27. Moreover, they showed in [8] that if $G$ is an insoluble CI-group, then $G = L \times N$ with $(|L|, |N|) = 1$, where $L \cong \mathrm{PSL}(2,5)$, $\mathrm{SL}(2,5)$, $\mathrm{PSL}(2,13)$ or $\mathrm{SL}(2,13)$, and $N$ is a direct product of elementary abelian groups.

However, the lead-up to Li's result was a comprehensive investigation of finite groups $G$ for which all Cayley graphs $\mathrm{Cay}(G, S)$ with $|S|$ bounded above by some constant $m$ ($m = 1, 2, 3, \ldots$) are CI-graphs. Such groups are called $m$-CI-*groups*. Similar questions have been posed and studied for Cayley di-graphs leading to investigations of so-called DCI-groups and $m$-DCI-groups (see [51, 54, 57, 58, 95]), but we shall confine our comments as far as possible to Cayley graphs rather than Cayley di-graphs. The study of $m$-CI-groups was suggested by M. Y. Xu [93] in 1988. Xu was influenced on the one hand by the results of Babai and Frankl, and on the other hand by work of Toida [88] on circulant graphs of small valency, proving that all Cayley graphs of valency 3 for cyclic groups are CI-graphs.

We shall give an indication of the basic framework for this study by considering further the property of being an $m$-CI-group, for $m$ small. Suppose that $G$ is a finite 2-CI-group. If $S = \{x, y\}$ is a Cayley subset of $G$, then $S = S^{-1}$ so either $y = x^{-1}$ or $x$ and $y$ are *involutions* (that is $o(x) = o(y) = 2$, where $o(x)$ the *order* of an element $x$). Note that $\mathrm{Cay}(G, S)$ has

$|G|/n$ connected components, each of which is a cycle $C_n$ of length $n$, where $n = o(x)$ if $y = x^{-1} \neq x$, and $n = 2o(xy)$ if $o(x) = o(y) = 2$. By considering Cayley subsets of the form $\{x, x^{-1}\}$ with $o(x) \geq 2$ (noting that such a set consists of the single element $x$ if $o(x) = 2$), we see that

(a) for each $n \geq 2$, and each pair $x, y$ (if any) of elements of $G$ of order $n$, $x$ can be mapped to either $y$ or $y^{-1}$ by some element of $\mathrm{Aut}(G)$.

If $x^\sigma = y$ for some $\sigma \in \mathrm{Aut}(G)$, then we say that $x$ and $y$ are *fused* under $\mathrm{Aut}(G)$, while if $x^\sigma = y^{-1}$ for some $\sigma \in \mathrm{Aut}(G)$, then we say that $x$ and $y$ are *inverse fused* under $\mathrm{Aut}(G)$. Thus, for each $n \geq 2$, each pair of elements of $G$ of order $n$ is either fused or inverse fused under $\mathrm{Aut}(G)$. Such groups are called FIF-*groups*. So $G$ is an FIF-group. Further, suppose that $G$ has two Cayley subsets $S_1 = \{x, y\}$, with $o(x) = o(y) = 2$ and $o(xy) = n$, and $S_2 = \{z, z^{-1}\}$, with $o(z) = 2n$, for some $n \geq 2$. Then $\mathrm{Cay}(G, S_i)$ is the disjoint union of $|G|/2n$ copies of $C_{2n}$, for $i = 1, 2$, but $S_1$ cannot be mapped to $S_2$ by any element of $\mathrm{Aut}(G)$, which is a contradiction. Thus $G$ cannot contain a Cayley subset of both kinds, and so

(b) for each $n \geq 2$, if $G$ contains a pair of involutions $x, y$ such that $o(xy) = n$, then $G$ contains no elements of order $2n$.

In the case where $r = 2$ this observation has the following important consequence. If $G$ contains no elements of order 4, then a Sylow 2-subgroup of $G$ is elementary abelian. On the other hand, if $G$ does not contain a pair of commuting involutions, then a Sylow 2-subgroup of $G$ is cyclic or generalised quaternion. Thus

(c) a Sylow 2-subgroup of $G$ is cyclic, elementary abelian, or generalised quaternion.

These restrictions on a finite group $G$ are very strong, and extra restrictions arise for $m$-CI-groups, with $m \geq 3$. Jiping Zhang [95] made a study of finite groups $G$ in which every pair of elements of the same order are fused under $\mathrm{Aut}(G)$. His techniques were extended by Li and the author and used to study finite FIF-groups (in [54] for simple groups, and in [55] for general finite groups). The following result for simple groups is the easiest to state. We note that it depends on the finite simple group classification.

**Theorem 14.2 ([54, Theorem 1.2])** *A finite nonabelian simple group $G$ is an FIF-group if and only if $G$ is one of the following:* $A_5$, $A_6$, $\mathrm{PSL}(2,7)$, $\mathrm{PSL}(2,8)$, $\mathrm{PSL}(3,4)$, $\mathrm{Sz}(8)$, $M_{11}$ *or* $M_{23}$.

A detailed description of finite FIF-groups is given in [55]. We state below a corollary of the main result of that paper, namely a description of the structure of finite insoluble FIF-groups.

**Theorem 14.3 ([55])** *Let $G$ be a finite insoluble FIF-group. Then $G = A \times B$, where $|A|, |B|$ are coprime, $A$ is a soluble FIF-group and $B$ either is one of the simple groups in Theorem* 14.2, *or is* $\mathrm{SL}(2, q)$ *for $q = 5, 7$ or 9, or is* $(C \times \mathrm{Sz}(8)) \rtimes \mathbb{Z}_{3^s m}$, *where $m, s \geq 1$.*

Using these results to study $m$-CI-groups for small values of $m \geq 2$, yielded the following theorem. We note that [56] contains a detailed description of the structure of arbitrary finite 4-CI-groups which, in particular, extends the results in [7, 8].

**Theorem 14.4 ([54, Theorem 1.3] and [56])** *Let $G$ be a finite insoluble group.*

(a) *If $G$ is simple, then $G$ is a 2-CI-group if and only if $G = \mathrm{A}_5$ or* $\mathrm{PSL}(2,8)$.

(b) *If $G$ is simple, then $G$ is a 3-CI-group if and only if $G = \mathrm{A}_5$.*

(c) *If $G$ is a 4-CI-group, then $G = U \times V$, where $(|U|, |V|) = 1$, $U$ is abelian, and $V$ is* $\mathrm{A}_5$ *or* $\mathrm{SL}(2,5)$.

Cai Heng Li was able to use this result, together with an ingenious graph construction, to show that all finite CI-groups are soluble. The proof strategy goes roughly as follows. Suppose that $G$ is a finite insoluble CI-group. Then by Theorem 14.4, $G = U \times V$ where $(|U|, |V|) = 1$, $U$ is abelian and $V = \mathrm{A}_5$ or $\mathrm{SL}(2,5)$. Since $V$ is a characteristic subgroup of $G$ it follows that $V$ is also a CI-group. Thus it is sufficient to show that neither $A_5$ nor $\mathrm{SL}(2,5)$ is a CI-group, and this is what Li did.

**Theorem 14.5 ([52])** *The simple group* $\mathrm{A}_5$ *is not a 29-CI-group, and the group* $\mathrm{SL}(2,5)$ *is not a 58-CI-group.*

**Corollary 14.6** *All finite CI-groups are soluble.*

The proof of Theorem 14.5 uses the criterion given in Theorem 14.1 for a Cayley graph to be a CI-graph. Li constructs Cayley graphs for $A_5$ and $\mathrm{SL}(2,5)$ of valencies 29 and 58 respectively, which are not CI-graphs. The former graph is constructed as an orbital graph of a certain transitive permutation representation of the group $\mathrm{PSL}(2,29)$. We outline the ideas below. The *lexicographic product* $\Gamma_1[\Gamma_2]$ of graphs $\Gamma_1 = (\Omega_1, E_1)$ by $\Gamma_2 = (\Omega_2, E_2)$ is the graph $(\Omega_1 \times \Omega_2, E)$ where $\{(\alpha_1, \alpha_2), (\beta_1, \beta_2)\} \in E$ (where $\alpha_i, \beta_i \in E_i$, for $i = 1, 2$) if and only if either $\{\alpha_1, \beta_1\} \in E_1$, or $\alpha_1 = \beta_1$ and $\{\alpha_2, \beta_2\} \in E_2$.

**Sketch of the proof of Theorem** 14.5: Let $G = \mathrm{PSL}(2,29)$. Then $G$ has subgroups $B$, $C$ isomorphic to $\mathrm{A}_5$ but not conjugate in $G$, see [23]. It was shown in [52, Theorem 1.1] that there exists an arc-transitive graph $\Gamma = \Gamma(G, H, HgH)$ of valency 29, where $H \cong \mathbb{Z}_{29} \rtimes \mathbb{Z}_7$ and $g$ is an involution, such that $\mathrm{Aut}\,\Gamma \cong \mathrm{PSL}(2,29) \times \mathbb{Z}_2$. (This used the construction method described in Section 11.) Therefore $B$ and $C$ are not conjugate in $\mathrm{Aut}\,\Gamma$.

Now $(|H|, \mathrm{A}_5|) = 1$ and $|H||\mathrm{A}_5| = |G|$. It follows that $H \cap B = H \cap C = 1$ and hence that $G = BH = CH$. Since $H$ is the stabiliser of a vertex of $\Gamma$, we deduce from $G = BH = CH$ that both $B$ and $C$ are transitive on the vertices of $\Gamma$, and we deduce from $H \cap B = H \cap C = 1$ that $B$ and $C$ are both regular on vertices. Thus, by Lemma 5.2, we may identify the vertex set of $\Gamma$ with $B$ in such a way that $\Gamma = \mathrm{Cay}(B, S)$ for some Cayley subset $S$ of $B$. Moreover, with this identification, we have both $B$ and $C$ as regular subgroups of $\mathrm{Sym}(B)$, contained in $\mathrm{Aut}\Gamma$. Since all regular permutation representations of any given group are permutationally equivalent, it follows that $B$ and $C$ are conjugate in $\mathrm{Sym}(B)$. It then follows from Theorem 14.1 that $\Gamma = \mathrm{Cay}(B, S)$ is not a CI-graph, and hence that $B \cong A_5$ is not a 29-CI-group.

Next assume that $F = \mathrm{SL}(2,5)$ and let $\overline{F} = F/Z(F) \cong \mathrm{A}_5$. Since $\mathrm{A}_5$ is not a 29-CI-group, there exist two Cayley subsets $\overline{S}, \overline{T}$ of $\overline{F}$ of cardinality 29 such that the graphs $\Gamma_1 = \mathrm{Cay}(\overline{F}, \overline{S})$ and $\Gamma_2 = \mathrm{Cay}(\overline{F}, \overline{T})$ are isomorphic, but $\overline{S}^\alpha \neq \overline{T}$ for any $\alpha \in \mathrm{Aut}(\overline{F})$. Let $Z(F) = \langle e \rangle \cong \mathbb{Z}_2$, and let $S$ and $T$ be the full preimages of $\overline{S}$ and $\overline{T}$ under $F \to \overline{F}$, respectively. (If $s \mapsto \overline{s}$

then the full preimage of $\overline{s}$ is $\{s, se\}$.) Then $|S| = 2|\overline{S}| = 58$. It is not difficult to show that $\mathrm{Cay}(F, S) = \Gamma_1[\overline{K}_2] \cong \Gamma_2[\overline{K}_2] = \mathrm{Cay}(F, T)$. If there is an element $\sigma \in \mathrm{Aut}(F)$ such that $S^\sigma = T$, then $\overline{S}^{\overline{\sigma}} = \overline{T}$, where $\overline{\sigma}$ is the automorphism of $\overline{F}$ induced by $\sigma$. Since the latter cannot hold for any $\overline{\sigma}$, it follows that $\mathrm{Cay}(F, S)$ is not a CI-graph of $F$, and so $F$ is not a 58-CI-group. □

## Postscript: A fable

Once upon a time, a few years ago, a small band of people from the land of Permutation Groups traveled to the land of Graph Theory to settle. They brought with them their latest technology and intended to set up business in their new country. This was not the first time people had travelled to the land of Graph Theory from the land of Permutation Groups, but the new wave of settlers was different. Their technology was powerful, and of a type not known widely in the new land. Their arrival was met with mixed responses. Some residents were welcoming while others were sceptical of either the value or the commitment of the new settlers. However the newcomers were not alone. They met small groups of settlers from other lands – the land of Model Theory, the Federation of Computer Science, and the region of Mathematical Statistics. They set up joint business enterprises with the young enterpreneurs from the land of Graph Theory and for a time these were extremely successful and the economy of the land of Graph Theory prospered from this injection of new technology.

However after some time the new enterprises became less profitable. Fewer new products were being developed and the novelty of those that came on the market decreased. Some of the settlers realised that their decision to settle in the new land was a mistake, and returned to their homeland of Permutation Groups. The resident sceptics felt justified in their mistrust of the newcomers. But there were some among the settlers who realised that it was not sufficient to offer their new technology, powerful though it was, as simply to be used unchanged to enhance the appropriate technology from the land of Graph Theory. Even though this had led to some wonderful successes something more was needed. In fact essentially two things were needed. First, the settlers had to learn to understand the culture and technology of the land of Graph Theory at a deeper and broader level; they had to discover where their own technology reqired refinement and development so as to be more appropriately and widely used in the new land. Secondly, and simultaneously, they had to realise that experts from the land of Graph Theory needed to visit the land of Permutation Groups to learn its culture and technology better and to promote their own. Both of these things began to happen, and the resulting new joint enterprises flourished to the benefit of both lands. The governments of both lands after negotiation, agreed to offer joint citizenship to those of their citizens who were indeed by this stage sharing the culture and sharing their time between them.

## References

[1] A. Adám, Research problem 2-10, *J. Combin. Theory* **2** (1967), 309.

[2] B. Alspach and T. D. Parsons, Isomorphisms of circulant graphs and digraphs, *Discrete Math.* **25** (1979), 97–108.

[3] M. Aschbacher, On the maximal subgroups of the finite classical groups, *Invent. Math.* **76** (1984), 469–514.

[4] M. Aschbacher and L. L. Scott, Maximal subgroups of finite groups, *J. Algebra* **92** (1985), 44–80.

[5] L. Babai, Isomorphism problem for a class of point-symmetric structures, *Acta Math. Acad. Sci. Hungar.* **29** (1977), 329–336.

[6] L. Babai, Arc transitive covering digraphs and their eigenvalues, *J. Graph Theory* **9** (1985), 363–370.

[7] L. Babai and P. Frankl, Isomorphisms of Cayley graphs I, in: *Combinatorics (Proc. Fifth Hungarian Colloq., Keszthely 1976)*, Colloq. Math. Soc. János Bolyai **18**, North-Holland, Amsterdam, 1978, 35–52.

[8] L. Babai and P. Frankl, Isomorphisms of Cayley graphs II, *Acta Math. Acad. Sci. Hungar.* **34** (1979), 177–183.

[9] R. W. Baddeley, Two-arc transitive graphs and twisted wreath products, *J. Alg. Combin.* **2** (1993), 215–237.

[10] R. W. Baddeley and C. E. Praeger, On classifying all full factorisations and multiple-factorisations of the finite almost simple groups, Technical Report No. 1996/13, University of Leicester, 1996.

[11] R. W. Baddeley and C. E. Praeger, On primitive overgroups of quasiprimitive permutation groups, Technical Report No. 1996/14, University of Leicester, 1996.

[12] R. W. Baddeley and C. E. Praeger, Expansion and blow-up decompositions of quasiprimitive permutation groups, in preparation.

[13] N. Biggs, *Algebraic Graph Theory*, Cambridge University Press, Cambridge, 1993.

[14] J. T. M. van Bon and A. E. Brouwer, The distance-regular antipodal covers of classical distance-regular graphs, in: *Combinatorics (Proc. Seventh Hungarian Colloq., Eger 1987)*, Colloq. Math. Soc. János Bolyai **52**, North-Holland, Amsterdam, 1988, 141–166.

[15] J. T. M. van Bon and A. M. Cohen, Prospective classification of distance-transitive graphs, in: *Combinatorics '88*, vol.1 (A. Barlotti et al., eds.), Mediterranean Press, Rende, 1991, 25-38.

[16] J. T. M. van Bon and A. M. Cohen, Linear groups and distance-transitive graphs, *European J. Combin.* **10** (1989), 399–412.

[17] W. Bosma and J. Cannon, *Handbook of* MAGMA *Functions*, School of Mathematics and Statistics, Sydney University, 1994.

[18] A. E. Brouwer, A. M. Cohen and A. Neumaier, *Distance-Regular Graphs*, Springer Verlag, Berlin-Heidelberg-New York, 1989.

[19] A. E. Brouwer and J. Hemmeter, A new family of distance-regular graphs and the $\{0,1,2\}$-cliques in dual polar graphs, *European J. Combin.* **13** (1992), 71-79.

[20] P. J. Cameron, Finite permutation groups and finite simple groups, *Bull. London Math. Soc.* **13** (1981), 1–22.

[21] P. J. Cameron, There are only finitely many finite distance-transitive graphs of given valency greater than 2, *Combinatorica* **2** (1982), 9–13.

[22] I. V. Chuvaeva and D. V. Pasechnik, Distance-transitive graphs of type $q \cdot K_{q,q}$ and projective planes, *European J. Combin.* **11** (1990), 341–346.

[23] J. H. Conway, R. T. Curtis, S. P. Norton, R. A. Parker and R. A. Wilson, *Atlas of Finite Groups*, Clarendon Press, Oxford, 1985.

[24] B. Elspas and J. Turner, Graphs with circulant adjacency matrices, *J. Combin. Theory* **9** (1970), 297-307.

[25] X. G. Fang, Construction and classification of some families of almost simple 2-arc transitive graphs, PhD Thesis, University of Western Australia, 1995.

[26] X. G. Fang and C. E. Praeger, Finite two-arc transitive graphs admitting a Suzuki simple group, Research Report, Department of Mathematics, University of Western Australia, 1996.

[27] X. G. Fang and C. E. Praeger, Finite two-arc transitive graphs admitting a Ree simple group, Research Report, Department of Mathematics, University of Western Australia, 1996.

[28] X. G. Fang and C. E. Praeger, On graphs admitting arc-transitive actions of almost simple groups, Research Report, Department of Mathematics, University of Western Australia, 1996.

[29] I. A. Faradzev and A. A. Ivanov, Distance-transitive representations of the groups $G$, $\mathrm{PSL}_2(q) \leq G \leq \mathrm{P\Gamma L}_2(q)$, *European J. Combin.* **11** (1990), 348–356.

[30] A. D. Gardiner, Antipodal covering graphs, *J. Combin. Theory Ser. B* **16** (1974), 255–273.

[31] A. D. Gardiner, Imprimitive distance-regular graphs and projective planes, *J. Combin. Theory Ser. B* **16** (1974), 274–281.

[32] A. D. Gardiner and C. E. Praeger, On 4-valent symmetric graphs, *European J. Combin.* **15** (1994), 375–381.

[33] A. D. Gardiner and C. E. Praeger, A geometrical approach to imprimitive symmetric graphs, *Proc. London Math. Soc. (3)* **71** (1995), 524–546.

[34] C. D. Godsil, R. A. Liebler, and C. E. Praeger, Antipodal distance transitive covers of complete graphs, in preparation.

[35] A. Hassani, L. Nochefranca and C. E. Praeger, Finite two-arc transitive graphs admitting a two-dimensional projective linear group, Research Report, Department of Mathematics, University of Western Australia, 1995.

[36] J. Hemmeter, Halved graphs, Hamming and Johnson graphs, *Utilitas Math.* **25** (1984), 115–118.

[37] J. Hemmeter, Distance-regular graphs and halved graphs, *European J. Combin.* **7** (1986), 119–129.

[38] J. Hemmeter, The large cliques in the graph of quadratic forms, *European J. Combin.* **9** (1988), 395–410.

[39] D. Holt, The mechanical computation of first and second cohomology groups, *J. Symbolic Comput.* **1** (1985), 351–361.

[40] N. F. J. Inglis, Multiplicity-free permutation characters, distance-transitive graphs, and classical groups, PhD Thesis, University of Cambridge, 1986.

[41] N. F. J. Inglis, M. W. Liebeck and J. Saxl, Multiplicity-free permutation representations of finite linear groups, *Math. Z.* **192** (1986), 329–337.

[42] I. M. Isaacs, *Character Theory of Finite Groups*, Academic Press, New York, 1976.

[43] A. A. Ivanov, Distance-transitive representations of the symmetric groups, *J. Combin. Theory Ser. B* **41** (1986), 255–274.

[44] A. A. Ivanov, Distance-transitive graphs and their classification, in: *Investigations in the Algebraic Theory of Combinatorial Objects* (I.A.Faradžev et al., eds.), Math. Appl. (Soviet Ser.) **84**, Kluwer, Dordrecht, 1994, 283–378.

[45] A. A. Ivanov, R. A. Liebler, T. Penttila, and C. E. Praeger, Antipodal distance-transitive covers of complete bipartite graphs, to appear in *European J. Combin.*

[46] A. A. Ivanov, S. A. Linton, K. Lux, J. Saxl and L. Soicher, Distance- transitive representations of the sporadic groups, *Comm. Algebra* **23** (1995), 3379–3427.

[47] A. A. Ivanov and C. E. Praeger, On finite affine 2-arc transitive graphs, *European J. Combin.* **14** (1993), 421–444.

[48] A. A. Ivanov and C. E. Praeger, Problem Session at ALCOM-91, *European J. Combin.* **15** (1994), 105–112.

[49] P. M. Kleidman and M. Liebeck, *The Subgroup Structure of the Finite Classical Groups*, London Math. Soc. Lecture Note Ser. **129**, Cambridge University Press, Cambridge, 1990.

[50] L. G. Kovacs, Maximal subgroups in composite finite groups, *J. Algebra* **99** (1986), 114–131.

[51] C. H. Li, Isomorphisms of finite Cayley graphs, PhD Thesis, University of Western Australia, 1996.

[52] C. H. Li, Finite CI-groups are solvable, submitted.

[53] C. H. Li, A family of quasiprimitive 2-arc transitive graphs which have non-quasi-primitive full automorphism groups, preprint 1996.

[54] C. H. Li and C. E. Praeger, The finite simple groups with at most two fusion classes of every order, *Comm. Algebra* **24** (1996), 3681–1704.

[55] C. H. Li and C. E. Praeger, Finite groups in which any two elements of the same order are either fused or inverse-fused, Research Report, Department of Mathematics, University of Western Australia, 1996.

[56] C. H. Li and C. E. Praeger, On the isomorphism problem for finite Cayley graphs of bounded valency, preprint.

[57] C. H. Li, C. E. Praeger and M. Y. Xu, Isomorphisms of finite Cayley digraphs of bounded valency, submitted.

[58] C. H. Li, C. E. Praeger and M. Y. Xu, Finite groups with the Cayley isomorphism property, submitted.

[59] M. W. Liebeck, The affine permutation groups of rank 3, *Proc. London Math. Soc. (3)* **54** (1987), 477–516.

[60] M. W. Liebeck, C. E. Praeger and J. Saxl, Distance transitive graphs with symmetric or alternating automorphism group, *Bull. Austral. Math. Soc.* **35** (1987), 1–25.

[61] M. W. Liebeck, C. E. Praeger and J. Saxl, On the ONan-Scott Theorem for finite primitive permutation groups, *J. Austral. Math. Soc. Ser. A* **44** (1988), 389–396.

[62] R. A. Liebler, The classification of distance-transitive graphs of type $q \cdot K_{q,q}$, *European J. Combin.* **12** (1991), 125–128.

[63] B. D. McKay, Transitive graphs with fewer than 20 vertices, *Math. Comp.* **33** (1979), 1101–1121.

[64] B. D. McKay, Topics in Computational Graph Theory, PhD Thesis, University of Melbourne, 1980.

[65] B. D. McKay, nauty Users guide (version 1.5), Technical Report TR-CS-90-02, Department of Computer Science, Australian National University, 1990.

[66] J. McCorkindale, The 2-modular representation theory of $PSU_3(q)$, $q \equiv 3 (\mathrm{mod}\ 4)$, D. Phil. Thesis, University of Oxford, 1990.

[67] B. D. McKay and G. F. Royle, The transitive graphs with at most 26 vertices, *Ars Combin.* **30** (1990), 161–176.

[68] M. Muzychuk, Ádáms conjecture is true in the square-free case, *J. Combin. Theory Ser. A* **72** (1995), 118–134.

[69] B. H. Neumann, Twisted wreath products of groups, *Arch. Math.* **14** (1963), 1–6.

[70] P. M. Neumann, Finite permutation groups, edge-coloured graphs, and matrices, in: *Topics in Group Theory and Computation* (M. P. J. Curran, ed.), Academic Press, London, 1977, 82–116.

[71] P. M. Neumann, G. A. Stoy and E. C. Thompson, *Groups and Geometry*, Oxford University Press, Oxford, (1994).

[72] P. P. Palfy, Isomorphism problem for relational structures with a cyclic automorphism, *European J. Combin.* **8** (1987), 35–43.

[73] C. E. Praeger, Imprimitive symmetric graphs, *Ars Combin.* **19A** (1985), 149–163.

[74] C. E. Praeger, The inclusion problem for finite primitive permtation groups, *Proc. London Math. Soc. (3)* **60** (1990), 68–88.

[75] C. E. Praeger, An ONan-Scott Theorem for finite quasiprimitive permutation groups and an application to 2-arc transitive graphs, *J. London Math. Soc. (2)* **47** (1993), 227–239.

[76] C. E. Praeger, On a reduction theorem for finite, bipartite, 2-arc transitive graphs, *Australas. J. Combin.* **7** (1993), 21–36.

[77] C. E. Praeger, Finite quasiprimitive graphs, submitted for the proceedings of the British Combinatorics Conference, 1997.

[78] C. E. Praeger, J. Saxl and K. Yokohama, Distance transitive graphs and finite simple groups, *Proc. London Math. Soc. (3)* **55** (1987), 1–21 .

[79] C. E. Praeger and J. Wang, On primitive representations of finite alternating and symmetric groups with a 2-transitive subconstituent, *J. Algebra* **180** (1996), 808–833.

[80] G. F. Royle, The transitive groups of degree 12, *J. Symbolic Comput.* **4** (1987), 255–268.

[81] G. F. Royle and C. E. Praeger , Constructing the vertex transitive graphs of order 24, *J. Symbolic Comput.* **8** (1989), 309–326.

[82] G. Sabidussi, Vertex-transitive graphs, *Monatsh. Math.* **68** (1964), 426-438.

[83] J. Saxl, On multiplicity-free permutation representaions, *Finite Geometries and Designs* (P. J. Cameron, J. W. P. Hirschfeld and D. Hughes, eds.), London Math. Soc. Lecture Note Ser. **49**, Cambridge University Press, Cambridge, 1981, 337–353.

[84] L. L. Scott, Representations in characteristic $p$, *Santa Cruz conference on finite groups, Proc. Sympos. Pure Math.* **37** (1980), 318–331.

[86] M. Suzuki, *Group Theory I*, Springer-Verlag, Berlin, 1982.

[87] D. E. Taylor, Two-graphs and doubly transitive groups, *J. Combin. Theory Ser. B* **16** (1992), 113–122.

[88] S. Toida, A note on Adáms conjecture, *J. Combin. Theory (B)* **23** (1977), 239-246.

[89] W. T. Tutte, A family of cubical graphs, *Proc. Cambridge Philos. Soc.* **43** (1947), 459–474.

[90] W. T. Tutte, On the symmetry of cubic graphs, *Canad. J. Math.* **11** (1959), 621–624.

[91] R. Weiss, $s$-Transitive graphs, *Colloq. Math. Soc. János Bolyai* **25** (1978), 827–847.

[92] R. Weiss, The non-existence of 8-transitive graphs, *Combinatorica* **1** (1983), 309–311.

[93] M. Y. Xu, Vertex-transitive graphs and finite simple groups, preprint (unpublished), 1988.

[94] H. P. Yap, Point symmetric graphs with $p \leq 13$ points, *Nanta Math.* **6** (1986), 8–10.

[95] J. P. Zhang, On finite groups all of whose elements of the same order are conjugate in their automorphism groups, *J. Algebra* **153** (1992), 22-36.

# Vertex-transitive graphs and digraphs

Raffaele SCAPELLATO

*Dipartimento di Matematica*
*Politecnico di Milano*
*20133 Milano*
*Italy*

Notes by

Walter PACCO

**Abstract**

The aim of the present paper is to illustrate concepts and methods employed while working in the area of vertex-transitive graphs (or VT-graphs for short), through recently obtained significant results.

Let $G$ be a permutation group acting on the non-empty set $V$. For a fixed $(a,b) \in V \times V$, all pairs $(g(a), g(b))$ with $g \in G$ form the arc set of a digraph $\Gamma$, such that $G \leq \mathrm{Aut}\,\Gamma$. Then $\Gamma$ is said to be an *orbital digraph*. This is a fundamental concept for the study of VT-digraphs, because they all arise as *generalized orbital digraphs*, i.e. edge-disjoint unions of orbital digraphs.

Letting $p$ and $q$ denote prime numbers, we will first present the classification theorem (or **CT**) for VT-graphs of order $pq$, due to Marušič, Praeger, Scapellato, Wang and Xu. It says that such a graph must be one of the following:

(i) a metacirculant;

(ii) a graph from an exceptional class (with $p$ a Fermat prime and $q \mid p-1$);

(iii) an orbital graph arising from certain rank 3 representations of $P\Omega^{\pm}(2d, 2)$ or $M_{22}$;

(iv) a generalized orbital graph associated with the action of $A_7$ on triples;

(v) a generalized orbital graph associated with a 2-dimensional projective special linear group.

Furthermore, the five classes of graphs above are mutually disjoint. These will be described in this paper, too. Marušič and Scapellato extended the **CT** to digraphs, by proving a similar statement.

Calling *semiregular* a non-trivial permutation whose orbits all have the same length, Marušič and Jordan have independently asked whether every VT-graph has an automorphism acting semiregularly on its vertex set, as happens for Cayley graphs. Marušič and Scapellato proved that this is the case for both cubic VT-graphs and VT-digraphs of order $2p^2$.

A somewhat related question was posed by Lovász: does every VT-graph have a hamiltonian path? We will survey the known results on this topic. Applying the **CT**, Marušič proved that every VT-graph $\Gamma$ of order $pq$ and having an imprimitive automorphism subgroup, is necessarily hamiltonian, unless $\Gamma$ is the Petersen graph.

The canonical double covering of a graph $\Gamma$ is $\mathbf{B}(\Gamma) = \Gamma \times K_2$. Pacco and Scapellato proved that for a fixed bipartite digraph $\widetilde{\Gamma}$, the isomorphism classes of digraphs $\Gamma$ such



*G. Hahn and G. Sabidussi (eds.), Graph Symmetry*, 319–378.

that $\mathbf{B}(\Gamma) \cong \tilde{\Gamma}$ are in 1-1 correspondence with the conjugacy classes in $\boldsymbol{Aut}\,\tilde{\Gamma}$ of the involutions that interchange the stable parts of $\tilde{\Gamma}$.

Pacco and Scapellato also introduced the following generalization of orbital digraphs. Replace $\boldsymbol{G} \leq \boldsymbol{S}_n$ by a sugroup $\boldsymbol{G}$ of $\boldsymbol{S}_n \times \boldsymbol{S}_n$, and take as arcs the pairs $(g(a), h(b))$ with $(g, h) \in \boldsymbol{G}$. Then $\Gamma$ is said to be a *two-fold orbital digraph.* There is a close relationship between $\boldsymbol{G}$ and the automorphism group of $\mathbf{B}(\Gamma)$. Every two-fold orbital digraph such that $\boldsymbol{Aut}\,\mathbf{B}(\Gamma) \cong \boldsymbol{Aut}\,\Gamma \times \mathbb{Z}_2$ is arc-transitive.

## Introduction

The restless growth of graph theory which we witness nowadays makes it difficult to confine this matter to a single book of reasonable length, covering enough material to provide a fair idea of how much was done on the subject. In particular, the results recently obtained in algebraic graph theory, in view of their connection with the classification of finite simple groups, deserve special attention.

However, the present paper does not aim to be a complete survey of the topic of vertex-transitive graphs (from now on called VT-*graphs*), but rather to give a few instances of concepts and methods employed while working in this area.

As one may expect, most of the material here is based on my own research. But many other related results are also discussed. It was out of the question to write down all proofs. Hence, I did insert only those taken from my articles, and replaced the others with references to the original papers. There are, however, a few exceptions, usually consisting of fairly short arguments that give a feeling of the corresponding topic. On the other hand, almost no proofs are given on matters like hamiltonicity, thus avoiding long computations needed to obtain long cycles, for which an extensive literature is available.

We will often deal with *digraphs* rather than with undirected graphs. Namely, in spite of the fact that most questions in this area were raised and studied only in an undirected setting, almost always their directed versions make sense, and are as interesting as the former. Furthermore, some definitions may become even simpler. For instance, for a Cayley digraph $\boldsymbol{Cay}(\boldsymbol{G}, E)$, no restriction has to be imposed upon the set $E$. Graph-theoretical operations will also be defined for digraphs. Thus, the canonical double covering (from now on also indicated by the acronym $\boldsymbol{CDC}$) of a digraph $\Gamma$ is the categorical product $\Gamma \times K_2$.

The interplay between graphs and permutation groups can be exploited via the following standard construction. Let $\boldsymbol{G}$ be a permutation group acting on the non-empty set $V$. For a fixed $(a, b) \in V \times V$, all pairs $(g(a), g(b))$ with $g \in \boldsymbol{G}$ form the arc set of a digraph $\Gamma$, such that $\boldsymbol{G} \leq \boldsymbol{Aut}\,\Gamma$. Then $\Gamma$ is said to be an *orbital digraph.* This is a fundamental concept for our purpose, because in view of Theorem 2.2.8 every VT-digraph arises as an edge-disjoint union of orbital digraphs. By slightly generalizing this definition, we will define and apply the concept of a two-fold orbital digraph. As we shall see, this construction allows us to obtain graphs having an edge-transitive $\boldsymbol{CDC}$.

It is well known that every VT-graph of prime order is a circulant, that is, a Cayley graph on a cyclic group. An interesting generalization was introduced by Alspach and Parsons [AP]. These graphs, called *metacirculants*, have a satisfactory algebraic description in terms of appropriate subsets of residue classes (as in the special case of circulants). They play a rôle in the problem of the classification of VT-graphs whose order is a product of two primes. Independent solutions to this question were recently found by two teams of mathematicians

(C. E. Praeger, R. J. Wang, M.-Y. Xu [PWX] and D. Marušič, R. Scapellato [MS5]). Their approaches were entirely different, but orbital digraphs turned out to be a basic tool for both. A statement of this result is presented here as Theorem 3.2.1, providing an explicit description of the graphs involved.

The aforementioned classification was generalized to vertex-transitive $pq$-digraphs (Theorem 3.3.3). On the other hand, even for the undirected case, the complete description is far beyond us. Namely, for graphs whose order is a product of three or more primes the situation is much more complex than that of the $pq$-case, where almost all VT-graphs having an imprimitive automorphism subgroup are metacirculants, the only exception being the class of Marušič-Scapellato graphs. The situation changes dramatically when the number of prime factors increases, which gives rise to several different cases to analyze. As shown by Marušič, Scapellato and Zgrablić [MSZ], even the case of three primes is affected by this phenomenon.

Such a ramification should not be surprising. It may be related with the fact that not that many properties are shared by all VT-graphs. For example, though every such graph is distance degree regular (see Definition 1.2.1), its distance degrees sequence may fail to be unimodal, contrary to the special case of distance-transitive graphs (Corollary 1.2.4). In fact, infinitely many counterexamples exist (see Example 1.2.5). Among possible properties common to all VT-graphs, two will be considered here. Many further problems and results can be found for instance in the survey of Cameron [Ca2] and in the book of Yap [Ya].

Calling *semiregular* a non-trivial permutation whose orbits all have the same length, Marušič [Ma1] and Jordan [Ma5] have independently asked whether every VT-graph has an automorphism acting semiregularly on its vertex set, which of course happens for all Cayley graphs. Such a digraph will be said to be *rotary*. A partial affirmative solution to this problem will be given here, by showing that all cubic VT-graphs are rotary (Theorem 4.1.5). The same property will be proved for VT-digraphs of order $2p^2$, where $p$ is a prime (Theorem 4.1.6). This is a complement to an earlier result of Marušič [Ma1], who proved that all digraphs of prime-power order are rotary (Theorem 4.1.4).

Another intriguing question was posed by Lovász [Lo]: does every VT-graph have a hamiltonian path? This is somewhat related to the preceeding question (see Alspach [Al2]). We will survey the known results on this topic, some of which concern Cayley graphs. A direct application of the classification theorem of vertex-transitive $pq$-graphs will be given. Namely, if such a $\Gamma$ has an imprimitive automorphism subgroup, it is necessarily hamiltonian, unless $\Gamma$ is the Petersen graph (Theorem 4.2.12).

The determination of the full automorphism group of a graph is definitively a difficult problem (see e.g. Godsil [Go]). In particular, it is not easy to decide whether the automorphism group of the ***CDC*** of a given graph $\Gamma$ is actually isomorphic to ***Aut*** $\Gamma \times \mathbb{Z}_2$. A graph satisfying this last property will be called *stable*. Nedela and Škoviera [NS2] have recently raised the problem of finding an edge-transitive unstable graph. This is still unsolved, but we have at least a non-trivial instance of an unstable Cayley graph (see Example 5.1.2 and the comments preceding it).

As known, a graph $\Gamma$ has a connected ***CDC*** $\widetilde{\Gamma}$ if and only if $\Gamma$ is connected and non-bipartite. This holds true also for the directed case. Given such a $\widetilde{\Gamma}$, it is then natural to ask whether $\Gamma$ is unique up to isomorphism. An answer is provided by Theorem 5.1.5: the isomorphism classes of digraphs $\Gamma$ whose ***CDC*** is $\widetilde{\Gamma}$ are in 1-1 correspondence with the conjugacy classes in ***Aut*** $\widetilde{\Gamma}$ of certain involutions. By the way, this is an instance of a question where symmetry plays a rôle in the solution of a problem, though it does not appear explicitly into the statement.

As a tool for the study of $\boldsymbol{CDC}$s, we will introduce here the natural generalization of orbital digraphs that we have mentioned earlier. Let us replace $\boldsymbol{G} \leq \boldsymbol{S}_n$ by a sugroup $\mathcal{G}$ of $\boldsymbol{S}_n \times \boldsymbol{S}_n$, and take as arcs the pairs $(g(a), h(b))$ with $(g, h) \in \mathcal{G}$. Then $\Gamma$ is said to be a two-fold orbital digraph (or $\boldsymbol{TOG}$ in short). Though the arc set of $\Gamma$ is still an orbit of a group $\mathcal{G}$ acting on $V \times V$, as in the case of orbital digraphs, its behaviour may be quite wild. For example, it could fail to be vertex-transitive, and may have loops, even for $a \neq b$. On the other hand, we shall demonstrate (Proposition 6.1.3) a close relationship between $\mathcal{G}$ and the automorphism group of $\Gamma \times K_2$ (rather than of $\Gamma$). Furthermore, we shall prove that every stable $\boldsymbol{TOG}$ is arc-transitive (Proposition 6.3.2).

A first study of $\boldsymbol{TOG}$s will be carried out here, focusing on properties that differ from those of orbital digraphs. A characterization will be given for undirected, disconnected $\boldsymbol{TOG}$s (Theorems 6.2.2 and 6.2.3).

This paper is organized as follows. It consists of six chapters, each divided into sections. The first two provide some background on graphs and on permutation groups. They are especially meant for the readers that are not experts in the area. A little familiarity with graph theory and elementary group theory is neverthless expected. Some definitions and results on graphs are given in Chapter 1. An introduction to vertex-transitive graphs is then provided in Chapter 2.

The classification problem for vertex-transitive $pq$-graphs (and digraphs) is treated in Chapter 3 in full detail. Further results on VT-graphs, concerning both semiregular automorphisms and hamiltonian paths and cycles, are presented in Chapter 4.

Canonical double coverings are dealt with in Chapter 5, followed by the related topic of two-fold orbital digraphs, which form the subject of Chapter 6.

I am grateful to the organizers of the NATO-ASI on Graph Symmetry and Applications, and especially to Geňa Hahn and Gert Sabidussi for their kind invitation. They did a very fine work, and succeeded in collecting a lot of young and *less* young researchers under the sunshine of a beautiful summer in Montreal (i.e., Montréal). I thank also all the people who attended my lectures there.

I am indebted to Dragan Marušič and Walter Pacco, as co-authors of the papers on which the present one is based. Moreover, thanks are due to Roman Nedela, for the inspiring discussions we had during his stay in Italy, on some of the topics treated here. Useful comments came also from Marko Lovrečič Saražin.

Finally, I am grateful to Walter Pacco for the work he has performed as my note-taker, and also for his skill in decrypting some of my hand-written notes.

# Chapter 1

# Graphs and their metric structure

A *directed graph* or *digraph* is an ordered pair $\Gamma = (V, A)$, where $V$ is a set and $A \subseteq V \times V$. The elements of $V$ are the *vertices* of $\Gamma$, those of $A$ are the *arcs*. If $u = v$, the arc $(u, v)$ is said to be a *loop*. We write $V(\Gamma) = V$ and $A(\Gamma) = A$. Letting $A^T = \{(u,v) \mid (v,u) \in A\}$, if $A = A^T$ then $\Gamma$ is called *undirected*. Then $E(\Gamma) = \{\{u,v\} \mid (u,v) \in A\}$ is the set of *edges* of $\Gamma$. If $(u,v) \in A(\Gamma)$, we often write $u \rightarrow_\Gamma v$ (or $u \rightarrow v$ for short). Similarly, if $\Gamma$ is undirected,

$u \to v$ stands for $\{u, v\} \in E(\Gamma)$.

For a fixed vertex $u$ of $\Gamma$, denote by $\mathrm{val}^+(u)$ and $\mathrm{val}^-(u)$ the out-valency and the in-valency of $u$. Of course in the undirected case the number $\mathrm{val}(u) = \mathrm{val}^+(u) = \mathrm{val}^-(u)$ is called the *valency* of $u$.

The notion of a bipartite graph will be dealt with both for the undirected and the directed case, as follows.

**Definition 1.0.1** A digraph $\Gamma$ is said to be *bipartite* if there exists a partition of the vertex set $V(\Gamma)$ into two subsets $V_0$ and $V_1$ such that for all arcs $u \to v$ either $u \in V_0$ and $v \in V_1$ or $u \in V_1$ and $v \in V_0$. Then $V_0$ and $V_1$ are said to be the *stable parts* of $\Gamma$.

Note that the stable parts of a bipartite digraph $\Gamma$ are uniquely determined if and only if $\Gamma$ is weakly connected.

We will often follow the usual terminology, to call an undirected graph simply a *graph*, and to identify it with the pair $(V, E)$. The rest of our notation is mostly standard and will not be specified.

## 1.1 Products and their by-products

Let $\Gamma_1$ and $\Gamma_2$ be digraphs. There are several 'natural' ways to define a new digraph $\Gamma$ such that $V(\Gamma) = V(\Gamma_1) \times V(\Gamma_2)$ and the adjacency

$$(u_1, u_2) \to (v_1, v_2) \tag{1}$$

is determined solely by the adjacency in $\Gamma_1$ and $\Gamma_2$, according to a given rule. The most common operations of this kind are the following three:

- the *cartesian product* $\Gamma = \Gamma_1 \square \Gamma_2$, where (1) holds whenever $u_1 \to v_1$ and $u_2 = v_2$, or else $u_1 = v_1$ and $u_2 \to v_2$;
- the *categorical product* $\Gamma = \Gamma_1 \times \Gamma_2$, where (1) holds whenever $u_1 \to v_1$ and $u_2 \to v_2$;
- the *strong product* $\Gamma = \Gamma_1 \boxtimes \Gamma_2$, where $A(\Gamma) = A(\Gamma_1 \square \Gamma_2) \cup A(\Gamma_1 \times \Gamma_2)$.

Of course, if $\Gamma_1$ and $\Gamma_2$ are undirected, then so is $\Gamma$. Moreover, letting $\circ$ denote any of these three operations, there is always a monomorphism

$$\phi : \boldsymbol{Aut}\,(\Gamma_1 \times \Gamma_2) \to \boldsymbol{Aut}\,(\Gamma_1 \circ \Gamma_2)$$

defined by $\phi : (f_1, f_2) \mapsto f$, where $f : (v_1, v_2) \mapsto (f_1(v_1), f_2(v_2))$. This is easy to see, because in any of the three products the adjacency is based on the relations $=$ and $\to$, which are preserved by the $f_i$.

For both the cartesian and the strong product of two undirected graphs $\Gamma_1$ and $\Gamma_2$, their connectedness is sufficient (besides being obviously necessary) in order to have connectedness for $\Gamma_1 \circ \Gamma_2$. Asking the same question for the categorical product, the answer is less categorical. As indicated below, this operation does not always provide a connected graph when $\Gamma_1$ and $\Gamma_2$ are. For a proof see [CM], p. 84, Theorem 5.53.

**Proposition 1.1.1** *The categorical product $\Gamma = \Gamma_1 \times \Gamma_2$ of the undirected graphs $\Gamma_1$ and $\Gamma_2$ is connected if and only if both $\Gamma_1$ and $\Gamma_2$ are connected, and at least one of them is not bipartite. Moreover, $\Gamma$ is bipartite if at least one of $\Gamma_1$ and $\Gamma_2$ is.* □

More information on these operations can be found e.g. in [CM] and [Sa].

**Definition 1.1.2** The categorical product $\Gamma \times K_2$ is called the *canonical double covering* (***CDC***) of $\Gamma$, and is denoted by $\mathbf{B}(\Gamma)$.

In view of Definition 1.1.2, $V(\mathbf{B}(\Gamma)) = V(\Gamma) \times \mathbb{Z}_2$ and $(u, \varepsilon) \to (v, \varepsilon + 1)$ $(\varepsilon \in \mathbb{Z}_2)$ if and only if $u \to v$ in $\Gamma$. The name *canonical double covering* will be justified later (see Example 1.1.5). From Proposition 1.1.1 we now get the following result.

**Corollary 1.1.3** *The* ***CDC*** *of $\Gamma$ is connected if and only if $\Gamma$ is connected and non-bipartite.* □

The construction that we are now going to introduce originated in algebraic topology (where coverings are looked at as to 'local homeomorphisms'). However, our point of view (as that of [Bi]) is purely algebraic. There is no need of topological background here.

**Definition 1.1.4** For a fixed integer $k$, consider the symmetric group $\boldsymbol{S}_k$, acting on the set $X = \{0, 1, \ldots, k-1\}$. Let $\Gamma$ be a digraph and $\sigma : A(\Gamma) \to \boldsymbol{S}_k$ be a map. The *$k$-covering* $\widetilde{\Gamma}$ associated with the *base digraph* $\Gamma$ and the *voltage* $\sigma$ is the digraph with vertex set $V(\Gamma) \times \{0, 1, \ldots, k-1\}$ and arc set consisting of the pairs $((u, i)(v, j))$ such that $(u, v) \in A(\Gamma)$ and $\sigma(u, v)(i) = j$.

For $k = 2$ one speaks usually of a (general) *double covering.* These graphs were characterized by Zelinka [Ze].

The above is still a sort of product, in the sense that we deal with a cartesian product of $V(\Gamma)$ with the set $X$ that can be looked at as $V(K_k)$. It differs from those mentioned earlier because, among other things, the valency of the vertex $(u, i)$ is *equal* to that of $u$. The other products produce usually an increment of the valency.

Clearly, if $\sigma = \mathrm{id}$, then the $k$-covering $\widetilde{\Gamma}$ is nothing else than a disjoint union of $k$ digraphs, each isomorphic with $\Gamma$. More interesting is the following special case.

**Example 1.1.5** Let $\sigma : A(\Gamma) \to \boldsymbol{S}_2$ be the map taking each arc into the transposition (0 1). Then the resulting 2-covering $\widetilde{\Gamma}$ is isomorphic to $\Gamma \times K_2$. The identity on the set $\Gamma \times \{0, 1\}$ is in fact an isomorphism, because $(u, i) \sim_{\widetilde{\Gamma}} (y, j)$ if and only if $\sigma(i) = j$, that is $i \neq j$. Hence this particular covering can be identified with the ***CDC***.

More information on this construction can be found for example in [Bi], [Mal] or [MM].

## 1.2 Metric properties

This is a short account about definitions and results on graphs that enjoy special properties with respect to distance, some of which will be recalled later. For more information, the reader may consult the references or the monograph [BCN].

All graphs in this section will be *undirected*, *loopless* and *connected*. Each such $\Gamma$ gives rise to a metric space $(V(\Gamma), \eth)$, where the distance $\eth(u,v)$ between two vertices of $\Gamma$ is the length of a shortest path joining $u$ and $v$. The maximum value taken by $\eth$ is called the *diameter* of $\Gamma$ and is denoted by $\operatorname{diam}\Gamma$. Note that $\boldsymbol{Aut}\,\Gamma = \boldsymbol{Aut}\,(V(\Gamma), \eth)$, i.e. that the automorphisms of $\Gamma$ are exactly the distance-preserving bijections of $V(\Gamma)$ onto itself.

Besides valency, other parameters can be associated to a vertex $v$. Such are the numbers usually called distance degrees (see e.g. [Hi]). Vertex-transitivity implies of course distance degree regularity, in the sense of the following.

**Definition 1.2.1** Given a graph $\Gamma$ of diameter $d$, and a vertex $v$ of $\Gamma$, let

$$\Gamma_i(v) = \{u \in V(\Gamma) \mid \eth(u,v) = i\}$$

for $i = 0, 1, \ldots, d$. The *distance degrees* of $v$ are the numbers $k_0, k_1, \ldots, k_d$ where $k_i = |\Gamma_i(v)|$. When the $k_i$ depend only on $i$, the graph is said to be *distance degree regular.*

We will use the abbreviation $\Gamma(v) = \Gamma_1(v)$. (In other papers, this set is often referred to as $N_\Gamma(v)$).

Let $\Gamma$ be a graph. For $(u,v) \in V(\Gamma) \times V(\Gamma)$ with $\eth(u,v) = m$, define:

$$\begin{aligned} A(u,v) &= \{w \mid w \sim u,\ \eth(w,v) = m\}; \\ B(u,v) &= \{w \mid w \sim u,\ \eth(w,v) = m+1\}; \\ C(u,v) &= \{w \mid w \sim u,\ \eth(w,v) = m-1\}. \end{aligned}$$

Clearly, $A(u,v)$, $B(u,v)$ and $C(u,v)$ are pairwise disjoint, and their union is $\Gamma_1(u)$. Note that $|A(u,v)|$, $|B(u,v)|$ and $|C(u,v)|$ depend on $(u,v)$, and some of these sets may be empty. For instance, bipartite graphs (and they only) satisfy $A = \emptyset$ for all $(u,v)$.

**Definition 1.2.2** Let $\Gamma$ be a graph, and let $u,v$ be vertices of $\Gamma$, with $\eth(u,v) = m$. Then $\Gamma$ is said to be *distance-regular* if the numbers $a_m = |A(u,v)|$, $b_m = |B(u,v)|$ and $c_m = |C(u,v)|$ depend only on $m$.

The numbers $a_m$, $b_m$ and $c_m$ as above are usually called the *parameters* of $\Gamma$.

Distance-regularity is a very strong property. Not surprisingly, it implies distance degree regularity. This fact, together with some relations among the parameters, is stated just below (for a proof see [Bi], Proposition 20.4).

**Proposition 1.2.3** *Let $\Gamma$ be a distance-regular graph with diameter $d$. Then $\Gamma$ is distance degree regular. Moreover, the following hold:*

(i) $k_i b_i = k_{i+1} c_{i+1}$ *for all* $i = 0, 1, \ldots, d-1$;

(ii) $k_1 = b_0 > b_1 \geq b_2 \geq \ldots \geq b_{d-1}$;

(iii) $1 = c_1 \leq c_2 \leq \ldots \leq c_k$. □

There are other known relations among distance degrees and parameters of a distance-regular graphs (see, for example, [Bi] or [BW]). The following (see [PSY], Lemma 1.1) proves that the distance degrees sequence is *unimodal* for such graphs.

**Corollary 1.2.4** *Let $\Gamma$ be a distance-regular graph of diameter $d$. Then there are two integers $h, l$ such that:*

$$k_1 < k_2 < \ldots < k_h = \ldots = k_l > \ldots > k_d.$$

**Proof** In view of Proposition 1.2.3, (i), we have $k_i b_i = k_{i+1} c_{i+1}$ for all $i = 0, 1, \ldots, d-1$. Thus we get

$$\frac{k_i}{k_{i+1}} = \frac{c_{i+1}}{b_i} \leq \frac{c_{i+2}}{b_{i+1}} = \frac{k_{i+1}}{k_{i+2}},$$

from which the statement follows. □

Note that in general vertex-transitivity does not imply unimodality, as shown below.

**Example 1.2.5** Consider the strong product $\Gamma = C_n \boxtimes K_2$ for $n \geq 7$. The constant value 4 is repeated in the degree sequence of $\Gamma$, even though it is not its maximum. Therefore the sequence is not unimodal.

Let us now introduce another remarkable class of graphs that essentially consists of the distance-regular graphs of diameter two (see Proposition 1.2.8).

**Definition 1.2.6** An undirected graph $\Gamma$ is said to be *strongly regular*, with parameters $k$, $\lambda$ and $\mu$, if it is regular of valency $k$, and the number of vertices adjacent to two fixed vertices $u, v \in V(\Gamma)$ is equal to $\mu$ or to $\lambda$, according as $u$ and $v$ are adjacent or not.

It is known that $\lambda m = k(k - \mu - 1)$, hence $m$ is obtained from this equation. If $\Gamma$ is connected, we have $|V| = 1 + k + m$; then the triple $(k, \lambda, \mu)$ gives also the order of $\Gamma$. Moreover, if $\Gamma$ is strongly regular, with parameters $(k, \lambda, \mu)$, the same holds for its complement $\Gamma^c$, whose parameters are

$$\overline{k} = m,\ \overline{\lambda} = m - k + \mu + 1,\ \overline{\mu} = m - k + \lambda - 1.$$

The results below characterize strongly regular graphs in both the disconnected and the connected case.

**Proposition 1.2.7** *Let $\Gamma$ be a strongly regular graph. Then the following are equivalent:*

(i) *$\Gamma$ is disconnected;*

(ii) *$\Gamma$ is a union of complete graphs;*

(iii) $\lambda = 0$. □

**Proposition 1.2.8** *A graph is distance-regular of diameter 2 if and only it is connected and strongly regular.* □

For more information about strongly regular graphs we strongly recommend Cameron [Ca1].

Because of the relevance of shortest paths in this context, it may seem natural to introduce the following further class of graphs.

**Definition 1.2.9** A graph $\Gamma$ is said to be *uniformly geodetic* if the number of shortest paths joining two vertices $u$, $v$ of $\Gamma$ depends only on $\partial(u, v)$. When this number is always equal to 1, we speak of a *geodetic graph.*

This means that there is a map $F: \{1, \ldots, \text{diam } \Gamma\} \to \mathbb{N}$ taking $m$ into the number $F(m)$ of shortest paths between a pair of vertices distance $m$ apart.

**Proposition 1.2.10** *A graph is uniformly geodetic if and only if $|C(u, v)|$ depends only on $\partial(u, v)$. In particular, every distance-regular graph is uniformly geodetic.* □

The hypercube $Q_n$ is known to be distance-regular, hence uniformly geodetic. The corresponding $F$ is given by $F(m) = m!$ for all $m$. Conversely (see e.g. [CS]), each bipartite uniformly geodetic graph with such $F$ is a hypercube.

However, the class of uniformly geodetic graphs is much wider than that of distance-regular graphs. It also contains several non-regular examples. An easy one is that of trees, all of which geodetic. A further instance is given below.

**Example 1.2.11** Let $\mathcal{A} = (P, R, I, \|)$ be an affine space. Define $\Gamma$ by $V(\Gamma) = P \cup R$, and $u \sim v$ if and only $u$ and $v$ are two parallel lines, or a point and a line that are incident. As remarked in [Sc], Proposition 11, if $\mathcal{A}$ is not a plane, then $\Gamma$ is uniformly geodetic, having diameter 3 and satisfying $F(1) = F(2) = 1$ and $F(3) = 2k$. On the other hand, if $\mathcal{A}$ is a plane, then $\Gamma$ is geodetic, of diameter 2.

# Chapter 2

# Permutation groups and vertex-transitive graphs

This introductory chapter starts by reviewing a few known facts and definitions on permutation groups. Then vertex-transitive digraphs are considered. As remarked already, we will mostly deal with digraphs, rather than graphs, especially in view of the increasing interest of graph-theorists for the directed case (see e.g. [CLP]).

## 2.1 Permutation groups

A *permutation group* on the non-empty set $X$ is a triple $(\boldsymbol{G}, X, a)$, where $\boldsymbol{G}$ is a group and $a : \boldsymbol{G} \to \boldsymbol{Sym} X$ a homomorphism (here $\boldsymbol{Sym} X$ is the symmetric group on $X$). The integer $|X|$ is called the *degree* of $(\boldsymbol{G}, X, a)$. We then say that $\boldsymbol{G}$ *acts* on $X$. This action is said to be *faithful* when $a$ is injective. In order to simplify the notation, the map $a$ will not appear explicitly later. Also, $\boldsymbol{G}$ will often stand for $(\boldsymbol{G}, X, a)$.

The relation: $x \cong y$ if and only if there is $g \in \boldsymbol{G}$ such that $g(x) = y$, is an equivalence on $X$. Its classes are the *orbits* of $\boldsymbol{G}$ in $X$. The group $\boldsymbol{G}$ is said to be *transitive* on $X$ when it has a unique orbit. That is, for all $x, y \in X$, there is *at least* one $g \in \boldsymbol{G}$ such that $g(x) = y$. Dually, $\boldsymbol{G}$ is called *semiregular* if, for all $x, y \in X$, there is *at most* one $g \in \boldsymbol{G}$ with $g(x) = y$. When both properties are satisfied, the permutation group is said to be *regular*. In other words, for all $x, y \in X$, there is *exactly* one $g \in \boldsymbol{G}$ such that $g(x) = y$.

For $x \in X$, the set of all elements of $\boldsymbol{G}$ such that $g(x) = x$ is a subgroup $\boldsymbol{G}_x$ of $\boldsymbol{G}$, called the *stabilizer* of $x$. When $\boldsymbol{G}$ is transitive, its stabilizers form a conjugacy class. If $\boldsymbol{G}$ is regular, then $\boldsymbol{G}_x = 1$, and conversely. An orbit of $\boldsymbol{G}_x$, different from the trivial one $\{x\}$, is said to be a *suborbit* of $\boldsymbol{G}$ *relative* to $x$. The number of elements of a suborbit is said to be a *subdegree* of $\boldsymbol{G}$.

The following is a very useful arithmetical condition.

**Theorem 2.1.1** *If $\boldsymbol{G}$ is a permutation group on $X$, then the order of the orbit of $x \in X$ is $|\boldsymbol{G} : \boldsymbol{G}_x|$. In particular, if $\boldsymbol{G}$ is transitive, then $|\boldsymbol{G} : \boldsymbol{G}_x| = |X|$.* □

As an example of an application of Theorem 2.1.1, we have the next result, which will be needed later.

**Lemma 2.1.2** *Let $\boldsymbol{G}$ be a transitive group on a set $V$ of degree $n$ and order $ns$. Let $\boldsymbol{H}$ be a subgroup of $\boldsymbol{G}$ such that $n$ divides $|\boldsymbol{H}|$ and $([\boldsymbol{G} : \boldsymbol{H}], |\boldsymbol{H}|) = 1$. Then $\boldsymbol{H}$ is transitive.*

**Proof** Let $|\boldsymbol{H}| = nt$. Then $(nt, s/t) = 1$. Let $x\boldsymbol{H}$ denote the orbit of $\boldsymbol{H}$ containing $x$. For any $x \in \boldsymbol{H}$, we must have $|x\boldsymbol{H}||\boldsymbol{H}_x| = |\boldsymbol{H}|$ (Theorem 2.1.1). Now $|\boldsymbol{H}_x|$ divides $|\boldsymbol{H}|$ as well as $s$ since $\boldsymbol{H}_x \subseteq \boldsymbol{G}_x$. So $|\boldsymbol{H}_x|$ divides $(s, nt)$, which is equal to $t$. But then $|x\boldsymbol{H}| \geq nt/t = n$. □

If $\boldsymbol{G}$ is transitive on $X$, an equivalence $\equiv$ on $X$ is said to be *compatible* with $\boldsymbol{G}$ when

$$\forall x, y \in X\ \forall g \in \boldsymbol{G}\ \ x \equiv y \ \ \Rightarrow \ \ g(x) \equiv g(y).$$

The elements of $X/_{\equiv}$ are called blocks, and the quotient $\overline{X} = X/_{\equiv}$ is said to be an *imprimitivity block system* for $\boldsymbol{G}$. Clearly, the elements of $\boldsymbol{G}$ take blocks into blocks: this means that there is another permutation group $(\boldsymbol{G}, \overline{X}, \overline{a})$, where $\overline{a}$ is the action of $\boldsymbol{G}$ on blocks. The situation may be represented diagrammatically as follows.

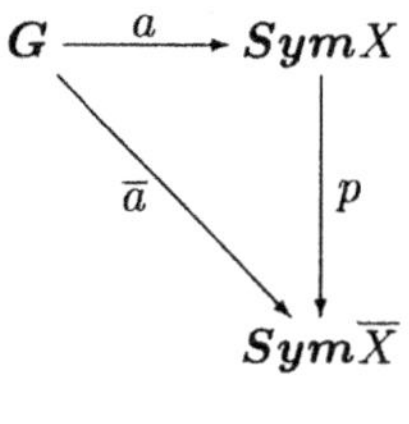


Conversely, one sees that a partition $\mathcal{P}$ of $X$ with the above property is the quotient of a suitable equivalence compatible with $\boldsymbol{G}$. For such a $\mathcal{P}$, we then have that $|\mathcal{P}|$ divides $|\boldsymbol{G}|$.

Of course, the *trivial* equivalences $\mathrm{id}_X$ and $X \times X$ corresponding to $\mathcal{P} = \{\{x\} | x \in X\}$ and $\mathcal{P} = \{X\}$, respectively, are *always* compatible. Now, a transitive group $\boldsymbol{G}$ is said to be *imprimitive* when there is a non-trivial equivalence compatible with $\boldsymbol{G}$, and *primitive* otherwise. Note that if $\Gamma$ is a connected and bipartite graph and $\boldsymbol{Aut}\,\Gamma$ is transitive, then the two stable parts of $\Gamma$ form an imprimitivity block system of $\boldsymbol{Aut}\,\Gamma$. A further instance is given below.

**Example 2.1.3** Given a graph $\Gamma$, let $\mathcal{R}$ be the relation on $V(\Gamma)$ defined by

$$u\mathcal{R}v \iff \Gamma(u) = \Gamma(v)$$

for $u, v \in V(\Gamma)$. Then $\mathcal{R}$ is an equivalence, compatible with $\boldsymbol{Aut}\,\Gamma$. Therefore, when $\mathcal{R}$ is non-trivial and the action of this group is transitive, it is also imprimitive.

A permutation group $\boldsymbol{G}$ on $X$ is said to be *doubly transitive* if, for all $x, y, x', y'$ with $x \neq y$ and $x' \neq y'$, there is $g \in \boldsymbol{G}$ such that $g(x) = x'$ and $g(y) = y'$. It is easily seen that each doubly transitive group is primitive. The converse does not hold: for instance, all groups of prime order are primitive. Hence we shall call *uniprimitive* a group that is primitive but not doubly transitive.

The properties we are dealing with depend on the action, not only on the group itself. As an example, the group $\boldsymbol{G} = \boldsymbol{GL}(n, \mathbb{K})$ acts both on the set $P$ of points and the set $L$ of lines of the affine space of dimension $n$ over $\mathbb{K}$. The former is doubly transitive (hence primitive). For the latter, the relation $\|$ is compatible, so this action is imprimitive: the blocks are the classes of parallel lines.

If $\Gamma$ is a graph, the double transitivity of $\boldsymbol{Aut}\,\Gamma$ forces $\Gamma$ to be trivial. However, the full automorphism group is primitive for many non-trivial graphs. The main object of our study is introduced below.

**Definition 2.1.4** A digraph $\Gamma$ is called *vertex-transitive* (or a VT-*digraph*) if $\boldsymbol{Aut}\,\Gamma$ is transitive on $V(\Gamma)$.

Note that, as we saw for affine spaces, the automorphism group of a structure acts not only on the elements of its carrier, but also on all other 'parts' (e.g., lines, in that case). For graphs, other than vertices, one can consider edges and arcs, which leads to the concepts of edge- and arc-transitivity. Other kinds of transitivity have also been considered. These include $t$-transitivity, $t \geq 2$, looking at the action on induced paths of length $t$, and distance-transitivity, where $\boldsymbol{Aut}\,\Gamma$ is transitive on each of the sets $S_i = \{(u,v) \mid \eth(u,v) = i\}$. More recently, special attention has been paid to graphs which are both vertex- and edge-transitive, but not arc-transitive (see e.g. [AMN]).

The various kinds of transitivity are often related to each other. For instance, arc-transitivity implies of course edge-transitivity, and also vertex-transitivity (unless there are isolated vertices). On the other hand, the following is known (see e.g. [Ya]).

**Theorem 2.1.5** *Let $\Gamma$ be un undirected graph. If $\Gamma$ is edge-transitive, then it is either vertex-transitive or bipartite.* □

## 2.2 Orbital digraphs

We will see soon how VT-digraphs arise from the action of a group (Definition 2.2.3).

**Definition 2.2.1** Let $\boldsymbol{G}$ be a transitive permutation group on the set $V$. The orbits of the action of $\boldsymbol{G}$ on $V \times V$ are called *orbitals.* One of them, $D_0 = \{(u,u) \mid u \in V\}$, is called the *trivial orbital.* The *transpose* $D^T$ of the orbital $D$ is defined by $D^T = \{(u,v) \mid (v,u) \in D\}$. We say that $D$ is *self-paired* when $D = D^T$. The number $r$ of orbitals of $\boldsymbol{G}$ (including $D_0$) is the *rank* of $\boldsymbol{G}$

Clearly the rank of $G$ satisfies $r \geq 2$, except for the trivial case $|V| = 1$. Moreover $r = 2$ means that $G$ is doubly transitive. As remarked already, such a $G$ is not interesting as an automorphism group acting on the vertices of a graph. From this point of view, the first nontrivial case is $r = 3$. It corresponds to the *strongly regular graphs* (see Proposition 2.2.9). From now on, all orbitals will be assumed to be non-trivial.

**Proposition 2.2.2** *If either $r$ or $|G|$ is even, then there is a self-paired orbital.*

**Proof** If $r$ is even, there is an odd number of orbitals. As they are pairwise transpose, at least one must be self-paired.

If $|G|$ is even, the group has an element $g$ of order 2. Hence there are $u, v \in V$ with $u \neq v$, $g(u) = v$ and $g(v) = u$. Thus $(u, v) \in D$ and $(v, u) \in D^T$ belong to the same orbital, so that $D = D^T$. □

**Definition 2.2.3** Let $G$ be a permutation group acting on the set $V$. If $D$ is an orbital of $G$, we call *orbital digraph* the digraph $\Gamma(D) = (V, D)$. If $D$ is a union of orbitals of $G$, then $\Gamma(D) = (V, D)$ it is said to be a *generalized orbital digraph.*

Clearly, $\Gamma$ is undirected exactly when $D$ is self-paired. Note that an orbital digraph is either undirected or without opposite arcs: in fact, $D \cap D^T \neq \emptyset$ implies $D = D^T$ for all orbitals $D$.

**Example 2.2.4** Consider the natural action of $G = \langle (12 \dots n) \rangle$ on the set $V = \{1, 2, \dots, n\}$. The orbitals containing $(1, 2)$, $(1, 3), \dots, (1, n)$ are all different, hence the rank is $n$. Each of the corresponding graphs is a directed cycle of length $n$, or else a union of cycles whose length is a fixed divisor of $n$. They are illustrated by Figure 1 for small values of $n$.

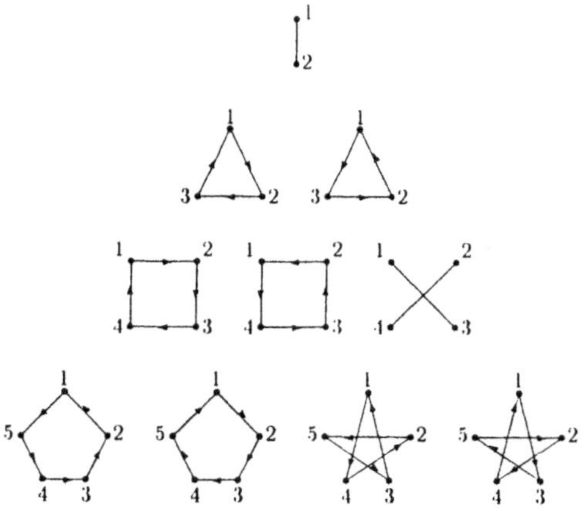

Figure 1: Orbital digraphs for small $n$

The following is a straightforward consequence of Definition 2.2.3.

**Proposition 2.2.5** *Each orbital digraph is arc-transitive.* □

Let $\Gamma = \Gamma(D, \boldsymbol{G})$ be an orbital digraph of $\boldsymbol{G}$. If $v$ is a vertex of $\Gamma$, then the vertices $w$ adjacent from $v$, that is, those such that $(v, w) \in D$, form an orbit of $\boldsymbol{G}_v$, that is, a *suborbit* of $\boldsymbol{G}$. Conversely, we can associate an orbital digraph to each suborbit of $\boldsymbol{G}$. This observation leads to the following alternative definition of orbital and generalized orbital graphs. It is easily checked that the digraphs concerned are in fact the same.

**Definition 2.2.6** Let $\boldsymbol{G}$ be a permutation group acting on the set $V$, and let $W$ be a suborbit of $\boldsymbol{G}$ relative to a fixed $v \in V$. We call *orbital digraph of* $\boldsymbol{G}$ *with respect to* $W$ the digraph $\Gamma(\boldsymbol{G}, W)$ which has vertex set $V$, and whose arcs are the pairs $(u, w)$ such that $g(u) = v$ and $g(w) \in W$ for some $g \in \boldsymbol{G}$. If $W$ is a union of suborbits $W_1$, $W_2$, ..., $W_m$ relative to $v$, the edge-disjoint union $\Gamma(\boldsymbol{G}, W)$ of the digraphs $\Gamma(\boldsymbol{G}, W_i)$ is called a *generalized orbital digraph.*

Note that the subdegrees of $\boldsymbol{G}$ are equal to the valencies of the corresponding orbital digraphs.

**Theorem 2.2.7** *The group* $\boldsymbol{G}$ *is primitive if and only if each orbital digraph of* $\boldsymbol{G}$ *is strongly connected.*

**Proof** Let $\Gamma = (V, D)$ be an orbital digraph. Let $\approx$ denote the equivalence relation defined by: $u \approx v$ if and only if there are directed paths both from $u$ to $v$ and from $v$ to $u$. If $u \approx v$ then

$$u = u_1 \to u_2 \to \cdots \to u_k = v;$$

then for $g \in \boldsymbol{G}$ we get

$$g(u) = g(u_1) \to g(u_2) \to \cdots \to g(u_k) = g(v),$$

and similarly for the path from $v$ to $u$. Hence $g(u) \approx g(v)$. Therefore $\approx$ is compatible with $\boldsymbol{G}$. Thus, if some orbital digraph is not strongly connected, $\boldsymbol{G}$ turns out to be imprimitive.

Conversely, let $\boldsymbol{G}$ be imprimitive, having $\{V_1, \ldots, V_k\}$ as an imprimitivity block system, and let $u, v, w \in V_1$, with $u \neq v$. The orbital digraph $\Gamma$ associated with $(u, v)$ is non-trivial, and is not strongly connected, because no element of $\boldsymbol{G}$ can take the arc $(u, v)$ into $(w, w')$, with $w' \in V_i$, $i \neq 1$. □

**Theorem 2.2.8** *Let* $\Gamma = (V, A)$ *be a digraph, and let* $\boldsymbol{G} \leq \boldsymbol{Aut}\,\Gamma$ *be vertex-transitive. Then* $A$ *is a union of orbitals of* $\boldsymbol{G}$ *on* $V$. *Conversely, if* $\boldsymbol{G}$ *is a transitive permutation group on* $V$ *and* $A$ *is a union of orbitals of* $\boldsymbol{G}$ *on* $V$, *then the digraph* $(V, A)$ *is vertex-transitive. Moreover,* $\boldsymbol{G}$ *is arc-transitive on* $\Gamma$ *if and only if* $\Gamma$ *is an orbital digraph on* $\boldsymbol{G}$. *Finally,* $\Gamma$ *is undirected if and only if* $D \subseteq A$ *implies* $D^T \subseteq A$.

**Proof** Let $\mathcal{D}$ be the set of all orbitals $D$ of $\boldsymbol{G}$ such that $D \cap A \neq \emptyset$. Clearly $D \in \mathcal{D}$ implies $D \subseteq A$, so that $A$ is the union of all $D \in \mathcal{D}$. The rest is straightforward. □

An alternative description available for all VT-digraphs is due to Sabidussi (see e.g. [Ya], Theorem 4.3, p. 102).

For edge-transitive graphs no analogue of Theorem 2.2.8 is known. But recall Theorem 2.1.5: these graphs cannot be obtained as unions of orbital graphs because they may fail to be vertex-transitive.

**Proposition 2.2.9** *If $\boldsymbol{G}$ has rank 3 and $D$ is a self-paired orbital of $\boldsymbol{G}$, then the graph $\Gamma(D)$ is strongly regular.*

**Proof** If $u \sim v$ and $u' \sim v'$, since $\Gamma(D)$ is an orbital graph, there is $g \in \boldsymbol{G}$ such that $g(u) = u'$ and $g(v) = v'$. If $w \sim u$ and $w \sim v$ we then have $g(w) \sim u'$ and $g(w) \sim v'$. It is now clear that $g$ induces a bijection from $\Gamma(u) \cap \Gamma(v)$ to $\Gamma(u') \cap \Gamma(v')$. Their numbers of elements are therefore the same.

Assume now that $u \not\sim v$ and $u' \not\sim v'$. The other orbital graph is $\Gamma^c$, and the pairs $u, v$ and $u', v'$ form edges in $\Gamma^c$. Then there is $\overline{g} \in \boldsymbol{G}$ such that $\overline{g}(u) = u'$ and $\overline{g}(v) = v'$. At this point we can repeat the same argument as above, concluding that $|\Gamma^c(u) \cap \Gamma^c(v)| = |\Gamma^c(u') \cap \Gamma^c(v')|$. □

Finally, let us give a primitivity criterion for rank 3 graphs. For a proof see [BW].

**Theorem 2.2.10** *Let $\Gamma$ be connected, with $\boldsymbol{G} \leq \boldsymbol{Aut}\,\Gamma$ of rank 3. The action of $\boldsymbol{G}$ is vertex-primitive if and only if $\lambda \neq k$.* □

## 2.3 Vertex-transitive digraphs

Let us now introduce the most remarkable subclass of VT-digraphs.

**Definition 2.3.1** Let $\boldsymbol{G}$ be a group and $E \subseteq \boldsymbol{G}$. The *Cayley digraph* $\Gamma = \boldsymbol{Cay}(\boldsymbol{G}, E)$, is defined by

$$V(\Gamma) = \boldsymbol{G}, \ A(\Gamma) = \{(u, ue) : \ u \in \boldsymbol{G}, e \in E\}.$$

According to Definition 2.3.1 we have $u \to v$ if and only if $u^{-1}v \in E$. All Cayley digraphs are of course regular, having in- and out-valency equal to $|E|$. But much more can be said about them. The following is well known (see [Bi], Lemma 16.3).

**Theorem 2.3.2** *A digraph $\Gamma$ has a regular automorphism group $\boldsymbol{G} \leq \boldsymbol{Aut}\,\Gamma$ if and only if there is $E \subseteq \boldsymbol{G}$ such that $\Gamma \cong \boldsymbol{Cay}(\boldsymbol{G}, E)$.* □

This means that Cayley digraphs are those having a vertex-regular automorphism group, namely the left regular representation of $\boldsymbol{G}$, consisting of the permutations $x \mapsto gx$, $g \in \boldsymbol{G}$. The next result is also well known. Here we use the notation: $E^{-1} = \{e^{-1} | e \in E\}$, $E^2 = \{ee' | e, e' \in E\}$.

**Proposition 2.3.3** *For $\Gamma = \boldsymbol{Cay}(\boldsymbol{G}, E)$ the following hold:*

(i) *$\Gamma$ is connected if and only if $\langle E \rangle = \boldsymbol{G}$;*

(ii) *$\Gamma$ is undirected if and only if $E = E^{-1}$, and loopless if and only if $1 \notin E$;*

(iii) $\Gamma$ *is bipartite if and only if* $|\langle E\rangle : \langle E^2\rangle| = 2$. □

Note that if $E$ is a subgroup of $\boldsymbol{G}$, the relation $\sim$ is nothing but the equivalence associated with the left cosets of $E$. The graph $\Gamma$ has $|\boldsymbol{G} : E|$ connected components, the cosets themselves, each isomorphic to complete graph with loops added to each vertex, where $m = |E|$. As a more significant instance, let us look at hypercubes.

**Example 2.3.4** Let $V$ be a vector space of dimension $n$ over the field $GF(2)$, and let $E$ be a basis of $V$. Letting $\boldsymbol{G} = (V, +)$, the graph $\boldsymbol{Cay}(\boldsymbol{G}, E)$ is isomorphic to the hypercube $Q_n$ of dimension $n$.

The class of Cayley digraphs is far from exhausting all VT-digraphs. The same holds even for the undirected case: the smallest example of a non-Cayley VT-graph is the *Petersen graph* $O_3$. This remarkable graph appears in so many different graph-theoretical questions that it seems worthy of a whole book. In fact, this was done by Holton and Sheehan [HS].

The four classes we are going to introduce (quasi-Cayley digraphs, Kneser graphs, metacirculants and generalized Petersen graphs) are essentially different. Each of them contains $O_3$, together with infinitely many further non-Cayley graphs and digraphs. Other examples can be found e.g. in [Ta3], [JS] and [MP].

In order to generalize the notion of a Cayley digraph, we may note that we do not need a group in the definition, but any groupoid (i.e., set with any binary operation) would work, except that not so much can be expected for the symmetry of such a graph. But special kinds of groupoids are still of interest. Recall that a *quasigroup* is a groupoid $(Q, \cdot)$, where for each choice of $a, b \in Q$ the equations $ax = b$ and $xa = b$ have both a unique solution in $Q$. In other words, the maps $x \mapsto ax$ and $x \mapsto xa$ are permutations, just as in the case of groups. If, furthermore, $E \subseteq Q$ satisfies $a(bE) = (ab)E$ for all $a, b \in E$, then the quasigroup is said to be *right-associative.*

**Definition 2.3.5** Let $Q$ be a quasigroup and $E \subseteq Q$. The *quasigroup digraph* $\Gamma = \boldsymbol{QG}(Q, E)$ is defined by

$$V(\Gamma) = Q, \quad A(\Gamma) = \{(u, ue) | u \in Q, e \in E\}.$$

In particular, if $E$ is right associative, $\Gamma$ is said to be a *quasi-Cayley digraph.*

The following result was shown by Dörfler [Dö].

**Theorem 2.3.6** *A digraph* $\Gamma$ *is regular if and only if it is a quasigroup graph. Moreover, if* $\Gamma = \boldsymbol{QG}(Q, E)$ *with* $E$ *right-associative, then* $\Gamma$ *is vertex-transitive.* □

A proof of the second part of Theorem 2.3.6 can be given by an explicit representation of a quasi-Cayley digraph as a union of orbitals. Letting $M$ be the permutation group generated by all $x \mapsto gx$ for $g \in Q$, the union of all orbital graphs individuated by the pairs $(1, e)$, $e \in E$, is exactly $\boldsymbol{QG}(Q, E)$: it suffices to remark that $(gu, g(ue)) = (gu, (gu)e)$ because of the right associativity of the subset $E$. Of course, this description applies as well to the special case of Cayley digraphs, where $M$ is the left regular representation of the underlying group.

Further properties of quasi-Cayley graphs were recently found and exploited in Gauyacq (see [Ga]), where several examples are provided of VT-graphs that are not quasi-Cayley. Among them, infinitely many belong to the class of the Kneser graphs (see Definition 2.3.7). In fact, in view of Theorem 2.3.8, these are seldom Cayley graphs.

**Definition 2.3.7** Let $k$ and $n$ be positive integers with $n \geq 2k+1$ and $k \geq 2$. The *Kneser graph* $\mathcal{K}(n,k)$ has the $k$-subsets of $\{1,2,\ldots,n\}$ as vertices, with $X \sim Y$ if and only if $X \cap Y = \emptyset$. In the special case $n = 2k+1$, $\mathcal{K}(n,k) = O_{k+1}$ is said to be an *odd graph.*

Note that $O_3$ is the Petersen graph, which justifies our notation for this one.

It transpires from the definition itself that each Kneser graph is vertex-transitive. The following result was obtained by Godsil [Go].

**Theorem 2.3.8** *The graph $\mathcal{K}(n,k)$ is Cayley if and only if one of the following holds:*

(i) $k = 2$, *$n$ is a prime-power and* $n \equiv 3 (\mathrm{mod}\, 4)$*;*

(ii) $k = 3$, $n = 8$ *or* $32$.

*In particular, all odd graphs are non-Cayley.* □

Here is a further class of VT-graphs.

**Definition 2.3.9** Let $m, n \geq 1$ be integers. A graph $\Gamma$ is said to be an *$(m,n)$-metacirculant* if it has an automorphism $f$ with a cycle decomposition

$$f = (v_0^0 v_1^0 \cdots v_{n-1}^0)(v_0^1 v_1^1 \cdots v_{n-1}^1) \cdots (v_0^{m-1} v_1^{m-1} \cdots v_{n-1}^{m-1})$$

and an automorphism $g$ which cyclically permutes the orbits

$$V_i = \{v_0^i, v_1^i, \ldots, v_{n-1}^i\} \quad (i = 0, 1, \ldots, m-1).$$

It can be shown that there exists $r \in \mathbb{Z}_m^*$ such that $g(v_j^i) = v_{rj}^{i+1}$ for all $i$ and $j$, forcing $r^h \equiv 1 (\mathrm{mod}\, n)$ for some multiple $h$ of $m$. This means that a metacirculant is a graph with a transitive cyclic or metacyclic automorphism subgroup. In the first case the term *circulant* is used. It is proved in [AP] that every $(m,n)$-metacirculant can be associated with an array $(m, n, r, S_0, S_1, \ldots, S_k)$ where $k = \lfloor \frac{m}{2} \rfloor$, $S_i = \{s \in \mathbb{Z}_n : v_0^0 v_s^i \text{ is an edge}\}$ and $r^m S_i = S_i$. Let $M(m, n, r, S_0, S_1, \ldots, S_k)$ denote the corresponding graph.

Metacirculants can thus be described by subsets of certain groups, in quite a similar way as Cayley graphs. In fact, circulants are nothing but Cayley graphs of cyclic groups. However, many metacirculants are non-Cayley (among others, the Petersen graph). A complete characterization of metacirculant Cayley graphs can be found in [AP], Theorem 12. Though these graphs were studied only in the undirected case, their straightforward generalization to digraphs is possible, as remarked in [MS7]. This will lead to Definition 3.3.2.

**Definition 2.3.10** Let $n > k \geq 1$. The graph $\Gamma = \mathrm{GPG}(n,k)$ having $V(\Gamma) = \{u_i \mid i \in \mathbb{Z}_n\} \cup \{v_i \mid i \in \mathbb{Z}_n\}$ as vertex set and exactly the following adjacencies for all $i$:

(i) $u_i \sim v_i$;

(ii) $u_i \sim u_{i+1}$;

(iii) $v_i \sim v_{i+k}$,

is called a *generalized Petersen graph.*

Clearly $\mathrm{GPG}(n,k) \cong \mathrm{GPG}(n,n-k)$ and $\mathrm{GPG}(n,1) \cong C_n \Box K_2$. Note that $\mathrm{GPG}(5,2)$ is the Petersen graph; that is, $\mathrm{GPG}(5,2) \cong O_3$.

In general, $\Gamma = \mathrm{GPG}(n,k)$ has the automorphisms $f$ and $g$ defined by:

$$f : u_i \mapsto u_{i+1}, \quad v_i \mapsto v_{i+1},$$

$$g : u_i \mapsto u_{-i}, \quad v_i \mapsto v_{-i}.$$

Henceforth, $\boldsymbol{Aut}\,\Gamma$ has a dihedral subgroup $\langle f, g\rangle \cong \boldsymbol{D}_n$. This is of course a metacyclic group, but never transitive: its orbits are $\{u_i \mid i \in \mathbb{Z}_n\}$ and $\{v_i \mid i \in \mathbb{Z}_n\}$. Unlike the Petersen graph, many $\mathrm{GPG}(n,k)$ have no further automorphisms, so that they are not vertex-transitive. Namely, if $(n,k) \neq 1$, the set of $v_i$ splits into $(n,k)$ cycles, each of length $n/(n,k)$, while the $u_i$ form a unique cycle.

The following statement (proved in [FGW]) characterizes the vertex-transitive generalized Petersen graphs.

**Theorem 2.3.11** *The graph* $\mathrm{GPG}(n,k)$ *is vertex-transitive if and only if* $k^2 \equiv \pm 1 (\bmod\, n)$, *or else* $(n,k) = (10,2)$. □

The exceptional case $\mathrm{GPG}(10,2)$ is the dodecahedron, which is also arc-transitive, but non-Cayley. In fact, the following holds (see [NS1] or [Lov]).

**Theorem 2.3.12** *The graph* $\mathrm{GPG}(n,k)$ *is Cayley if and only if* $k^2 \equiv 1 (\bmod\, n)$. □

Further properties, like edge-transitivity, hold only for a finite number of pairs $(n,k)$: see [FGW].

As just seen, the usual description of the generalized Petersen graphs is explicit: there are these vertices and these edges, etc. An alternative one is provided here. Namely, Corollary 2.3.14 characterizes them among all cubic graphs.

Let us recall that, given a graph $\Gamma = (V,E)$, a set $X \subseteq V$ is called *dominating* if for each vertex $u \in V \setminus X$ there exists $v \in X$ such that $u \sim v$.

**Proposition 2.3.13** *Let* $\Gamma$ *be a cubic graph, having a dominating cycle* $C = (u_0, \dots, u_{m-1})$, $3 \leq m < |V(\Gamma)|$, *and an automorphism* $f$ *taking each* $u_i$ *into* $u_{i+1}$. *Then there are vertices* $v_j$ *such that* $V(\Gamma) = \{u_i | i \in \mathbb{Z}_m\} \cup \{v_j | j \in \mathbb{Z}_n\}$, *and* $\Gamma$ *satisfies one of the following, with no other edge than the listed ones:*

(i) $u_iu_{i+1}$, $u_iv_i$, $v_iv_{i+k}$, *with* $m = n$ *and* $k \leq n$ *fixed;*

(ii) $u_iu_{i+1}$, $u_jv_j$, $v_ju_{j+n}$, $v_jv_{j+k}$, *with* $m = 2n$, $n = 2k$*;*

(iii) $u_iu_{i+1}$, $u_iv_{[i]}$, *and* $m = 3n$, *where* $[i]$ *is* $i \bmod n$.

*Conversely, each graph described by one of* (i)-(iii) *satisfies the above condition with* $C = (u_0, \dots, u_{m-1})$.

**Proof** Let $v_0$ be a vertex outside of $C$, satisfying $v_0 \sim u_0$. Under the action of $f$, we obtain vertices $v_i$ such that $u_i v_i$ are edges, $i \in \mathbb{Z}_m$.

Since $C$ is dominating, we have $m \geq |V(\Gamma)|/2$. If $m = |V(\Gamma)|/2$, the $v_i$ are all distinct. Now $v_0 \sim v_k$ for exactly one $k$, so that $v_i v_{i+k} \in E(\Gamma)$ for all $i$. Thus (i) holds, with $n = m$.

Assume now that $m > |V(\Gamma)|/2$. Let $n$ be the smallest integer such that $v_j = v_{j+n}$ for some $j$. Clearly, this equality holds for *all* $j$. If some $v_j$, and so every $v_j$, is adjacent only to vertices of $C$, these are $u_j$, $u_{j+n}$ and $u_{j+2n}$. Therefore $m = 3n$, and (iii) follows.

Otherwise, there is a pair of adjacent vertices not belonging to $C$. The subgraph induced by $V(\Gamma) \setminus C$ is then a disjoint union of $K_2$'s, say $nK_2$. This easily leads to (ii). □

Note that for $k = 1$ we get from (ii) the Kuratowski graph $K_{3,3}$. The smallest instances of graphs corresponding to (iii) are $K_4$ and $Q_3$, obtained for $n = 1$ and $n = 2$, respectively.

**Corollary 2.3.14** *A cubic graph with* $2n$ *vertices* ($n \geq 3$) *is a generalized Petersen graph if and only if it has a dominating cycle* $C = (u_0, \ldots, u_{n-1})$ *and an automorphism taking each* $u_i$ *into* $u_{i+1}$.

**Proof** The graphs described by Proposition 2.3.13 (i) are exactly the generalized Petersen graphs. □

We conclude this section with some informal information about *Frucht notation*. A more detailed description can be found in the original paper [Fr] or in [CM], pp. 211-216.

Apart from very small graphs, to draw a graph is by no means trivial. To make a picture representing a large one may be time-consuming and almost useless for the purpose of checking the actual properties of the graph. The idea of Frucht notation, that will be illustrated by Example 2.3.15, is to choose an appropriate partition $\{V_1, \ldots, V_s\}$ of the vertex set, so that each $V_i$ induces some 'elementary' graph (e.g., a cycle), and links between different $V_i$ can also be described in a most easy way (e.g., with a labeling as in the generalized Petersen graphs). The output will be a diagram with circles (representing $V_i$) linked by edges and arcs. Inside the former and above the latter, symbols are written, representing which kind of graph are inside the circles and how they are related.

**Example 2.3.15** Let us consider the graph $\Gamma$ drawn in Figure 2, that is represented by Frucht notation as explained above.

There are six vertices, $u_i$ ($i = 0, \ldots, 5$), on the outer cycle, and another six, $v_j$ ($j = 0, \ldots, 5$), on the inner cycle of $\Gamma$. These will be represented by the left and the right circle of the diagram, respectively. Moreover, $u_i$, $u_j$ are adjacent if and only if $i - j = \pm 1$, where addition is taken modulo 6. This situation is represented by the symbol 6(1) inside the left circle of the diagram. Similarly, $v_i \sim v_j$ if and only if $i - j = \pm 1$ or $i - j = \pm 2$. This corresponds to the symbol 6(1, 2) inside the right circle. The edge between the two circles means that $u_i \sim v_j$ whenever $i \equiv j \pmod 6$. The arrow from left to right, labeled by 1, stands for $u_i \sim v_j$ whenever $j \equiv i + 1 \pmod 6$.

A graph lends itself particularly well to the above kind of description when a semiregular automorphism $f$ can be found. Then take $\{V_1, \ldots, V_s\}$ as the set of all orbits of $f$. Now each $V_i$ induces a circulant (for the restriction of $f$ to this graph is transitive), and $f$ gives the description of edges between different $V_i$.

**Example 2.3.16** The graph $\mathrm{GPG}(n, k)$ has the semiregular automorphism $f$ consisting of the two cycles $(u_0 \cdots u_{n-1})$, $(v_0 \cdots v_{n-1})$. Hence Figure 3 represents $\mathrm{GPG}(n, k)$.

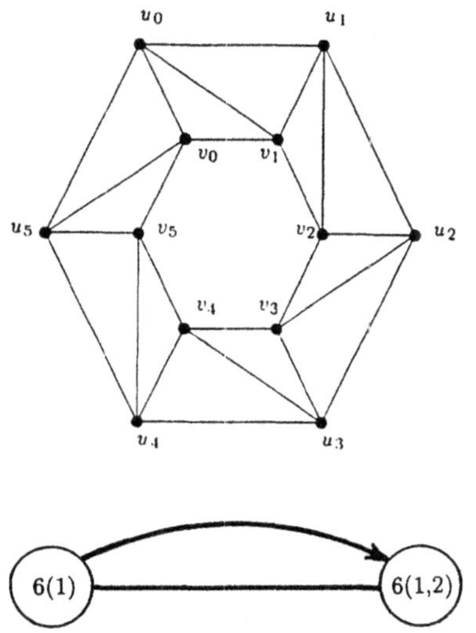


Figure 2: A graph and its Frucht diagram

Figure 3: The Generalized Petersen graph

# Chapter 3

## The classification of vertex-transitive $pq$-graphs

When describing the way we followed towards the classification of VT-graphs of order $pq$, we find it advisable to adopt the following (non-standard) terminology.

Letting $p \geq q$, a graph $\Gamma$ as above will be called:

- *p-imprimitive* if there is a transitive $\boldsymbol{G} \leq \boldsymbol{Aut}\,\Gamma$ having a subgroup with an imprimitivity block system with blocks of order $p$;

- *q-imprimitive* if it is not $p$-imprimitive but there is a transitive $\boldsymbol{G} \leq \boldsymbol{Aut}\,\Gamma$ having a subgroup with an imprimitivity block system with blocks of order $q$;

- *primitive* if it is neither $p$-imprimitive nor $q$-imprimitive, i.e. $\boldsymbol{Aut}\,\Gamma$ has definitively no imprimitive subgroup.

Hence the word 'primitive' is used in a more restrictive sense than usual. For example, the Petersen graph is 5-imprimitive, even though its full automorphism group is primitive.

But our choice is reasonable, in view of the fact that the existence of an imprimitive subgroup $\boldsymbol{G}$ can be exploited in order to describe the graph, no matter if $\boldsymbol{G}$ is contained in a larger, primitive subgroup.

The full proof of our Classification Theorem is very long. The first step was [Ma4], which takes care of the $p$-imprimitive graphs: they are all metacirculants. We will then describe here the remaining cases. Note that the definition of a metacirculant implies that every Cayley $pq$-graph is a metacirculant, so all further graphs that we will come across will be non-Cayley VT-graphs.

## 3.1 The imprimitive case

This section is based on [MS1], and takes care of $q$-imprimitive $pq$-graphs.

It is a straightforward consequence of Definition 2.3.9 that every Cayley $pq$-graph is a metacirculant. Moreover, Alspach and Parsons [AP] gave conditions on $p$, $q$ for the existence of non-Cayley metacirculants. Therefore, we have the following.

**Proposition 3.1.1** *Let $\Gamma$ be a metacirculant with $pq$ vertices ($p > q$ primes). If $p \not\equiv 1 \pmod q$, then $\Gamma$ is a circulant. If $p \not\equiv 1 \pmod{q^2}$, then $\Gamma$ is a Cayley graph. If $p \equiv 1 \pmod{q^2}$, then there are non-Cayley metacirculants with $pq$ vertices. Moreover every Cayley $pq$-graph is a metacirculant.* □

Let $k$ be a positive integer greater than 1, and $m$ a divisor of $2^k - 1$ greater than 1. Then there exists a unique $m$-imprimitive representation of the group $\boldsymbol{SL}(2, 2^k)$ with $PG(1, 2^k)$ as the imprimitivity block system [MS3]. This representation has $m$ suborbits of size $2^k$ and $m$ trivial suborbits. It is proved in [MS3], Theorem 3.4, that the corresponding nontrivial orbital graphs give rise to a unique $m$-fold covering, denoted by $\Gamma(k, m)$, whose base graph is the complete graph $\Gamma$ with vertex set $PG(1, 2^k)$. Let $\rho \in \boldsymbol{S}_m$ be the $m$-cycle $(01 \ldots m-1)$, $\tau \in \boldsymbol{S}_m$ the involution that fixes 0 and interchanges $r$ with $-r$ ($r \in \mathbb{Z}_m$), and let $\tau_r = \rho^r \tau \rho^{-r}$ ($r \in \mathbb{Z}_m$). With a fixed generator $w$ of $GF(2^k)^*$ and the usual identification of $PG(1, 2^k)$ with $GF(2^k) \cup \{\infty\}$, we have that $\Gamma(k, m)$ is the covering graph of $\Gamma$ with the voltage defined as follows:

$$\sigma(x, y) = \begin{cases} \text{id}, & \text{if } x = \infty \text{ or } y = \infty; \\ \tau_r, & \text{if } x + y = w^r. \end{cases} \tag{2}$$

More generally, a graph $\Gamma$ admitting an $m$-imprimitive representation of $\boldsymbol{SL}(2, 2^k)$ with $PG(1, 2^k)$ as the imprimitivity block system must be an edge-disjoint union of the corresponding orbital graphs, and so it is necessarily isomorphic to one of the graphs $\Gamma(k, m, S, T)$ defined below (see [MS3], Corollary 3.5). Recall that a subset $S$ of $\mathbb{Z}_n$ is said to be *symmetric* when $-S = S$.

**Definition 3.1.2** Let $m$ be a divisor of $2^k - 1$, $S$ a symmetric subset of $\mathbb{Z}_m \setminus \{0\}$ and $T$ a subset of $\mathbb{Z}_m$. The *Marušič-Scapellato graph* (or MS-*graph*) $\Gamma(k, m, S, T)$ has vertex set $PG(1, 2^k) \times \mathbb{Z}_m$ such that, for each $r \in \mathbb{Z}_m$ and each $x \in PG(1, 2^k)$, the neighbours of $(x, r)$ are all the vertices of the form

$$(\infty, r + s), s \in S, \text{ and } (y, r + t), y \in GF(2^k), t \in T$$

if $x = \infty$; and all vertices of the form

$$(x, r+s), s \in S(\infty, r-t), t \in T, \quad \text{and } (x+w', -r+t+2i), i \in \mathbb{Z}_m, t \in T$$

if $x \neq \infty$.

For example, in this notation $\Gamma(k,m) = \Gamma(k,m,\emptyset,\{0\})$.

The above name was given to these graphs in [PWX], while in our papers [MS1] and [MS5] they are called Fermat graphs. Namely, the case that is of particular interest to us arises when $2^k+1$ is a prime number $p$. This may only happen when $k = 2^a$ and so $p = 2^{2^a}+1$ is a Fermat prime. Then, for any prime $q$ that divides $2^{2^a}-1$, the construction in Definition 3.1.2 gives rise to $q$-imprimitive $pq$-graphs. The following is a direct consequence of [MS3], Corollary 3.5.

**Proposition 3.1.3** *Let $p = 2^{2^a}+1$ be a Fermat prime, $q$ a divisor of $2^{2^a}-1$, and $\Gamma$ a vertex-transitive $pq$-graph that admits a $q$-imprimitive representation of $\boldsymbol{SL}(2, 2^{2^a})$, where the corresponding imprimitivity block system is identified with $PG(1, 2^{2^a})$. Then there are a symmetric subset $S$ of $\mathbb{Z}_q^*$ and a subset $T$ of $\mathbb{Z}_q$ such that $\Gamma \cong \Gamma(2^a, q, S, T)$.* □

From now on, $p > q$ will be odd primes and $\Gamma$ a vertex-transitive $pq$-graph. Let $\boldsymbol{G}$ be a $q$-imprimitive subgroup of $\boldsymbol{Aut}\,\Gamma$ and $\mathcal{B}$ the corresponding imprimitivity block system of $\boldsymbol{G}$. For a block $B \in \mathcal{B}$, let $\boldsymbol{G}_B$ denote the subgroup of all those automorphisms in $\boldsymbol{G}$ that fix $B$ setwise. If $\boldsymbol{G}_v$ denotes the stabilizer of a vertex $v \in B$, we have:

$$[\boldsymbol{G} : \boldsymbol{G}_B] = p, \quad [\boldsymbol{G}_B : \boldsymbol{G}_v] = q. \tag{3}$$

The restriction of $\boldsymbol{G}_B$ to $B$ defines a homomorphism $\phi_B : \boldsymbol{G}_B \to \boldsymbol{S}_q$. The following result is straightforward.

**Proposition 3.1.4** *$\boldsymbol{G}_B/\ker\phi_B$ is isomorphic to some transitive subgroup of $\boldsymbol{S}_q$.* □

Let $\widetilde{\boldsymbol{G}}$ denote the homomorphic image of $\boldsymbol{G}$ acting on $\mathcal{B}$.

**Proposition 3.1.5** *Let $\widetilde{\boldsymbol{G}}$ be doubly transitive. If there are distinct blocks $B, B' \in \mathcal{B}$ and automorphisms $g, g' \in \boldsymbol{G}_B \cap \boldsymbol{G}_{B'}$ such that the restrictions of $g$ to $B$ and $B'$ are a $q$-cycle and* id, *respectively, and the restrictions of $g'$ to $B$ and $B'$ are a $q$-cycle and* id, *respectively, then $\Gamma$ is a metacirculant.*

**Proof** The existence of the automorphisms $g$ and $g'$ implies that the bipartite subgraph of $\Gamma$ induced by $B$ and $B'$ is either $K_{q,q}$ or totally disconnected. Moreover, since $\widetilde{\boldsymbol{G}}$ is doubly transitive, any two bipartite graphs induced by pairs of distinct blocks in $\mathcal{B}$ are isomorphic. Therefore either $\Gamma$ or $\Gamma^c$ is disconnected and therefore a union of isomorphic circulants. Hence $\Gamma$ must be a metacirculant. □

The next Proposition 3.1.6, proved in [Ma4], Theorem 3.4, plays an important role in the characterization of imprimitive $pq$-graphs.

**Proposition 3.1.6** *Let $\Gamma$, $G$ and $\tilde{G}$ have the meaning described above. Then either $\Gamma$ is a metacirculant or $\tilde{G}$ is nonsolvable and* $\ker(G \to \tilde{G}) = 1$. □

**Proposition 3.1.7** *If $G$ is a nonsolvable doubly transitive permutation group of degree $m$ that contains an $m$-cycle, then one of the following holds:*

(i) $G \cong A_m$ *or* $S_m$;

(ii) $m = 11$ *and* $G \cong PSL(2, 11)$ *or* $M_{11}$;

(iii) $m = 23$ *and* $G \cong M_{23}$;

(iv) $m = (k^n - 1)/(k - 1)$ *for some prime power $k$, and $G$ is isomorphic to a subgroup of* $Aut\, PSL(n, k)$ *containing* $PSL(n, k)$.

**Proof** This is a direct consequence of the classification of simple groups (see [Gor], Theorem 1.49). □

An application of Proposition 3.1.7 to the case we are interested in leads us to the following result.

**Proposition 3.1.8** *Let $p > q$ be odd primes, $\Gamma$ a $q$-imprimitive $pq$-graph, and $G$ a minimal transitive $q$-imprimitive subgroup of $Aut\,\Gamma$. Then one of the following holds:*

(i) $G \cong A_p$, *of degree* $p \geq 5$;

(ii) $p = 11$ *and* $G \cong PSL(2, 11)$ *or* $M_{11}$;

(iii) $p = 23$ *and* $G \cong M_{23}$;

(iv) $p = (k^n - 1)/(k - 1)$ *for some prime $n$ and a prime power $k$ such that* $(n, k - 1) = 1$, *and* $G \cong PSL(n, k) = SL(n, k)$.

**Proof** A nonsolvable permutation group of prime degree is by [Wi], Theorem 11.7, necessarily doubly transitive. On the other hand, it is easy to see that a vertex-transitive $pq$-graph whose automorphism group has a transitive subgroup with an intransitive normal subgroup must be a metacirculant. Note that the number $(k^n - 1)/(k - 1)$ is not a prime unless $n$ is a prime and $(n, k - 1) = 1$. Hence, by Propositions 3.1.6 and 3.1.7, the assertion follows. □

The next result brings us a step closer to the characterization of imprimitive $pq$-graphs.

**Proposition 3.1.9** *Let $p > q$ be odd primes, $\Gamma$ a $q$-imprimitive $pq$-graph, and $G$ a minimal $q$-imprimitive subgroup of $Aut\,\Gamma$. Then $p = (k^n - 1)/(k - 1)$ for some prime $n$ and a prime power $k$ such that $(n, k - 1) = 1$, and $G \cong PSL(n, k) = SL(n, k)$.*

**Proof** Let $\mathcal{B}$ be the corresponding imprimitivity block system of $G$. We can apply Proposition 3.1.8 and distinguish the corresponding cases (i)-(iv).

Consider first (i), and take $G = A_p$. If $p \geq 7$, then the stabilizer of $G_B$ of a block $B \in \mathcal{B}$ is the simple group $A_{p-1}$, and so $\ker \phi_B$ is either $A_{p-1}$ or id, both contradicting Proposition 3.1.4. Thus $p = 5$ and so $G \cong PSL(2, 4) = SL(2, 4)$.

In case (ii), $G \cong PSL(2,11)$, then the block stabilizer $G_B$ is isomorphic to $A_5$. Since $|G| = 2^2 \cdot 3 \cdot 5 \cdot 11$, and by (3), we have $|G_B| = 2^2 \cdot 3 \cdot 5$. Hence, it follows that $q \in \{3,5\}$. But $A_5$ is simple and so $\ker \phi_B = \mathrm{id}$. Thus $q = 5$. Since $PSL(2,11)$ contains a subgroup of order 55 (see [C-W], p. 7) and, for $v \in B$, the number $|G_v| = 12$ is prime to 55, the fact that every Cayley $pq$-graph is a metacirculant and Lemma 2.6 in [MS1] together imply that $\Gamma$ is a metacirculant and thus 11-imprimitive, a contradiction.

In case (iii) we have $G \cong M_{11}$. Then the block stabilizer $G_B$, for $B \in \mathcal{B}$, must be isomorphic to $M_{10}$, an extention of $A_6$ by $\mathbb{Z}_2$. Since $|M_{10}| = 2^4 \cdot 3^2 \cdot 5$, it follows that $q \in \{3,5\}$. Because of the simplicity of $A_6$, a subgroup isomorphic to $A_6$ is the only proper nontrivial normal subgroup of $M_{10}$. But the relative quotient group $\mathbb{Z}_2$ is not isomorphic to a transitive subgroup of $S_q$. Therefore $\ker \phi_B = 1$ and so $G_B \cong M_{10}$ and $|G_B| > |S_q|$, which contradicts Proposition 3.1.4.

Finally, in case (iv) we have $G \cong M_{23}$. Then a block stabilizer of $G$ is isomorphic to $M_{22}$ and so contains no subgroup of prime index (see [C-W], p. 39), contradicting (3). This completes the proof. □

In view of the above results, we may restrict our discussion to linear groups $SL(n,k)$. Following a sequence of lemmas that analyze the structure of vertex-transitive $pq$-graphs admitting a $q$-imprimitive representation of $SL(n,k)$, where $n$ is an odd prime, we show that these graphs are necessarily metacirculants (see Proposition 3.1.14). This result together with Propositions 3.1.3 and 3.1.9 provides the basis for the proof of Theorem 3.1.15.

Assume that $n$ is an odd prime, $k$ is a prime power, $(n, k-1) = 1$, $p = (k^n - 1)/(k-1)$, $G = SL(n,k)$, $\mathbb{F} = GF(k)$, and that $\Gamma$ is a vertex-transitive $pq$-graph admitting a $q$-imprimitive representation of $G$. Then since $SL(n,k) = PSL(n,k)$, the blocks of $G$ may be identified with the points of the projective space $PG(n-1,k)$.

Let $H$, $K$ and $J$ denote the following groups of matrices, where $U$ is an $(n-1)$-row vector:

$$H = \left\{ \begin{bmatrix} \lambda & U \\ 0 & M \end{bmatrix} : M \in GL(n-1,k), \lambda \in \mathbb{F}^*, \lambda \det M = 1 \right\},$$

$$K = \left\{ \begin{bmatrix} \lambda^q & U \\ 0 & M \end{bmatrix} : M \in GL(n-1,k), \lambda \in \mathbb{F}^*, \lambda^q \det M = 1 \right\},$$

$$J = \left\{ \begin{bmatrix} 1 & U \\ 0 & M \end{bmatrix} : M \in SL(n-1,k) \right\}.$$

Note that $J \subseteq K \subseteq H \subseteq G$. We specify that the matrix action is on the left. Let $B$ denote the block $[1,0,\ldots,0]^T$. Then clearly,

$$H = G_B. \tag{4}$$

Furthermore, let $B' = [0,0,\ldots,0,1]^T$, and define the following groups of matrices:

$$H' = \left\{ \begin{bmatrix} M & 0 \\ U & \lambda \end{bmatrix} : M \in GL(n-1,k), \lambda \in \mathbb{F}^*, \lambda \det M = 1 \right\},$$

$$K' = \left\{ \begin{bmatrix} M & 0 \\ U & \lambda^q \end{bmatrix} : M \in GL(n-1,k), \lambda \in \mathbb{F}^*, \lambda^q \det M = 1 \right\},$$

$$J' = \left\{ \begin{bmatrix} M & 0 \\ U & 1 \end{bmatrix} : M \in \boldsymbol{SL}(n-1,k) \right\}.$$

Then

$$\boldsymbol{H}' = \boldsymbol{G}_{B'}. \tag{5}$$

This notation will be used throughout the rest of the section.

**Lemma 3.1.10** *$\boldsymbol{J}$ is transitive on $B$ if and only if $\boldsymbol{J}'$ is transitive on $B'$.*

**Proof** It suffices to show that $\boldsymbol{J}$ and $\boldsymbol{J}'$ are conjugate in $\boldsymbol{G}$. Let

$$g = \begin{bmatrix} 0 & 0 & 1 \\ 0 & -I & 0 \\ 1 & 0 & 0 \end{bmatrix},$$

where $I$ is the identity matrix of order $n-2$. Then $g$ takes $B$ into $B'$ and $g^{-1}\boldsymbol{J}g = \boldsymbol{J}'$. □

**Lemma 3.1.11** *If $\boldsymbol{J}$ is intransitive on $B$, then $\Gamma$ is a metacirculant.*

**Proof** Let $\boldsymbol{C}$ be the cyclic group generated by the matrix

$$\begin{bmatrix} w & U \\ 0 & w^{-1}I \end{bmatrix},$$

where $w$ is a generator of $\mathbb{F}^*$. It is easy to see that

$$\boldsymbol{H} = \boldsymbol{J} \cdot \boldsymbol{C} \tag{6}$$

and

$$\boldsymbol{K} = \boldsymbol{J} \cdot \boldsymbol{C}^q. \tag{7}$$

Of course, $\boldsymbol{J}$ is normal in $\boldsymbol{H}$, and therefore the intransitivity of $\boldsymbol{J}$ implies that

$$\boldsymbol{J} \subseteq \boldsymbol{H}_v \text{ for all } v \in B. \tag{8}$$

Therefore, by (4) and (6), $\boldsymbol{C}$ is transitive on $B$. Hence $q = |B|$ must divide $k-1 = |\boldsymbol{C}|$. But $\boldsymbol{C}$ is cyclic and so, for each $v \in B$, the stabilizer $\boldsymbol{C}_v$ is equal to $\boldsymbol{C}^q$, the unique subgroup of index $q$ in $\boldsymbol{C}$. In view of (7) and (8), we then have

$$\boldsymbol{K} \subseteq \boldsymbol{H}_v \text{ for all } v \in B. \tag{9}$$

Using Lemma 3.1.10, a similar argument as above shows that

$$\boldsymbol{K}' \subseteq \boldsymbol{H}_{v'} \text{ for all } v' \in B'. \tag{10}$$

Consider a matrix of the form

$$\begin{bmatrix} \mu & U & 0 \\ 0 & A & 0 \\ 0 & W & \lambda^q \end{bmatrix}$$

where $\mu \notin (\mathbb{F}^*)^q$, and $A \in \boldsymbol{GL}(n-2,k)$ with $\mu\lambda^q \det A = 1$. This matrix belongs to $(\boldsymbol{H} \setminus \boldsymbol{K}) \cap \boldsymbol{K}'$. By (4), (9) and (10), its restrictions to the blocks $B$ and $B'$ are a $q$-cycle and id, respectively. Similarly, a matrix of the form

$$\begin{bmatrix} \lambda^q & W & 0 \\ 0 & A & 0 \\ 0 & U & \mu \end{bmatrix}$$

with $\lambda, \mu$, and $A$ as above, belongs to $(\boldsymbol{H}' \setminus \boldsymbol{K}') \cap \boldsymbol{K}$. By (5), (9) and (10), its restrictions to the blocks $B'$ and $B$ are a $q$-cycle and id, respectively. Proposition 3.1.5 then implies that $\Gamma$ is a metacirculant. □

In order to deal with the case where the group $\boldsymbol{J}$ is transitive on $B$, we introduce the following notation:

$$\boldsymbol{A} = \left\{ \begin{bmatrix} 1 & U \\ 0 & I \end{bmatrix} : I \in \boldsymbol{SL}(n-1,k), U \text{ is an } (n-1)\text{-row vector} \right\},$$

$$\boldsymbol{L} = \left\{ \begin{bmatrix} 1 & 0 \\ 0 & M \end{bmatrix} : M \in \boldsymbol{SL}(n-1,k) \right\}.$$

**Lemma 3.1.12** *If $\boldsymbol{J}$ is transitive on $B$ and $\phi_B : \boldsymbol{J} \to \boldsymbol{S}_q$ is the restriction of $\boldsymbol{J}$ to $B$, then $\boldsymbol{A} \subseteq \ker \phi_B$ and $\boldsymbol{L}$ is transitive on $B$.*

**Proof** Note that $\boldsymbol{A}$ is an elementary abelian group of order $k^{n-1}$ and that $\boldsymbol{J}$ is the semidirect product of $\boldsymbol{A}$ by $\boldsymbol{L}$. Thus

$$\boldsymbol{A} \lhd \boldsymbol{J}. \tag{11}$$

Let us assume that $\boldsymbol{A}$ is transitive on $B$. Then $q$ must divide $k^{n-1}$, and so $k$ is a power of $q$. Since $\boldsymbol{A}$ is abelian and transitive on $B$, it follows that $\boldsymbol{A}$ is regular on $B$, and so the restriction of an element of $\boldsymbol{A}$ to $B$ is either a $q$-cycle or id. Let $\boldsymbol{D} = \boldsymbol{A} \cap \ker \phi_B$. Then

$$[\boldsymbol{A} : \boldsymbol{D}] = q. \tag{12}$$

Moreover, since both $\boldsymbol{A}$ and $\ker \phi_B$ are normal in $\boldsymbol{J}$, we have

$$\boldsymbol{D} \lhd \boldsymbol{J}. \tag{13}$$

Now $\boldsymbol{A}$ can be seen as a vector space over the field $\mathbb{F}$. The action of $\boldsymbol{L}$ on $\boldsymbol{A}$ by conjugation is transitive on hyperplanes of the vector space $\boldsymbol{A}$. In view of (12), $\boldsymbol{D}$ must contain at least one hyperplane but not all of them. Therefore $\boldsymbol{D}$ is not fixed by the action of $\boldsymbol{L}$ defined above, contradicting (13). Hence $\boldsymbol{A}$ must be intransitive and so, by (11), $\boldsymbol{A} \subseteq \boldsymbol{J}_v$ for each $v \in B$, that is, $\boldsymbol{A} \subseteq \ker \phi_B$. Therefore $\boldsymbol{L}$ must be transitive on $B$. □

**Lemma 3.1.13** *If $n = 3$ and $q$ divides $k$, then $\Gamma$ is a metacirculant.*

**Proof** In view of Lemmas 3.1.11 and 3.1.12 we may assume that $\boldsymbol{L}$ is transitive on $B$. Let $\phi_B : \boldsymbol{L}_B \to B$ denote the restriction of $\boldsymbol{L}_B$ to $B$. Letting $g$ denote the matrix

$$\begin{bmatrix} 1 & 0 & 0 \\ 0 & 1 & 0 \\ 0 & 1 & 1 \end{bmatrix},$$

it is easy to see that

$$g \text{ has order } q = \operatorname{char} \mathbb{F}. \tag{14}$$

Moreover, $g$ belongs to no proper normal subgroup of $\boldsymbol{L} \cong \boldsymbol{SL}(2,k)$. For $k = 3$, since $\boldsymbol{PSL}(2,3) \cong \boldsymbol{A}_4$, all normal subgroups of $\boldsymbol{L}$ lie in the unique 2-Sylow subgroup of $\boldsymbol{L}$, but $g$ has order 3. For $k > 3$, all normal subgroups of $\boldsymbol{L}$ are contained in $Z(\boldsymbol{L})$, and consist of diagonal matrices alone, but $g$ is not a diagonal matrix. Thus, in particular, $g \notin \ker \phi_B$ and so, by (14),

$$\text{the restriction of } g \text{ to } B \text{ is a } q\text{-cycle.} \tag{15}$$

Let

$$\boldsymbol{A}' = \left\{ \begin{bmatrix} I & 0 \\ U & 1 \end{bmatrix} : I \in \boldsymbol{SL}(n-1,k), U \text{ is an } (n-1)\text{-row vector} \right\}.$$

Then $g \in \boldsymbol{A}'$ and so, by Lemmas 3.1.10 and 3.1.12, the restriction of $g$ to $B'$ is id. Therefore, by (15), $g$ satisfies the assumptions of Proposition 3.1.5. A similar argument shows that the restrictions of the matrix

$$g' = \begin{bmatrix} 1 & 1 & 0 \\ 0 & 1 & 0 \\ 0 & 1 & 1 \end{bmatrix}$$

to $B'$ and $B$ are a $q$-cycle and id, respectively. By Proposition 3.1.5, we conclude that $\Gamma$ is a metacirculant. □

**Proposition 3.1.14** *Let $n$ be an odd prime, $k$ a prime power with $(n, k-1) = 1$, and $\boldsymbol{G} = \boldsymbol{SL}(n,k)$. If $q < p = (k^n - 1)/(k-1)$ are primes and $\Gamma$ is a vertex-transitive pq-graph admitting a q-imprimitive representation of $\boldsymbol{G}$ with $PG(n-1,k)$ as the imprimitivity block system, then $\Gamma$ is a metacirculant.*

**Proof** If $n = 3$ and $k = 2$, it follows that $p = 7$ and $q = 3$. But $\boldsymbol{G} = \boldsymbol{SL}(3,2)$ contains a subgroup of order 21 (see [C-W], p.3), which is, by Lemma 2.6 in [MS1], regular, and so $\Gamma$ is a Cayley $pq$-graph, hence a metacirculant. If $n = k = 3$, then $p = 13$ and $q = 3$, and so $\Gamma$ is a metacirculant by Lemma 3.1.13. Therefore we may assume that $\boldsymbol{SL}(n-1,k)$ is a nonsolvable group. Hence, with the notation of this section, we may, in view of Lemmas 3.1.11 and 3.1.12, assume that $\boldsymbol{L} \cong \boldsymbol{SL}(n-1,k)$ is nonsolvable and transitive on the block $B$. Let $\phi_B : \boldsymbol{L} \to \boldsymbol{S}_q$ be the restriction of $\boldsymbol{L}$ to $B$, and let $\boldsymbol{R} = \phi_B(\boldsymbol{L})$. Then $\boldsymbol{L}$ is also nonsolvable and so by [Wi], Theorem 11.7, doubly transitive. Since $\boldsymbol{R}$ contains only abelian proper normal subgroups and no projective linear group is isomorphic to $\boldsymbol{M}_{11}$ or $\boldsymbol{M}_{23}$, Proposition 3.1.7 implies that $\boldsymbol{R} = \boldsymbol{PSL}(n-1,k)$ is one of the following groups:

(i) $\boldsymbol{A}_q$;

(ii) $\boldsymbol{PSL}(2,11)$, with $q = 11$;

(iii) $\boldsymbol{PSL}(m,h)$, where $q=(h^m-1)/(h-1)$, $m$ is a prime and $(m-1,h)=1$.

In case (i), $\boldsymbol{A}_q \cong \boldsymbol{PSL}(n-1,k)$, and so, since $q$ is a prime, it follows by [BW], p. 43, that $q=5$, $n=3$, and $k\in\{4,5\}$. For $k=4$, we have $(n,k-1)\neq 1$ and so $k=5$ and $\boldsymbol{G}=\boldsymbol{SL}(3,5)$. Therefore $\Gamma$ is a metacirculant by Lemma 3.1.13. Case (ii) is not possible since we have $n=3$ and $k=11$, and so $p=11^2+11+1=133$ is not a prime number. Finally, in case (iii), suppose first that $m\neq n-1$. Then by [BW], pp. 43-44, either $(m,h)=(3,2)$ and $(n-1,k)=(2,7)$, or $(m,h)=(2,7)$ and $(n-1,k)=(3,2)$, both of which are impossible. Hence $m=n-1$. For $m$ to be a prime we must have $n=3$ and so, for $q$ to be a prime, $h=k=2^{2^a}$ for some $a$. But then $p=2^{2^a}+2^a+1$ is a prime only for $a=1$. Hence $\boldsymbol{G}=\boldsymbol{SL}(3,2)$, which has already been covered. This completes the proof. □

We are now ready to characterize all imprimitive $pq$-graphs.

**Theorem 3.1.15** *Let $\Gamma$ be a vertex-transitive pq-graph such that $\boldsymbol{Aut}\,\Gamma$ contains an imprimitive subgroup. Then either $\Gamma$ is a metacirculant or $\Gamma$ is q-imprimitive and $p=2^{2^a}+1$ is a Fermat prime, q divides $2^{2^a}-1$, and there are a symmetric subset $S$ of $\mathbb{Z}_q^*$ and a proper, nonempty subset $T$ of $\mathbb{Z}_q$ such that $\Gamma\cong\Gamma(2^a,q,S,T)$.*

**Proof** In view of Propositions 3.1.3, 3.1.9 and 3.1.14, it suffices to show that given a Fermat prime $p=2^{2^a}+1$, a prime $q$ dividing $2^{2^a}-1$, a symmetric subset $S$ of $\mathbb{Z}_q^*$ and a subset $T$ of $\mathbb{Z}_q$, it follows that

$$\Gamma(2^a,q,S,T) \text{ is a metacirculant if and only if } T\in\{\emptyset,\mathbb{Z}_q\}. \qquad (16)$$

If $T\in\{\emptyset,\mathbb{Z}_q\}$, then either $\Gamma(2^a,q,S,T)$ or its complement is disconnected, and so $\Gamma(2^a,q,S,T)$ is a metacirculant (see [Ma4], Lemma 3.1).

Assume now that $\Gamma=\Gamma(2^a,q,S,T)$ is a metacirculant and let $\Gamma'$ be the corresponding base graph of $PG(1,2^{2^a})$. Since $p\not\equiv 1(\bmod q)$, Proposition 3.1.1 implies that $\Gamma$ is a circulant. Let $g$ be an automorphism of $\Gamma$ with $V(\Gamma)$ as its unique orbit and let $f=g^q$. Furthermore, let $x\in PG(1,2^a)$ and let $(x,r)$, $r\in\mathbb{Z}_q$, be the vertices of the corresponding block in $\Gamma$. Then the orbits of $f$ are precisely the sets $V_r=\{f^j(x,r): j\in\mathbb{Z}_p\}$ $(r\in\mathbb{Z}_q)$ and, in particular, each $V_r$ intersects each block in $\Gamma$ in exactly one vertex. Therefore, for each $r\in\mathbb{Z}_q$, the symbol of the circulant $[V_r]$, the subgraph of $\Gamma$ induced by $V_r$, depends solely on the structure of the set $T$. We call it $A_r(T)$.

Suppose first that $T=\{0\}$. For each $r\in\mathbb{Z}_q$, let $A_r=A_r(\{0\})$. We claim that

$$\{A_0,A_1,\dots,A_{q-1}\} \text{ partitions } \mathbb{Z}_p^*. \qquad (17)$$

Let $j\in\mathbb{Z}_p^*$ be arbitrary and let $C$ be the cycle $xf^j(x)\cdots f^{(p-1)j}(x)x$ in $\Gamma'$. Let $\rho$ be the $q$-cycle $(01\dots q-1)$, $\tau$ the involution that fixes 0 and interchanges $r$ with $-r$ $(r\in\mathbb{Z}_q)$, and let $\tau_r=\rho^r\tau\rho^{-r}$ $(r\in\mathbb{Z}_q)$. Recalling the definition of the voltage $\sigma$ in (2), we can see that in the product $\Pi_{k=0}^{p-1}(\sigma(f^{kj}(x),f^{(k+1)j}(x))$ two of the factors are id. The remaining factors, an odd number of them, belong to the set $\{\tau_0,\tau_1,\dots,\tau_{q-1}\}$. Therefore this product must be some $\tau_r$. Since $r$ is the only fixed point of the involution $\tau_r$, it follows that the cycle $C$, lifted to the covering graph $\Gamma$, gives rise to $(q-1)/2$ cycles of length $2p$ and a unique cycle of length $p$, the latter being contained in the circulant $[V_r]$. Thus $j\in A_s$ if and only if $s=r$. This proves (17).

If $T = \{t\}$, then a similar argument as above shows that $A_r(\{t\}) = A_{r+t}$ for each $r \in \mathbb{Z}_q$. We may therefore conclude that, in the case of an arbitrary subset $T$ of $\mathbb{Z}_q$,

$$A_r(T) = \bigcup_{t \in T} A_{r+t} \quad (r \in \mathbb{Z}_q). \tag{18}$$

Moreover, let $r$ and $s$ be arbitrary elements of $\mathbb{Z}_q$, and let $j \in A_r(T)$. Then $v \sim f^j(v)$ for any $v \in V_r$. There exists $h \in \langle g \rangle$ such that $g(v) \in V_s$. Of course, $g(v) \sim gf^j(v) = f^j g(v)$ and so $j \in A_s(T) = A_r(T)$. In view of (17) and (18), it follows that $T \in \{\emptyset, \mathbb{Z}_q\}$. This gives us (16) and concludes the proof. □

## 3.2 The primitive case

Our main result (Theorem 3.2.1) will be proved here, followed by a detailed description of primitive $pq$-graphs. We shall make heavy use of generalized orbital digraphs, following Definition 2.2.6.

The main tool in our classification is the Liebeck-Saxl's list of all non-abelian socles of primitive groups of degree $mp$, where $m < p$ (see[LS], p.239, Table 3). Namely, if such a group $\boldsymbol{G}$ has no imprimitive subgroups, its socle cannot be abelian, otherwise it would be transitive, hence regular. This is however impossible, since $mp$ is not a prime power. Therefore, the automorphism group of a primitive $pq$-graph must necessarily contain a subgroup from the list mentioned above. Besides, the extra conditions that the group is not doubly transitive and that $m = q$ is a prime, reduces the list to the one given in Table 1.

**Theorem 3.2.1** *A vertex-transitive pq-graph must be one of the following:*

(i) *a metacirculant;*

(ii) *an* MS-*graph with* $p$ *a Fermat prime and* $q \mid p-1$*;*

(iii) *an orbital graph arising from certain rank* 3 *representations of* $\boldsymbol{P\Omega}^{\pm}(2d,2)$ *or* $\boldsymbol{M}_{22}$*;*

(iv) *a generalized orbital graph associated with the action of* $\boldsymbol{A}_7$ *on triples;*

(v) *a generalized orbital graph associated with a 2-dimensional projective special linear group.*

*Furthermore, the five classes of graphs above are mutually disjoint.*

**Proof** We shall first show that the groups in Table 1 not appearing in Table 2 either contain imprimitive subgroups, and so the corresponding generalized orbital graphs are metacirculants and MS-graphs, or give rise to generalized orbital graphs associated with some groups in Table 2.

Metacirculant arise precisely in these cases: from groups $\boldsymbol{PSL}(2,19)$, $\boldsymbol{PSL}(2,29)$, $\boldsymbol{PSL}(2,59)$ in row 11 – in this proof 'row $j$' will always stand for the $j$-th row of Table 1– and from all the groups in rows 1, 5, 9, 12, 13, 14 and 16. To see this we first show that in each of these cases the normalizer of a Sylow $p$-subgroup acts transitively and hence imprimitively.

| row | *soc* $\boldsymbol{G}$ | $(p,q)$ | *action* | *comment* |
|---|---|---|---|---|
| 1 | $\boldsymbol{A}_p$ | $(p, \frac{p-1}{2})$ | pairs | $p \geq 5$ |
| 2 | $\boldsymbol{A}_{p+1}$ | $(p, \frac{p+1}{2})$ | pairs | $p \geq 5$ |
| 3 | $\boldsymbol{A}_7$ | $(7,5)$ | triples | |
| 4 | $\boldsymbol{PSL}(4,2)$ | $(7,5)$ | 2-spaces | |
| 5 | $\boldsymbol{PSL}(5,2)$ | $(31,5)$ | 2-spaces | |
| 6 | $\boldsymbol{PSp}(4,k)$ | $(k^2+1, k+1)$ | 1-spaces | $p$ and $q$ Fermat primes |
| 7 | $\boldsymbol{P\Omega}^{\varepsilon}(2d,2)$ | $(2^d-\varepsilon, 2^{d-1}+\varepsilon)$ | singular | $\varepsilon = +1$ : $d$ Fermat prime |
| | | | 1-spaces | $\varepsilon = -1$ : $d-1$ Mersenne prime |
| 8 | $\boldsymbol{PSL}(2,p)$ | $(p, \frac{p+1}{2})$ | coset of $\boldsymbol{D}_{p-1}$ | $p \equiv 1 (mod\, 4)$ $p \geq 13$ |
| 9 | $\boldsymbol{PSL}(2,p)$ | $(p, \frac{p-1}{2})$ | coset of $\boldsymbol{D}_{p+1}$ | $p \equiv 3 (mod\, 4)$ |
| 10 | $\boldsymbol{PSL}(2,q^2)$ | $(\frac{q^2+1}{2}, q)$ | coset of $\boldsymbol{PGL}(2,q)$ | |
| 11 | $\boldsymbol{PSL}(2,p)$ | $(19,3)$, $(29,7)$ $(59,29)$, $(61,31)$ | coset of $\boldsymbol{A}_5$ | |
| 12 | $\boldsymbol{PSL}(2,23)$ | $(23,11)$ | coset of $\boldsymbol{S}_4$ | |
| 13 | $\boldsymbol{PSL}(2,11)$ | $(11,5)$ | coset of $\boldsymbol{A}_4$ | |
| 14 | $\boldsymbol{M}_{23}$ | $(23,11)$ | see [C-W] | |
| 15 | $\boldsymbol{M}_{22}$ | $(11,7)$ | see [C-W] | |
| 16 | $\boldsymbol{M}_{11}$ | $(11,5)$ | see [C-W] | |

Table 1: Non-abelian socles of uniprimitive groups of degree $pq$

Consider the group $\boldsymbol{A}_p$ acting on pairs. If $f$ denotes the $p$-cycle $(01\ldots p-1)$ and $g$ is given by the formula $g(i) = ri$ $(i \in GF(p))$, where $r$ has order $\frac{p-1}{2}$ in $GF(p)^*$, then the normalizer of $\langle f \rangle$ in $\boldsymbol{A}_p$ is $\langle f, g \rangle$ which is easily seen to be transitive on pairs.

In all other cases the transitivity of the normalizer follows by Lemma 2.1.2. With the exception of the group $\boldsymbol{PSL}(2,19)$ in row 11, all these normalizers contain a regular subgroup, and so the corresponding graphs are Cayley. On the other hand, the generalized orbital graphs of the rank 4 group $\boldsymbol{PSL}(2,19)$ are non-Cayley. In fact, apart from the Petersen graph and its complement, these are the only non-Cayley metacirculants with a primitive automorphism group.

Moreover, none of the graphs in the remaining cases of Table 1 is a metacirculant. Namely, we have $p \not\equiv 1 (\mathrm{mod}\, q)$ and so by [AP] to be a metacirculant such a graph would have to be a circulant, that is, a Cayley graph of a cyclic group of composite order.

Moving on to MS-graphs, the necessary condition is of course that $p$ be Fermat prime and that $q$ divide $p-2$. This is only satisfied for the rank 3 action of $\boldsymbol{A}_6$ on pairs in row 2, as well as for the group $\boldsymbol{PSL}(2,9)$ – which is isomorphic to $\boldsymbol{A}_6$ – in row 10, and for the rank 3 action of the group $\boldsymbol{PSp}(4,k)$ in row 6 with $p = k^2 + k$ and $q = k+1$ being Fermat primes. Surprisingly enough, the two orbital graphs in row 6 are in fact MS-graphs. Namely, the group $\boldsymbol{PSp}(4,k)$ contains an isomorphic copy of $\boldsymbol{PSL}(2,k^2)$ acting imprimitively on 1-spaces. This follows in view of the fact that

$$\boldsymbol{SL}(2,k^2) \subseteq \boldsymbol{Sp}(4,k) \quad \text{for all prime powers } k. \tag{19}$$

To see this let $I$ denote the identity $2\times 2$-matrix, $E$ the $4\times 4$-matrix $\begin{bmatrix} 0 & I \\ -I & 0 \end{bmatrix}$, $\mathbb{K} = GF(k)$ and $\mathcal{V}$ a 4-dimensional vector space of columns over $\mathbb{K}$. Now a symplectic form over $\mathcal{V}$ is given by $(X,Y) = X^T EY$. The group $\boldsymbol{Sp}(4,k)$ then consists precisely of the $4\times 4$-matrices $S$ such that

$$S^T ES = E. \tag{20}$$

Let $f(x) = x^2 + rx + s$ be an irreducible polynomial over $\mathbb{K}$. Then $f(x)$ is the characteristic polynomial of the matrix $M = \begin{bmatrix} -r & -s \\ 1 & 0 \end{bmatrix}$. By the Hamilton-Cayley theorem we get $f(M) = 0$. Let $\mathbb{K}'$ be the field of order $k$ which consists of all 2-dimensional scalar matrices $aI$, where $a \in \mathbb{K}$. Then $M \notin \mathbb{K}'$ and $[\mathbb{K}'(M) : \mathbb{K}'] = 2$. We may thus consider $\boldsymbol{SL}(2,k^2)$ as the group of the $2\times 2$-matrices $\begin{bmatrix} A & B \\ C & D \end{bmatrix}$, where $A,B,C,D \in \mathbb{K}'(M)$ and $AD - BC = I$, the unit of $\mathbb{K}'$. It is then not difficult to see that these matrices satisfy (20), obtaining thus an embedding of $\boldsymbol{SL}(2,k^2)$ inside $\boldsymbol{Sp}(4,k)$ and proving (19). A computation shows that this group is transitive and – since it does not appear in the list of Liebeck and Saxl [LS] – imprimitive on the 1-spaces of $\mathcal{V}$. This gives us an indirect proof of the fact that the two orbital graphs in row 6 are MS-graphs, as they certainly cannot be metacirculants. In the smallest admissible case $(p,q) = (5,3)$, the two graphs are $L(K_6)$ and $L(K_6)^c$ which also turn out to be the orbital graphs for the group $\boldsymbol{A}_6 = \boldsymbol{PSL}(2,9)$ in rows 2 and 10. In Figure 4 the graph $L(K_6)^c$ is shown using the notation of Frucht and emphasizing the three orbits of a semiregular automorphism of order 5.

We are now left with primitive groups without imprimitive subgroups. To conclude the proof we have to consider all those cases in which two different groups give rise to the same

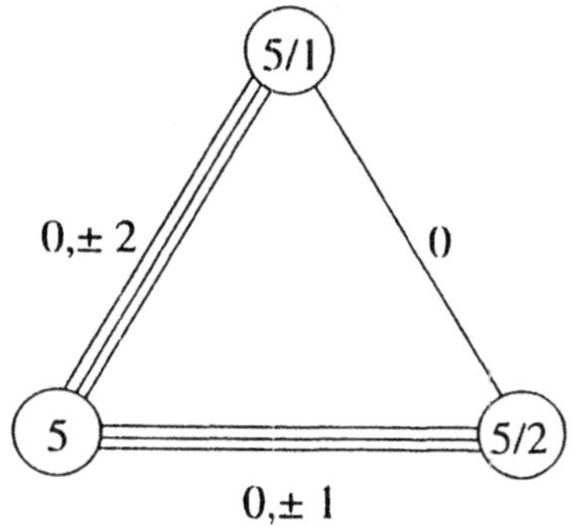


Figure 4: $L(K_6)^c$

| $soc\, G$ | $(p,q)$ | action | comment | case |
|---|---|---|---|---|
| $P\Omega^{\varepsilon}(2d,2)$ | $(2^d-\varepsilon, 2^{d-1}+\varepsilon)$ | singular 1-spaces | $\varepsilon=+1$ : $d \geq 5$ Fermat prime | (iii) |
| | | | $\varepsilon=-1$ : $d-1$ Mersenne prime | |
| $M_{22}$ | $(11,7)$ | see [C-W] | | (iii) |
| $A_7$ | $(7,5)$ | triples | | (iv) |
| $PSL(2,p)$ | $(p, \frac{p+1}{2})$ | coset of $D_{p-1}$ | $p \equiv 1 (mod\, 4)$ $p \geq 13$ | (v) |
| $PSL(2,q^2)$ | $(\frac{q^2+1}{2}, q)$ | coset of $PGL(2,q)$ | $q \geq 5$ | (v) |
| $PSL(2,61)$ | $(61,31)$ | coset of $A_5$ | | (v) |

Table 2: Uniprimitive groups of degree $pq$ without imprimitive subgroups and with non-isomorphic generalized orbital graphs

generalized orbital graph. Of course, this situation can only occur when the degrees of action are equal. Using this fact, some tedious computations reduce the list of possible overlappings to three pairs of groups corresponding respectively to: rows 2 and 8, rows 3 and 4, and for $p = 61$ rows 2, 8 and 11.

The rank 3 action of $\boldsymbol{A}_{p+1}$ on pairs of $\{1, 2, \ldots, p+1\}$ contains the action of $\boldsymbol{PSL}(2,p)$ on cosets of $\boldsymbol{D}_{p-1}$. This is seen by identifying $\{1, 2, \ldots, p+1\}$ with the point set of the projective line over $GF(p)$ and observing that $\boldsymbol{D}_{p-1}$ is the stabilizer of a pair of points. As a consequence, the two orbital graphs associated with this action of $\boldsymbol{A}_{p+1}$ can be found among the generalized orbital graphs of the relative action of $\boldsymbol{PSL}(2,p)$. This justifies the omission of row 2 of Table 1 from Table 2.

Next, as $\boldsymbol{PSL}(4,2)$ is isomorphic to $\boldsymbol{A}_8$, it certainly contains a copy of $\boldsymbol{A}_7$ as a subgroup. It follows from [Hup], Satz 2.6, p. 157, that every $\boldsymbol{A}_7$ inside $\boldsymbol{PSL}(4,2)$ acts doubly transitively on the 15 non-zero vectors in $V(4,2)$ and so it acts transitively on 2-subspaces. This gives a transitive representation of $\boldsymbol{A}_7$ of degree 35. But $\boldsymbol{A}_7$ has only one conjugacy class of maximal subgroups of index 35, and so this representation is equivalent to the action of $\boldsymbol{A}_7$ on triples. Note also that the groups $\boldsymbol{P\Omega}^+(6,2)$ and $\boldsymbol{PSL}(4,2)$ are isomorphic and the action of $\boldsymbol{P\Omega}^+(6,2)$ on singular 1-spaces in row 7 of Table 1 is equivalent to the above action of $\boldsymbol{PSL}(4,2)$. This justifies the additional condition $d \geq 5$ in row 1 of Table 2.

Let us now consider the possible overlapping of rows 2 and 11 for $p = 61$. There is a unique permutation representation of $\boldsymbol{PSL}(2,61)$ of degree 62, the natural action on the projective line. Also, the normalizer of this group in $\boldsymbol{S}_{62}$ contains an odd order element, for example the one corresponding to the matrix $\begin{bmatrix} 1 & -5 \\ 0 & 11 \end{bmatrix}$. Hence all subgroups of $\boldsymbol{A}_{62}$ isomorphic to $\boldsymbol{PSL}(2,61)$ are conjugate in $\boldsymbol{A}_{62}$. Therefore considering the action of $\boldsymbol{A}_{62}$ on pairs, the subgroups isomorphic to $\boldsymbol{PSL}(2,61)$ inside $\boldsymbol{A}_{62}$ give rise to an action on cosets of $\boldsymbol{D}_{60}$. This shows that rows 2 and 11 give rise to disjoint classes of graphs.

Finally, suppose that there is a graph $\Gamma$ which is a generalized orbital graph of $\boldsymbol{H} = \boldsymbol{PSL}(2,61)$ acting on cosets of $\boldsymbol{D}_{60}$, as well as of $\boldsymbol{H}' = \boldsymbol{PSL}(2,61)$ acting on cosets of $\boldsymbol{A}_5$. Let $\boldsymbol{G} = \langle \boldsymbol{H}, \boldsymbol{H}' \rangle$. The socle $\boldsymbol{S}$ of $\boldsymbol{G}$ is a primitive group which must appear in the Liebeck-Saxl's list ([LS], Table 3, p. 239). But apart from the two actions of $\boldsymbol{PSL}(2,61)$, the only other such group is $\boldsymbol{A}_{62}$ acting on pairs. However, the previous argument excludes the embedding of $\boldsymbol{H}'$ in $\boldsymbol{A}_{62}$. Thus rows 8 and 11 give rise to disjoint classes of graphs. □

The rest of this section is devoted to the description of primitive graphs arising from groups in Table 2. We begin discussing cases (iii) and (iv) of Theorem 3.2.1, that is, the generalized orbital graphs associated with groups in rows 1, 2 and 3 of Table 2. With the exception of four graphs arising from the action of $\boldsymbol{A}_7$ on triples, the rest all have rank 3. Those appearing in Table 3 and their complements are precisely all of these graphs. (For the arithmetic conditions on $p$ and $q$ see Table 2).

The rank 3 graphs from the first three rows correspond to cases C1, C4 and S2 in Hubaut's list ([Hu], pp. 366-373). A straightforward computation shows that the action of $\boldsymbol{A}_7$ on the set of triples of $\{1, 2, \ldots, 7\}$ gives rise to the three orbital graphs of valency 4, 12 and 18, with the respective adjacencies defined according to whether the intersection of two triples has no elements, two elements or one element. This justifies the last two rows in Table 3 (see also the proof of Theorem 3.2.1).

Note that the smallest admissible pair $(p,q)$ of prime numbers for the second row in Table

| group | vertex set | adjacency | valency |
|---|---|---|---|
| $\boldsymbol{PSL}(4,2)$ | 2-spaces | intersecting spaces | 18 |
| $\boldsymbol{P\Omega}^{\pm 1}(2d,2)$ | singular 1-spaces | orthogonal spaces | $2(2^{d-1} \mp 1)(2^{d-2} \pm 1)$ |
| $\boldsymbol{M}_{22}$ | blocks of $S(3,6,22)$ | disjoint blocks | 16 |
| $\boldsymbol{A}_7$ | triples | empty intersection | 4 |
| $\boldsymbol{A}_7$ | triples | 2-element intersection | 12 |

Table 3: Generalized orbital graphs of low rank groups

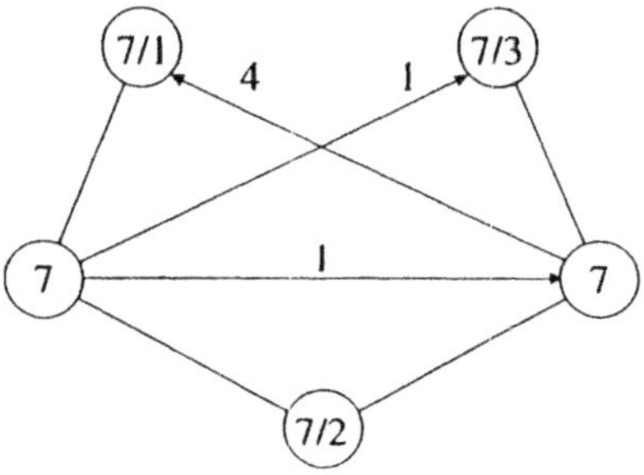


Figure 5: The odd graph $O_4$

3 is $(17,7)$. Let us also mention that the rank 3 graphs in Table 3 and their complements together with the orbital graphs of the action of $\boldsymbol{A}_{p+1}$ on pairs – which correspond to case K2 in Hubaut's list – exhaust the class of all rank 3 primitive $pq$-graphs.

Finally, observe that the graph in row 4 of Table 3 is $O_4$. A neat representation of this graph may be given using the notation of Frucht which emphasizes the five orbits $\{f^i(135) : i = 0,1,\ldots,6\}$, $\{f^i(467) : i = 0,1,\ldots,6\}$, $\{f^i(125) : i = 0,1,\ldots,6\}$, $\{f^i(346) : i = 0,1,\ldots,6\}$, and $\{f^i(127) : i = 0,1,\ldots,6\}$ of the cycle $(1234567)$ (see Figure 5).

The generalized orbital $pq$-graphs arising from the action of projective linear special groups will now be classified, by giving an explicit description of suborbits. In particular, we compute the subdegrees of the corresponding actions. Let us also remark that the subdegrees of all primitive permutation representations of $\boldsymbol{PSL}(2,k)$ were calculated in [Tc]. Some excerpts of this thesis appeared in [FI].

The graphs arising from the action of $\boldsymbol{G} = \boldsymbol{PSL}(2,p)$ on the set $\mathcal{H}$ of right cosets of $\boldsymbol{H} = \boldsymbol{D}_{p-1}$ can be described in the following way. (For further details as well as the proof see [MS2].)

First, note that in order for $q = \frac{p+1}{2}$ to be a prime we must have $p \equiv 1(\mathrm{mod}\, 4)$. Let $\mathbb{F} = GF(p)$ and $\mathbb{F}^* = \mathbb{F} \setminus \{0\}$. For reasons of simplicity we refer to the elements of $\boldsymbol{G}$ as matrices: this should cause no confusion. Let $\boldsymbol{H}$ consist of all matrices of the form

$$\begin{bmatrix} x & 0 \\ 0 & x^{-1} \end{bmatrix} \text{ and } \begin{bmatrix} 0 & -x \\ x^{-1} & 0 \end{bmatrix} \quad (x \in \mathbb{F}^*).$$

For a typical element $g = \begin{bmatrix} a & b \\ c & d \end{bmatrix}$ of $\boldsymbol{G}$ we let $\xi(g) = ad$ and $\eta(g) = a^{-1}b$. Further, we call $\chi(g) = (\xi(g), \eta(g))$ the *character* of $g$. This has nothing to do with the usual meaning of 'character' in group theory. Let $\sim$ be the equivalence relation on $\mathbb{F} \times \mathbb{F}^*$ defined by $(\xi, \eta) \sim (1-\xi, \frac{\xi\eta}{\xi-1})$ for $\xi \neq 0, 1$. There is then a natural identification of the sets $\mathcal{H}$ and $((\mathbb{F} \times \mathbb{F}^*)/\sim) \cup \{\infty\}$, where $\infty$ corresponds to $\boldsymbol{H}$ and $(\xi, \eta)$ corresponds to the coset $\boldsymbol{H}g$ satisfying $\chi(g) = (\xi, \eta)$.

Let $S^*$ denote the set of all non-zero squares in $\mathbb{F}$ and let $N^* = \mathbb{F}^* \setminus S^*$. For each $\xi \in S^*$ define the following subsets of $\mathcal{H}$. Let $\mathcal{S}_\xi^+ = \{(\xi, \eta) : \eta \in S^*\}$, $\mathcal{S}_\xi^- = \{(\xi, \eta) : \eta \in N^*\}$, and $\mathcal{S}_\xi = \mathcal{S}_\xi^+ \cup \mathcal{S}_\xi^-$. We observe that, for $\xi \neq 0, 1$, the sets $\{\mathcal{S}_\xi^+, \mathcal{S}_\xi^-\}$ and $\{\mathcal{S}_{1-\xi}^+, \mathcal{S}_{1-\xi}^-\}$ coincide. Moreover, since $(\frac{1}{2}, \eta) \sim (\frac{1}{2}, -\eta)$ it follows that the cardinality of $\mathcal{S}_\xi$ is $p-1$, except for $\xi = \frac{1}{2}$ where the cardinality is $\frac{p-1}{2}$. Similarly, the cardinalities of $\mathcal{S}_\xi^+$ and $\mathcal{S}_\xi^-$ are $\frac{p-1}{2}$, except for $\xi = \frac{1}{2}$ where the cardinalities are $\frac{p-1}{4}$.

**Theorem 3.2.2** *The action of $\boldsymbol{G} = \boldsymbol{PSL}(2,p)$ on $\mathcal{H}$ has*

(i) $(p+7)/4$ *suborbits of length* $p-1$, *all of them self-paired. These are* $\mathcal{S}_0^+ \cup \mathcal{S}_1^+$, $\mathcal{S}_0^- \cup \mathcal{S}_1^-$ *and* $\mathcal{S}_\xi$ *for all those* $\xi$ *which satisfy* $\xi^{-1} - 1 \in N^*$.

(ii) $(p-5)/2$ *suborbits of length* $(p-1)/2$, *namely* $\mathcal{S}_\xi^+$ *and* $\mathcal{S}_\xi^-$, *where* $\xi^{-1} - 1 \in S^*$. *Among them the self-paired suborbits correspond to all those* $\xi$ *for which* $\xi$ *and* $\xi - 1$ *belong to* $N^*$. *Their number is* $(p-9)/4$ *if* $p \equiv 1 (\mathrm{mod}\, 8)$ *and* $(p-5)/4$ *if* $p \equiv 5 (\mathrm{mod}\, 8)$.

(iii) 2 *suborbits of length* $(p-1)/4$, *namely* $\mathcal{S}_{\frac{1}{2}}^+$ *and* $\mathcal{S}_{\frac{1}{2}}^-$, *which are self-paired if and only if* $p \equiv 1 (\mathrm{mod}\, 8)$.

**Proof** See the Theorem in [MS2]. □

It was noted in the proof of Theorem 3.2.1 that the rank 3 action of $\boldsymbol{A}_{p+1}$ on pairs from $\{1, 2, \ldots, p+1\}$ contains the action of $\boldsymbol{PSL}(2,p)$ on cosets of $\boldsymbol{D}_{p-1}$. Therefore the two orbital graphs can be found among the generalized orbital graphs $\Gamma(\boldsymbol{G}, \boldsymbol{H}, W)$, where $W$ is a self-paired union of certain suborbits described in Theorem 3.2.2. In fact, it can be seen that the two orbital graphs are $\Gamma(\boldsymbol{G}, \boldsymbol{H}, \mathcal{S}_0 \cup \mathcal{S}_1)$ and its complement.

With the explicit description of the suborbits of $\boldsymbol{G}$ on $\mathcal{H}$ the construction of the relative generalized orbital graphs $\Gamma = \Gamma(\boldsymbol{G}, \boldsymbol{H}, W)$, where $W$ is a self-paired union of suborbits of $\boldsymbol{G}$, is relatively simple. Namely, the edge set of $\Gamma$ is $\{\{\boldsymbol{H}g, \boldsymbol{H}wg\} : g \in \boldsymbol{G}, w \in W\}$, and hence in order to determine the structure of $\Gamma$ a rule for the multiplication of characters is applied.

The description of these graphs is best done via a factorization modulo the Sylow $p$-subgroup generated by the matrix $\begin{bmatrix} 1 & 1 \\ 0 & 1 \end{bmatrix}$. For example, the smallest admissible pair of primes $p = 13$ and $q = 7$ gives rise to the following suborbits: $\mathcal{S}_0^+ \cup \mathcal{S}_1^+$, $\mathcal{S}_0^- \cup \mathcal{S}_1^-$, $\mathcal{S}_2$, $\mathcal{S}_3$, $\mathcal{S}_5$ of size 12, and $\mathcal{S}_4^+$, $\mathcal{S}_4^-$ of size 6, all of them self-paired; and $\mathcal{S}_6^+$, $\mathcal{S}_6^-$ of size 6, and $\mathcal{S}_7^+$, $\mathcal{S}_7^-$ of size 3, which are not self-paired. Therefore each of the corresponding generalized orbital graphs is a union of the graphs $\Gamma(\boldsymbol{G}, \boldsymbol{H}, W)$ with $W = \{\mathcal{S}_0, \mathcal{S}_1, \mathcal{S}_2, \mathcal{S}_3, \mathcal{S}_5, \mathcal{S}_6, \mathcal{S}_4^+, \mathcal{S}_4^-, \mathcal{S}_7\}$. As an illustration, using the notation of Frucht, the generalized orbital graph corresponding to $W = \mathcal{S}_7$ is given in Figure 6.

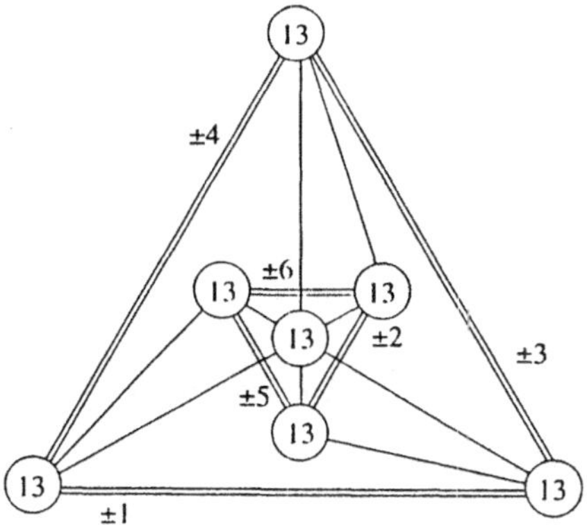


Figure 6: $\Gamma(\boldsymbol{PSL}(2,13), \boldsymbol{D}_{12}, \mathcal{S}_7)$

Finally, the subdegrees of the action of $\boldsymbol{G} = \boldsymbol{PSL}(2,61)$ on the set $\mathcal{A}$ of cosets of $\boldsymbol{A}_5$ can be computed by the method of [IKTS]. But we refer the reader also to [MS6], where a criterion for self-pairedness of suborbit is given. We obtain the following result.

**Theorem 3.2.3** *The action of $\boldsymbol{G} = \boldsymbol{PSL}(2,61)$ on the set $\mathcal{A}$ of cosets of $\boldsymbol{A}_5$ gives rise to*

(i) *one suborbit each of length* 1, 6 *and* 10, *respectively;*

(ii) *two suborbits of length* 12; (iii) *four suborbits of length* 20;

(iv) *five suborbits of length* 30; (v) 27 *suborbits of length* 60.

*Except for* 8 *suborbits of length* 60, *all other suborbits are self-paired.* □

A description of the relative generalized orbital graphs is possible with methods mentioned before the statement of Theorem 3.2.3. Namely, the stabilizers of suborbits of length 1, 6, 10, 12, 20, 30 and 60 are respectively $\boldsymbol{A}_5$, $\boldsymbol{D}_{10}$, $\boldsymbol{D}_6$, $\mathbb{Z}_5$, $\mathbb{Z}_3$, $\mathbb{Z}_2$ and 1. So we choose a particular subgroup $\boldsymbol{A}_5$ inside $\boldsymbol{PSL}(2,61)$ – for example, the one generated by the elements $\alpha = \begin{bmatrix} 3 & 0 \\ 0 & -20 \end{bmatrix}$ and $\beta = \begin{bmatrix} 8 & 22 \\ 22 & -8 \end{bmatrix}$ – as well as copies of the corresponding stabilizers, and perform the necessary computations to obtain a more detailed information about the suborbits and the generalized orbital graphs.

The graphs arising from the action of $\boldsymbol{G} = \boldsymbol{PSL}(2,q^2)$ on the set $\mathcal{H}$ of right cosets of $\boldsymbol{H} = \boldsymbol{PGL}(2,q)$ can be described as follows. (For further details as well as the proof see [MS4].)

Let $\mathbb{F} = GF(q)$. Fixing a non-square $\beta \in \mathbb{F}$, let $\alpha$ be a solution of the equation $x^2 = \beta$. The extension $\mathbb{V} = \mathbb{F}(\alpha)$ of $\mathbb{F}$ is then isomorphic to $GF(q^2)$. We may let $\boldsymbol{G} = \boldsymbol{PSL}(2,\mathbb{V})$ and $\boldsymbol{H} = \langle \boldsymbol{K}, \tilde{a} \rangle$, where $\boldsymbol{K} = \boldsymbol{PSL}(2,\mathbb{F})$ and $\tilde{a} = \begin{bmatrix} \alpha & 0 \\ 0 & \alpha^{-1} \end{bmatrix}$. It can be shown that $\boldsymbol{H}$ is isomorphic to $\boldsymbol{PGL}(2,\mathbb{F})$. For reasons of simplicity we refer to the elements of $\boldsymbol{G}$ as matrices.

Let $\Omega$ be the set of all elements $x \in \mathbb{F}$ such that $x = y^2\beta + y$ for some $y \in \mathbb{F}$. Let $\Omega^+$ and $\Omega^-$ be the set of all non-zero squares and non-squares in $\Omega$ respectively. Every coset $\boldsymbol{H}g \neq \boldsymbol{H}$ in $\mathcal{H}$ contains a canonical representative $\begin{bmatrix} 1 & y\alpha \\ z\alpha & 1+yz\beta \end{bmatrix}$, $y \in \mathbb{F}, z \in \mathbb{F}^*$. Let us call $\omega = y^2z^2\beta + yz \in \Omega$ the *character* of $\boldsymbol{H}g$. It can be shown that $\omega$ is independent of the choice of the representatives. One can then also see that each suborbit contains cosets with equal characters. More precisely, letting $\mathcal{S}_\omega$ be the set of all cosets $\boldsymbol{H}g$ whose characters are equal to a given $\omega \in \Omega$, the suborbits of the action of $\boldsymbol{G}$ on $\mathcal{H}$ are described by the following result ([MS4], Theorem 1.1).

**Theorem 3.2.4** *The action of $\boldsymbol{G} = \boldsymbol{PSL}(2, q^2)$ on $\mathcal{H}$ has the following suborbits:*

(i) $\mathcal{S}_0$, *the only suborbits of length $(q^2 - 1)$;*

(ii) $\mathcal{S}_{-(4\beta)^{-1}}$, *the only suborbits of length $q(q-1)/2$ if $q \equiv 1 (\mathrm{mod}\, 4)$ and of length $q(q+1)/2$ if $q \equiv 3 (\mathrm{mod}\, 4)$;*

(iii) $\mathcal{S}_\omega$, $\omega \in \Omega \setminus \{(-4\beta)^{-1}\}$, *of length $q(q-1)$; their number is $(q-5)/4$ if $q \equiv 1 (\mathrm{mod}\, 4)$ and $(q-3)/4$ if $q \equiv 3 (\mathrm{mod}\, 4)$;*

(iv) $\mathcal{S}_\omega$, $(\omega \in \Omega^+ \setminus \{(-4\beta)^{-1}\}$, *of length $q(q+1)$; their number is $(q-1)/4$ if $q \equiv 1 (\mathrm{mod}\, 4)$ and $(q-3)/4$ if $q \equiv 3 (\mathrm{mod}\, 4)$.*

*Furthermore, all suborbits are self-paired.* □

Let us mention that the lengths of these suborbits can also be deduced from formula (7) of [FI]. Moreover, an alternative way is followed by [PX].

With the explicit description of the suborbits of $\boldsymbol{G}$ on $\mathcal{H}$ the construction of the corresponding generalized orbital graphs $\Gamma(\boldsymbol{G}, \boldsymbol{H}, W)$, where $W$ is a union of suborbits, is relatively simple. The structure of these graphs is best understood via a factorization modulo a cyclic group of order $(q^2+1)/2$. The computations however are rather tedious. To illustrate the ideas we therefore choose to consider the smallest case $q = 3$ which gives rise to the graphs $L(K_6)$ and $L(K_6)^c$, the MS-graphs arising also in rows 2, 7 of Table 1.

The two non-trivial $\boldsymbol{H}$-suborbits are $\mathcal{S}_1$ of length 6 containing the cosets with representatives

$$\begin{bmatrix} 1+\alpha & 0 \\ 0 & -1+\alpha \end{bmatrix}, \begin{bmatrix} -1+\alpha & 1 \\ 1-\alpha & \alpha \end{bmatrix}, \begin{bmatrix} -1-\alpha & 1 \\ 1+\alpha & -\alpha \end{bmatrix},$$

$$\begin{bmatrix} 1 & 1-\alpha \\ \alpha & -1+\alpha \end{bmatrix}, \begin{bmatrix} 1 & -1-\alpha \\ \alpha & -1-\alpha \end{bmatrix}, \begin{bmatrix} 1 & -\alpha \\ \alpha & -1 \end{bmatrix},$$

and $\mathcal{S}_0$ of length 8 containing the cosets with representatives

$$\begin{bmatrix} 0 & 1 \\ -1 & -\alpha \end{bmatrix}, \begin{bmatrix} 1 & 0 \\ \alpha & 1 \end{bmatrix}, \begin{bmatrix} 1 & 1 \\ \alpha & 1+\alpha \end{bmatrix}, \begin{bmatrix} 1 & -1 \\ \alpha & 1-\alpha \end{bmatrix},$$

$$\begin{bmatrix} 0 & 1 \\ -1 & \alpha \end{bmatrix}, \begin{bmatrix} 1 & 0 \\ -\alpha & 1 \end{bmatrix}, \begin{bmatrix} 1 & 1 \\ -\alpha & 1-\alpha \end{bmatrix}, \begin{bmatrix} 1 & -1 \\ -\alpha & 1+\alpha \end{bmatrix}.$$

Two complementary graphs having 15 vertices are obtained. The graph $\Gamma(\boldsymbol{G}, S_1) = L(K_6)^c$ is depicted in Figure 4, where the notation of Frucht is used to emphasize the three orbits of the cyclic group of order 5 generated by the element $\begin{bmatrix} 1 & -1 \\ \alpha & 1-\alpha \end{bmatrix}$.

## 3.3 A generalization to digraphs

In this section we give a classification of vertex-transitive digraphs whose order is a product of two primes. As it is known that each vertex-transitive $p^2$-digraph is a Cayley digraph of one of the two nonisomorphic groups of order $p^2$ (see [Ma3]), we may restrict ourselves to VT-digraphs whose order is a product of two distinct primes.

The object of this section is therefore to classify vertex-transitive $pq$-digraphs (Theorem 3.3.3), thus extending the results of Theorem 3.2.1. In order to emphasize the existence or non-existence of directed graphs in our extended classification we find it convenient to use the following terminology.

**Definition 3.3.1** Digraphs that are not undirected graphs are called *genuine* digraphs.

The class of vertex-transitive $pq$-graphs coincides with that of all generalized orbital graphs arising from transitive permutation groups of degree $pq$. Such groups are of three essentially different types: $p$-imprimitive, $q$-imprimitive or primitive. We base our classification on a detailed description of digraphs belonging to each of these three disjoint classes. Before doing this, we make a simple observation which allows us to easily extend Theorem 3.2.1 to the directed case (Theorem 3.3.3).

If $\Gamma$ is a digraph and $\widetilde{\Gamma}$ is the underlying graph of $\Gamma$, obtained by adding $(u,v)$ whenever $(v,u) \in A(\Gamma)$, then clearly $\boldsymbol{Aut}\,\Gamma \subseteq \boldsymbol{Aut}\,\widetilde{\Gamma}$. In particular, this means that every VT-digraph $\Gamma$ gives rise to a VT-graph $\widetilde{\Gamma}$. Therefore non-existence results for VT-graphs of certain kinds will imply analogous results for VT-digraphs. Hence a classification of VT-graphs of a given order may be used to obtain classification results for VT-digraphs of that order. This is precisely what we do here for vertex-transitive $pq$-digraphs.

Note that we can generalize the terminology of [AP] and [MS5], introducing the concept of $(q,p)$-*dimetacirculant* digraph (see Definition 3.3.2). In fact, by [Ma1] (Theorem 3.6), the classes of $(q,p)$-dimetacirculants and $p$-imprimitive $pq$-digraphs coincide.

**Definition 3.3.2** Let $m, n \geq 1$ be integers. A digraph $\Gamma$ is said to be an $(m,n)$-*dimetacirculant* if it has an automorphism $f$ with a cycle decomposition

$$f = (v_0^0 v_1^0 \cdots v_{n-1}^0)(v_0^1 v_1^1 \cdots v_{n-1}^1) \cdots (v_0^{m-1} v_1^{m-1} \cdots v_{n-1}^{m-1})$$

and an automorphism $g$ which cyclically permutes the orbits

$$V_i = \{v_0^i, v_1^i, \ldots, v_{n-1}^i\} \quad (i = 0, 1, \ldots, m-1).$$

**Theorem 3.3.3** *The vertex-transitive pq-digraphs are precisely the ones listed in Table 4. Also, the rows of Table 4 correspond to mutually disjoint families of digraphs, whose description is incorporated in the proof of this theorem.*

| class of digraphs | description | genuine digraphs |
|---|---|---|
| $p$-imprimitive | dimetacirculants | yes |
| $q$-imprimitive | MS-digraph | yes |
| primitive | $\boldsymbol{P\Omega}^{\pm}(2d,2)$ acting on singular 1-spaces | no |
| | $\boldsymbol{M}_{22}$ with the natural action of degree 77 | no |
| | $\boldsymbol{A}_7$ acting on triples | no |
| | $\boldsymbol{PSL}(2,p)$ acting on cosets of $\boldsymbol{D}_{p-1}$ | yes |
| | $\boldsymbol{PSL}(2,q^2)$ acting on cosets of $\boldsymbol{PGL}(2,q)$ | no |
| | $\boldsymbol{PSL}(2,61)$ acting on cosets of $\boldsymbol{A}_5$ | yes |

Table 4: Vertex-transitive $pq$-digraphs

**Proof** Let $\Gamma$ be a vertex-transitive $pq$-digraph. We have to distinguish three cases, depending on the particular class to which $\Gamma$ belongs.

Suppose first that $\Gamma$ is $p$-imprimitive. A combined application of Theorems 3.4 and 3.6 of [Ma1] implies the existence of an automorphism $f$ of $\Gamma$ with $q$ orbits of length $p$ such that the subgroup of $\boldsymbol{Aut}\,\Gamma$, which contains the automorphisms of $\Gamma$ mapping the orbits of $f$ among themselves, is transitive. Letting the cycle decomposition of $f$ be

$$f = (v_0^0 v_1^0 \dots v_{p-1}^0)(v_0^1 v_1^1 \dots v_{p-1}^1) \cdots (v_0^{q-1} v_1^{q-1} \dots v_{p-1}^{q-1})$$

we may therefore assume the existence of an automorphism $g$ that cyclically permutes the sets

$$V_i = \{v_0^i, v_1^i, \dots, v_{p-1}^i\} \;\; (i = 0, 1, \dots, q-1),$$

which means that there exists $r \in \mathbb{Z}_m^*$ such that $g(v_j^i) = v_{rj}^{i+1}$ for all $i$ and $j$, forcing $r^h \equiv 1(\bmod p)$ for some multiple $h$ of $q$. Hence $\boldsymbol{Aut}\,\Gamma$ contains a transitive metacyclic subgroup, and so $\Gamma$ is a $(q,p)$-dimetacirculant. Now by the same argument as in [AP] it can be shown that $\Gamma$ is completely determined by an array $(q, p, r, S_0, S_1, \dots, S_{q-1})$, where $r^q \equiv 1(\bmod p)$ and $S_i = \{s \in \mathbb{Z}_p : v_0^0 v_s^i \text{ is an arc}\}$. On the other hand, every array of this form gives rise to a $(q,p)$-dimetacirculant and $p$-imprimitive $qp$-digraph. Of course, there are genuine digraphs among dimetacirculants. This justifies the first row of Table 4.

Suppose now that $\Gamma$ is $q$-imprimitive. Using the Theorem of [MS3], we have that $p = 2^k - 1$ is a Fermat prime and $q$ divides $p-2$, and that the underlying graph $\widetilde{\Gamma}$ is a generalized orbital digraph of the group $\boldsymbol{SL}(2,2^k)$ acting on the set $PG(2,2^k-1) \times \mathbb{Z}_q$, hence an MS-*graph*. Therefore $\Gamma$ must be a generalized orbital digraph of the same group action. Combining Lemma 2.1 and the Theorem of [MS3] we have that this action has $q$ suborbits of length 1 and $q$ suborbits of length $p-1$, of which precisely the latter are self-paired. Also, each of the nontrivial suborbits of length 1 gives rise to a union of $p$ directed cycles of length $q$, the corresponding sets of vertices being the blocks of size $q$. Therefore $\Gamma$ must be an edge-disjoint union of an MS-graph and a genuine digraph of the form $pC$, where $C$ is a dicirculant with $q$ vertices. This justifies the second row of Table 4.

Suppose finally that $\Gamma$ is a primitive digraph. The results on primitive $pq$-digraphs proved in previous sections are now of crucial importance. In particular, Table 2 gives us the complete list of primitive $pq$-graphs. So $\widetilde{\Gamma}$ is one of these graphs and therefore $\Gamma$ must be precisely one of the digraphs in the last six rows of Table 4. Next, we note that all of the relative primitive groups except for the action of $\boldsymbol{PSL}(2,p)$ on cosets of $\boldsymbol{D}_{p-1}$, and the action of $\boldsymbol{PSL}(2,61)$ on cosets of $\boldsymbol{A}_5$, give rise to self-paired suborbits (see Theorem 3.2.3). This justifies the last column in each of these rows. As a final remark, the fact that the rows of Table 4 give rise to mutually disjoint families of digraphs is a direct consequence of the corresponding statement in Theorem 3.2.1. □

The following is now immediate.

**Corollary 3.3.4** *A vertex-transitive genuine pq-digraph must be one of the following:*

(i) *a dimetacirculant;*

(ii) *an* MS-*digraph;*

(iii) *a generalized orbital digraph of* $\boldsymbol{PSL}(2,p)$ *acting on cosets of* $\boldsymbol{D}_{p-1}$, *or of* $\boldsymbol{PSL}(2,61)$ *acting on cosets of* $\boldsymbol{A}_5$.

□

# Chapter 4

# Two problems on vertex-transitive graphs

Among all questions that arose in the area of vertex-transitive graphs, besides the 'basic' one concerning the classification, we have chosen two unsettled existence problems. One concerns semiregular automorphisms, and the other hamiltonian paths.

The results presented in the first section were proved in [MS8].

## 4.1 Semiregular automorphisms

A single permutation $f$ may generate a transitive group: the graphs with such an automorphism are of course circulants.

**Definition 4.1.1** Let $f$ be a fixed-point-free permutation, with $r$ orbits of the same length $s \geq 2$. Then $f$ is said to be $(r,s)$-*semiregular*. A permutation $f$ is then called *semiregular* when it is $(r,s)$-semiregular for some $r$, $s$. Accordingly, a digraph $\Gamma$ is said to be $(r,s)$-*rotary* or *rotary* if there is $f \in \boldsymbol{Aut}\,\Gamma$ that is $(r,s)$-semiregular or semiregular, respectively.

In view of Definition 4.1.1, a permutation $f$ is semiregular if and only if $\langle f \rangle$ is itself semiregular.

In spite of their name, rotary graphs do not form a restricted club. They are most common, and in particular all known VT-digraphs seem to satisfy this property. Therefore, Marušič [Ma1] and Jordan [Jo], independently raised the following question.

**Problem 4.1.2** *Does every* VT-*graph have a semiregular automorphism?*

This is a very natural problem indeed: the orbits of a semiregular automorphism $f$ can be used in order to describe the given graph. This is clear when looking at metacirculants and generalized Petersen graphs, both of which are rotary. More generally, if $f$ is a semiregular automorphism of $\Gamma$, it is often useful to consider the quotient $\Gamma/_f$, defined by taking $V(\Gamma/_f)$ as the set of orbits of $\langle f \rangle$ and letting $U \sim W$ if and only if $\exists\, u \in U$, $\exists\, w \in W$ such that $u \sim w$. Of course, $\Gamma/_f$ is nothing but the 'underlying graph' of the Frucht diagram associated with $f$. Recently, Lovrečič Saražin and Pacco [LP] have exploited semiregular automorphisms in order to describe certain graphs of valency 4.

Although Problem 4.1.2 is clearly related to permutation groups, it is not equivalent to asking about transitive groups without semiregular automorphisms. Namely, such groups exist, but possibly never coincide, as permutation groups, with the full automorphism group of a digraph. Infinitely many examples of transitive permutation groups having no semiregular elements were given in [FKS]. They have all degree $p(p+1)$.

Of course, it is a direct consequence of the definition that every Cayley digraph has a semiregular automorphism. For non-Cayley digraphs the problem seems to be a difficult one. Further families of VT-digraphs with semiregular automorphisms include VT-digraphs of orders $mp$, where $m \leq p$, and $p^k$ (see [Ma1], Theorems 3.3, 3.4). On the other hand, no VT-digraph without a semiregular automorphism is known to us.

We will give here a partial affirmative answer to Problem 4.1.2, by proving that for both cubic graphs (Theorem 4.1.5). and digraphs of order $2p^2$, where $p$ is a prime, vertex-transitivity is a sufficient condition to be rotary.

**Lemma 4.1.3** *Let $\Gamma$ be a connected* VT-*graph of prime valency $p$, let $v \in V(\Gamma)$ and $\boldsymbol{G}$ a transitive subgroup of $\boldsymbol{Aut}\,\Gamma$. Then $|\boldsymbol{G}_v|$ is neither divisible by $p^2$ nor by any prime greater than $p$.*

**Proof** Let us first show that $|\boldsymbol{G}_v|$ is not divisible by any prime greater than $p$. Namely, if there exists a prime $q > p$ dividing $|\boldsymbol{G}_v|$, then there is an automorphism $a \in \boldsymbol{G}_v$ of order $q$, moving at least one neighbour $u$ of $v$. But then, since $\Gamma(v)$ is a union of nontrivial orbits of $a$, we get $|\Gamma(v)| \geq q > p$, a contradiction because $p$ is the valency of $\Gamma$.

Of course, $p$ divides $|\boldsymbol{G}_v|$. We now prove that $p^2$ does not divide $|\boldsymbol{G}_v|$. Let $\boldsymbol{P}$ be a Sylow $p$-subgroup of $\boldsymbol{G}_v$. We shall prove that an element $a \in \boldsymbol{G}_v$ of order a power of $p$, fixing some vertex in $\Gamma(v)$, must necessarily be the identity.

Let $u \in V(\Gamma)$, with $\partial(v,u) = i$. By induction on $i$ we now prove that $a(u) = u$. This is clearly true for $i = 0, 1$. Assume that $i \geq 2$ and that the statement is true for all vertices $u'$ such that $\partial(v,u') \leq i-1$. Let $z$ and $w$ be the two vertices immediately preceding $u$ in a path of length $i$ from $v$ to $u$. Then $\partial(v,w)$ and $\partial(v,z)$ are both less than $i$. By induction hypothesis, $a(w) = w$ and $a(z) = z$. It follows that $a \in \boldsymbol{G}_w$ and that $a$ has a fixed point in $\Gamma(w)$, that is, the vertex $z$. But then, since the valency of $\Gamma$ is prime, it follows that $a$ fixes $\Gamma(w)$ pointwise. In particular, $a(u) = u$. We conclude that $a = 1$. In view of this, the homomorphism associating to an element $a \in \boldsymbol{P}$ its restriction to $\Gamma(v)$ must be injective. Hence $\boldsymbol{P}$ is isomorphic to some subgroup of the symmetric group $\boldsymbol{S}_p$. But $p^2$ does not divide $p! = |\boldsymbol{S}_p|$ and so $|\boldsymbol{P}| = p$. □

**Theorem 4.1.4** *Every* VT-*digraph of prime-power order is rotary.*

**Proof** See [Mal], Theorem 3.3. □

**Theorem 4.1.5** *Every cubic* VT-*graph is rotary.*

**Proof** Let $\Gamma$ be a cubic VT-graph, and let $\boldsymbol{G} = \boldsymbol{Aut}(\Gamma)$. We may assume that $\Gamma$ is connected. Now Lemma 4.1.3 implies that each element of $\boldsymbol{G}$ of order $p \neq 2, 3$ is semiregular. We may therefore assume that $|\boldsymbol{G}| = 2^h \cdot 3^k$ for some $h, k \geq 0$. In view of the Burnside $p$-$q$-Theorem ([Su], pp. 104-105), the group $\boldsymbol{G}$ is solvable. Hence a minimal normal subgroup $\boldsymbol{K}$ of $\boldsymbol{G}$ is elementary abelian.

Suppose now that $\boldsymbol{K}$ contains no elements of order 3. Then $\boldsymbol{K}$ is clearly a 2-group. In view of Theorem 4.1.4, we may assume that the order of $\Gamma$ is a multiple of 3. As $\boldsymbol{G}$ is transitive, the same holds for the order of $\boldsymbol{G}$. Further we may suppose that $\boldsymbol{G}$ has no semiregular elements of order 3. Henceforth, there is an element $g \in \boldsymbol{G}$ of order 3, fixing some vertex $v$ of $\Gamma$ and cyclically permuting the neighbours of $v$. Because of vertex-transitivity, this happens to each vertex of $\Gamma$.

Let $a \in \boldsymbol{K}$ be an involution that is not semiregular. The connectedness of $\Gamma$ then implies that there are two adjacent vertices $v, w$ of $\Gamma$ such that $a(v) = v$ and $a(w) \neq w$. Let $b \in \boldsymbol{G}_v$ be an element of order 3 which cyclically permutes the neighbours of $v$. Then $a^b \in \boldsymbol{K}^b \cap \boldsymbol{G}_v = \boldsymbol{K} \cap \boldsymbol{G}_v$. Clearly, $aa^b \in \boldsymbol{K}$. Besides, $a^b \neq a$ and, more precisely, the restriction of $aa^b$ to $\Gamma(v)$ is a 3-cycle, contradicting the fact that $\boldsymbol{K}$ is a 2-group. Hence our assumption about nonexistence of elements of order 3 in $\boldsymbol{K}$ was false. In view of the fact that $\boldsymbol{K}$ is elementary abelian, this forces $\boldsymbol{K}$ to be a 3-group. Now Lemma 4.1.3 implies that $|\boldsymbol{K} \cap \boldsymbol{G}_v| = 3$ for all vertices $v$ of $\Gamma$. It follows that $\{\mathrm{Fix}(a) : a \in \boldsymbol{K} \setminus \{1\}\}$ is a partition of $V(\Gamma)$. Namely, if $u \in \mathrm{Fix}(a) \cap \mathrm{Fix}(b)$ for $a, b \in \boldsymbol{K}$, then $b \in \{a, a^{-1}\}$, and so clearly $\mathrm{Fix}(a) = \mathrm{Fix}(b)$.

Now, since $\Gamma$ is connected, there exist two adjacent vertices $u$ and $v$ belonging to different blocks of this partition. Considering the action of the elements of $\boldsymbol{G}$ fixing precisely one of $u$ and $v$, we may easily obtain that the subgraph of $\Gamma$ induced by $\Gamma(u) \cup \Gamma(v)$ is $K_{3,3}$. But then, since $\Gamma$ is connected and of valency 3, we have $\Gamma = K_{3,3}$, which is of course a rotary graph. □

**Theorem 4.1.6** *A* VT-*digraph of order* $2p^2$ *is* $(2p, p)$-*rotary.*

**Proof** In view of [Mal], we may assume that $\Gamma$ is connected. Let $\boldsymbol{P}$ be a Sylow $p$-subgroup of $\boldsymbol{Aut}\,\Gamma$. In view of [Wi], Theorem 3.4′, it follows that $\boldsymbol{P}$ has two orbits of length $p^2$, say $A$ and $B$. Let $a \in Z(\boldsymbol{P})$ be of order $p$. Assume that $a$ is not semiregular. Then, without loss of generality, $a^A$ is semiregular and $a^B = 1$. By [Mal], Proposition 3.2, every transitive permutation group of degree $p^2$ has a regular subgroup. Hence let $\boldsymbol{Q} \leq \boldsymbol{P}$ be such that $|\boldsymbol{Q}^B| = p^2$.

Assume first that $\boldsymbol{Q}^B = Z_{p^2}$. Then let $f \in \boldsymbol{Q}$ be such that $f^B$ generates $\boldsymbol{Q}^B$, that is, $f^B$ is a $p^2$-cycle. If $f^A$ is also a $p^2$-cycle, then $f^p$ is a $(2p, p)$-semiregular automorphism of $\Gamma$. Otherwise the above holds for $f^p a$.

We may now assume that $\boldsymbol{Q}^B = Z_p \times Z_p$. Then there are elements $g, h \in \boldsymbol{Q}$ such that $\langle g, h\rangle^B = \boldsymbol{Q}^B$ and both $g^B, h^B$ are semiregular of order $p$. Suppose first that one of $g^A$ and $h^A$, say $g^A$, is not a $p^2$-cycle. If $g^A = 1$ then $f = ga$ is a $(2p, p)$-semiregular automorphism of $\Gamma$. Hence we may assume that $g^A$ has order $p$ and moreover that $g$ has some fixed points in $A$. But then $g^A$ must necessarily fix each of the orbits $A_0, A_1, \ldots, A_{p-1}$ of $a$ in $A$. Let $B_0, B_1, \ldots, B_{p-1}$ be the orbits of $g^A$ in $B$. Note that $a^{B_j} = 1$ for each $j$. But $B_j$ is an orbit of $g$ and $g(A_i) = A_i$ for each $i$.

In order to describe the graph at this point, we adopt the following notation. If $\Gamma$ is a digraph and $U, W$ are disjoint subsets of $V(\Gamma)$ we let $\Gamma[U, W]$ denote the bipartite digraph induced by all those arcs of $\Gamma$ of the form $(u, w)$ and $(w, u)$ with $u \in U$ and $w \in W$.

From the last formula we have proved, it follows that for each $i, j$ the arc set of $\Gamma[A_i, B_j]$ is either empty or one of $A_i \times B_j$, $B_j \times A_i$ and $A_i \times B_j \cup B_j \times A_i$. It is then clear that the permutation $f$, defined by the rule

$$f(v) = \begin{cases} a(v), & v \in A \\ g(v), & v \in B \end{cases}$$

is the desired automorphism of $\Gamma$.

We are now left with the case where both $g^A$ and $h^A$ are $p^2$-cycles. If $\langle g^A \rangle = \langle h^A \rangle$ then say $g^A = (h^i)^A$ for some $i$, and so $(gh^{-i})^A = 1$. Since $(gh^{-i})^B$ has order $p$, the automorphism $f = agh^{-i}$ has the desired form. Assume now that $\langle g^A \rangle \neq \langle h^A \rangle$. Note that the orbits of $a^A$ in $A$ form an imprimitivity block system of $\boldsymbol{P}^A$. On the other hand, a cyclic group of order $p^2$ has only one imprimitivity block system $P$. More precisely, if $g^A = (0\,1 \ldots p^2 - 1)$, then $P$ consists of the residue classes mod $p$ in $\{0, 1, \ldots, p^2 - 1\}$. Of course, the same must be true for $h^A$. Without loss of generality we may assume that $h^A(0) = 1$. Then $g^A(h^A)^{-1}$ fixes 0, and certainly $g^A(h^A)^{-1} \neq 1$. Hence $g^A(h^A)^{-1}$ has order $p$. We may now apply the argument in the case where not both $g^A$ and $h^A$ were $p^2$-cycles, by replacing $g$ with $gh^{-1}$. Therefore, the permutation $f$, defined by the rule

$$f(v) = \begin{cases} a(v), & v \in A \\ gh^{-1}(v), & v \in B \end{cases},$$

is the desired automorphism of $\Gamma$, which concludes the proof. □

## 4.2 Hamiltonian paths and cycles in VT-graphs

Let us now give a short survey, with almost no proofs, of known results on the existence of hamiltonian paths and cycles in VT-graphs. Details and further information can be found in the original papers, or in [Ya]. We just mention that the techniques used to construct hamiltonian paths and cycles often involve semiregular automorphisms. The underlying idea is to look at the quotient $\Gamma/_f$ with respect to such an $f$, and to lift paths of this graph in a suitable way (see e.g. Alspach [Al2]). This provides one more motivation to study Problem 4.1.2.

All graphs are assumed to be connected.

To begin with, the following question, raised by Lovász [Lo], is still open.

**Problem 4.2.1** *Does every* VT*-graph have a hamiltonian path?*

Several constructions solving Problem 4.2.1 for special cases provide also partial answers to the related question below.

**Problem 4.2.2** *Which* VT*-graphs have a hamiltonian cycle?*

There are indeed four graphs for which the answer to Problem 4.2.2 is negative. One is our professional counterexample $O_3$. Another is the *Coxeter graph* $Y$. The remaining two are obtained from $O_3$ and $Y$ by replacing vertices with triangles (see [Ya]). None of these is a Cayley graph. Since they are the only known exceptions to Problem 4.2.2, the following has also been asked.

**Problem 4.2.3** *Does every Cayley graph have a hamiltonian cycle?*

Before going on and state the available results on Problems 4.2.1, 4.2.2, 4.2.3, let us mention a result (due to Babai: see [Ca2], Theorem 4.5) that answers the existence of fairly 'long' cycles in all VT-graphs.

**Theorem 4.2.4** *Every* VT-*graph of order* $n$ *has a cycle of length* $\geq \sqrt{3n}$. □

For the more general case of regular graphs, Jackson [Ja] proved that every 2-connected regular graph of order $n$ and valency $\geq \frac{n}{3}$ is hamiltonian. As VT-graphs are necessarily 2-connected, this implies the following.

**Proposition 4.2.5** *A* VT-*graph of order* $n$ *and valency* $\geq \frac{n}{3}$ *has a hamiltonian cycle.* □

This result was used by Marušič (see [Ma4]) as a step towards Theorem 4.2.6, a statement that puts together several results (see [Ma1] and [Ma2]).

**Theorem 4.2.6** *Except for the Petersen graph, every* VT-*graph of order* $p$, $p^2$, $2p$, $3p$ *or* $2p^2$ *(*$p$ *prime) has a hamiltonian cycle.* □

**Theorem 4.2.7** *Every* VT-*graph of order* $4p$ *or* $5p$ *(*$p$ *prime) has a hamiltonian path.*

The following result was proved in [Ma2].

**Theorem 4.2.8** *Every connected Cayley graph of a semidirect product of a cyclic group of prime order by an abelian group of odd order is hamiltonian.* □

More can be said on hamiltonian cycles in Cayley graphs. In particular, it is known that they always exist if the underlying group is abelian (see [GW]). Alspach conjectured that every Cayley graph of even valency $2k$ has a decomposition into $k$ edge-disjoint hamiltonian cycles. Baumann, Lesch and Schmeichel [BLS] proved that this is true for $k = 2$.

Concerning metacirculants, the following result was established independently by Alspach and Parsons ([AP], Theorem 1) and by Marušič ([Ma1], Theorem 7.4).

**Proposition 4.2.9** *Let* $p$ *and* $q$ *be primes. Then every connect metacirculant with* $pq$ *vertices, other than the Petersen graph, is hamiltonian.* □

Marušič conjectured that the Petersen graph is the only connected metacirculant without a Hamilton cycle. This was proved by Alspach [Al1], under the additional assumption that the number of vertices in a block is a prime power. More recently, Tan considered cubic metacirculant graphs, obtaining the following result (see [Ta1] and [Ta2]).

**Theorem 4.2.10** *A connected cubic $(m,n)$-metacirculant $\Gamma \neq O_3$ has a hamiltonian cycle provided that one of the following holds:*

(i) *$m$ is a multiple of 4;*

(ii) *$n$ is even and every odd prime divisor of $n$ divides also $m$;*

(iii) *$n = 2^h p^k$, $p$ odd prime, $h > 0$, $k \geq 0$.* □

The next two results concern imprimitive graphs. They were established by Marušič (see [Ma5]).

**Lemma 4.2.11** *Let $k$ and $m$ be a positive integers greater that 1, and let $m$ divide $2^k - 1$. If $m < 2^k - 1$ then $\Gamma(k,m)$ is hamiltonian.*

**Proof** Let $\Gamma = \Gamma(k,m)$ and let $\widetilde{\Gamma}$ be the base graph of $\Gamma$ with the corresponding labeling function $\sigma$. Moreover, let $\rho$, $\tau$ and $\tau_r$ ($\tau \in \mathbb{Z}_m$) have the meaning described in the paragraph preceding Definition 3.1.2. Fix a generator $w$ in $GF(2^k)^*$ and let $C = 01ww^2 \ldots w^n 0$, with $n = 2^k - 2$, be a cycle in $\widetilde{\Gamma}$ which leaves out the vertex $\infty$. Then the respective labels of the corresponding arcs $(0,1), (1,w), (w,w^2), \ldots, (w^{n-1}, w^n)(w^n, 0)$ of $C$ are $\tau_0, \tau_\alpha, \tau_{\alpha+1}, \ldots, \tau_{\alpha+n-1}, \tau_n$, where $\alpha$ satisfies the equation $1 + w = w^\alpha$. Clearly, $m < n$ and so

$$\text{each } \tau_r (r \in \mathbb{Z}_m) \text{ is a label of at least one arc in } C. \tag{21}$$

Moreover, since $\tau_b \tau_c = \rho^{2(b-c)}$ for all $b, c \in \mathbb{Z}_m$, it follows that the value of the labeling $\sigma$ computed on the cycle $C$ equals

$$\sigma(C) = \tau_0 \overbrace{\rho^{-2} \cdots \rho^{-2}}^{n/2} \tau_n = \tau_0 \rho^{-n} \tau_n = \rho^{-n}.$$

But $\rho^m = \mathrm{id}$ and so, since $m$ divides $n+1$, it follows that $\sigma(C) = \rho^{-(n+1)}\rho = \rho$. This implies that the cycle $C$ lifts to a Hamilton cycle, call it $H$, in the graph $\Gamma - \{(\infty, r) : r \in \mathbb{Z}_m\}$.

The hamiltonicity of $\Gamma$ will now follow if we show that, for each $r \in \mathbb{Z}_m$, the vertex $(\infty, r)$ is a common neighbour of two successive vertices in $H$. Fix $r \in \mathbb{Z}_m$. Then, by (21), there exist $x, y \in GF(2^k)$ such that $\sigma(x,y) = \tau_r$. Of course, $\tau_r(r) = r$ and so $(y,r)$ is a successor of $(x,r)$ in $H$. But, by the definition of $\Gamma(k,m)$ , both $(x,r)$ and $(y,r)$ are neighbours of $(\infty, r)$ and so the cycle $H$ can be extended by inserting $(\infty, r)$ between $(x,r)$ and $(y,r)$. Doing this for each $r \in \mathbb{Z}_m$ we obtain a Hamilton cycle in $\Gamma$. □

Using Lemma 4.2.11 we may now establish the hamiltonicity of imprimitive $pq$-graphs.

**Theorem 4.2.12** *Let $\Gamma \neq O_3$ be a connected imprimitive $pq$-graph, where $p$ and $q$ are primes. Then $\Gamma$ is hamiltonian.*

**Proof** If $\Gamma$ is a metacirculant, then it is hamiltonian by Proposition 4.2.9. In view of Theorem 3.1.15 we may therefore assume that $p = 2^{2^a} - 1$ is a Fermat prime and $q$ divides $2^{2^a} - 1$, and that $\Gamma$ contains a subgraph isomorphic to $\Gamma(2^a, q)$.

Since 3 divides $p = 2^{2^a} - 1$, it follows that the latter is a prime if and only if $p = 5$ and $q = 3$. The graph $\Gamma(2,3)$ turns out to be $L(O_3)$, the line graph of the Petersen graph, and has a Hamilton cycle (see [Ma4], Theorem 5.6). We may therefore assume that $q < p = 2^{2^a} - 1$, and so $\Gamma$ is hamiltonian by Lemma 4.2.11. This completes the proof. □

# Chapter 5

# Canonical double coverings

A characterization of canonical double coverings is presented here, together with further developements towards the idea of a two-fold orbital graph. The two sections of this chapter are based on [PS1] and [PS2], respectively. Proofs not provided here are to be found there.

## 5.1 Isomorphism classes

Two graphs $\Gamma$ and $\Gamma'$ will be said to be **B**-*isomorphic* if $\mathbf{B}(\Gamma) \cong \mathbf{B}(\Gamma')$.

Let $\boldsymbol{Aut}_0\Gamma$ be the stabilizer in $\boldsymbol{Aut}\,\mathbf{B}(\Gamma)$ of the stable parts. If $\Gamma$ is connected and non-bipartite, then

$$|\,\boldsymbol{Aut}\,\mathbf{B}(\Gamma) : \boldsymbol{Aut}_0\mathbf{B}(\Gamma)\,| = 2.$$

Each automorphism $f \in \boldsymbol{Aut}\,\Gamma$ always induces two automorphisms $\mathbf{f}^+$ and $\mathbf{f}^-$ of $\mathbf{B}(\Gamma)$: the *liftings* of $f$. The former fixes the components, and the latter interchanges them. More precisely

$$\mathbf{f}^+(u,\varepsilon) = (f(u),\varepsilon), \quad \mathbf{f}^-(u,\varepsilon) = (f(u), 1+\varepsilon).$$

for all $u \in V(\Gamma)$ and $\varepsilon \in \mathbb{Z}_2$.

**Definition 5.1.1** A graph $\Gamma$ is said to be *stable* if each automorphism of $\mathbf{B}(\Gamma)$ is a lifting.

Clearly, a graph $\Gamma$ is stable if and only if $\boldsymbol{Aut}\,\mathbf{B}(\Gamma) \cong \boldsymbol{Aut}\,\Gamma \times \mathbb{Z}_2$.

But there are also connected, non-bipartite graphs with $\mathcal{R} = \mathcal{R}' = \mathrm{id}$ (see Example 2.1.3), and still unstable. A circulant with these properties is exhibited below.

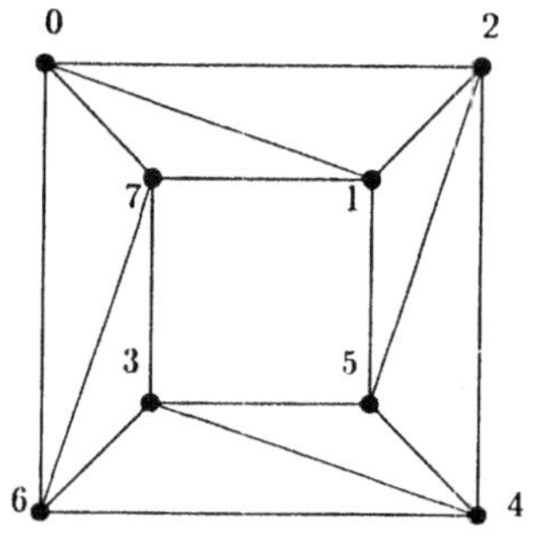


Figure 7: An unstable circulant

**Example 5.1.2** Let $\Gamma = \boldsymbol{Cay}(\mathbb{Z}_8, \{\pm1, \pm2\})$ be the graph drawn in Figure 7, and let $\boldsymbol{H}$ be the subgroup of index 2 of $\mathbb{Z}_8$. For a vertex $(x,\varepsilon)$ of $\mathbf{B}(\Gamma)$ we let $f(x,\varepsilon) = (x+4,\varepsilon)$ if either $x \in \boldsymbol{H}$ and $\varepsilon = 0$, or $x \notin \boldsymbol{H}$ and $\varepsilon = 1$, and $f(x,\varepsilon) = (x,\varepsilon)$ otherwise. It is easily checked that $f \in \boldsymbol{Aut}\,\mathbf{B}(\Gamma)$. Since clearly $f$ is not a lifting, $\Gamma$ is unstable. However, this graph is not edge-transitive, because no automorphism of $\Gamma$ can take $\{0,1\}$ into $\{0,2\}$.

**Definition 5.1.3** Let $\widetilde{\Gamma}$ be a bipartite digraph. An involution $a \in \boldsymbol{Aut}\,\widetilde{\Gamma}$ is said to be a *polarity* if $a$ interchanges the classes of $\widetilde{\Gamma}$, and a *strong polarity* if furthermore no arc of the form $u \to a(u)$ exists.

Of course, if $a$ is a (strong) polarity, the conjugacy class of $a$ in $\boldsymbol{Aut}\,\widetilde{\Gamma}$ consists only of (strong) polarities.

**Proposition 5.1.4** *Let $\widetilde{\Gamma}$ be a connected digraph. There is $\Gamma$ such that $\widetilde{\Gamma} = \mathbf{B}(\Gamma)$ if and only if $\widetilde{\Gamma}$ is bipartite and has a polarity. Loops of $\Gamma$ correspond to vertices $u$ such that $u \to a(u)$ in $\widetilde{\Gamma}$. In particular, $\Gamma$ is loopless if and only if $a$ is a strong polarity.*

**Proof** Let $\widetilde{\Gamma} = \mathbf{B}(\Gamma)$, and take $V(K_2) = \mathbb{Z}_2$. Now $(u, \varepsilon) \mapsto (u, \varepsilon + 1)$ is a polarity of $\widetilde{\Gamma}$.

Conversely, if $a \in \boldsymbol{Aut}\,\widetilde{\Gamma}$ is a polarity of the bipartite digraph $\widetilde{\Gamma}$, with classes $V_0$, $V_1$, define $\Gamma$ by $V(\Gamma) = V_0$ and $u \to_\Gamma v$ if and only if $u \to_{\widetilde{\Gamma}} a(u)$. We claim that the map $f : V(\widetilde{\Gamma}) \to V(\Gamma) \times \mathbb{Z}_2$, defined by $f : u \mapsto (u, 0)$ and $f : a(u) \mapsto (u, 1)$ is an isomorphism from $\widetilde{\Gamma}$ to $\Gamma \times K_2$. Namely, since $a$ is an involution, we have

$$u \to_{\widetilde{\Gamma}} v \Leftrightarrow u \to_\Gamma a(v) \Leftrightarrow (u, 0) \to (a(v), 1) \Leftrightarrow f(u) \to_{\Gamma \times K_2} f(v). \qquad \square$$

The digraph $\Gamma$ obtained from $a$ as in Proposition 5.1.4 will be denoted by $\Gamma_a$. When dealing with two digraphs $\Gamma$ and $\Gamma'$ with $\mathbf{B}(\Gamma) \cong \mathbf{B}(\Gamma') \cong \widetilde{\Gamma}$, we can identify the vertex set of both $\Gamma$ and $\Gamma'$ with a class $V_0$ of the bipartite graph $\widetilde{\Gamma}$, and write $\Gamma = \Gamma_a$ and $\Gamma' = \Gamma_b$ for suitable involutions $a, b \in \boldsymbol{Aut}\,\widetilde{\Gamma}$. Since the vertex set is the same, to avoid any risk of confusion, we will write $u \to_a v$ for $u \to v$ and $u \sim_a v$ for $u \sim v$ in $\Gamma_a$.

If $\boldsymbol{Aut}\,\widetilde{\Gamma}$ has several polarities, then $\widetilde{\Gamma} \cong \mathbf{B}(\Gamma)$ for several graphs $\Gamma$. These graphs may not be isomorphic. Namely, we have the following result.

**Theorem 5.1.5** *The number of non-isomorphic graphs $\Gamma$ such that $\widetilde{\Gamma} \cong \mathbf{B}(\Gamma)$ is equal to that of conjugacy classes of polarities in $\boldsymbol{Aut}\,\widetilde{\Gamma}$. The number of non-isomorphic loopless graphs $\Gamma$ such that $\widetilde{\Gamma} \cong \mathbf{B}(\Gamma)$ is equal to that of conjugacy classes of strong polarities in $\boldsymbol{Aut}\,\widetilde{\Gamma}$.*

**Proof** Let $a, b$ be polarities of $\widetilde{\Gamma}$. If they are conjugate, there exists $g \in \boldsymbol{Aut}\,\widetilde{\Gamma}$ such that $ga = bg$. Let $\Gamma_a$ and $\Gamma_b$ be such that $\widetilde{\Gamma} \cong \mathbf{B}(\Gamma_a) \cong \mathbf{B}(\Gamma_b)$. We can always assume that $g$ fixes the classes of $\widetilde{\Gamma}$. In fact, from $ga = bg$ it follows that $(bg)a = b(bg)$. Hence $a$ and $b$ are conjugate through $g' = bg \in \boldsymbol{Aut}\,\widetilde{\Gamma}$, and either $g$ or $g'$ fixes the classes of $\widetilde{\Gamma}$. Now, we have:

$$\begin{aligned} u \to_a w &\Longleftrightarrow u \to a(w) \\ &\Longleftrightarrow g(u) \to ga(w) \\ &\Longleftrightarrow g(u) \to bg(w) \\ &\Longleftrightarrow g(u) \to_b g(w). \end{aligned}$$

Therefore, the restriction of $g$ to $V_0 = V(\Gamma_a) = V(\Gamma_b)$ induces an isomorphism of $\Gamma_a$ onto $\Gamma_b$.

Conversely, assume $\Gamma_a \cong \Gamma_b$, and let $g_0 : \Gamma_a \longrightarrow \Gamma_b$ be an isomorphism. Then

$$u \to a(w) \Leftrightarrow u \to_a w \Leftrightarrow g_0(u) \to_b g_0(w) \Leftrightarrow g_0(u) \to bg_0(w) \qquad (22)$$

Define $g : V(\widetilde{\Gamma}) \longrightarrow V(\widetilde{\Gamma})$ by letting $g(u) = g_0(u)$ if $u \in V_0$, and $g(u) = bg_0a(u)$ otherwise. Let $u \to v$. If $u \in V_0$, then, by (22), $g_0(u) \to bg_0a(v)$, that is, $g(u) \to g(v)$. If $u \in V_1$, then $a(u) \to a(v)$. In view of (22), this implies $g_0a(u) \to bg_0(v)$, that is, $bg_0a(u) \to g_0(v)$, so $g(u) \to g(v)$. Therefore $g \in \boldsymbol{Aut}\,\widetilde{\Gamma}$. Let $u \in V(\widetilde{\Gamma})$. If $u \in V_0$ then $ga(u) = bg_0a^2(u) = bg_0(u) = bg(u)$. If $u \notin V_0$ then $ga(u) = g_0a(u) = bg(u)$. In fact, from the definition of $g$, we have $g(u) = bg_0a(u)$, so that $bg(u) = b^2g_0a(u) = g_0a(u)$. Hence $ga = bg$, and $a$ and $b$ are conjugate. □

A necessary condition that for a graph $\Gamma$ there exists a non-isomorphic $\Gamma'$ with $\mathbf{B}(\Gamma) \cong \mathbf{B}(\Gamma')$ is provided by the following statement.

**Theorem 5.1.6** *Let* $\Gamma \cong \Gamma_a$ *and* $\Gamma' \cong \Gamma_b$ *be two non-isomorphic digraphs with* $\mathbf{B}(\Gamma) \cong \mathbf{B}(\Gamma') \cong \widetilde{\Gamma}$. *Then there exists* $f \in \boldsymbol{Aut}\,\Gamma_a \cap \boldsymbol{Aut}\,\Gamma_b$ *such that* $f^2 = 1 \neq f$ *and* $f = c^m \mid V_0$ *for some* $m$, *where* $c = ab$. *Furthermore, letting* $j = \lfloor \frac{m}{2} \rfloor$, *if* $u \to_a f(u)$ *then there is a loop at* $c^j(u)$ *in the digraph* $\Gamma_a$ *or* $\Gamma_b$ *according as* $m$ *is even or odd.* □

**Corollary 5.1.7** *Under the hypothesis of Theorem* 5.1.6, *if both* $\Gamma$ *and* $\Gamma'$ *are loopless, then there is no vertex* $u$ *such that* $u \to_\Gamma f(u)$. □

Note that if $f$ is an automorphism of $\Gamma$ as in Corollary 5.1.7, then $v$ and $f(v)$ are always adjacent in $\Gamma^c$ Then $f$ can be called an *adjacency-automorphism*. Zgrablić [Zg] proved that if $\Gamma = \boldsymbol{Cay}(\boldsymbol{G}, E)$, where $E$ is a union of conjugacy classes of $\boldsymbol{G}$, then its adjacency-automorphisms generate a vertex-transitive group.

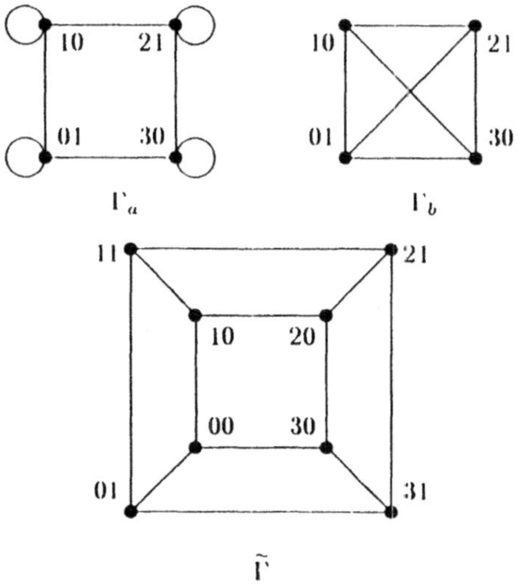


Figure 8: The cube as a canonical double covering

**Example 5.1.8** Let $\widetilde{\Gamma}$ be a cube with $V(\widetilde{\Gamma}) = \mathbb{Z}_4 \times \mathbb{Z}_2$ and with automorphisms $a\colon (x,\varepsilon) \mapsto (x, \varepsilon + 1)$, $b : (x,\varepsilon) \mapsto (x+2, \varepsilon+1)$ (see Figure 8). In this case we have $m = 1$ and $f = ab : (x,\varepsilon) \mapsto (x+2,\varepsilon)$. Note that $u \not\to_a f(u)$ for all $u \in V(\Gamma_a)$.

Allowing loops, Corollary 5.1.7 is no longer true. Namely, the following holds.

**Proposition 5.1.9** *Let $\Gamma$ be a digraph with loops such that there exists an involutory automorphism $f \in \mathbf{Aut}\,\Gamma$ with $f \neq 1$ and $u \not\to_\Gamma f(u)$ for all $u \in V(\Gamma)$. Then there exists a digraph $\Gamma'$ with $\Gamma \not\cong \Gamma'$ and $\mathbf{B}(\Gamma) = \mathbf{B}(\Gamma')$.*

**Proof** By definition, $\widetilde{\Gamma} = \mathbf{B}(\Gamma_a)$, where $a : (u, \varepsilon) \mapsto (u, \varepsilon + 1)$. Letting $b : (x, \varepsilon) \mapsto (f(x), \varepsilon + 1)$, we have a polarity of $\widetilde{\Gamma}$. If $\Gamma_b$ has a loop in $u$, then $(u, \varepsilon) \to b(u, \varepsilon)$. As $b(u, \varepsilon) = (f(u), \varepsilon + 1) = a(f(u), \varepsilon)$, we get $u \to_a f(u)$, a contradiction. □

## 5.2 A category of digraphs

As a background for ***TOG***, which will be defined in the next chapter, we introduce here a category $\mathcal{G}^*$. It will play for ***TOG*** the same rôle as the usual category $\mathcal{G}$ of digraphs plays for orbital digraphs. Henceforth, instead of $\mathbf{Aut}\,\Gamma$, we shall deal with $\mathbf{Aut}^*\,\Gamma$, the group of $\mathcal{G}^*$-automorphisms of $\Gamma$.

**Definition 5.2.1** Let $\mathcal{G}^*$ be the category whose objects are the digraphs $\Gamma$, while the arrows from $\Gamma_1$ to $\Gamma_2$ are the pairs $(g, h)$, where $g, h : V(\Gamma_1) \to V(\Gamma_2)$ are maps such that $u \to_{\Gamma_1} v \iff g(u) \to_{\Gamma_2} h(v)$.

If $(g, h) : \Gamma_1 \to \Gamma_2$ is invertible, then $(g, h)$ is said to be a *$\mathcal{G}^*$-isomorphism*. In this case, we say that $\Gamma_1$ and $\Gamma_2$ are *$\mathcal{G}^*$-isomorphic*, and write $\Gamma_1 \approx \Gamma_2$.

Of course, arrows include full digraph homomorphisms, up to an identification of $(f, f)$ with $f$. In particular, $\mathcal{G}^*$-isomorphisms include isomorphisms. For a digraph $\Gamma$, we will denote by $\mathbf{Aut}^*\,\Gamma$ the group of invertible $\mathcal{G}^*$-isomorphisms, or $\mathcal{G}^*$-automorphisms of $\Gamma$.

Two digraphs $\Gamma$ and $\Gamma'$ such that $\mathbf{B}(\Gamma) \cong \mathbf{B}(\Gamma')$ are always isospectral, but it is known that the converse is not true (see below).

**Example 5.2.2** The following two graphs are isospectral.

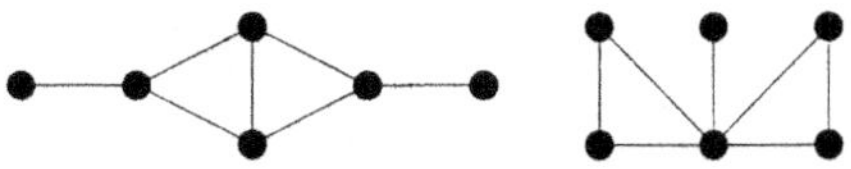

Two $\mathcal{G}^*$-isomorphic digraphs have the same degree sequence. On the other hand, there are many examples of digraphs with the same degree sequences that are not $\mathcal{G}^*$-isomorphic. In fact, if $\Gamma_1$ and $\Gamma_2$ are bipartite and $\Gamma_1 \not\cong \Gamma_2$, then $\mathbf{B}(\Gamma_1) \not\cong \mathbf{B}(\Gamma_2)$, because $\mathbf{B}(\Gamma_i) \cong 2\Gamma_i$ $(i = 1, 2)$. But there are also non-bipartite examples (see [PS2]). Furthermore, two $\mathcal{G}^*$-isomorphic digraphs are not necessarily isospectral.

**Proposition 5.2.3** *Two digraphs $\Gamma_1$ and $\Gamma_2$, such that $\mathbf{B}(\Gamma_1) \cong \mathbf{B}(\Gamma_2)$, are $\mathcal{G}^*$-isomorphic. The converse holds in the undirected case.*

**Proof** Let $\Gamma_1$ and $\Gamma_2$ such that $\mathbf{B}(\Gamma_1) \cong \widetilde{\Gamma} \cong \mathbf{B}(\Gamma_2)$. By definition there exists two maps $g_i, h_i : V(\Gamma_i) \to V(\widetilde{\Gamma})$ $(i = 1, 2)$ such that $uv \in A(\Gamma_i) \Leftrightarrow g_i(u)h_i(v) \in A(\widetilde{\Gamma})$. Letting $g = g_1 g_2^{-1}$ and $h = h_1 h_2^{-1}$, we have $g, h : V(\Gamma_1) \to V(\Gamma_2)$. Moreover:

$$u \to_{\Gamma_1} v \quad \Leftrightarrow \quad g_1(u) \to_{\widetilde{\Gamma}} h_1(v) \quad \Leftrightarrow \quad g_2^{-1} g_1(u) \to_{\Gamma_2} h_2^{-1} h_1(v),$$

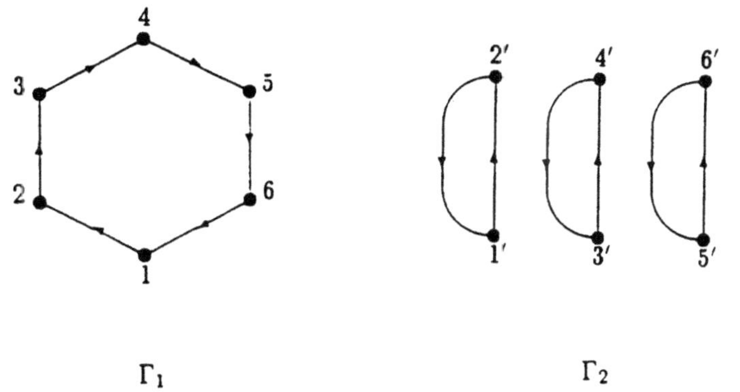


Figure 9: Two $\mathcal{G}^*$-isomorphic digraphs

that is, $u \to_{\Gamma_1} v \Leftrightarrow g(u) \to_{\Gamma_2} h(v)$. Hence $(g,h)$ is a invertible arrow in $\mathcal{G}^*$ from $\Gamma_1$ to $\Gamma_2$. Therefore, the digraphs $\Gamma_i$ $(i = 1,2)$ are $\mathcal{G}^*$-isomorphic.

Conversely, assume that $(g,h) : \Gamma_1 \to \Gamma_2$ is a $\mathcal{G}^*$-isomorphism, with $\Gamma_i$ $(i = 1,2)$ undirected, and define $f : V(\mathbf{B}(\Gamma_1)) \to V(\mathbf{B}(\Gamma_2))$ by:

$$f(w,\varepsilon) = \begin{cases} (g(w),0) & \text{if } \varepsilon = 0 \\ (h(w),1) & \text{if } \varepsilon = 1. \end{cases}$$

Then, for each arc $(u,0) \to (v,1)$ of $\mathbf{B}(\Gamma_1)$, we have $f(u,0) = (g(u),0)$ and $f(v,1) = (g(v),1)$. Therefore $(g(u),0)(h(v),1)$ is an arc of $\mathbf{B}(\Gamma_2)$. Clearly, $f : \mathbf{B}(\Gamma_1) \to \mathbf{B}(\Gamma_2)$ is invertible, hence $\mathbf{B}(\Gamma_1) \cong \mathbf{B}(\Gamma_2)$. □

Note that the map $f : \mathbf{B}(\Gamma_1) \to \mathbf{B}(\Gamma_2)$ preserves only the arcs going from $V(\Gamma_1) \times \{0\}$ to $V(\Gamma_2) \times \{1\}$. Hence it cannot help to extend Proposition 5.2.3 to the directed case. In fact, that result does not hold in this more general situation, as shown below.

**Example 5.2.4** For the digraphs drawn in Figure 9 we clearly have $\mathbf{B}(\Gamma_1) \not\cong \mathbf{B}(\Gamma_2)$. However, there is a $\mathcal{G}^*$-isomorphism $(g,h) : \Gamma_1 \to \Gamma_2$, defined by:

$$g = \begin{pmatrix} 1 & 2 & 3 & 4 & 5 & 6 \\ 1' & 2' & 3' & 4' & 5' & 6' \end{pmatrix}, \quad h = \begin{pmatrix} 1 & 2 & 3 & 4 & 5 & 6 \\ 5' & 2' & 1' & 4' & 3' & 6' \end{pmatrix}.$$

**Definition 5.2.5** Let $\Gamma$ be a digraph. The *transpose* of $\Gamma$, denoted by $\Gamma^T$, is the digraph whose arcs are the pairs $(v,u)$, where $(u,v) \in A(\Gamma)$. If $\Gamma^T = \Gamma$, the digraph is said to be *self-paired.*

Of course, $\Gamma^T$ is usually called the *inverse* of $\Gamma$. Our choice of the term 'transpose' is consistent with its use in the context of the two-fold orbitals (cf. Definition 2.2.1). The term 'self-paired' for a digraph (instead of the usual term 'undirected') is likewise chosen by analogy with orbitals. In what follows, 'self-paired' and 'undirected' will be used interchangeably.

Let $V$ be an $n$-set. We will write $\mathcal{S} = \boldsymbol{Sym}V \times \boldsymbol{Sym}V$. Let us consider the *natural* action of $\mathcal{S}$ on $V \times V$ defined as follows:

$$(g,h)(u,v) = (g(u), h(v)),$$

with $(g,h) \in \mathcal{S}$ and $(u,v) \in V \times V$. A special case is the *full diagonal subgroup* $\mathcal{D}$ of $\mathcal{S}$, defined by:

$$\mathcal{D} = \{(g,g) \mid g \in \boldsymbol{Sym}V\}.$$

Each subgroup of $\mathcal{D}$ is said to be *diagonal.* Denote by $\pi_1$ and $\pi_2$ the canonical projections of $\mathcal{S}$ into $\boldsymbol{Sym}V$, that is $\pi_1 : (g,h) \mapsto g$ and $\pi_2 : (g,h) \mapsto h$. For a subgroup $\boldsymbol{G} \leq \boldsymbol{Sym}V$ we will denote with $\mathcal{D}(\boldsymbol{G})$ the diagonal subgroup of $\mathcal{S}$ built on $\boldsymbol{G}$, i.e.:

$$\mathcal{D}(\boldsymbol{G}) = \{(g,g) | g \in \boldsymbol{G}\}.$$

**Definition 5.2.6** Let $\mathcal{G} \leq \mathcal{S}$. The *transpose* of $\mathcal{G}$, written as $\mathcal{G}^T$, is the set of ordered pairs $(h,g)$ such that $(g,h) \in \mathcal{G}$. If $\mathcal{G}^T = \mathcal{G}$, the group is said to be *self-paired.*

If $\mathcal{G}$ is self-paired, its projections $\pi_i(\mathcal{G})$ $(i = 1,2)$ are, of course, equal.

**Proposition 5.2.7** *If* $\Gamma$ *is self-paired, then* $\boldsymbol{Aut}^* \Gamma$ *is self-paired.*

**Proof** Let $(g,h) \in \boldsymbol{Aut}^* \Gamma$. For all $uv \in A(\Gamma)$, we have $vu \in A(\Gamma)$, therefore $g(v)h(u) \in A(\Gamma)$ and $h(u)g(v) \in A(\Gamma)$. Hence $(h,g) \in \boldsymbol{Aut}^* \Gamma$. □

Note that $\boldsymbol{Aut}^* \Gamma \cong \boldsymbol{Aut}\, \Gamma$ means that $\boldsymbol{Aut}^* \Gamma = \mathcal{D}(\boldsymbol{Aut}\, \Gamma)$, i.e. that there does not exist $(g,h) \in \boldsymbol{Aut}^* \Gamma$ with $g \neq h$. Hence, if $\boldsymbol{Aut}^* \Gamma \cong \boldsymbol{Aut}\, \Gamma$, then $\Gamma$ is stable. Conversely, if $\boldsymbol{Aut}^* \Gamma$ is self-paired, then $\Gamma$ is stable if $\boldsymbol{Aut}^* \Gamma \cong \boldsymbol{Aut}\, \Gamma$.

# Chapter 6

# Two-fold orbital digraphs

Our last chapter presents a first study of this new concept of a two-fold covering graph. It is a shortened version of [PS1], where the reader can find more information on this topic.

## 6.1 Two-fold orbital digraphs

Let us now introduce our generalization of orbital digraphs.

**Definition 6.1.1** Let $\mathcal{G} \leq \mathcal{S}$. For a fixed element $(a,b)$ in $V \times V$, let:

$$\mathcal{G}(a,b) = \{(g(a), h(b)) \mid (g,h) \in \mathcal{G}\}.$$

The set $\mathcal{G}(a,b)$ is called a *two-fold orbital,* and it is the set of the arcs of a digraph, which will be still denoted by $\mathcal{G}(a,b)$, having $V$ as vertex set. The graph $\Gamma = (V, \mathcal{G}(a,b))$ is said to be a *two-fold orbital digraph* (***TOG***) or a *$\mathcal{G}$-orbital graph.* The pair $(a,b)$ is said to be a *representative* of $\Gamma$.

Though a two-fold orbital digraph is usually directed, we will not adopt the acronym ***TOD***, which would sound quite strange to a German reader.

Orbitals and orbital graphs are obtained from the action of a transitive group $(G, V)$ on $V \times V$. It is apparent that these are special cases of two-fold orbitals and two-fold orbital digraphs, corresponding to the case where $\mathcal{G}$ is diagonal, that is, $\mathcal{G} = \mathcal{D}(\mathcal{G})$.

While orbital graphs are arc-transitive, usually this is not true for ***TOG***s, which may not be vertex-transitive, even when the projections $\pi_i(\mathcal{G})$ are transitive. Namely, $a \neq b$ does not forbid that $g(a) = h(b)$ for suitable $g$ and $h$. Hence we may have loops on some (but not all) vertices. Actually, many groups $\mathcal{G}$ admit at least one ***TOG*** having loops at some vertex, but not at each. In fact, if $(g, h) \in \mathcal{G}$ and $g \neq h$, let $a$ be such that $g(a) \neq h(a)$. Then $\mathcal{G}(a, a)$ has a loop at $a$ but possibly not at $g(a)$.

For a $\mathcal{G}$-orbital graph, the elements of $\mathcal{G}$ are not isomorphisms, but are $\mathcal{G}^*$-isomorphisms (see Proposition 6.1.3). Proposition 6.3.2 will give a condition for the transitivity of a ***TOG***.

Note that, even for a self-paired group $\mathcal{G}$, the graph $\mathcal{G}(a, b)$ may not be self-paired: this is known to happen even for diagonal $\mathcal{G}$ (orbital digraphs). On the other hand, $\mathcal{G}(a, b)$ may be self-paired even if $\mathcal{G}$ is not self-paired (see Example 6.1.2). Nevertheless, its $\mathcal{G}^*$-automorphism group is self-paired (see Proposition 5.2.7).

**Example 6.1.2** Let $\mathcal{G} = \boldsymbol{D}_4 \times \boldsymbol{S}_4 \leq \boldsymbol{S}_4 \times \boldsymbol{S}_4$. Then $\mathcal{G}$ is not self-paired, but $\mathcal{G}(1, 2)$ is a complete digraph.

The following result is a straightforward consequence of the definition of a ***TOG***.

**Proposition 6.1.3** *If* $\Gamma$ *is* $\mathcal{G}$*-orbital, then* $\mathcal{G} \leq \boldsymbol{Aut}^* \Gamma$. □

The following result allows us to compute the in- and out-valencies of a vertex in a ***TOG***. The regular ***TOG***s will be characterized by Proposition 6.1.5.

**Proposition 6.1.4** *Let* $\Gamma = \mathcal{G}(a, b)$. *For any* $u \in V(\Gamma)$ *we have:*

(i) $\mathrm{val}^+(u) = |\mathcal{G}_{\{u\} \times V(\Gamma)} : \mathcal{G}_{(u,v)}|$, *where* $(u, v) \in \mathcal{G}(a, b)$;

(ii) $\mathrm{val}^-(u) = |\mathcal{G}_{V(\Gamma) \times \{u\}} : \mathcal{G}_{(v,u)}|$, *where* $(v, u) \in \mathcal{G}(a, b)$.

**Proof** We will prove only (i). Assume, without loss of generality, that $(u, v) = (a, b)$, as $\mathcal{G}(u, v) = \mathcal{G}(a, b)$. We know that $\mathrm{val}^+(u)$ is equal to the number of elements of an orbit of $\mathcal{G}_{\{a\} \times V(\Gamma)}$ on $\{a\} \times V(\Gamma)$. Then (i) follows by a simple counting principle (see [BW], p. 3). □

The next result was proved in [PS2].

**Proposition 6.1.5** *A* $\mathcal{G}$*-orbital graph is regular if the projections* $\pi_i(\mathcal{G})$ $(i = 1, 2)$ *are transitive.* □

## 6.2 The disconnected case

Clearly, a disconnected orbital digraph is always the union of a set of disjoint isomorphic digraphs, and conversely. This fact is no longer true for a ***TOG***. As we shall see (Theorems 6.2.3 and 6.2.4), a disconnected ***TOG*** need not be vertex-transitive.

The undirected case is simpler: these graphs will be characterized by Theorems 6.2.2 and 6.2.3.

**Lemma 6.2.1** *The* ***CDC*** *of a* ***TOG*** *is still a* ***TOG***.

**Proof** Let $\Gamma$ be a $\mathcal{G}$-orbital graph, and let $\tilde{\Gamma} = \mathbf{B}(\Gamma)$. Define the following isomorphism:

$$\begin{array}{rcl} \mathcal{G} & \to & \tilde{\mathcal{G}} \\ (g,h) & \mapsto & (\tilde{g},\tilde{h}), \end{array}$$

where

$$\begin{array}{rcl} \tilde{g}(u,i) & = & (g(u),i) \\ \tilde{h}(u,i) & = & (h(u),i). \end{array}$$

Then $\tilde{\Gamma}$ is a $\tilde{\mathcal{G}}$-orbital graph. □

**Theorem 6.2.2** *Let $\Gamma$ be an undirected, disconnected* ***TOG***, *with connected components $\Gamma_1, \ldots, \Gamma_k$, and:*

$$|V(\Gamma_1)| \geq |V(\Gamma_2)| \geq \cdots \geq |V(\Gamma_k)|.$$

*Then each $\Gamma_i$ $(i = 1, \ldots, k)$ is still a* ***TOG***. *Moreover:*

(i) *if $|V(\Gamma_1)| = |V(\Gamma_k)|$, then $\Gamma_1, \Gamma_2, \ldots, \Gamma_k$ are pairwise $\mathcal{G}^*$-isomorphic;*

(ii) *otherwise, there exists a unique index $r \in \{1, \ldots, k-1\}$ such that*

$$\Gamma_1 \cong \Gamma_2 \cong \cdots \cong \Gamma_r \not\approx \Gamma_{r+1} \approx \cdots \approx \Gamma_k$$

*and $\Gamma_1 \cong \mathbf{B}(\Gamma_k)$.*

**Proof** By the definition of a ***TOG***, for all pairs $i, j$ such that $1 \leq i < j \leq k$, there exists a permutation $(g,h) \in \mathcal{G}$ taking an arc of $\Gamma_i$ to an arc of $\Gamma_j$. Let us set

$$A = \{u \in V(\Gamma_i) \mid g(u) \in V(\Gamma_j)\},$$

$$B = \{u \in V(\Gamma_i) \mid h(u) \in V(\Gamma_j)\}.$$

If $u \in A$, from $u \sim_{\Gamma_i} v$, it follows that $g(u) \sim_{\Gamma_j} h(v)$, hence $h(v) \in V(\Gamma_j)$, that is $v \in B$. Conversely, if $v \in B$ and $u \sim_{\Gamma_i} v$, then $g(u) \sim_{\Gamma_j} h(v)$, and $u \in A$. Proceeding in the same way on $B$, we get

$$u \in A \iff \Gamma_i(u) \subseteq B,$$

$$u \in B \iff \Gamma_i(u) \subseteq A.$$

Since the $\Gamma_i$ are connected, $V(\Gamma_i) = A \cup B$. Moreover, if $u \in A \cap B \neq \emptyset$, we have $A = B = V(\Gamma_i)$. As $\Gamma_i$ and $\Gamma_j$ are connected components of $\Gamma$, $g$ and $h$ take vertices of $\Gamma_i$ to $\Gamma_j$, without involving other components. Thus, if $A = B$, we have $|V(\Gamma_i)| \leq |V(\Gamma_j)|$, hence $|A| = |V(\Gamma_i)| = |V(\Gamma_j)|$. Furthermore $u \to_{\Gamma_i} v \iff g(u) \to_{\Gamma_j} h(v)$. Then $\Gamma_i \approx \Gamma_j$. If $|V(\Gamma_1)| = |V(\Gamma_k)|$, we have (i).

When $A \cap B = \emptyset$, $g$ fixes $B$ and takes $A$ to $V(\Gamma_j)$. Similarly, $h$ fixes $A$ and takes $B$ to $V(\Gamma_j)$. Hence $|A| \leq |V(\Gamma_j)|$ and $|B| \leq |V(\Gamma_j)|$. This means that $g^{-1}$ and $h^{-1}$ take $V(\Gamma_j)$ to $A$ and $B$, respectively. Then $|V(\Gamma_j)| \leq |A|$ and $|V(\Gamma_j)| \leq |B|$. Hence $|A| = |B| = |V(\Gamma_j)|$ and $|V(\Gamma_i)| = 2|V(\Gamma_j)|$. Moreover $g^{-1}(u) \to_{\Gamma_i} h^{-1}(v) \iff u \sim_{\Gamma_j} v \iff g^{-1}(v) \to_{\Gamma_i} h^{-1}(u)$, and so $\Gamma_i \cong \mathbf{B}(\Gamma_j)$.

Assume that there exist $r, s$ such that $1 \leq r < s \leq k-1$ and

$$|V(\Gamma_r)| > |V(\Gamma_{r+1})| = |V(\Gamma_s)| > |V(\Gamma_{s+1})|.$$

Now, we have $\Gamma_r \cong \mathbf{B}(\Gamma_{r+1})$, $\Gamma_{r+1} \approx \Gamma_s$ and $\Gamma_s \cong \mathbf{B}(\Gamma_{s+1})$. Then, in view of Proposition 5.2.3, $\Gamma_r \cong \mathbf{B}(\Gamma_s)$. Therefore $\Gamma_r \cong \mathbf{B}(\mathbf{B}(\Gamma_{s+1}))$, but this is a contradiction, because the latter graph is disconnected, and $\Gamma_r$ is not. Hence there exists a unique $r \in \{1, \ldots, k-1\}$ such that

$$|V(\Gamma_1)| = \cdots = |V(\Gamma_r)| > |V(\Gamma_{r+1})| = \cdots = |V(\Gamma_k)|,$$

which yields (ii).

Consider $\Gamma_i$ and $\Gamma_j$ such that $|V(\Gamma_i)| = |V(\Gamma_j)|$. The set $\{A(\Gamma_i) \mid 1 \leq i \leq k\}$ is an imprimitivity block system of $(\mathcal{G}, A(\Gamma))$. Then, letting $\mathcal{G}_i$ be the stabilizer of $\Gamma_i$ in $\mathcal{G}$, the graphs $\Gamma_i$ are $\mathcal{G}_i$-orbital. If $|V(\Gamma_i)| = 2|V(\Gamma_j)|$, then $\Gamma_i$ is a ***CDC*** of the ***TOG*** $\Gamma_j$. By Lemma 6.2.1, $\Gamma_i$ is a ***TOG***, too. □

In view of Theorem 6.2.2, the set $\{\Gamma_1, \ldots, \Gamma_k\}$ of the connected components of an undirected ***TOG*** splits into two classes (the first one could be empty) $\{\Gamma_1, \ldots, \Gamma_r\}$ and $\{\Gamma_{r+1}, \ldots, \Gamma_k\}$, each consisting of pairwise $\mathcal{G}^*$-isomorphic graphs. The elements of the former are isomorphic to the ***CDC*** (unique up to isomorphisms) of the elements of the latter.

For example, let $\Gamma$ be a disconnected undirected regular graph of valency 2, that is a union of separate cycles. In view of Theorem 6.2.2, if $\Gamma$ is a ***TOG***, then $\Gamma \cong kC_n$, or else $\Gamma \cong rC_{2n} \cup (k-r)C_n$ with $n$ odd.

**Theorem 6.2.3** *Let $\Gamma_1, \ldots, \Gamma_k$ be undirected graphs and $r \in \{1, \ldots, k-1\}$ such that*

(i) *for all $r+1 \leq h \leq k$, the graphs $\Gamma_j$ are $\mathcal{G}_h$-orbital;*

(ii) *$\Gamma_p \approx \Gamma_q$ for all $r+1 \leq p < q \leq k$;*

(iii) *for each $i \in \{1, \ldots, r\}$ there exists $j \in \{r+1, \ldots, k\}$ such that $\Gamma_i \cong \mathbf{B}(\Gamma_j)$.*

*Then $\overline{\Gamma} = \bigcup_{i=1}^{k} \Gamma_i$ is a **TOG**.* □

Theorems 6.2.2 and 6.2.3 together cover the undirected case. A counterexample for the directed one is provided by Example 6.2.5. On the other hand, for the class of regular digraphs of valency one (i.e. such that $\mathrm{val}^+(u) = \mathrm{val}^-(u) = 1$ for all $u \in V(\Gamma)$) we have Theorem 6.2.4 below.

**Theorem 6.2.4** *Each regular digraph $\Gamma$ of valency one is a* ***TOG***.

**Proof** The result is trivial if $\Gamma$ is connected, as $\Gamma$ is then arc-transitive. Otherwise, $\Gamma$ is a union of disjoint, strongly connected cycles. Assume first that $\Gamma$ has only two components $\Gamma_1$ and $\Gamma_2$. Denote the vertices in $\Gamma_1$ by $u_i$ $(i = 1, \ldots, n)$, and those in $\Gamma_2$ by $v_j$ $(j = 1, \ldots, m)$. Hence, the arcs in $\Gamma_1$ and $\Gamma_2$ will be $u_i u_{i+1}$ $(i \in \mathbb{Z}_{n+1})$ and $v_j v_{j+1}$ $(j \in \mathbb{Z}_{m+1})$, respectively.

Consider the following permutations on $V(\Gamma)$:

$$g = (u_0 v_0 v_1 \ldots v_m u_1 u_2 \ldots u_n),$$

$$h = (u_1 v_1 v_2 \ldots v_0 u_2 u_3 \ldots u_0).$$

Each element of $\mathcal{G} = \langle (g, h) \rangle$ takes $u_0 u_1$ to an arc in $\Gamma$, since the image of $u_i u_{i+1}$ (resp. $v_i v_{i+1}$) under the action of $(p, q) \in \mathcal{G}$ is either $u_{i+1} u_{i+2}$ or $v_j v_{j+1}$ (resp. $v_{i+1} v_{i+2}$ or $u_j u_{j+1}$). Therefore $(p, q)$ takes arcs of $\Gamma$ into arcs of $\Gamma$. On the other hand, each arc in $\Gamma$ can be obtained as image of $u_0 u_1$, under the action of a suitable power of $(g, h)$. In fact, if the arc is $v_i v_{i+1}$ or $u_i u_{i+1}$, we can use $(g, h)^{i+1}$ $(i \in \mathbb{Z}_{m+1})$ or $(g, h)^{m+i+1}$ $(i \in \mathbb{Z}^*_{n+1})$, respectively. Hence $\Gamma$ is a $\mathcal{G}$-orbital graph.

Examine now the general case, where the connected components of $\Gamma$ are $\Gamma_i$ for $i = 1, \ldots, k$:

$$\Gamma = \bigcup_{i=1}^{k} \Gamma_i.$$

For each $r$, $1 \leq r \leq k - 1$, define the arrow $(g_r, h_r)$, acting as above on vertices of the components $\Gamma_r$ and $\Gamma_{r+1}$, and fixing all remaining vertices of $\Gamma$. Then $\Gamma$ is $\mathcal{G}$-orbital with $\mathcal{G} = \langle (g_r, h_r) \mid 1 \leq r \leq k - 1 \rangle$. □

Let us now illustrate Theorem 6.2.4.

**Example 6.2.5** Consider the following disconnected graph $\Gamma$.

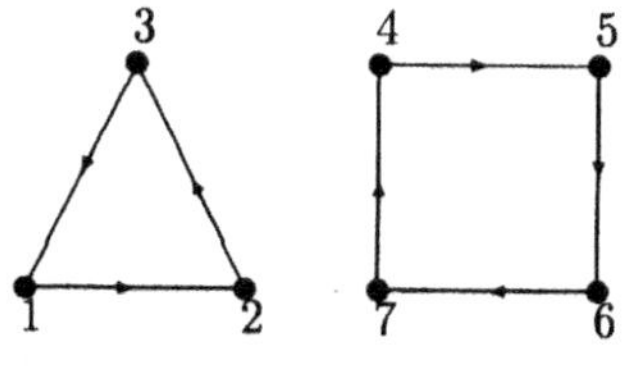


Set $g = (1456723)$ and $h = (1256743)$. Then $\Gamma$ is $\mathcal{G}$-orbital, where $\mathcal{G} = \langle (g, h) \rangle$.

Recall that a group $\boldsymbol{G}$, transitive on $X$, is imprimitive if and only if it has a disconnected orbital graph $\Gamma$ (see Theorem 2.2.7). The subgroup $\boldsymbol{K}$ of $\boldsymbol{G}$, fixing set-wise the connected components of $\Gamma$, is an intransitive normal subgroup, hence $\mathcal{D}(\boldsymbol{K}) \lhd \mathcal{D}(\boldsymbol{G})$. For a $\mathcal{G}$-orbital graph $\Gamma$, the group $\boldsymbol{K}$ as above may not satisfy this property.

**Corollary 6.2.6** *Let $\Gamma$ be an undirected disconnected $\mathcal{G}$-orbital graph corresponding to case* (ii) *of Theorem* 6.2.2. *Then $\mathcal{D}(\boldsymbol{K})$ is not normal in $\mathcal{G}$.*

**Proof** The graph $\Gamma$ has a connected component $\Gamma_0$ such that $\mathbf{B}(\Gamma_0)$ is also a connected component (up to isomorphisms). Define two self-maps $g, h$ of the set $V(\Gamma_0) \cup V(\mathbf{B}(\Gamma_0))$ as follows. For all $u \in V(\Gamma_0)$, $g$ interchanges $(u,0)$ and $u$ and fixes $(u,1)$, while $h$ interchanges $(u,1)$ and $u$ and fixes $(u,0)$. Then let $f \in \boldsymbol{Aut}\,\Gamma$ be the automorphism interchanging the stable parts of $\mathbf{B}(\Gamma_0)$ and fixing all remaining vertices of $\Gamma$. The map

$$(f,f)^{-1}(g,h)(f,f)$$

interchanges arcs $uv$ and $(u,1)(v,0)$, fixes arcs $(u,0)(v,1)$ between $\Gamma_0$ and $\mathbf{B}(\Gamma_0)$, and fixes the eventual other components of $\Gamma$. Therefore $(f,f)$ does not normalize $\mathcal{D}(\boldsymbol{K})$. □

## 6.3 Applications to canonical double coverings

As a consequence of the remarks below, several questions concerning ***CDC*** s may be looked at in the setting of ***TOG***.

**Lemma 6.3.1** *If $\mathcal{G}$ is self-paired but not diagonal, then $\mathcal{G}(a,b)$ is stable for no pair $(a,b)$.*

**Proof** As $\mathcal{G}$ is self-paired, each of its elements induces an automorphism of the ***CDC*** of $\mathcal{G}(a,b)$. On the other hand, since $\mathcal{G}$ is not diagonal, there exist automorphisms $(g,h)$ with $g \neq h$, hence which are not the liftings of any automorphism. In view of Definition 5.1.1, the graph $\mathcal{G}(x,y)$ cannot be stable. □

**Proposition 6.3.2** *Every stable **TOG** is arc-transitive.*

**Proof** Let $\Gamma$ be a $\mathcal{G}$-orbital graph. It is known that $\mathcal{G} \leq \boldsymbol{Aut}^*\,\Gamma$, so clearly $\Gamma$ is $\boldsymbol{Aut}^*\,\Gamma$-orbital. Moreover (Proposition 6.1.3), $\boldsymbol{Aut}^*\,\Gamma$ is self-paired. Hence, Lemma 6.3.1 forces $\boldsymbol{Aut}^*\,\Gamma$ to be diagonal. □

In view of Proposition 6.3.2, if $\Gamma$ is a stable graph which is not arc-transitive, then $\Gamma$ never arises as a $\mathcal{G}$-orbital graph.

**Lemma 6.3.3** *Let $\mathfrak{b}$ be the automorphism of $\mathbf{B}(\Gamma)$ swapping each $(u,0)$ with $(u,1)$. The involutions of $\mathbf{B}(\Gamma)$ interchanging the stable parts are exactly all the elements of the form $\mathfrak{b}(p,p^{-1})$. Moreover, such an element is conjugate to $\mathfrak{b}$ if and only if $p = f^{-1}e$, where $(e,f) \in \boldsymbol{Aut}\,\mathbf{B}(\Gamma)$.*

**Proof** Clearly, the elements, which we are dealing with, take the form $\mathfrak{b}(p,q)$. Moreover, $(\mathfrak{b}(p,q))^2(u,0) = (u,0)$ if and only if $qp = 1$. The conjugate of $\mathfrak{b}$ by $(e,f)$ is

$$(e^{-1},f^{-1})\mathfrak{b}(e,f) : \begin{cases} (u,0) \mapsto (e(u),0) \mapsto (e(u),1) \mapsto (f^{-1}e(u),1) \\ (u,1) \mapsto (f(u),1) \mapsto (f(u),0) \mapsto (e^{-1}f(u),0). \end{cases}$$

Therefore $(e^{-1},f^{-1})\mathfrak{b}(e,f) = \mathfrak{b}(f^{-1}e, e^{-1}f)$. □

**Proposition 6.3.4** *Let $\mathcal{G} \leq \mathcal{S}$ and $\mathcal{K} = \mathcal{G} \cap \mathcal{D}$. The map*

$$\begin{array}{rcl}\phi : \mathcal{G} & \to & \boldsymbol{Sym}V \\ (g,h) & \mapsto & gh^{-1}\end{array}$$

*satisfies*

$$\phi(g,h) = \phi(g',h') \iff (g',h') \in (g,h)\mathcal{K}.$$

**Proof** Let $\phi(g,h) = \phi(g',h')$. If $g' = gf$, then $f^{-1}h^{-1} = h'^{-1}$, that is, $h' = hf$. Hence $(g',h') = (g,h)(f,f)$. It follows that $(f,f) \in \mathcal{G}$, therefore $(f,f) \in \mathcal{K}$ and $(g',h') \in (g,h)\mathcal{K}$. The converse is similar. □

Clearly, $|\mathrm{Im}\phi| = |\mathcal{G} : \mathcal{K}|$. In general $\mathcal{K}$ is not normal in $\mathcal{G}$ because the conjugate of $(g,h)$ by $(f,f) \in \mathcal{K}$ is $(g^{-1}fg, h^{-1}fh)$. Hence $\mathcal{K} \lhd \mathcal{G}$ if and only if $g^{-1}h$ centralizes $f$, for each $(f,f) \in \mathcal{K}$.

It could happen that $\mathcal{K} \lhd \mathcal{G}$ and $\phi$ is not a homomorphism. For example, if $G \leq \boldsymbol{Sym}V$ has odd order and $\mathcal{G} = \{(g,g^{-1}) \mid g \in G\}$, we have $\mathcal{K} = 1$, but $\phi(g,g^{-1}) = g^2$ is not a homomorphism.

Assume that $\mathcal{G}$ is not diagonal with abelian projections $\pi_i(\mathcal{G})$ $(i = 1,2)$, and denote by $\mathcal{A}$ the set $\{(g,h) \in \mathcal{G} : h = g^{-1}\}$. If $\Gamma$ is a $\mathcal{G}$-orbital graph, then $|\mathcal{A}|$ is the number of polarities of $\mathbf{B}(\Gamma)$.

**Lemma 6.3.5** *Let $\mathcal{G} \leq \mathcal{S}$ and $\Gamma = \mathcal{G}(a,b)$. Then there exists $\Gamma' \ncong \Gamma$ such that $\mathbf{B}(\Gamma) \cong \mathbf{B}(\Gamma')$ if and only if*

$$|\mathcal{A}| > |\mathcal{G} : \ker\phi|. \tag{23}$$

**Proof** If $\pi_i(\mathcal{G})$ are abelian, then $\phi$ is a homomorphism. Then, there exist involutions of $\mathbf{B}(\Gamma)$ not conjugate to $\mathfrak{b}$ if and only if (23) holds. □

Infinitely many pairs $\Gamma_1$, $\Gamma_2$ of graphs having isomorphic $\boldsymbol{CDC}$ are provided by the following.

**Theorem 6.3.6** *Let $G$ be an abelian group, $\iota : g \longmapsto g^{-1}$ and let $\Gamma_1 = \mathcal{D}_\iota(G)(a,b)$. Then there exists a graph $\Gamma_2 \ncong \Gamma_1$, for some $(a,b)$, such that $\mathbf{B}(\Gamma_1) \cong \mathbf{B}(\Gamma_2)$.*

**Proof** Clearly $|\mathcal{A}| = |G|$. Denote by $m$ the number of involutions of $G$. Thus we have:

$$|\mathcal{D}_\iota(G) : \ker\phi| = \frac{|\mathcal{D}_\iota(G)|}{m+1}.$$

But $|\mathcal{D}_\iota(G)| = |G|$, then $|\mathcal{A}| = |G| > |\mathcal{D}_\iota(G) : \ker\phi|$. The result follows now from Lemma 6.3.5. □

# References

[Al1] Alspach, B., Hamilton cycles in metacirculant graphs with prime cardinal blocks, in: *Proc. Graph Theory in Memory of G.A. Dirac, Sandbjerg (1985)* (L.D. Andersen et al., eds.), *Ann. Discrete Math.* **41** (1989), 7-16.

[Al2] Alspach, B., Lifting Hamilton cycles of quotient graphs, *Discrete Math.* **78** (1989), 25-36.

[AMN] Alspach, B., Marušič, D. and Nowitz, L., Constructing graphs which are 1/2-transitive, *J. Austral. Math. Soc. Ser. A* **56** (1994), 391-402.

[AP] Alspach, B. and Parsons, T.D., A construction for vertex-transitive graphs, *Canad. J. Math.* **34** (1982), 307-318.

[BLS] Baumann, U., Lesch, M. and Schmeichel, I., A note on Hamiltonian decompositions of Cayley graphs, *Abh. Math. Sem. Univ. Hamburg* **65** (1995), 105-111.

[Bi] Biggs, N.L., *Algebraic Graph Theory*, 2nd edition, Cambridge University Press, Cambridge, 1993.

[BW] Biggs, N.L. and White, A.T., *Permutation Groups and Combinatorial Structures*, London Math. Soc. Lecture Note Ser. **33**, Cambridge University Press, Cambridge, 1979.

[BCN] Brouwer, A.E., Cohen, A.M. and Neumaier, A., *Distance-Regular Graphs*, Springer-Verlag, Berlin, 1989.

[Ca1] Cameron, P.J., Strongly regular graphs, in: *Selected Topics in Graph Theory* (L.W. Beineke and R.J. Wilson, eds.), vol. 1, Academic Press, New York, 1979.

[Ca2] Cameron, P.J., Automorphism groups of graphs, in: *Selected Topics in Graph Theory* (L.W. Beineke and R.J. Wilson, eds.), vol. 2, Academic Press, New York, 1983.

[CM] Capobianco, M. and Molluzzo, J.C., *Examples and Counterexamples in Graph Theory*, North-Holland, New York, 1978.

[CS] Ceccherini, P.V. and Sappa, A., $F$-binomial coefficients and related combinatorial topics: perfect matroid designs, partially ordered sets of full binomial type and $F$-geodetic graphs, *Combinatorics '84 (Bari, 1984)* (A. Barlotti et al., eds.), *Ann. Discrete Math.* **30** (1986), 143-158.

[CLP] Conder, M., Lorimer, P. and Praeger, C.E., Constructions for arc-transitive digraphs, *J. Austral. Math. Soc. Ser. A* **59** (1995), 61-80.

[C-W] Conway, J.H. et al., *Atlas of Finite Groups*, Clarendon Press, Oxford, 1985.

[Dö] Dörfler, W., Every regular graph is a quasigroup graph, *Discrete Math.* **10** (1974), 181-183.

[FI] Faradzev, I.A. and Ivanov, A.A., Distance-transitive representation of groups with $\boldsymbol{PSL}_2(q) \leq G \leq \boldsymbol{P\Gamma L}_2(q)$, *European J. Combin.* **11** (1990), 347-356.

[FKS] Fein, B., Kantor, W.M. and Schacher, M., Relative Brauer groups II, *J. Reine Angew. Math.* **328** (1981), 39-57.

[Fr] Frucht, R., How to describe a graph, *Ann. New York Acad. Sci.* **175** (1970), 159-167.

[FGW] Frucht, R., Graver, J.E. and Watkins, M.E., The groups of generalized Petersen graphs, *Proc. Cambridge Philos. Soc.* **70** (1971), 211-218.

[GW] Gallian, J.A. and Witte, D., A survey: Hamiltonian cycles in Cayley graphs, *Discrete Math.* **51** (1984), 293-304.

[Ga] Gauyacq, G., *Routages uniformes dans les graphes sommet-transitifs*, Thèse, Université Bordeaux I, 1995.

[Go] Godsil, C.D., More odd graph theory, *Discrete Math.* **32** (1980), 205-207.

[Gor] Gorenstein, D., *Finite Simple Groups*, Plenum Press, New York, 1982.

[Hi] Hilano, T., Distance degrees of vertex-transitive graphs, *Graphs Combin.* **5** (1989), 223-228.

[HS] Holton, M. and Sheehan, J., *The Petersen Graph*, Cambridge University Press, Cambridge, 1993.

[Hu] Hubaut, X.L., Strongly regular graphs, *Discrete Math.* **13** (1975), 357-381.

[Hup] Huppert, B., *Endliche Gruppen*, I, Springer-Verlag, Berlin, 1967.

[IKTS] Ivanov, A.A., Klin, M.Kh., Tsaranov, S.V. and Shpektorov, S.V., On the problem of computation of subdegrees of transitive permutation groups, *Russian Math. Surveys* **38** no. 6 (1983), 123-124.

[Ja] Jackson, B., Hamilton cycles in regular graphs, *J. Graph Theory* **2** (1978), 363-365.

[JS] Jajcay, R. and Širáň, J., A construction of vertex-transitive non-Cayley graphs, *Australas. J. Combin.* **10** (1994), 105-114.

[Jo] Jordan, D., Eine Symmetrieeigenschaft von Graphen, in: *Graphentheorie und ihre Anwendungen (Stadt Wehlen, 1988)*, Dresdner Reihe Forsch. **9**, Dresden, 1988, 17-20.

[LS] Liebeck, M.W. and Saxl, J., Primitive permutation groups containing an element of large prime order, *J. London Math. Soc. (2)* **31** (1985), 237-249.

[Lo] Lovász, L., Problem 11, in: *Combinatorial Structures and Their Applications*, (R. Guy et al., eds.), Gordon and Breach, New York, 1979, 497.

[Lov] Lovrečič Saražin, M., A note on the generalized Petersen graphs that are also Cayley graphs, *J. Combin. Theory Ser. B*, to appear.

[LP] Lovrečič Saražin, M. and Pacco, W., I grafi di Petersen generalizzati: un ampliamento del concetto, preprint (1996).

[Mal] Malnič, A., Group actions, coverings and lifts of automorphisms, preprint (1996).

[MM] Malnič, A. and Marušič, D., Imprimitive groups and graph coverings, in: *Coding Theory, Design Theory, Group Theory*, (O. Jungnickel and S.A. Vandstone, eds.), Wiley, New York, 1993, 227-235.

[Ma1] Marušič, D., On vertex symmetric digraphs, *Discrete Math.* **36** (1981), 69-81.

[Ma2] Marušič, D., Hamiltonian circuits in Cayley graphs, *Discrete Math.* **46** (1983), 49-54.

[Ma3] Marušič, D., Vertex-symmetric graphs and digraphs of order $p^k$, *Ann. Discrete Math.* **27** (1985), 115-128.

[Ma4] Marušič, D., On vertex-transitive graphs of order $qp$, *J. Combin. Math. Combin. Comput.* **4** (1988), 97-114.

[Ma5] Marušič, D., Hamiltonicity of vertex-transitive $pq$-graphs, in: *Fourth Czechoslovakian Symposium on Combinatorics, Graphs and Complexity (Prachatice 1990)*, (J. Nešetřil, and M. Fiedler, eds.), *Ann. Discrete Math.* **51** (1992), 209-212.

[MS1] Marušič, D. and Scapellato, R., Characterizing vertex-transitive $pq$-graphs with an imprimitive automorphism subgroup, *J. Graph Theory* **16** (1992), 375-387.

[MS2] Marušič, D. and Scapellato, R., A class of non-Cayley vertex-transitive graphs associated with $PSL(2,p)$, *Discrete Math.* **109** (1992), 161-170.

[MS3] Marušič, D. and Scapellato, R., Imprimitive representations of $SL(2,2^k)$, *J. Combin. Theory Ser. B* **58** (1993), 46-57.

[MS4] Marušič, D. and Scapellato, R., A class of graphs arising from the action of $PSL(2,q^2)$ on cosets of $PGL(2,q)$, *Discrete Math.* **134** (1994), 99-110.

[MS5] Marušič, D. and Scapellato, R., Classifying vertex-transitive graphs whose order is a product of two primes, *Combinatorica* **14** (1994), 187-201.

[MS6] Marušič, D. and Scapellato, R., Permutation groups with conjugacy complete stabilizer, *Discrete Math.* **134** (1994), 93-98.

[MS7] Marušič, D. and Scapellato, R., Classification of vertex-transitive $pq$-digraphs, *Istit. Lombardo Accad. Sci. Lett. Rend. A* **128** (1994), 31-36.

[MS8] Marušič, D. and Scapellato, R., Permutation groups, vertex-transitive digraphs and semiregular automorphisms, submitted.

[MSZ] Marušič, D., Scapellato, R. and Zgrablić, B., On quasiprimitive $pqr$-graphs, *Algebra Colloq.* **2** (1995), 295-314.

[MP] McKay, B. and Praeger, C.E., Vertex-transitive graphs that are not Cayley graphs. II, *J. Graph Theory* **22** (1996), 321-334.

[NS1] Nedela, R. and Škoviera, M., Which generalized Petersen graphs are Cayley graphs?, *J. Graph Theory* **19** (1995), 1-11.

[NS2] Nedela, R. and Škoviera, M., Regular embeddings of canonical double coverings of graphs, *J. Combin. Theory Ser. B* **67** (1996), 249-277.

[PS1] Pacco, W. and Scapellato, R., Digraphs having the same canonical double covering, *Discrete Math.*, to appear.

[PS2] Pacco, W. and Scapellato, R., Two-fold orbital graphs, preprint, Politecnico di Milano (1996).

[PSY] Praeger, C.E., Saxl, J. and Yokoyama, K., Distance transitive graphs and finite simple groups, *Proc. London Math. Soc. (3)* **55** (1987), 1-21.

[PX] Praeger, C.E. and Xu, M.-Y., Vertex-primitive graphs of order a product of two distinct primes, *J. Combin. Theory Ser. B* **59** (1993), 245-266.

[PWX] Praeger, C., Wang, H.J., and Xu, M.-Y., Symmetric graphs of order a product of two distinct primes, *J. Combin. Theory Ser. B* **58** (1993), 299-316.

[Sa] Sabidussi, G., Graph multiplication, *Math. Z.* **72** (1960), 446-457.

[Sc] Scapellato, R., On $F$-geodetic graphs, *Discrete Math.* **80** (1990), 313-325.

[Su] Suzuki, M., *Group Theory I*, Springer-Verlag, Berlin, 1982.

[Ta1] Tan, N.D., On Hamilton cycles in cubic $(m, n)$-metacirculant graphs, *Austral. J. Combin.* **8** (1993), 211-232.

[Ta2] Tan, N.D., Hamilton cycles in cubic $(m, n)$-metacirculant graphs with $m$ divisible by 4, *Graphs Combin.* **10** (1994), 67-73.

[Ta3] Tan, N.D., Cubic $(m, n)$-metacirculant graphs which are not Cayley graphs, *Discrete Math.* **154** (1996), 237-244.

[Tc] Tchuda, T., Combinatorial-geometric characterization of some primitive representations of the groups $PSL_n(q)$, $n = 2, 3$, Ph.D. Thesis, Kiev, 1986 (in Russian).

[Wa] Watkins, M.E., Vertex-transitive graphs that are not Cayley graphs, in: *Cycles and Rays* (G. Hahn, G. Sabidussi and R.E. Woodrow, eds.), NATO ASI Ser. C 301, Kluwer Academic Publishers, Dordrecht, 1990, 243-256.

[Wi] Wielandt, H., *Finite Permutation Groups*, Academic Press, New York, 1966.

[Ya] Yap, H.P. (ed.), *Some Topics in Graph Theory*, London Math. Soc. Lecture Note Ser. **108**, Cambridge University Press, Cambridge, 1986.

[Ze] Zelinka, B., On double covers of graphs, *Math. Slovaca* **32** (1982), 49-54.

[Zg] Zgrablić, B., On adjacency-transitive graphs, preprint (1996).

# Ends and automorphisms of infinite graphs

Mark E. WATKINS
*Mathematics Department*
*Syracuse University*
*Syracuse, NY 13244-1150*
*USA*

Notes by
Sanghamitra CHAKRABARTY

**Abstract**

These lectures introduce the finite graph theorist to a medley of topics and theorems in infinite graphs theory. Section 1: three graph theoretical notions required for a study of infinite graphs, namely end-equivalence (as developed by R. Halin), a refinement of the notion of connectivity, and growth. Section 2: an extension to infinite graphs (by C. Thomassen and the author) of W.T. Tutte's thoerem on arc-transitivity. Section 3: two-ended graphs, especially various characterizations of strips. Section 4: rays, double rays, quasi-axes, and the automorphism group action upon them. Section 5: joint work by P. Niemeyer and the author on fiber-equivalence, which is a refinement of end-equivalence. Section 6: the classification of locally finite, edge-transitive planar graphs by J.E. Graver and the author in terms of the number of ends, their Petrie walks, and the local behavior of their automorphism groups.

## Section 1
## Prerequisites

This article is aimed at the reader who is familiar with most of the basic notions of finite graphs but has been unexposed, whether by chance or the reader's own reticence, to infinite graphs. For you, the demystification of infinite graphs can begin HERE.

All graphs and their subgraphs herein will be denoted by capital Greek letters. Thus we say that $\Gamma$ is a graph with vertex set $V(\Gamma)$ and edge set $E(\Gamma)$. For $x \in V(\Gamma)$, the valence of $x$ will be denoted by $\rho(x)$. If $\rho(x) = n$ for some integer $n \geq 0$, then $x$ is $n$-valent. If $\rho(x) = n$ for all $x \in V(\Gamma)$, then $\Gamma$ is $n$-valent, and we may write $\rho(\Gamma) = n$. A one-way infinite path is a *ray* and a two-way infinite path is a *double ray.* Thus when we use the term *path*, we understand it to refer to a path of finite length. The set (or additive group or ring) of the integers is denoted by $\mathbb{Z}$; the set of positive integers is denoted by $\mathbb{Z}^+$.

If $\rho(x)$ is finite for all $x \in V(\Gamma)$, then $\Gamma$ is *locally finite.* Most but not all of this article is concerned with locally finite graphs. Locally finite graphs have a certain intuitive appeal. For if a myopic person were to stand somewhere in an infinite, locally finite graph and look all about, the view would be exactly the same as if this person were standing in the midst of a finite graph of large diameter. Thus many of the "local" notions for infinite, locally finite



*G. Hahn and G. Sabidussi (eds.), Graph Symmetry,* 379–414.

graphs coincide with analogous notions for finite graphs; the theories diverge only when one looks at infinite graphs "globally".

This point is particularly well illustrated with the notion of connectivity. For an infinite graph $\Gamma$, we define its *finite connectivity* by

$$\kappa_f(\Gamma) := \inf\{|S| : S \subset V(\Gamma); \Gamma - S \text{ has at least one finite component}\}.$$

If $\Gamma$ has at least one vertex of finite valence, then $\kappa_f(\Gamma) < \infty$. A *$\kappa_f$-separator* of $\Gamma$ is a set $S \subset V(\Gamma)$ such that $|S| = \kappa_f(\Gamma)$ and $\Gamma - S$ has at least one finite component. In infinite, locally finite graphs, the parameter of finite connectivity has many of the properties of connectivity in finite graphs. The "global" counterpart of $\kappa_f$ is *infinite conectivity*, defined as follows:

$$\kappa_\infty(\Gamma) := \inf\{|S| : S \subset V(\Gamma); \Gamma - S \text{ has at least two infinite components}\},$$

which makes sense except when $\Gamma$ consists of infinitely many finite components. If there is no finite subset of $V(\Gamma)$, including $\emptyset$, whose deletion leaves two infinite components, then we write $\kappa_\infty(\Gamma) = \infty$. A *$\kappa_\infty$-separator* of $\Gamma$ is a set $S \subset V(\Gamma)$ such that $|S| = \kappa_\infty(\Gamma)$ and $\Gamma - S$ has at least two infinite components. (These terms first appeared in [JW1].)

For example, let $\Delta$ denote a double ray, let $\Delta_n$ denote an $n$-circuit where $n \geq 3$, and let $\Box$ denote the binary operation of cartesian product of graphs. Then $\kappa_f(\Delta\Box\Delta_n) = \rho(\Delta\Box\Delta_n) = 4$, while $\kappa_\infty(\Delta\Box\Delta_n) = n$, showing that either of these connectivities may be the larger one. The $\kappa_f$-separators are the neighborhoods of the vertices, and the $\kappa_\infty$-separators are the copies of $\Delta_n$. The *connectivity* of $\Gamma$ is defined, naturally, by:

$$\kappa(\Gamma) := \min\{\kappa_f(\Gamma), \kappa_\infty(\Gamma)\}.$$

The overall "shape" of an infinite graph (we're speaking very globally now) is described more effectively with the aid of the concept of the "ends" of the graph. This notion, previously used in describing infinite groups, was first applied to graphs by Rudolf Halin [Hal]. Given a graph $\Gamma$, let $\mathcal{R}(\Gamma)$ denote the set of all rays in $\Gamma$, and let $\Pi_1, \Pi_2 \in \mathcal{R}(\Gamma)$. We say that $\Pi_1$ and $\Pi_2$ are *end-equivalent* and write $\Pi_1 \sim \Pi_2$ if there exists $\Pi_3 \in \mathcal{R}(\Gamma)$ such that both $V(\Pi_3 \cap \Pi_1)$ and $V(\Pi_3 \cap \Pi_2)$ are infinite. It is not hard to see that $\sim$ is an equivalence relation on the set $\mathcal{R}(\Gamma)$. The equivalence classes with respect to $\sim$ are the *ends* of $\Gamma$. The set of ends of $\Gamma$ will be denoted by $\mathcal{E}(\Gamma)$, and we define the cardinal $\epsilon(\Gamma) = |\mathcal{E}(\Gamma)|$.

If $\Theta$ is a subgraph of $\Gamma$ and $\Pi \in \mathcal{R}(\Gamma)$, we say that $\Pi$ *is contained in* $\Theta$ if $\Theta$ contains a subray of $\Pi$. If $\Pi_1, \Pi_2 \in \mathcal{R}(\Gamma)$, we say that $\Theta$ *separates* $\Pi_1$ and $\Pi_2$ if these two rays are contained in distinct components of $\Gamma - \Theta$ in the manner just defined. The following useful observation is immediate:

**Proposition 1.1 ([Hal, ¶1.1])** *Let* $\Pi_1, \Pi_2 \in \mathcal{R}(\Gamma)$. *The following are equivalent:*

(a) $\Pi_1 \sim \Pi_2$;

(b) *No finite subgraph of* $\Gamma$ *separates* $\Pi_1$ *and* $\Pi_2$.

(c) $\Gamma$ *contains infinitely many pairwise-disjoint (finite)* $\Pi_1\Pi_2$*-paths.*

If $\Gamma$ contains no ray at all, then clearly $\Gamma$ either is finite or is infinite but not locally finite, and $\epsilon(\Gamma) = 0$. Also $\epsilon(\Delta\Box\Delta) = 1$ while $\epsilon(\Delta\Box\Delta_n) = 2$. For any cardinal $\aleph$, the graph consisting of a vertex $v_0$ together with $\aleph$ rays emanating from $v_0$ but being otherwise vertex-disjoint has exactly $\aleph$ ends. We note in particular that any infinite, locally finite tree having all valences at least 3 has exactly $2^{\aleph_0}$ ends; the rays in any finite subset of an end intersect in their unique common subray.

When $\Gamma$ is locally finite, then an equivalent definition of $\epsilon(\Gamma)$ holds, namely

$$\epsilon(\Gamma) = \sup\{\#\text{infinite components of } \Gamma - S : S \subset V(\Gamma); |S| < \infty\}.$$

Given a graph $\Gamma$, a permutation $\alpha$ of $V(\Gamma)$ is an *endomorphism* (resp., *automorphism*) of $\Gamma$ if for any $u$, $v \in V(\Gamma)$, $[u,v] \in E(\Gamma)$ implies (resp., if and only if) $[\alpha(u), \alpha(v)] \in E(\Gamma)$. If $\Gamma$ is finite, then clearly every endomorphism is also an automorphism. The set of all automorphisms of $\Gamma$ forms a group, denoted by $\mathrm{Aut}(\Gamma)$. A group $G \leq \mathrm{Aut}(\Gamma)$ acts *vertex-transitively* (resp., *edge-transitively*) if for all $a$, $b$ in $V(\Gamma)$ (resp., in $E(\Gamma)$), there exists $\alpha \in G$ such that $\alpha(a) = b$. Clearly every endomorphism of $\Gamma$ acts on the set $\mathcal{E}(\Gamma)$. Let us now begin bringing together the notions of ends and morphisms of graphs.

**Theorem 1.2 ([Ha3, Theorem 4])** *Every endomorphism of a connected graph fixes either an end or a nonempty finite subgraph.*

An automorphism of $\Gamma$ is called a *translation* if it has an infinite orbit. When $\Gamma$ is locally finite, it is not difficult to show that for any $\alpha \in \mathrm{Aut}(\Gamma)$, either every orbit of $\alpha$ is finite or every orbit of $\alpha$ is infinite. (In [Ha3] Halin calls automorphisms of the former kind "type 1" and of the latter kind "type 2".) Halin went on to show:

**Theorem 1.3 ([Ha3, Corollary 15])** *If $\Gamma$ is connected and locally finite and admits a translation, then $\epsilon(\Gamma) = 1$, 2, or infinity.*

Note that the term "infinity" in Theorem 1.3 may stand for either $2^{\aleph_0}$, as in the case of the infinite $r$-valent tree ($r \geq 3$), or simply $\aleph_0$, as in the case of the tree consisting of a double ray $\Delta$ together with a countable set of pairwise-disjoint rays, one of which emanates from each vertex of $\Delta$. Here the stabilizer of $\Delta$ in $\mathrm{Aut}(\Gamma)$ is the infinite dihedral group $D_\infty$. Now it is time to add some connectivity notions to the soup.

For a subset $F \subset V(\Gamma)$, its *boundary* $\partial F$ is the set of vertices not in $F$ which are adjacent to at least one vertex in $F$. If $F$ and $\overline{F} := V(\Gamma) \backslash (F \cup \partial F)$ are infinite and $|\partial F| = \kappa_\infty(\Gamma) < \infty$, then $F$ is a *fragment* of $\Gamma$. The following important result is due to H. A. Jung.

**Theorem 1.4 ([J, Theorem 1])** *Let $\kappa_\infty(\Gamma) < \infty$ and let $G \leq \mathrm{Aut}(\Gamma)$ act vertex-transitively. Then for each fragment $F$ of $\Gamma$, there exists $\tau \in \mathrm{Aut}(\Gamma)$ such that $\tau(F \cup \partial F) \subsetneq F$.*

Jung's result holds also under a slightly weaker hypothesis, namely that $\mathrm{Aut}(\Gamma)$ act *almost transitively*, or more briefly, that $\Gamma$ be *almost transitive*; that is, that $\mathrm{Aut}(\Gamma)$ act with finitely many orbits. It is not hard to see that when $\Gamma$ is locally finite, then the automorphism $\tau$ of Theorem 1.4 must be a translation. Combining Theorems 1.3 and 1.4, we obtain the following very useful triage for the classes of infinite graphs in which we are frequently most interested:

**Corollary 1.5** *If $\Gamma$ is infinite, locally finite, and almost transitive, then $\epsilon(\Gamma) = 1$, 2 or $2^{\aleph_0}$.*

In order to introduce one remaining notion, that of "growth", in this introductory chapter, it is necessary to assume that $\Gamma$ is connected and locally finite. We fix $v_0 \in V(\Gamma)$, and for $n \geq 0$, we define the *n-ball* $B(n, v_0)$ about $v_0$ by

$$B(n, v_0) := \{v \in V(\Gamma) : d(v_0, v) \leq n\}$$

where $d(-,-)$ denotes the usual distance metric. It can be shown (as an exercise for the reader) that the choice of $v_0$ does not affect the following definitions; therefore it will be suppressed and we may write $B(n)$ in place of $B(n, v_0)$. We say that $\Gamma$ has *polynomial growth* of degree $d$ if there exist constants $C_1$ and $C_2$ such that

$$C_1 n^d \leq |B(n)| \leq C_2 n^d$$

holds for all $n \in \mathbb{Z}^+$. (The fact that, if $\Gamma$ is connected and almost transitive, then $d$ cannot be other than a positive integer can be deduced by catenating two theorems: [ST, Proposition 1] and a theorem of M. Gromov [Gr].) The growth of $\Gamma$ is *subexponential* if for any $a > 1$, one has $\liminf_{n\to\infty} |B(n)|/a^n = 0$. Otherwise $\Gamma$ has *exponential growth.*

Among almost transitive graphs, it is the 2-ended graphs – and only the 2-ended graphs – that have growth of degree 1, also called *linear growth.* The 1-ended graphs display all possible rates of growth $\geq 2$. For example, if $\Delta$ denotes a double ray, then for $n \geq 2$, the 1-ended graph $\Delta^n = \Delta\Box\Delta\Box\cdots\Box\Delta$ ($n$ times) has polynomial growth of degree $n$. Regular tessallations of the hyperbolic plane provide one-ended examples of graphs with exponential growth and will play a large role in Sections 5 and 6. Graphs with non-polynomial subexponential growth derive from Cayley graphs of groups with non-polynomial subexponential growth. As the author is unfamiliar with these groups, these graphs will not be discussed in this article. Finally, all infinitely-ended, almost transitive graphs have exponential growth.

## Section 2
## Vertex-, edge-, and arc-transitivity

This short section serves as a bridge to the theory of infinite graphs by showing by example how a now classical theorem from finite graph theory may – and may not – be generalized to infinite, locally finite graphs.

A group $G \leq \mathrm{Aut}(\Gamma)$ acts *arc-transitively* if given any $[u_1, v_1], [u_2, v_2] \in E(\Gamma)$, there exists some $\alpha \in G$ such that $\alpha(u_i) = v_i$, $i = 1, 2$. $\Gamma$ is *arc-transitive* if $\mathrm{Aut}(\Gamma)$ acts arc transitively. Thus a graph is arc-transitive if every edge can be "turned around", i.e., for any edge $[u, v]$ there is some automorphism that interchanges $u$ and $v$. Arc-transitive graphs are clearly both vertex- and edge-transitve, but it is the converse of this statement that concerns us for infinite graphs as it did W. T. Tutte for finite graphs.

**Theorem 2.1 ([Tu, ¶7.53])** *Every vertex-transitive, edge-transitive, finite graph of odd valence is arc-transitive.*

Tutte then inquired whether his theorem still held if the hypothesis of "odd valence" were dropped. A negative answer was provided by I. Z. Bouwer [Bo], who constructed a sequence of vertex-transitive, edge-transitive, non-arc-transitive graphs on $9 \cdot 6^n$ vertices and valence $2(n+1)$ ($n \in \mathbb{Z}^+$). Subsequently, D. F. Holt [Ho] constructed a 4-valent example on 27 vertices, half the size of the first graph in Bouwer's sequence. It is now known that there is none smaller. But we digress; we are concerned with graphs that are a great deal larger.

**Theorem 2.2 ([TW])** *Every connected vertex-transitive, edge-transitive graph of odd valence and subexponential growth is arc-transitive.*

In order to appreciate the leap from finite to infinite graphs, let us salvage as much as one can from Tutte's original proof. Let us assume only that $\Gamma$ is vertex-transitive and edge-transitive but not arc-transitive. We choose an arbitrary edge $[u, v]$. Since $\Gamma$ is edge-transitive but no edge can be "turned around", the orientation $(u, v)$ of this edge determines a unique orientation of each edge of $\Gamma$, yielding a directed graph $\vec{\Gamma}$ in which, because of vertex-transitivity, each vertex has a constant in-valence $\rho^-$ and a constant out-valence $\rho^+$, where $\rho := \rho(\Gamma) = \rho^- + \rho^+$.

So far, so good. But at this point Tutte invokes finiteness to conclude that clearly $\rho^-|V(\Gamma)| = |E(\Gamma)| = \rho^+|V(\Gamma)|$, whence $\rho^- = \rho^+$, and $\Gamma$ is even-valent.

To resume the proof of Theorem 2.2, let us assume further that $\Gamma$ is odd-valent and show that it must have exponential growth. Actually we show equivalently that $\vec{\Gamma}$ has exponential growth. For definiteness we suppose that $1 \leq \rho^- < \rho^+$, for if this is not so, we merely assign $[u, v]$ instead the orientation $(v, u)$. The remainder of the proof now follows from two technical lemmas. (For $S \subset V(\vec{\Gamma})$, we use the notation $\partial S$ as defined in Section 1 without regard to the orientation.)

**Lemma 2.3 ([TW, Lemma 1])** *For any finite subset $S \subset V(\vec{\Gamma})$,*

$$|\partial S| \geq \frac{2|S|}{\rho - 1}$$

*where $\rho = \rho^- + \rho^+$.*

**Proof** The number of arcs of $\vec{\Gamma}$ with initial vertex in $S$ is $\rho^+|S|$. Among these arcs, at most $\rho^-|S|$ also have their terminal vertex in $S$. Hence at least $(\rho^+ - \rho^-)|S|$ arcs have their initial vertex in $S$ and their terminal vertex in $\partial S$. Since each vertex of $\partial S$ is the terminal vertex of at most $\rho^-$ of these arcs, we have

$$|\partial S| \geq \frac{\rho^+ - \rho^-}{\rho^-}|S| \geq \frac{1}{\rho^-}|S| \geq \frac{2}{\rho - 1}|S|$$

as required. □

**Lemma 2.4 ([TW, Lemma 2])** *Let $v_0 \in V(\vec{\Gamma})$. For each nonnegative integer $k$, let $S_k := \{v \in V(\vec{\Gamma}) : d(v_0, v) = k\}$. Then*

$$|S_k| \geq 2\left(\frac{\rho + 1}{\rho - 1}\right)^{k-1}$$

*for all $k \in \mathbb{Z}^+$.*

**Proof** We note that $|S_0| = 1$ and $|S_1| = \rho$ and proceed by induction. Let $n \in \mathbb{Z}^+$ and assume that the above inequality holds whenever $1 \leq k \leq n$. Since $S_{n+1} = \partial(S_0 \cup S_1 \cup \cdots \cup S_n)$, we have by Lemma 2.4,

$$\begin{aligned}
|S_{n+1}| &\geq \frac{2}{\rho - 1}\sum_{k=0}^{n}|S_k| \\
&\geq \frac{2}{\rho - 1}\left[1 + \rho + 2\sum_{k=2}^{n}\left(\frac{\rho + 1}{\rho - 1}\right)^{k-1}\right] \\
&= \frac{2}{\rho - 1}\left[\rho - 1 + 2\frac{\left(\frac{\rho+1}{\rho-1}\right)^n - 1}{\frac{\rho+1}{\rho-1} - 1}\right] \\
&= 2\left(\frac{\rho + 1}{\rho - 1}\right)^n.
\end{aligned}$$

□

As $|B(n, v_0)| > |S_n|$ for $n \in \mathbb{Z}^+$, Theorem 2.2 follows.

Let the graph $\Gamma_{m,n}$ for $2 \le m < n$ be constructed as follows. Each *lobe* (i.e., maximal biconnected subgraph) of $\Gamma_{m,n}$ is isomorphic to the complete bipartite graph $K_{m,n}$. Each vertex of $\Gamma_{m,n}$ is incident with exactly two lobes in such a way that it is on the "$m$-side" in one lobe and on the "$n$-side" in the other lobe. Thus $\rho(\Gamma_{m,n})$ is odd when $m$ and $n$ have opposite parity. Since $\Gamma_{m,n}$ is infinitely-ended, it has exponential growth. It is easy to see that $\Gamma_{m,n}$ is both vertex- and edge-transitive. If an automorphism maps an edge $e$ onto itself, then it must map the unique lobe to which $e$ belongs onto itself. Hence no edge may be turned around, and so $\Gamma_{m,n}$ is not arc-transitive. We see that the hypothesis of subexponential growth may not be dropped from the hypothesis of Theorem 2.2.

Examples with only one end and odd valence may be obtained by taking the cartesian product of $\Gamma_{m,n}$ with itself an odd number $2k + 1$ times ($k \in \mathbb{Z}^+$), as $\rho((\Gamma_{m,n})^{2k+1}) = (2k+1)(m+n)$. One is certainly not limited to using $K_{m,n}$ as the starting point in the above examples. Any edge-transitive, non-arc-transitive graph would do.

No two-ended counterexamples are forthcoming; as noted in Section 1, all almost transitive, two-ended graphs have linear growth. It is of some interest, however, to see how Bouwer's graphs, which, in the case of finite graphs, prohibited extension of Tutte's theorem for finite graphs to graphs of even valence, may be adapted to construct two-ended, vertex-transitive, edge-transitive, non-arc-transitive graphs. For simplicity, we will adapt the first and smallest graph in Bouwer's sequence. It will be even-valent; in the next section we will see that this *must* be so. A portion of this graph $\Gamma_1$ is sketched in Figure 2.1.

Let $V(\Gamma_1) = \mathbb{Z}_9 \times \mathbb{Z}$, where $\mathbb{Z}_9$ denotes the ring of integers modulo 9. For $n \in \mathbb{Z}$, let $r(n)$ denote the residue of $n$ modulo 6. The set $E(\Gamma_1)$ consists of all 2-subsets of $\mathbb{Z}_9 \times \mathbb{Z}$ of the following two kinds: $[(a, n), (a, n + 1)]$ and $[(a, n), (a + 2^{r(n)}, n + 1)]$. The quotient graph of $\Gamma_1$ obtained by identifying all vertices $(a, m)$ and $(a, n)$ if and only if $m \equiv n \pmod 6$ is precisely Bouwer's graph on 54 vertices, and Bouwer's proof of vertex- and edge-transivity suffices as well for $\Gamma_1$.

To show that $\Gamma_1$ is not arc-transitive, suppose that some automorphism $\alpha$ turns around the edge $[(0, 1), (0, 2)]$. Since the only $\kappa_\infty$-separators of $\Gamma_1$ are the sets $T_n := \{(a, n) : a \in \mathbb{Z}_9\}$, $\alpha$ would have to transpose the sets $T_1$ and $T_2$, and hence must transpose the sets $T_0$ and $T_3$. The 6-circuit with vertices

$$(0, 1), (0, 2), (4, 3), (4, 2), (2, 1), (2, 2)$$

would be mapped into a 6-circuit with just one vertex in $T_0$, three vertices in $T_1$, and two vertices in $T_2$. But one easily verifies by inspection that no such 6-circuit exists.

# Section 3
# Strips

It was suggested in Section 1 that a study of many aspects of infinite, locally finite, almost transitive graphs may "decompose" into three such studies. The consequences may be very different depending upon whether the graphs in question have one, two, or infinitely many ends. Because the 2-ended graphs are usually the most tractable (a euphemistic way of saying that they are the least interesting, because they offer the smallest variety of weird things that can happen), we begin our study with these objects.

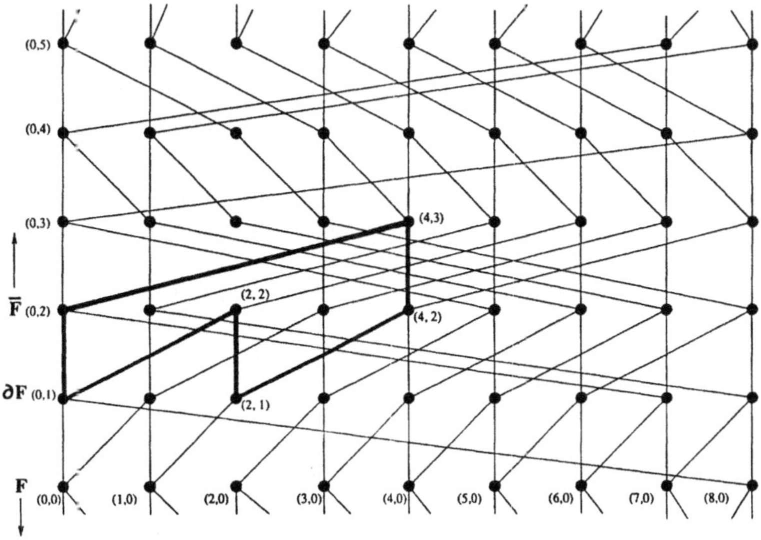


Figure 2.1: A two-ended vertex-transitive, edge-transitive, non-arc-transitive graph

In [JW2] a deliberately naive approach to almost transitive, 2-ended graphs was taken. The point of departure was the following notion.

**Definition** A *strip* $\Sigma$ is a connected graph that contains a connected subgraph induced by a set $F \subset V(\Sigma)$ and admits an automorphism $\tau$ such that $\tau[F \cup \partial F] \subseteq F$ and both $\partial F$ and $F \backslash \tau[F]$ are finite.

The idea intuitively is that a strip should consist of a countably infinite set of finite subgraphs isomorphic to $F \backslash \tau[F]$ that are somehow joined by edges end-to-end to form, well, a strip. Powers of the automorphism $\tau$ are supposed to translate these finite subgraphs up and down the strip. And this is exactly what does happen.

In order to discuss strips (and other graphs, too) we define an automorphism $\beta \in \mathrm{Aut}(\Gamma))$ to be *bounded* if there exists some $M > 0$ such that $d(v, \beta(v)) < M$ for all $v \in V(\Gamma)$. We let $B_0(\Gamma)$ denote the set of all bounded automorphisms of $\Gamma$.

**Lemma 3.1 ([JW2, Lemma 5.9])** *For any graph* $\Gamma$, $B_0(\Gamma)$ *is a normal subgroup of* $\mathrm{Aut}(\Gamma)$.

**Proof** Let $\beta_1, \beta_2 \in B_0(\Gamma)$ and pick $M_i > 0$ such that $d(v, \beta_i(v)) < M_i$ for all $v \in V(\Gamma)$ and $i = 1, 2$. For any $v \in V(\Gamma)$ we have $d(v, \beta_2\beta_1(v)) \leq d(v, \beta_1(v)) + d(\beta_1(v), \beta_2\beta_1(v)) < M_1 + M_2$. Hence $B_0(\Gamma)$ is a group. If $\alpha \in \mathrm{Aut}(\Gamma)$ and $v \in V(\Gamma)$, then $d(v, \alpha^{-1}\beta_1\alpha(v)) = d(\alpha(v), \beta_1\alpha(v)) < M_1$, as required. □

These automorphisms $\tau$ that satisfy the definition of a strip are indeed translations. It is shown in [JW2] that if $\Sigma$ is a strip, then all the translations of $\Sigma$ have finitely many orbits, and hence strips are almost transitive. One goes on to show that then all translations in $\mathrm{Aut}(\Sigma)$ belong to $B_0(\Sigma)$ and that the index $[\mathrm{Aut}(\Sigma) : B_0(\Sigma)] = 1$ or 2. (A more general underlying principle is that translations are bounded automorphisms in graphs with polynomial growth.) Finally strips are characterized as follows.

**Theorem 3.2 ([JW2, Theorem 5.13])** *A necessary and sufficient condition for an infinite connected graph to be a strip is that it be locally finite and admit an automorphism with finitely many orbits.*

It was stated in Section 1 that 2-ended, almost transitive graphs have linear growth. The condition of almost transitivity cannot be dropped. For consider a double ray $\Delta$ with $V(\Delta) = \{x_k : k \in \mathbb{Z}\}$, labeled in the natural way, and adjoin new edges to $\Delta$ so that $k$ new edges are attached at $x_{|k|}$ for each $k \in \mathbb{Z}$. Then for each $k \in \mathbb{Z}^+$, there are exactly $2k$ vertices at distance $k$ from $x_0$. One straightforwardly obtains $|B(n, x_0)| = n^2+n+1$, giving quadratic growth. If we replace $2k$ by $k^2$ in the above construction, one obtains a 2-ended graph with exponential growth.

There seems to be something "unfair" about the examples in the preceding paragraph. While they are locally finite, their valences are unbounded. We leave it as one of those exercises for the reader to determine the growth possibilities of 2-ended graphs with bounded valence. Meanwhile, we state two theorems about vertex-transitive strips.

**Theorem 3.3 ([IS])** *A necessary and sufficient condition for a locally finite, connected, vertex-transitive graph to be a strip is that it have linear growth.*

**Theorem 3.4 ([JW2, Theorem 5.14])** *Let $\Gamma$ be an infinite, connected graph, and suppose that $\mathrm{Aut}(\Gamma)$ contains a transitive abelian subgroup $G$. Then $\epsilon(\Gamma) = 1$ or 2. If $\epsilon(\Gamma) = 2$, then $\Gamma$ is a strip and $G \cong \mathbb{Z} \times T$, where $T$ is finite.*

At this point we turn to edge-transitive strips, which are characterized by the following theorem. Recall that every edge-transitive, non-vertex-transitive graph is bipartite. For brevity, for the strip $\Sigma$ let us temporarily write: $A := \mathrm{Aut}(\Sigma)$ and $B := B_0(\Sigma)$. For $x \in V(\Sigma)$, $A_x$ (resp., $B_x$) denotes the stabilizer of $x$ in $A$ (resp., $B$).

**Theorem 3.5 ([Wa2, Theorem 3.5])** *Let $\Sigma$ be an edge-transitive strip. Then all vertices of $\Sigma$ have even valence. More specifically, suppose that $F$ and $\overline{F}$ are connected fragments of $\Sigma$ and let $x \in \partial F$.*

*Case* 1. *If $\Sigma$ is arc-transitive or not vertex-transitive, then:*

(a) $[A : B] = 2$;

(b) *$\partial x$ is an $A_x$-orbit;*

(c) *$\partial x \cap F$ and $\partial x \cap \overline{F}$ are $B_x$-orbits of equal size.*

*Case* 2. *If $\Sigma$ is vertex-transitive but not arc-transitive, then either:*

(a) *$A = B$, in which case $\partial x \cap F$ and $\partial x \cap \overline{F}$ are $A_x$-orbits of equal size,*

*or*

(b) *$[A : B] = 2$ and*

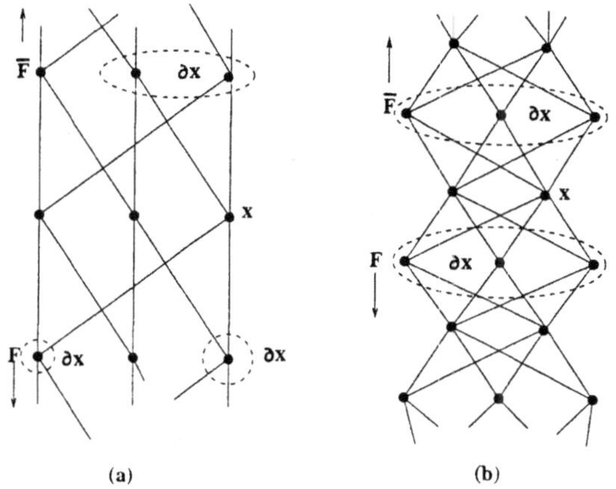


Figure 3.1: Examples of Theorem 3.5, Case 1: (a) an arc-transitive graph; (b) a non-vertex-transitive graph

(i) *Each of* $\partial x \cap F$ *and* $\partial x \cap \overline{F}$ *comprises exactly two* $B_x$*-orbits, all of the same size (thus* $4|\rho(\Sigma)$*), and*

(ii) $\partial x$ *comprises exactly two* $A_x$*-orbits, each being the union of one* $B_x$*-orbit in* $\partial x \cap F$ *and one* $B_x$*-orbit in* $\partial x \cap \overline{F}$.

A lemma needed for the proof of Theorem 3.5 showed that $\partial F$ is always an independent set. This anticipated the following more general result.

**Proposition 3.6 ([JW3, Corollary 3.3])** *Let* $\Gamma$ *be locally finite and edge-transitive with* $0 < \kappa_\infty < \infty$. *Then every* $\kappa_\infty$*-separator is an independent set.*

Examples of the two possibilities of Case 1 of Theorem 3.5 are illustrated in Figure 3.1; an arc-transitive example is shown in Figure 3.1(a), and a non-vertex-transitive example is shown in Figure 3.1(b). Subase 2(a) is illustrated by the graph shown in Figure 2.1, where $x = (4,1), F = \bigcup_{-\infty}^{0} T_k$ and $\partial F = T_1$.

A strip $\Sigma_1$ illustrating Subase 2(b) is more complicated and may be constructed in the following way. In [F], Jonathan Folkman presented a family of finite edge-transitive, non-vertex-transitive graphs of constant valence. The smallest member of this family is a 4-valent graph on 20 vertices (see Figure 3.2); let $\Phi$ denote this graph. Let $\{A, B\}$ denote the bipartition of $V(\Phi)$, where $A = \{a_1, \ldots, a_{10}\}$ and $B = \{b_1, \ldots, b_{10}\}$. Note that $A$ and $B$ are also the two vertex-orbits of $\mathrm{Aut}(\Phi)$.

Let $T$ be the set $\{x_1, \ldots, x_{10}, y_1, \ldots, y_{10}\}$ and let $V(\Sigma_1) = T \times \mathbb{Z}$. For each edge $[a_i, b_j] \in E(\Phi)$ and each $n \in \mathbb{Z}$, we have the following four edges in $E(\Sigma_1)$:

$$[(x_i, n), (x_j, n+1)], \quad [(y_i, n), (y_j, n+1)]$$
$$[(x_i, n), (y_j, n-1)], \quad [(y_i, n), (x_j, n-1)].$$

A portion of $\Sigma_1$, (where only the edges incident with $(x_1, n)$ or $(y_1, n)$ have been drawn in) is shown in Figure 3.3(a). In Figure 3.3(b) we see a quotient graph of $\Sigma_1$, where $X =$

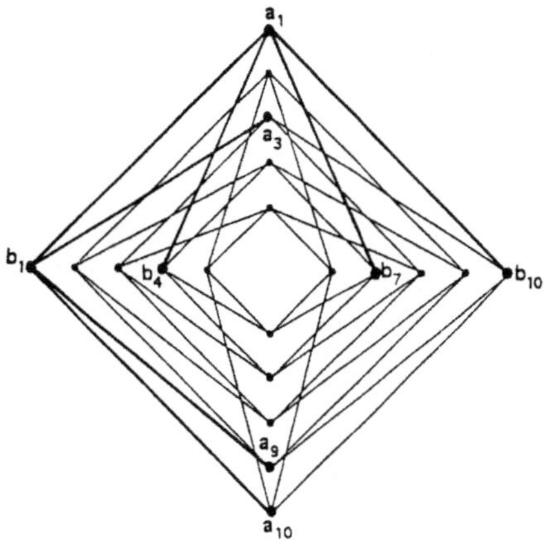


Figure 3.2: J. Folkman's 4-valent, edge-transitive, non-vertex-transitive graph on 20 vertices

$\{x_1, \ldots, x_{10}\}$ and $Y = \{y_1, \ldots, y_{10}\}$. All automorphisms must preserve the orientations of the arrowheads on the edges.

If $\Sigma$ is any planar strip, then $\Sigma$ admits an imbedding on an infinite cylinder in such a way that its two ends are separated by every circumference of the cylinder. Let $\tau$ be an automorphism of $\Sigma$ of the kind described in the definition of a strip, and let $n \in \mathbb{Z}^+$. Let $\Sigma/\tau^n$ be the quotient graph whose vertices are the $\tau^n$-orbits in $V(\Sigma)$. Thus $\Sigma/\tau^n$ is a finite graph. One may choose $n$ to be large enough so that if $[x]$ is the $\tau^n$-orbit containing $x$, then $\rho([x])$ in $\Sigma/\tau^n$ equals $\rho(x)$ in $\Sigma$, and $\Sigma/\tau^n$ inherits a natural imbedding in the torus. From this argument we deduce the following result.

**Lemma 3.7 ([Wa2, Lemma 4.1])** *Every planar strip contains vertices of valence at most 6.*

This lemma provides the first step in deducing from Theorem 3.5 that planar, edge-transitive strips are all 4-valent and, in fact, have the following very restrictive characterization. When we speak of the $n$-dimensional *lattice* "with the usual labeling", we mean the cartesian product $\Delta^n$, where $\Delta$ is the double ray with

$$V(\Delta) = \mathbb{Z} \quad \text{and} \quad E(\Delta) = \{[k, k+1] : k \in \mathbb{Z}\}.$$

**Theorem 3.8 ([Wa2, Theorem 4.2])** *For each $k \geq 2$ there exists a unique 2-ended, edge-transitive planar graph with $\kappa_\infty = k$. It is the quotient graph of $\Delta^2$ modulo the identifications $(x, y) \equiv (x + mk, y + mk)$ for all $(x, y) \in \mathbb{Z} \times \mathbb{Z}$ and all $m \in \mathbb{Z}$.*

Intuitively these graphs are obtained by "rolling up" $\Delta^2$ into an infinite cylinder, but the "rolling" must be done at a 45° angle to the coordinate axes. Any other angle results in identifications that yield a graph that is not edge-transitive.

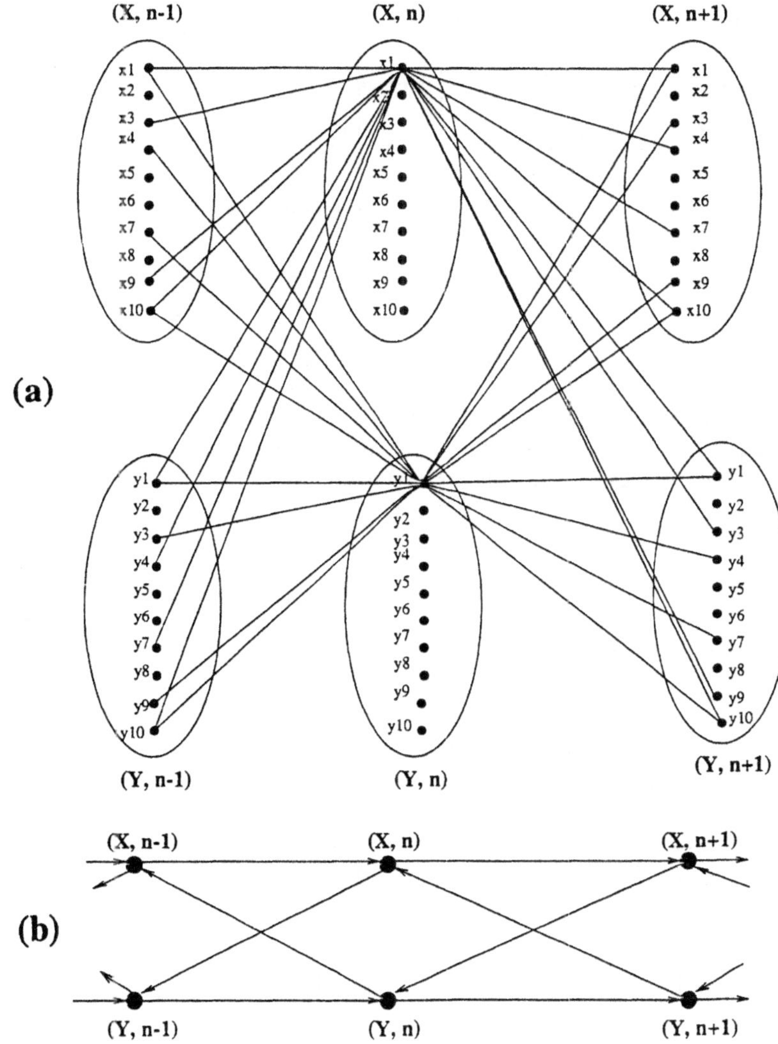


Figure 3.3: Example of Theorem 3.5, Case 2(b)

In Figure 3.4 we see a portion of the edge-transitive strip with $\kappa_\infty = 3$. Note how vertices on the upper perimeter are identified with vertices on the lower perimeter.

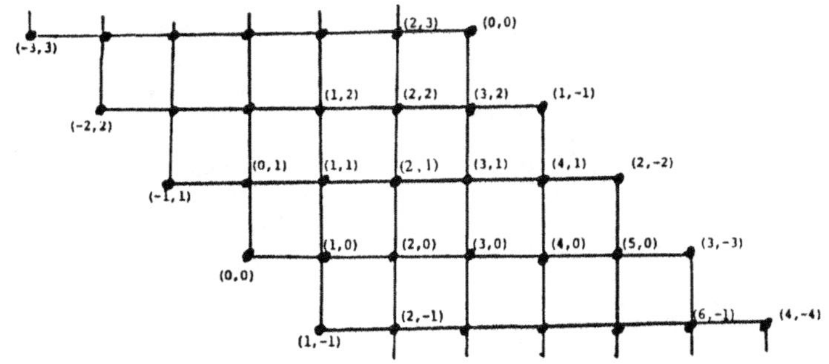


Figure 3.4: The planar edge-transitive strip with $\kappa_\infty = 3$

A path or ray or double ray $\Pi$ is *geodetic* if $d(x,y) = d_\Pi(x,y)$ holds for all $x, y \in V(\Pi)$ where $d_\Pi$ denotes the distance metric in $\Pi$ . A geodetic ray (resp., double ray) is called a *half-axis* (resp., an *axis*). Let $\Delta$ be a double ray and define its *straightness* to be

$$\sigma(\Delta) := \liminf_{d_\Delta(x,y)\to\infty} \frac{d(x,y)}{d_\Delta(x,y)}$$

where $x, y \in V(\Delta)$. Thus $0 \leq \sigma(\Delta) \leq 1$. $\Delta$ is a *quasi-axis* if $\sigma(\Delta) > 0$. By definition, the straightness of every axis is 1, but the converse is false.

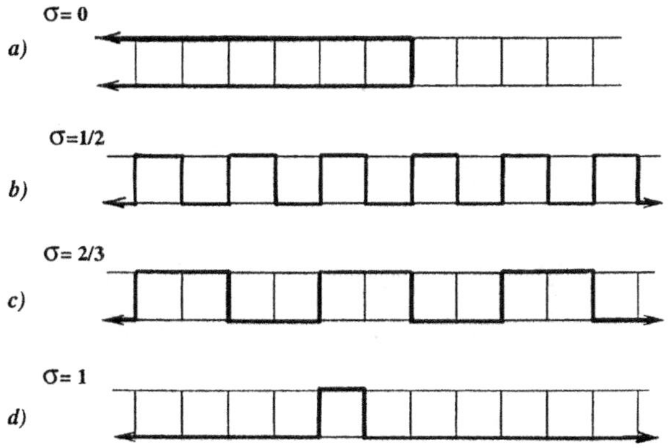


Figure 3.5: Four rays of increasing staightness in $\Delta \square K_2$

In Figure 3.5 are four examples of double rays in the strip $\Delta \square K_2$, where $\Delta$ is a double ray. Only the double ray in Figure 3.5(a) is not a quasi-axis. Assume that the patterns in Figure 3.5(b) and (c) repeat indefinitely in both directions. The quasi-axis in Figure 3.5(d) has straightness 1, but it is not an axis. Much more can and will be said about these geodetic objects in the next two sections when we consider graphs with one or infinitely many ends.

For graphs with exactly two ends, we offer only the following result. The author thanks David Witte for some suggestions that shorten the proof.

**Theorem 3.9** *Let $\Gamma$ be a locally finite graph, and let there be a quasi-axis $\Xi$ in $\Gamma$ such that for some $M > 0$, $d(v, \Xi) < M$ for all $v \in V(\Xi)$. Then $\Gamma$ has exactly two ends.*

**Proof** Clearly $\Gamma$ must be connected. Let $V(\Xi) = \{x_i : i \in \mathbb{Z}\}$ and $E(\Xi) = \{[x_i, x_{i+1}] : i \in \mathbb{Z}\}$. Since

$$\liminf_{d_\Xi(x_i,x_j)\to\infty} \frac{d(x_i, x_j)}{d_\Xi(x_i, x_j)} > 0,$$

we may choose $K > 0$ so that for all $i, j \in \mathbb{Z}$, if $|j - i| > 2K$, then $d(x_i, x_j) > 2M + 1$.

Let

$$U^- = \{v \in V(\Gamma) : d(v, x_p) < M \text{ for some } p < -K\}$$

and

$$U^+ = \{v \in V(\Gamma) : d(v, x_q) < M \text{ for some } q > K\}.$$

Thus $\langle U^-\rangle$ and $\langle U^+\rangle$ contain disjoint subrays of $\Xi$. Let $W = V(\Gamma)\backslash(U^+ \cup U^-)$. Note that $W$ is finite because $W \subset \bigcup_{i=-K}^{K} B(x_i, M)$.

We first show that if $v \in U^-$ and $w \in U^+$, then $v$ and $w$ are not adjacent. By definition, $d(v, x_p) < M$ and $d(w, x_q) < M$ for some $p < -M$ and some $q > M$. But if $[v, w]$ is an edge, then $d(x_p, x_q) \leq d(x_p, v) + d(v, w) + d(w, x_q) < M + 1 + M = 2M + 1$, contrary to the fact that $q - p > 2K$.

Similarly one can show that $U^- \cap U^+ = \emptyset$ by setting $v = w$ in the previous paragraph, obtaining a fortiori $d(x_p, x_q) < 2M$. From this it follows that $W \neq \emptyset$, or else $\Gamma$ would not be connected. Thus $U^-$ and $U^+$ induce distinct infinite components of $\Gamma - W$, from which it follows that $\Gamma$ has at least two ends, one containing $\Xi^+ := \Xi[x_K, \infty)$ and one containing $\Xi^- := \Xi(-\infty, x_{-K}]$.

Since $W$ is finite, given any ray $\Pi$ in $\Gamma$, some subray $\Pi'$ of $\Pi$ is contained in $\langle U^+\rangle$ or in $\langle U^-\rangle$. For definiteness assume the former. But then from every vertex $y \in V(\Pi')$ there exists a $y\Xi^+$-path of length less than $M$. Because of local finiteness, it is possible to select an infinite family of pairwise disjoint $\Pi\Xi^+$-paths, implying (cf. Proposition 1.1) that $\Pi$ and $\Xi^+$ belong to the same end. □

# Section 4
# Axes and quasi-axes

While Leonhard Euler is generally accorded credit for the first theorem in finite graph theory in 1736 (the characterization of graphs that admit an "Euler circuit"), it is the *Unendlichkeitslemma* of Dénes König, nearly two hundred years later, that may have been the first important theorem in infinite graph theory. We say "nearly" rather than "exactly ", because the same idea appeared in set theoretical clothing ten years earlier (cf. [K1, §3,¶E]).

**König's "Unendlichkeitslemma" ([K2,p.81])** *Let $S_1, S_2, \ldots$ be a (countably infinite) sequence of finite, nonempty, pairwise-disjoint subsets of vertices of a graph $\Gamma$. Suppose that* *for each* $n \in \mathbb{Z}^+$, *each vertex in* $S_{n+1}$ *is adjacent to some vertex in* $S_n$. *Then* $\Gamma$ *contains a ray with vertices* $x_1, x_2, \ldots$, *where* $x_n \in S_n$.

If $\Gamma$ is connected and locally finite, then one may choose $x_0 \in V(\Gamma)$ and let $S_n = \{y \in V(\Gamma) : d(x, y) = n\}$ for each $n \in \mathbb{Z}^+$. From the *Unendlichkeitslemma* we immediately obtain the following.

**Lemma 4.1** *Given any vertex* $x$ *of a connected, locally finite graph, there exists a half-axis that emanates from* $x$.

This lemma can be extended in several ways as follows.

**Theorem 4.2 ([Wa1, Theorems 3.1, 3.3, 4.1])** *Let* $\Gamma$ *be connected, infinite, and locally finite.*

(a) *Given any vertex* $x_0$ *and any end* $\mathcal{E}$ *of* $\Gamma$, *there exists a half-axis emanating from* $x_0$ *that belongs to* $\mathcal{E}$.

(b) *Given any two distinct ends* $\mathcal{E}_1$ *and* $\mathcal{E}_2$ *of* $\Gamma$, *there exists an axis which is the union of two half-axes, one belonging to* $\mathcal{E}_1$ *and the other to* $\mathcal{E}_2$.

(c) *If* $\Gamma$ *is vertex-transitive, then every vertex lies on an axis.*

To suggest some of the "flavor" of these extensions, we include a

**Proof of 4.2(a)** Let $x_0$ and $\mathcal{E}$ be given, and let $\Pi$ be a ray in $\mathcal{E}$ that emanates from $x$. Label $V(\Pi) = \{x_0, x_1, x_2 \ldots\}$ in the obvious way, and for each $n \in \mathbb{Z}^+$, let $S_n := \{v \in V(\Gamma) : d(v, x_0) = n\}$. Because $\Gamma$ is locally finite, each set $S_n$ is finite. For each $n \in \mathbb{Z}^+$, let $\Sigma_n$ be a geodesic joining $x_0$ and $x_n$. Thus $|S_k \cap V(\Sigma_k)| = 1$ for each $k = 1, \ldots, n$.

Some vertex $z_1 \in S_1$ lies on infinitely many of the geodesics $\Sigma_n$. Let

$$\mathbb{N}_1 := \{n \in \mathbb{Z}^+ : z_1 \in V(\Sigma_n)\}.$$

There exists a vertex $z_2 \in S_2$ that lies on infinitely many of the geodesics $\Sigma_n$ for which $n \in \mathbb{N}_1$. Clearly $[z_1, z_2] \in E(\Gamma)$. We let

$$\mathbb{N}_2 := \{n \in \mathbb{N}_1 : z_2 \in V(\Sigma_n)\}.$$

Continuing in this way, one constructs a half-axis $\Xi$ with $V(\Xi) = \{x_0, z_1, z_2, z_3, \ldots\}$, where for each $n \in \mathbb{Z}^+$, we have $z_n \in S_n$ and $n \in \mathbb{N}_n$.

It remains only to show that $\Xi \in \mathcal{E}$. Referring to Proposition 1.1(c), we do this by using segments of the paths $\Sigma_n$ to construct an infinite pairwise-disjoint family of $\Pi\Xi$-paths. Let $\nu(1) = \min \mathbb{N}_1$, and let $\Phi_1 = \Sigma_{\nu(1)}[z_1, x_{\nu(1)}]$. Note that $\Phi_1 \subset \bigcup_{k=1}^{\nu(1)} S_k$.

Inductively, for each $n \geq 2$ define $\nu(n) := \min \mathbb{N}_{\nu(n-1)+1}$ and let

$$\Phi_n = \Sigma_{\nu(n)}[z_{\nu(n-1)+1}, x_{\nu(n)}].$$

Note that if $k \in \mathbb{N}_m$ for some $k$ and $m$, then $k \geq m$. Thus $\Phi_n \subset \bigcup_{k=\nu(n-1)+1}^{\nu(n)} S_k$, and the family $\{\Phi_n : n \in \mathbb{Z}^+\}$ is the required family of paths. □

To prove 4.2(b), the point of departure is 4.2(a). From a common vertex there emanate quasi-axes $\Xi_1$ and $\Xi_2$ in $\mathcal{E}_1$ and $\mathcal{E}_2$, respectively. By truncating these quasi-axes if necessary, one obtains a double ray $\Delta$ that contains a subray of $\Xi_1$ and a subray of $\Xi_2$. But $\Delta$ might not be an axis, and so one replaces the "middle" of $\Delta$ by running successive short-cuts between two disjoint geodetic subrays. What is important is that because $\Gamma$ is locally finite and $\mathcal{E}_1 \neq \mathcal{E}_2$, this process must terminate after finitely many steps. This is shown using the following lemma that turns out also to be useful in another context in the next section.

**Lemma 4.3 ([Wal, Lemma 3.2])** *Let $\Gamma$ be connected and locally finite. Given any vertex $y \in V(\Gamma)$ and any half-axis $\Xi$ in $\Gamma$, there exists a half-axis emanating from $y$ that contains a subray of $\Xi$.*

From 4.2(b), we see that every connected, locally finite graph with two or more ends contains an axis, and so the contribution of 4.2(c) is really for one-ended graphs. To see that a non-vertex-transitive, one-ended graph $\Gamma$ may contain no axis, let $V(\Gamma) = \mathbb{Z}^+ \times \mathbb{Z}^+$, and join by edges all pairs of vertices of the forms

$$[(m,n),(m+1,n)], \quad [(m,n),(m,n+1)], \quad [(m+1,n),(m,n+1)]$$

for $m, n \in \mathbb{Z}^+$. Since $\mathrm{Aut}(\Gamma)$ in this example has infinitely many orbits, the reader is challenged to determine whether almost-transitivity is sufficient for a one-ended, connected, locally finite graph to contain an axis.

With local finiteness playing such a big role in the proof of 4.2(a), we might rightly suspect that this condition cannot be dropped. Indeed, consider the graph $\Gamma$ formed from a double ray $\Delta$ together with a vertex $y$ not on $\Delta$ that is adjacent to infinitely many vertices of $\Delta$ in each direction of $\Delta$. Then $\epsilon(\Gamma) = 2$, but $\Gamma$ contains not even a quasi-axis. Although local finiteness cannot be dropped, it can nonetheless be modified.

An end $\mathcal{E}$ of a graph $\Gamma$ is *dominated* by a vertex $x \in V(\Gamma)$ if no finite subgraph $\Phi$ of $\Gamma - x$ separates $x$ from $\mathcal{E}$; that is, $\mathcal{E}$ always includes a ray in $\Gamma - \Phi$ that emanates from $x$. If no such vertex $x$ exists, then $\mathcal{E}$ is *non-dominated.* Clearly $y$ dominates both ends of the graph in the previous example.

**Theorem 4.4 ([PW, Theorem 2.1])** *Given any vertex $x$ and any non-dominated end $\mathcal{E}$ of a connected graph $\Gamma$, there exists a half-axis in $\Gamma$ emanating from $x$ that belongs to $\mathcal{E}$.*

Suppose that $\tau$ is a translation of a connected graph $\Gamma$, and let $x \in V(\Gamma)$. Let $\Sigma$ be a geodetic $x\tau(x)$-path in $\Gamma$. The subgraph $\bigcup_{i=0}^{\infty} \tau^i(\Sigma)$ clearly contains a ray that emanates from $x$, and all such rays belong to the same end $\mathcal{E}$ of $\Gamma$. One can show (see [Ha3, Proposition 4]) that $\mathcal{E}$ is independent of the initially chosen vertex $x$ and thus is determined uniquely by the translation $\tau$. We call $\mathcal{E}$ the *direction* of $\tau$ and write $\mathcal{E} = \mathcal{D}(\tau)$.

Continuing the terminology of [Ha3], a ray or double ray $\Pi$ is *invariant* under $\tau$ (or *$\tau$-invariant*) if $\tau^k(\Pi) \subseteq \Pi$ for some $k > 0$. For any end $\mathcal{E}$, let $m_1(\mathcal{E})$ denote the supremum of the number of rays in all families of pairwise-disjoint rays in $\mathcal{E}$. More recent terminology is to say that $\mathcal{E}$ is *thin* if $m_1(\mathcal{E}) < \infty$; otherwise $\mathcal{E}$ is *thick.* A number of facts about these notions are lumped together in a marvelous catch-all theorem by Halin which should have been labeled *Hauptsatz* but wasn't.

**Theorem 4.5 ([Ha3, Theorem 9])** *Let $\tau$ be a translation of a connected, locally finite graph $\Gamma$. Then the following hold.*

(a) $m_1(\mathcal{D}(\tau)) = m_1(\mathcal{D}(\tau^{-1}))$.

(b) $\mathcal{D}(\tau) \neq \mathcal{D}(\tau^{-1})$ *if and only if $\mathcal{D}(\tau)$ is a thin end.*

(c) *There exists a system of* $m_1(\mathcal{D}(\tau))$ *pairwise-disjoint,* $\tau$*-invariant double rays (each of which contains a subray in* $\mathcal{D}(\tau)$ *and a subray in* $\mathcal{D}(\tau^{-1})$*).*

(d) *If* $\mathcal{D}(\tau)$ *is thick, then every finite family of pairwise-disjoint,* $\tau$*-invariant double rays can be extended to an infinite family of pairwise-disjoint,* $\tau$*-invariant double rays.*

(e) *If* $\mathcal{D}(\tau)$ *is thin, say* $m_1(\mathcal{D}(\tau)) = m < \infty$*, then* $\Gamma$ *contains an* $m$*-family* $\{\Delta_1, \ldots, \Delta_m\}$ *of* $\tau$*-invariant double rays (fixed setwise by* $\tau^n$ *for some* $n \in \mathbb{Z}^+$*) and an* $m$*-subset* $T$ *of* $V(\Gamma)$ *such that, for all* $k \in \mathbb{Z}$ *and each* $i \in \{1, \ldots, m\}$*,*

$$|\tau^{kn}(T) \cap V(\Delta_i)| = 1.$$

*In particular,* $\tau^{kn}(T)$ *separates the ends* $\mathcal{D}(\tau)$ *and* $\mathcal{D}(\tau^{-1})$*.*

Let us make a few remarks about this theorem. The first point is that *translatable* ends, that is, those that are the direction of some translation, have interesting properties (more of which will be presented in the next section). We digress for a moment to present an example of a non-translatable end. Consider an infinite graph whose lobes are copies of the 2-dimensional lattice $\Delta^2$ and the 3-dimensional lattice $\Delta^3$, and let each vertex be incident with exactly one lobe of each kind. Let $\alpha$ be the binary representation of some irrational number in $(0,1)$. Construct a ray emanating from a vertex $x_0$ with arbitrary initial edge $e_0 = [x_0, x_1]$. For each $n \in \mathbb{Z}^+$, adjoin edge $e_n = [x_n, x_{n+1}]$ in the same lobe as $e_{n-1}$ if and only if the $n$th binary digit of $\alpha$ is a 0. The end containing such a ray is not translatable. It is also thin. The graph in this example also has countably many translatable thick ends and thin ends.

Returning to the above theorem, note that part (d) is a real *tour de force.* An arbitrary finite family of pairwise-disjoint, $\tau$-invariant double rays in a translatable thick end can be extended not only to an arbitrarily large finite family, but indeed to an infinite family.

Part (e) of Theorem 4.5 (only part of which is paraphrased here) tells a great deal about the "geography" of translatable thin ends. Earlier in Halin's paper [Ha3, Theorem 8], we read that a translation $\tau$ fixes no end other than $\mathcal{D}(\tau)$ and $\mathcal{D}(\tau^{-1})$. Refining Theorem 1.3 above, we have:

**Theorem 4.6 ([Ha3, Theorem 11])** *If a connected, locally finite graph* $\Gamma$ *has two distinct non-inverse directions, then* $\epsilon(\Gamma) = 2^{\aleph_0}$*.*

With Halin's work as background, Norbert Polat and the author [PW] investigated whether some geodetic properties could be imposed upon the double rays in Halin's results.

**Theorem 4.7 ([PW, Theorem 2.4])** *Let* $\tau$ *be a translation of a connected graph* $\Gamma$ *such that* $m := m_1(\mathcal{D}(\tau)) < \infty$*. Then for some* $r$*,* $1 \le r \le m$*,* $\Gamma$ *contains an axis fixed by* $\tau^r$ *which is the union of two half-axes, one belonging to* $\mathcal{D}(\tau)$ *and the other to* $\mathcal{D}(\tau^{-1})$*.*

Thus if the direction of a translation is thin, there is an axis invariant under that translation; furthermore, there is a cap on the power of that translation needed to fix the axis. As in Theorem 4.4, note the absence of any assumption of local finiteness. The assumption that $\mathcal{D}(\tau)$ be thin is indeed necessary, however, as illustrated by the following example (cf. [PW, Example 4.2]).

We construct a graph $\Gamma$ that is locally finite, vertex-transitive, and one-ended. (Thus by [Th2, Proposition 5.6], that one end will be a thick end.) First let $T$ be a directed tree such that at each vertex there are exactly two incoming arcs and one outgoing arc. Next let $\Theta$ be the graph that consists of two "parallel" double rays $\Delta_0$ and $\Delta_1$ with $V(\Delta_0) = \{x_i : i \in \mathbb{Z}\}$ and $V(\Delta_1) = \{y_i : i \in \mathbb{Z}\}$ labeled in the natural way, together with all edges $[x_i, y_{2i}]$ and $[x_i, y_{2i+1}]$ for $i \in \mathbb{Z}$. (See Figure 4.1(a), and note, by the way, that $\Delta_1$ has straightness $\frac{1}{2}$ in $\Theta$.) We form $\Gamma$ in the following way from a countably infinite collection of copies of $\Theta$ that are in one-to-one correspondence with the arcs of $T$. Let $t \in V(T)$, let copies $\Theta_1$ and $\Theta_2$ correspond respectively to the two incoming arcs at $t$, and let copy $\Theta_3$ correspond to the ougoing arc at $t$. For each $i \in \mathbb{Z}$, we identify the three vertices $y_i \in V(\Theta_1)$, $y_{i-1} \in V(\Theta_2)$, and $x_i \in V(\Theta_3)$, and let us name this new vertex $w_{t,i}$. (See Figure 4.1(b).) This pattern of identification is repeated at each vertex of $T$ to construct the 6-valent graph $\Gamma$.

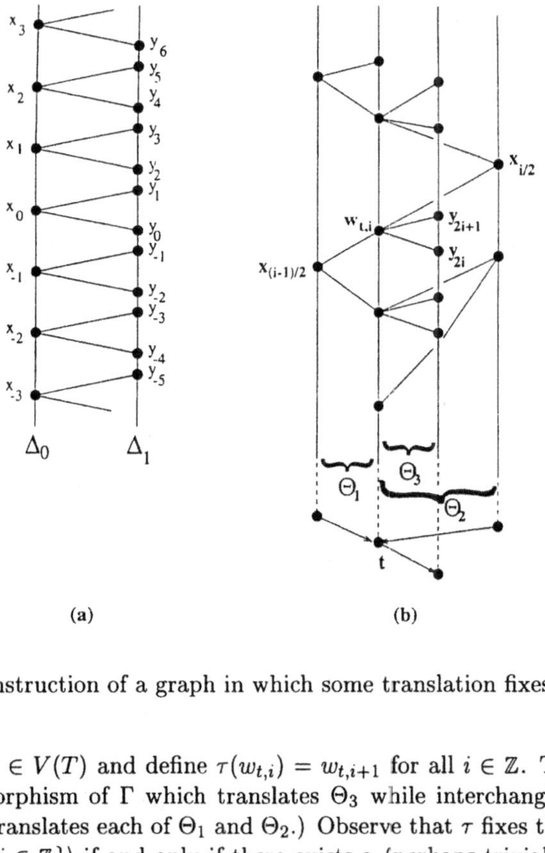


Figure 4.1: Construction of a graph in which some translation fixes no quasi-axis

Let us now fix $t \in V(T)$ and define $\tau(w_{t,i}) = w_{t,i+1}$ for all $i \in \mathbb{Z}$. Then $\tau$ is extendable to a unique automorphism of $\Gamma$ which translates $\Theta_3$ while interchanging $\Theta_1$ and $\Theta_2$. (In particular, $\tau^2$ also translates each of $\Theta_1$ and $\Theta_2$.) Observe that $\tau$ fixes the double rays of the form $\Delta_s = \langle\{w_{s,i} : i \in \mathbb{Z}\}\rangle$ if and only if there exists a (perhaps trivial) directed path in $T$ from $t$ to $s$. However,

$$\sigma(\Delta_s) = \lim_{|i-j|\to\infty} \frac{\log_2 |i-j|}{|i-j|} = 0,$$

and so none of these double rays $\Delta_s$ is even a quasi-axis, let alone an axis. It will follow from part (a) of the next theorem that no quasi-axis is invariant under the translation $\tau$.

**Theorem 4.8 ([PW, Theorems 3.1 and 3.2])** *Let $\tau \in \mathrm{Aut}(\Gamma)$ be a translation that fixes some quasi-axis $\Xi$ of a connected graph $\Gamma$.*

(a) *Every $\tau$-invariant double ray in $\Gamma$ is a quasi-axis.*

(b) *If $\Xi$ is not an axis, then there exists a $\tau$-invariant quasi-axis $\Xi'$ such that $\sigma(\Xi') > \sigma(\Xi)$.*

Let $\Xi$ be a double ray with $\sigma(\Xi) = 1$. In the previous section we saw that $\Xi$ need not be an axis. Suppose, however, that $\Xi$ is invariant under some translation. By Theorem 4.8(b), if $\Xi$ were not an axis, we could construct a quasi-axis $\Xi'$ such that $\sigma(\Xi) < \sigma(\Xi') \leq 1$, a contradiction. We have proved:

**Corollary 4.9 ([PW, Corollary 3.3])** *If $\sigma(\Xi) = 1$ and $\Xi$ is invariant under some translation, then $\Xi$ is an axis.*

If $\tau$ is a translation under which some quasi-axis is invariant, then Theorem 4.8(b) suggests the possibility of an infinite sequence of $\tau$-invariant quasi-axes with strictly increasing straightness. We are led to pose the following question:

**Question** If a translation $\tau$ fixes some quasi-axis, is it always the case that

$$\sup\{\sigma(\Xi) : \Xi \text{ is a quasi-axis invariant under } \tau\} = 1?$$

Jung & Niemeyer (see [JN, Theorem 6.4]) gave an affirmative answer to this question in the case of locally finite graphs, but the general question still remains open.

The complement of a double ray in a graph could conceivably be connected or have any number of components, either finite or infinite. If, however, a straightness constraint is imposed on the double ray or if some symmetry condition is imposed upon the graph, then the mischief of this double ray may be severely limited.

**Theorem 4.10 ([Wa1, Theorem 4.2])** *If $\Xi$ is an axis in a locally finite, vertex-transitive graph $\Gamma$, then $\Gamma - \Xi$ has no finite component.*

This earliest theorem on the subject leaves entirely open the question of the number of infinite components in the complement of even the most well-behaved double rays in the most symmetric of graphs. We therefore retreat to the more limited consideration of planar graphs. Intuitively, if one thinks of a fairly straight double ray in a locally finite planar graph, one thinks of drawing a line across the sheet of paper. The line will divide the graph into two infinite pieces, one on each side of the line. This is precisely the notion that we now wish to refine.

For important reasons, the planar graphs we consider are 3-connected. When a planar graph is 3-connected, the cyclic order of the edges incident with each vertex when that graph is embedded in the plane becomes an intrinsic property of the graph and is the same for all planar embeddings. This classic result of Whitney [Wh] for finite graphs has been generalized to infinite graphs in [I] and [Th1]. Also independent of the embedding of a graph $\Gamma$ is the collection of subsets of $E(\Gamma)$ that form the boundaries of the *faces*, i.e., the connected components of the complement of the embedded graph in the plane. Thus the *covalence* function $\rho^*$ is a well-defined function on the set of faces of a planar, 3-connected graph, counting the number of edges (also vertices) incident with each face. The set of faces of $\Gamma$ will be denoted by $F(\Gamma)$, and we will use the same symbols for both $\Gamma$ and its planar embedding. The number $\rho^*(f)$ may be infinite for some $f \in F(\Gamma)$, in which case $f$ will be called an *infinite face* and its boundary is a double ray.

**Theorem 4.11 ([BIW, Theorem 4.1])** *Let $\Gamma$ be locally finite, planar, 3-connected, and 1-ended with bounded covalences. Let $\Xi$ be a quasi-axis in $\Gamma$. Then $\Gamma - \Xi$ has exactly two infinite components, one on each side of $\Xi$.*

An immediate corollary is that the condition that $\Gamma$ have bounded covalences may be replaced by the condition that $\Gamma$ be almost transitive. Also the condition that $\Gamma$ be 3-connected may be replaced by the condition that $\Gamma$ be vertex-transitive. However, the condition that $\Gamma$ be one-ended is essential. (After all, wasn't $\Gamma$ one-ended in our intuitive mental picture of that piece of paper?)

In Figure 4.2 an infinitely-ended example is represented. It is even vertex-transitive. The quasi-axis $\Xi_0$ separates it into an arbitrary number of infinite components, while the axis $\Xi_1$ does not separate the graph at all. Figure 4.2(a) shows the 2-dimensional lattice in which certain faces are distinguished by having the symbol $\times$ placed in them. The same graph is drawn in Figure 4.2(b) but with one of the distinguished faces as its exterior face. Now place a copy of the graph in Figure 4.2(b) in each distinguished face of the graph in Figure 4.2(a). This creates infinitely more new distinguished faces. We repeat this process, placing a copy of the graph in Figure 4.2(b) in each new distinguished face, creating more distinguished faces. Repeat the process *ad infinitum*. The resulting graph is vertex-transitive, 6-valent, and infinitely ended. (Each end of $\Xi_1$ is understood to penetrate ever deeper into successive new "implants" of Figure 4.2(b).)

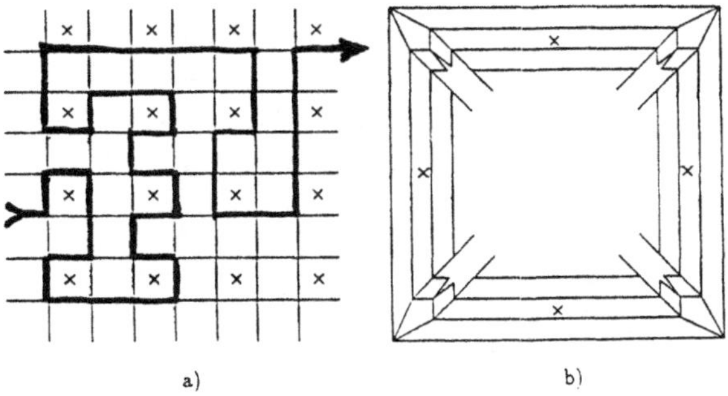


Figure 4.2 Infinitely-ended planar graph with quasi-axes whose complements do not have exactly two infinite components

One is led to inquire whether for one-ended, 3-connected, almost transitive graphs, this two-infinite-component-property is actually equivalent to planarity. The following example shows that we are not quite there. Consider the triangular (6-valent) regular tessellation of the plane. Inside each triangle, place three new vertices and adjoin 12 new edges to create a copy of $K_6$. This graph is obviously nonplanar, but it is still one-ended and 3-connected, and it has exactly two vertex-orbits. Moreover, the complement of every quasi-axis has exactly two infinite components.

A graph $\Gamma$ is *almost* 4*-connected* if $\kappa_f(\Gamma) \geq 3$ and for every $\kappa_f$-separator $T$, $\Gamma - T$ has exactly two components, one component consisting of a single vertex. This notion is due to Carsten Thomassen [Th2]. (The graph described in the preceding paragraph is not almost 4-connected. Although $\kappa_f = 3$, the deletion of any $\kappa_f$-separator leaves one infinite component and one copy of $K_3$.) Note that, by definition, an almost 4-connected graph need not be 3-connected.

**Theorem 4.12 ([BIW, Theorem 5.3])** *Let $\Gamma$ be locally finite,* 1*-ended, almost* 4*-connected, and almost transitive. Then $\Gamma$ is planar if and only if for every quasi-axis $\Xi$, $\Gamma - \Xi$ has exactly two infinite components.*

One may not replace "quasi-axis" by "axis" in Theorem 4.12, as the almost transitive graph in Figure 4.3 demonstrates. All axes in this nonplanar graph separate it into two infinite components, but some quasi-axes do not separate the graph at all. The authors were then led to the following conjecture.

**Conjecture 4.13 ([BIW, Conjecture 5.4])** *Let $\Gamma$ be locally finite,* 1*-ended, and vertex-transitive. Then $\Gamma$ is planar if and only if for every axis $\Xi$, $\Gamma - \Xi$ has exactly two infinite components.*

With Watkins out and Norbert Seifter in, Conjecture 4.13 was proved for graphs with polynomial growth. (See [BIS, Theorem 8.4]. The proof of [BIS, Theorem 8.4] uses [BIS, Theorem 6.1], but there is an error in the proof of the latter result. An independent proof of [BIS, Theorem 8.4] has been recently done by Imrich and J. Bornhöft and, as of the time of this writing, will be submitted for publication.)

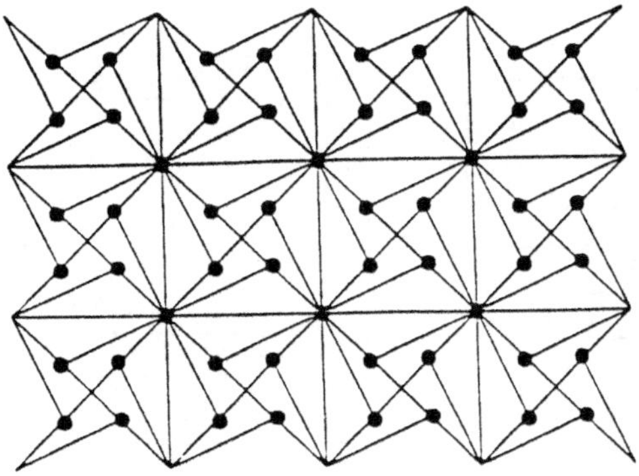

Figure 4.3: "Axis" may not replace "quasi-axis" in Theorem 4.12

# Section 5
# Fibers

As many of the concepts in these last two sections will be applied to locally finite planar graphs, and especially to one-ended graphs, we begin this section with a particularly useful way of representing these graphs in the plane. We will call these representations "Bilinski maps". They were first brought to the author's attention by Joseph Zaks, who learned them from his "mathematical father", Branko Grünbaum, who learned them in turn from a "mathematical uncle", S. Bilinski. (Examples can be seen in [GS1]. An earliest source is probably [B].)

Let $\Gamma$ be an infinite, locally finite, 1-ended, planar, 3-connected graph, and let $x \in V(\Gamma)$. $\Gamma$ may be regarded as the underlying graph of a *Bilinski map* $M$ in the light of the following notation (see Figure 5.1):

$U_0 = \{x\}$; $x$ will be called the *center* of $M$.

$F_1$ is the set of faces incident with $x$.

For $r \geq 1$, $U_r$ is the set of those vertices not in $U_{r-1}$ that are incident with a face in $F_r$.

For $r \geq 1$, $F_{r+1}$ is the set of faces not in $F_r$ that are incident with a vertex in $U_r$.

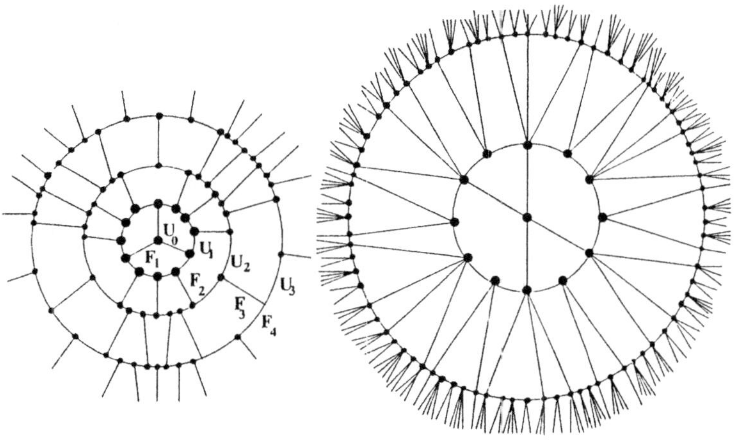


(a) (b)

Figure 5.1: Bilinski maps: (a) the hexagonal tessellation; (b) alternate valences and covalences of 4 and 6

Infinite faces can often be a nuisance. Due to the following result, this concern can sometimes be avoided.

**Theorem 5.1 ([BIW, Theorem 2.3])** *If a graph is locally finite, planar, 2-connected, one-ended, and almost transitive, then it has no infinite faces in any proper embedding.*

By a *proper embedding* we mean a planar embedding in which the boundary of every face is either a circuit or a double ray. This final phrase appears in the statement of Theorem 5.1 because planar graphs with $\kappa < 3$ generally do not necessarily have a unique planar embedding. Halin has characterized [Ha2] the forbidden subgraphs of a locally finite planar graphs that force all of its planar embeddings to have at least one point of accumulation (other than at $\infty$). All of these forbidden subgraphs entail a finite subgraph that separates two infinite subgraphs. It follows that every one-ended, locally finite planar graph has an embedding without accumulation points, that is, a proper embedding.

On the other hand, without the assumption of almost transitivity, infinite faces can be banned only by definition, as follows. We let $\mathcal{G}(a, a^*)$ denote the class of all planar, 3-connected, 1-ended graphs whose valences and covalences are finite and at least $a$ and at least $a^*$, respectively. Clearly the planar duals of graphs in $\mathcal{G}(a, a^*)$ belong to $\mathcal{G}(a^*, a)$. All the graphs in $\mathcal{G}(4,5) \cup \mathcal{G}(5,4)$ have exponential growth. Graphs in $\mathcal{G}(4,4)$ have quadratic growth provided that vertices of valence greater than 4 and faces of covalence greater than 4 are sufficiently sparse and sporadic, in a matter not yet made precise. (Another problem for the reader! For example, it is necessary but not sufficient to require that for any vertex $x$, if $\phi(n)$ is the number of vertices in $B(n, x)$ of valence at least 5, then $\lim_{n\to\infty} \phi(n)/|B(n,x)| = 0$, together with a dual statement for faces of covalence $\geq 5$.)

**Lemma 5.2 ([NW1, Theorem 3.2])** *Every graph in $\mathcal{G}(4,4)$ is the underlying graph of a Bilinski map with the properties that for each $r \geq 1$, the subgraph $\langle U_r \rangle$ is a circuit and every vertex in $U_r$ has at most one neighbor in $U_{r-1}$.*

This technical lemma has a long, inelegant proof, but is an indispensible tool in proving the following useful theorem, that also has a long, inelegant proof.

**Theorem 5.3 ([NW1, Theorem 4.1])** *Let $\Gamma \in \mathcal{G}(4,4)$ and let $\Pi := \Pi[x, y]$ be any path in $X$ with the property that for any face $f$, the number of edges incident with $f$ that lie on $\Pi$ is at most $\lfloor \frac{1}{2}(\rho^*(f) - 2) \rfloor$. Then $\Pi$ is the unique geodetic path joining $x$ and $y$.*

A walk in an embedded planar graph is a *Petrie walk* if every two consecutive edges are incident with a common face, but no three consecutive edges have this property. This term is adapted from H.S.M. Coxeter (see [C, pp. 24 and 32]). The following is an immediate consequence of Theorem 5.3.

**Corollary 5.4 ([NW1, Corollary 4.5])** *If $\Gamma \in \mathcal{G}(4,6)$, then all Petrie walks in $\Gamma$ are axes.*

A Petrie walk $\Pi^*$ in the planar dual $\Gamma^*$ of $\Gamma$ uses exactly the dual edges of a Petrie walk $\Pi$ in $\Gamma$, which leads one to inquire about the geodetic properties of Petrie walks in graphs in $\mathcal{G}(6,4)$, another open question for the reader.

In [BIS, p.17] it was conjectured that every edge of a vertex-transitive, one-ended graph lies on an axis. Since vertex-transitive, one-ended graphs are 3-connected, the conjecture applies (by Theorem 5.1) to the vertex-transitive members of $\mathcal{G}(3,3)$. The authors proved their conjecture under the additional hypothesis that the graph is bipartite [BIS, Theorem 3.2], and credited C. Thomassen with a proof when the graph is 3-valent.

**Corollary 5.5 ([NW1, Corollary 4.4])** *If $\Gamma \in \mathcal{G}(4,4)$, then every edge of $\Gamma$ lies on an axis.*

**Proof** Let $e \in E(\Gamma)$; say $e = [x, y]$. Regard $\Gamma$ as the underlying graph of a Bilinski map with center $x$. Since $\rho(x) \geq 4$, there exists another edge $e'$ incident with $x$ such that $e$ and $e'$ are incident with no common face. By Lemma 5.2 there exists a double ray $\Delta$ passing through $e$ and $e'$ which is the union of two half axes emanating from $x$, each of which contains exactly one vertex from $U_r$ for each $r \geq 0$. Since all valences and covalences are at least 4, these two half-axes never can approach each other sufficiently to be incident with any common face. By Theorem 5.3, every subpath of $\Delta$ is a geodetic path. Hence $\Delta$ is an axis. □

Thus the conjecture by Bonnington, Imrich, and Seifter is shown to be true, even without their transitivity assumption, except when some valence or covalence equals 3. In the presence of 3-valent vertices or 3-covalent faces, the condition of vertex-transitivity cannot be weakened to almost transitivity. Consider the graph obtained by inserting one vertex in the interior of each face of the infinite triangular tessellation of the plane and then joining each new vertex by an edge to each of the three vertices incident with the face in which it has been placed. There are now two vertex orbits, and none of the new edges is an interior edge of a geodetic path of length 3.

Consider the following equivalence relation on the set of rays of $\Gamma$: For rays $\Pi_1$ and $\Pi_2$, define $\Pi_1 \cong \Pi_2$ if for some $m \in \mathbb{Z}^+$ each of $\Pi_1$ and $\Pi_2$ is contained in an $m$-neighborhood of the other. Clearly $\cong$ is an equivalence relation; the equivalence classes with respect to this relation are called *fibers*. If $\Pi_1 \cong \Pi_2$ then, provided that $\Gamma$ is locally finite, one can easily construct an infinite family of pairwise-disjoint $\Pi_1\Pi_2$-paths. Thus fiber-equivalence is a refinement of end-equivalence in locally finite graphs. We would be correct to anticipate that fiber-equivalence might be especially useful in studying the structure of one-ended graphs.

The notion of fiber-equivalence is due to H. A. Jung and Peter Niemeyer (see [JN]) who called these objects "$d$-fibers". That was to distinguish them from "$b$-fibers". For rays $\Pi_1$ and $\Pi_2$, define $\Pi_1 \cong_b \Pi_2$ if there exist sequences $x_1, x_2, \ldots \in V(\Pi_1)$ and $y_1, y_2, \ldots \in V(\Pi_2)$ such that all three of the following sets of numbers are finite:

$$\{d(x_n, y_n) : n \in \mathbb{Z}^+\}, \quad \{d(x_n, x_{n+1}) : n \in \mathbb{Z}^+\}, \quad \{d(y_n, y_{n+1}) : n \in \mathbb{Z}^+\}.$$

The relation $\cong_b$ on the set of rays of a graph was shown in [N1] to be an equivalence relation, and $b$-fibers are the equivalence classes with respect to $\cong_b$. In locally finite graphs $\cong_b$ is in turn a refinement of fiber-equivalence.

A ray $\Pi$ is a *metric ray* if there exists $M > 0$ such that $d_\Pi(x,y) < Md(x,y)$ holds for all $x$, $y \in V(\Pi)$. Thus every subray of a quasi-axis is a metric ray.

**Theorem 5.6 ([JN, Theorems 3.2 and 4.1, Proposition 4.3])** *Let $\Gamma$ be a connected, locally finite graph.*

(a) *Every end of $\Gamma$ either is a fiber or is the union of uncountably many fibers.*

(b) *For any translation $\tau \in \mathrm{Aut}(\Gamma)$, all $\tau$-invariant rays belong to the same fiber.*

(c) *Each fiber of $\Gamma$ contains either only metric rays or only non-metric rays.*

While every end of a locally finite graph contains a half-axis, this does not hold for fibers. Thus we define a *geodetic fiber* to be a fiber that contains at least one half-axis. For one-ended planar graphs we have the following fibrous(?) analogue of Theorem 4.2(b).

**Theorem 5.7 ([NW1, Theorem 5.2])** *Let $\Gamma \in \mathcal{G}(4,5) \cup \mathcal{G}(5,4)$ and assume that the covalences in $\Gamma$ are bounded. Then for any half axes $\Xi_1$ and $\Xi_2$ that belong to distinct fibers of $\Gamma$, there exists an axis that contains both a subray of $\Xi_1$ and a subray of $\Xi_2$.*

A geodetic fiber $\mathcal{F}$ is of finite, countable, or uncountable *type* if every maximal set $\mathcal{S}$ of half-axes in $\mathcal{F}$ such that no two members of $\mathcal{S}$ contain a common subray is, respectively, finite, countable, or uncountable. Suppose that *some* maximal set $\mathcal{S}$ of half-axes in a fiber $\mathcal{F}$ has the property that no two members of $\mathcal{S}$ contain a common subray. If $\mathcal{S}$ is maximal with repect to this property and is finite (respectively, countable, uncountable), then by Lemma 4.3, we see immediately that *every* subset of $\mathcal{F}$ that is maximal with respect to this property is finite (respectively, countable, uncountable). Thus every geodetic fiber is of exactly one of these three types. In particular, if some vertex is the initial vertex of only finitely many half-axes belonging to a given fiber $\mathcal{F}$, then $\mathcal{F}$ is of finite type.

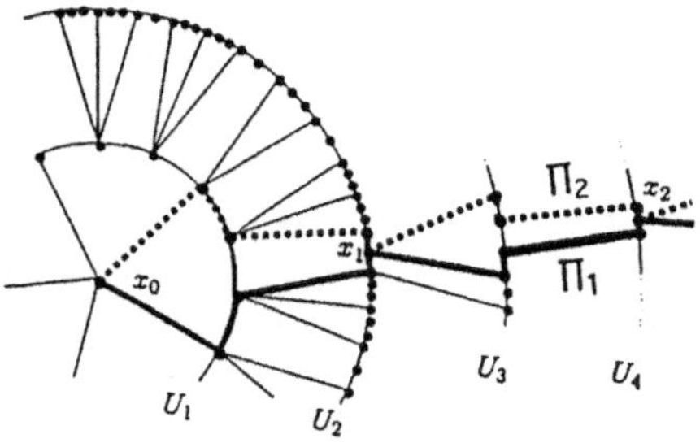


Figure 5.2: A pair of "defining half-axes" of a geodetic fiber of uncountable type

For example, consider the 5-valent, 5-covalent tessellation of the hyperbolic plane, represented as the underlying graph of a Bilinski map in Figure 5.2. Consider half-axes $\Pi_1$ and $\Pi_2$ that emanate from $x_0$ and meet only at every fourth vertex $x_1, x_2, \ldots$. There are uncountably many half-axes belonging to the same fiber that emanate from $x_0$ and contain exactly one of $\Pi_1[x_{n-1}, x_n]$ or $\Pi_2[x_{n-1}, x_n]$ for each $n \in \mathbb{Z}^+$. Hence this fiber is of uncountable type. Since each vertex is the initial vertex of exactly five pairs of "defining half-axes" such as $\Pi_1$ and $\Pi_2$, there are countably many geodetic fibers formed in this way.

Any ray that includes no more than one edge from any face is a half-axis (Theorem 5.3) and can be shown to belong to a fiber of finite type. It is clear from Figure 5.2 that each vertex $x$ is the root of five binary trees formed by the rays of this kind emanating from $x$, no two of which belong to the same fiber. Thus there are uncountably many such fibers. There also exist uncountably many half-axes that contain no subray fitting either of the above two descriptions. The reader can verify that they, too, belong to fibers of finite type. In summary, the graph in this example admits countably many geodetic fibers of uncountable type and uncountably many geodetic fibers of finite type. The next result confirms these findings.

**Theorem 5.8 ([NW1, Corollaries 5.4 and 5.5])**

(a) *Every graph in* $\mathcal{G}_{(}4,5) \cup \mathcal{G}(5,4)$ *contains uncountably many geodetic fibers of finite type.*

(b) *Every geodetic fiber of a graph in* $\mathcal{G}(4,5) \cup \mathcal{G}(5,4)$ *contains at most* 3 *pairwise-disjoint half-axes.*

**Theorem 5.9 ([NW1, Theorem 5.7])** *If* $\Gamma \in \mathcal{G}(5,5)$, *then its subgroup* $B_0(\Gamma)$ *of bounded automorphisms consists only of the identity automorphism.*

As remarked earlier, the one-ended planar graphs described in Theorems 5.8 and 5.9 have exponential growth. We will presently see that a class of graphs with polynomial growth has notably different properties. If $\Gamma$ is connected and locally finite, then $\Gamma$ is *periodic of rank* $d$ if some finitely generated abelian group of rank $d$ acts almost transitively on $V(\Gamma)$. Thus a periodic graph of degree $d$ has polynomial growth of degree $d$. A fiber $\mathcal{F}$ is *translatable* if some ray (hence, by Theorem 4.8(a), a metric ray) in $\mathcal{F}$ is invariant under some non-identity automorphism (hence by a translation) of $\Gamma$.

In constrast to the previous 5-valent, 5-covalent example of Figure 5.2, consider the (4-valent, 4-covalent) 2-dimensional lattice $\Delta^2$, drawn with the usual orientation. (We presume that no figure is needed.) Clearly $\mathbb{Z} \oplus \mathbb{Z}$ acts transitively on $V(\Delta^2)$. In any half-axis, all the horizontal edges must point in the same direction away from the root (right or left), and the same holds for all vertical edges (up or down).

There are essentially only two kinds of half-axes in $\Delta^2$. One kind contains finitely many vertical edges or finitely many horizontal edges, in which case the half-axis contains a subray all of whose edges point in the same one of four directions. Any two half-axes containing a subray pointing in the same direction belong to the same fiber, and from each vertex there emanate half-axes that join up eventually with each of the countably many rays pointing in each of the four directions. The half-axes of this kind give rise to exactly four translatable fibers of countable type.

On the other hand, if a half-axis $\Pi$ in $\Delta^2$ contains infinitely many of both vertical and horizontal edges, then on $\Pi$ there are countably many instances of a horizontal edge followed immediately by a vertical edge. Half-axes belonging to the same fiber as $\Pi$ are obtainable in the following uncountably many ways: for any subset of the set of these instances, replace these two edges in question by a vertical edge followed immediately by a horizontal edge. Thus $\Pi$ belongs to a fiber of uncountable type. Clearly there exist in $\Delta^2$ uncountably many such fibers. Finally, $\Pi$ belongs to a translatable fiber if and only if, with $\Delta^2$ regarded as a cartesian coordinate system, $\Pi$ lies within a neighborhood of a line drawn on $\Delta^2$ with rational slope.

The following two results, in particular Theorem 5.10(c), bear out the findings of this example.

**Theorem 5.10 ([NW2])** *Every periodic graph $\Gamma$ has either finitely many (but $\geq 2$) or uncountably many geodetic fibers, of which only finitely many are of countable type. Furthermore:*

(a) *$\Gamma$ has exactly two geodetic fibers if and only if $\Gamma$ is periodic of rank* 1. *In this case $\Gamma$ is a strip, both fibers are translatable, and neither is of countable type.*

(b) *If $\Gamma$ has finitely many but more than* 2 *geodetic fibers, then they are all translatable and none is of finite type.*

(c) *If $\Gamma$ has uncountably many geodetic fibers, then only finitely many of them are of countable type and the rest are of uncountable type. Moreover, only countably many goedetic fibers, including all those of countable type, are translatable.*

**Corollary 5.11 ([NW2])** *Every planar,* 3*-connected periodic graph has either two or uncountably many geodetic fibers.*

We conclude this section with an example of a periodic graph $\Gamma$ of rank $d = 2$ described by Theorem 5.10(b). Consistent with Theorem 5.11, it will be nonplanar. (See Figure 5.3.) Let $V(\Gamma) = \{x_{i,j}, y_{i,j} : (i,j) \in \mathbb{Z} \times \mathbb{Z}\}$ and let $E(\Gamma)$ include all edges of the three forms:

$$[x_{i,j}, x_{i+1,j}], \quad [y_{i,j}, y_{i,j+1}], \quad [x_{i,j}, y_{i,j}].$$

In effect, each vertex $(i,j)$ of $\Delta^2$ has been split into two adjacent vertices $x_{i,j}$ and $y_{i,j}$. Note that $\Gamma$ is a Cayley graph of the group

$$G = \langle g, h \mid h^2 = (hg)^2(hg^{-1})^2 = 1 \rangle,$$

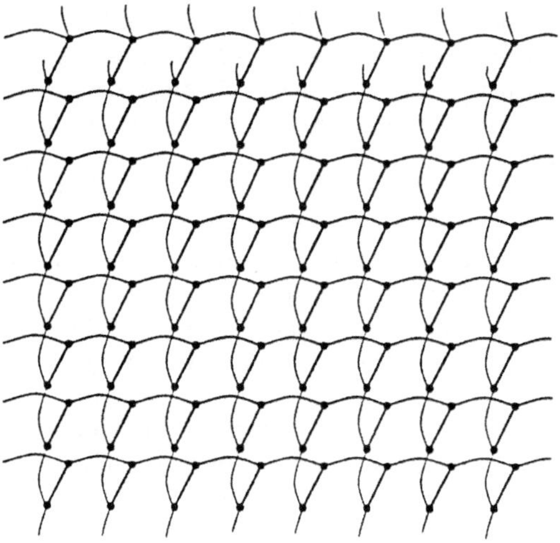

Figure 5.3: A graph with four geodetic fibers

where the edges colored $h$ are exactly those of the form $[x_{i,j}, y_{i,j}]$.

Observe that every half-axis in $\Gamma$ uses at most two of the edges colored $h$. Consequently every half-axis contains either finitely many edges of the form $[x_{i,j}, x_{i+1,j}]$ or finitely many edges of the form $[y_{i,j}, y_{i,j+1}]$, as every switch from one to the other requires traversing exactly one $h$-colored edge. Thus $\Gamma$ contains exactly four geodetic fibers, and by the same argument used earlier for $\Delta^2$, these fibers are all of countable type. Clearly they are translatable, as Theorem 5.10 requires.

A similar example, with six fibers of countable type, can be constructed by splitting each vertex of $\Delta^3$ into a copy of $K_3$, but generalizations of this kind appear to fail for $\Delta^d$ when $d \geq 4$.

# Section 6
# Edge-transitive planar graphs

Unless otherwise noted, the material in this section comes from [GW].

The *symbol* of an edge $e$ of an embedded locally finite, planar graph is a 4-tuple $\langle \rho_0, \rho_1; \rho_0^*, \rho_1^* \rangle$, where $\rho_0$ and $\rho_1$ are the valences of its two incident vertices and $\rho_0^*$ and $\rho_1^*$ are the covalences of its two incident faces. Such a graph $\Gamma$ is *edge-homogeneous* if all of its edges have the same symbol; in this case we may speak of $\langle \rho_0, \rho_1; \rho_0^*, \rho_1^* \rangle$ as the *edge-symbol* of $\Gamma$.

Suppose that $\Gamma$ is an edge-homogeneous, 3-connected planar graph with edge-symbol $\langle \rho_0, \rho_1; \rho_0^*, \rho_1^* \rangle$. Let $x \in V(\Gamma)$, let $r = \rho(x)$, and let $e_0, e_1, \ldots, e_{r-1}$ denote the edges incident with $x$ in counterclockwise cyclic order about $x$. Let $f_i$ be the (unique) face incident with edges $e_{i-1}$ and $e_i$, where all subscripts are understood modulo $r$. (See Figure 6.1(a).) If $\rho^*(f_i) = \rho_0^*$, then clearly $\rho^*(f_{i-1}) = \rho^*(f_{i+1}) = \rho_1^*$. It follows that if $r$ is odd, then $\rho_0^* = \rho_1^*$. Dually, we may consider the edges $e_0, e_1, \ldots, e_{r-1}$ which form the boundary of a face $f$ of covalence $r$, where $e_i = [x_{i-1}, x_i]$, and again all subscripts are understood modulo $r$. By the same reasoning, if $f$ has odd covalence, then $\rho_0 = \rho_1$. While it is elementary that these parity conditions are necessary for a 3-connected planar graph to be edge-homogeneous, Grünbaum and Shephard's surprising result is that the one-ended, edge-homogeneous graphs are characterized completely by their edge-symbol.

**Theorem 6.1 ([GS2, Theorems 1 and 2])** *There exists a locally finite, planar, 3-connected, edge-homogeneous graph with edge-symbol $\langle \rho_0, \rho_1; \rho_0^*, \rho_1^* \rangle$ if and only if all of $\rho_0, \rho_1, \rho_0^*, \rho_1^*$ are integers $\geq 3$ and exactly one of the following holds:*

(1) *all of $\rho_0, \rho_1, \rho_0^*, \rho_1^*$ are even;*

(2) *$\rho_0^* = \rho_1^*$ is even and at least one of $\rho_0, \rho_1$ is odd;*

(3) *$\rho_0 = \rho_1$ is even and at least one of $\rho_0^*, \rho_1^*$ is odd;*

(4) *$\rho_0 = \rho_1$, $\rho_0^* = \rho_1^*$, and all are odd.*

*If such a graph $\Gamma$ is 1-ended, then it is edge-transitive and is uniquely determined by its edge-symbol. If $\Gamma$ is 1-ended and $\rho_0 = \rho_1$, then $\Gamma$ is arc-transitive.*

Let us denote by $\mathcal{G}$ the class of all locally finite, edge-transitive, 3-connected planar graphs. There are nine finite members of $\mathcal{G}$, which include the five Platonic graphs. Let $\Gamma \in \mathcal{G}$ with edge-symbol $\langle \rho_0, \rho_1; \rho_0^*, \rho_1^* \rangle$. It follows from Euler's Theorem that $\Gamma$ is finite if and only if $\frac{1}{\rho_0} + \frac{1}{\rho_1} + \frac{1}{\rho_0^*} + \frac{1}{\rho_1^*} > 1$ holds. Moreover, $\Gamma$ has polynomial (linear or quadratic) growth or has exponential growth according as $\frac{1}{\rho_0} + \frac{1}{\rho_1} + \frac{1}{\rho_0^*} + \frac{1}{\rho_1^*}$ equals 1 or is $< 1$, respectively.

Theorem 6.1 gives one description of the one-ended members of $\mathcal{G}$. Theorem 3.8 above tells us that all of the two-ended members of $\mathcal{G}$ have edge-symbol $\langle 4,4;4,4 \rangle$. However, the one-ended graph $\Delta^2$ also has edge-symbol $\langle 4,4;4,4 \rangle$. It seems desirable, therefore, to establish further criteria to distinguish the members of $\mathcal{G}$.

Grünbaum and Shephard's proof of Theorem 6.1 relies heavily upon the geometry of the hyperbolic plane, in which all of the one-ended members of $\mathcal{G}$ with exponential growth have regular embeddings. (Those with quadratic growth have regular embeddings in the Euclidean plane.) The approach in [GW], on the other hand, is a more algebraic one.

For $\Gamma \in \mathcal{G}$, a *flag* of $\Gamma$ is a triple $(x, e, f)$, where $x \in V(\Gamma)$, $e \in E(\Gamma)$, $f \in F(\Gamma)$, and all three are mutually incident.

**Lemma 6.2** *If $\Gamma \in \mathcal{G}$, then each automorphism of $\Gamma$ is uniquely determined by its action on any flag of $\Gamma$.*

**Proof** Suppose that $\alpha, \beta \in \mathrm{Aut}(\Gamma)$ agree at each term of the flag $(x, e, f)$ of $\Gamma$, and let $\delta = \alpha\beta^{-1}$. We must show that $\delta$ is the identity automorphism. Clearly if $\delta$ fixes some flag on a given vertex, then it must fix every flag on that vertex. Thus if $y$ is adjacent to $x$, then $\delta$ fixes $y$, the edge $[x, y]$, and both faces incident with $[x, y]$. But then $\delta$ fixes all flags on the vertex $y$. Since $\Gamma$ is connected, we conclude that $\delta$ fixes each flag of $\Gamma$. □

Let $e = [x, y] \in E(\Gamma)$ and let $f$ and $g$ denote the two faces incident with $e$. Let $\lambda_e, \tau_e \in \mathrm{Aut}_e(\Gamma)$ be defined by $\lambda_e(x, e, f) = (y, e, f)$ and $\tau_e(x, e, f) = (x, e, g)$. (Abusing notation, we may regard $\mathrm{Aut}(\Gamma)$ as also acting on flags.) In the light of Lemma 6.2, these maps – if they are extendable at all – are uniquely defined and

$$\tau^2 = \lambda^2 = (\tau\lambda)^2 = \iota, \tag{6.1}$$

where $\iota$ denotes the identity of $\mathrm{Aut}(\Gamma)$. We define the group $K_e := \langle \lambda_e, \tau_e \rangle \cong \mathbb{Z}_2 \oplus \mathbb{Z}_2$, the Klein 4-group. Each of the five subgroups of $K_e$ is a candidate for the edge-stabilizer $\mathrm{Aut}_e(\Gamma)$. Note that by edge-transitivity, all edge-stabilizers are isomorphic as permutation groups. The role of the planar dual $\Gamma^*$ of $\Gamma$ must not be overlooked. If $e^* = [f, g]$ is the dual edge of $e$, then $\tau_{e^*} = \lambda_e$, $\lambda_{e^*} = \tau_e$, and if $\phi_e = \tau_e \lambda_e$, then $\phi_{e^*} = \phi_e$.

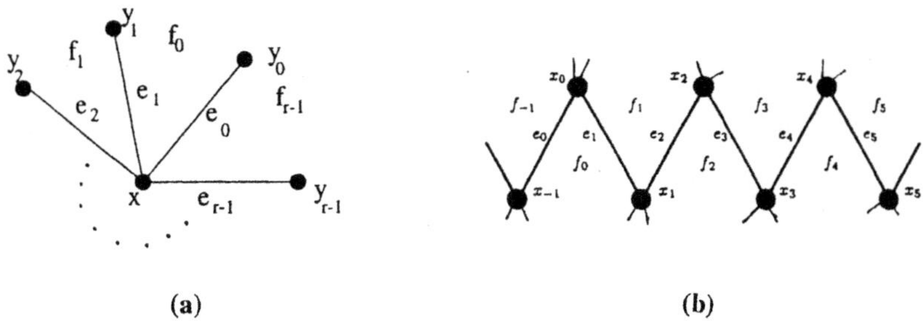


Figure 6.1: Local structures: (a) around a vertex; (b) along a Petrie walk

Let us now determine the stabilizers of vertices and faces and consider once again the situation in Figure 6.1(a), where $e_i = [x, y_i]$. By Lemma 6.2 we may define $\theta_{xf_0}, \sigma_x$, and $\sigma_{f_0}$ by

$$\theta_{xf_0}(x, e_0, f_0) = (x, e_1, f_0),\ \sigma_x(x, e_0, f_0) = (x, e_1, f_1),\ \sigma_{f_0}(x, e_1, f_0) = (y_0, e_0, f_0).$$

One straightforwardly verifies the following identities:

$$\theta_{xf_0}^2 = (\lambda_{e_0}\theta_{xf_0})^{\rho^*(f_0)} = (\tau_{e_0}\theta_{xf_0})^{\rho(x)} = \iota. \tag{6.2}$$

We also have

$$\theta_{xf_0}^2 = \sigma_x^{\rho(x)} = (\sigma_x \theta_{xf_0})^2 = \iota \tag{6.3}$$

and

$$\theta_{xf_0}^2 = \sigma_{f_0}^{\rho^*(f_0)} = (\sigma_{f_0}\theta_{xf_0})^2 = \iota. \tag{6.4}$$

The possibility of infinite faces has not been totally eliminated. If $\rho^*(f_0) = \infty$, let us agree to ignore the second terms of (6.2) and (6.4).

**Remark** In the particular case that $\Gamma$ is finite or that $\Gamma$ is a one-ended tessellation of the Euclidean or hyperbolic plane with edge-symbol $\langle \rho(x), \rho(x); \rho^*(f_0), \rho^*(f_0) \rangle$, a complete presentation of $\mathrm{Aut}(\Gamma)$ exists in terms of $\lambda_{e_0}, \tau_{e_0}$, and $\theta_{xf_0}$ together with the six relations given in (6.1) and (6.2). (An alternative presentation can be extracted from [C, Chapter 5].) Under other than these ideal circumstances, a presentation of $\mathrm{Aut}(\Gamma)$ may require many more, perhaps infinitely many relations. A local rather than global study of the action of $\mathrm{Aut}(\Gamma)$ is therefore not only easier but also sufficient for the broader class of graphs in $\mathcal{G}$.

The duality between vertices and faces is "old hat", but in the present context, we have in fact a "triality". Consider the portion of a Petrie walk $\Pi$ shown in Figure 6.1(b), and define the map $\gamma_\Pi$ determined by $\gamma_\Pi(x_0, e_0, f_0) = (x_1, e_1, f_1)$. Note that $\gamma_\Pi$ belongs to the stabilizer

$\mathrm{Aut}_\Pi(\Gamma)$ of $\Pi$. When $\Pi$ is a double ray (cf. Corollary 5.4 above), then $\gamma_\Pi$ is a translation. This gives rise to the relations

$$\theta_{xf_0}^2 = (\gamma_\Pi \theta_{xf_0})^2 = \iota. \tag{6.5}$$

Summarizing (6.1), (6.3), (6.4), and (6.5), we have:

**Lemma 6.3** *Let* $\Gamma \in \mathcal{G}$, $x \in V(\Gamma)$, $e \in E(\Gamma)$, $f \in F(\Gamma)$, *and let* $\Pi$ *be any Petrie walk in* $\Gamma$. *Then*

(a) $\mathrm{Aut}_e(\Gamma)$ *is a subgroup of the Klein* 4*-group* $K_e$;

(b) $\mathrm{Aut}_x(\Gamma)$ *is a subgroup of the dihedral group* $D_{\rho(x)}$;

(c) $\mathrm{Aut}_f(\Gamma)$ *is a subgroup of the* (*possibly infinite*) *dihedral group* $D_{\rho^*(f)}$;

(d) $\mathrm{Aut}_\Pi(\Gamma)$ *is a subgroup of the dihedral group* $D_m$, *where* $m$ *is the number* (*possibly infinite*) *of edges in* $\Pi$.

Because there are only four ways that a flag can be rotated or reflected, it is not difficult to show that $\sigma_x^4 \in \mathrm{Aut}(\Gamma)$ for every $x \in V(\Gamma)$, even though $\sigma_x$ or $\sigma_x^2$ may fail to be extendable to all of $\Gamma$. Similarly, $\sigma_f^4$ and $\gamma_\Pi^4$ are automorphisms of $\Gamma$ for every face $f$ and every Petrie walk $\Pi$. Thus not every subgroup of the dihedral groups in parts (b), (c), and (d) of Lemma 6.3 can actually be a stabilizer. In fact, in each case only six of the subgroups are realizable, the same six, up to permutation group isomorphism, namely the (full) dihedral group itself, three subgroups of index 2, and two subgroups of index 4. Choosing one stabilizer for each of the four kinds of objects (edges, vertices, faces, and Petrie walks) we obtain $5 \times 6^3 = 2160$ combinations of stabilizers. However, only 14 of these many combinations are *algebraically* compatible. They are listed in Table 6.1. Of these 14, only eight combinations of stabilizers are *graph theoretically* realizable.

**Remarks** 1. There is a rational method to the naming of the 14 types in Table 6.1. For $n = 2, 4$, or 5, the stabilizers the edges are isomorphic to the stabilizers of the Petrie walks as permuation groups. Type $n^*$ describes the automorphism group of the planar dual of a graph whose group is of type $n$; its vertex-stablizers are isomorphic to the stabilizers of its Petrie walks as permutation groups. In a graph whose group is of type $n^{\mathrm{P}}$, the Petrie walks play the role analogous to that of the vertices (respectively, faces) of a graph whose group is of type $n$ (respectively, $n^*$), while the stabilizers of the vertices and the stabilizers of the faces form isomorphic permutation groups. This holds with or without the suffix "ex" when $n = 2$.

2. When $n = 1$ or 3, there is a symmetry among the roles played by vertices, faces, and Petrie walks.

3. Two different vertex-stabilizers are listed for the graphs whose automorphism groups are of type 4. These graphs are bipartite. The vertices on one side of the bipartition have a stabilizer of one of the two forms listed while the vertices on the other side have a stabilizer of the other form listed. In this case, $\theta_{xf}$ is an automorphism for all vertices $x$ in exactly one of the two sides, which is implied by the way that edge- and Petrie-stabilizers are presented. The planar duals of these graphs are of type $4^*$. Those of type $4^{\mathrm{P}}$ have two orbits of Petrie walks; the members of different orbits have different stabilizers.

4. Graphs in $\mathcal{G}$ for which $\theta_{xf}$ is an automorphism for every incident vertex-face pair $x, f$ are called *ordinary*. Graphs that are not ordinary are called *extraordinary*. The extraordinary graphs do indeed display some unusual behavior.

| Type | $G_e$ | $G_x$ | $G_f$ | $G_\Pi$ | Remarks |
|---|---|---|---|---|---|
| 1 | $K_e$ | $\langle\sigma_x, \theta_{xf}\rangle$ | $\langle\sigma_f, \theta_{xf}\rangle$ | $\langle\gamma_\Pi, \theta_{xf}\rangle$ | $V, F, P$ |
| 2 | $\langle\tau_e\rangle$ | $\langle\sigma_x, \theta_{xf}\rangle$ | $\langle\sigma_f^2, \theta_{xf}\rangle$ | $\langle\gamma_\Pi^2, \theta_{xf}\rangle$ | $F, P;\ 2\vert\rho^*$ |
| 2ex | $\langle\tau_e\rangle$ | $\langle\sigma_x^2, \tau_e\rangle$ | $\langle\sigma_f\rangle$ | $\langle\gamma_\Pi\rangle$ | $V, F, P$ |
| $2^*$ | $\langle\lambda_e\rangle$ | $\langle\sigma_x^2, \theta_{xf}\rangle$ | $\langle\sigma_f, \theta_{xf}\rangle$ | $\langle\gamma_\Pi^2, \theta_{xf}\rangle$ | $V, P;\ 2\vert\rho$ |
| $2^*$ex | $\langle\lambda_e\rangle$ | $\langle\sigma_x\rangle$ | $\langle\sigma_f^2, \lambda_e\rangle$ | $\langle\gamma_\Pi\rangle$ | $V, F, P$ |
| $2^{\mathrm{P}}$ | $\langle\phi_e\rangle$ | $\langle\sigma_x^2, \theta_{xf}\rangle$ | $\langle\sigma_f^2, \theta_{xf}\rangle$ | $\langle\gamma_\Pi, \theta_{xf}\rangle$ | $V, F;\ 2\vert\rho;\ 2\vert\rho^*$ |
| $2^{\mathrm{P}}$ex | $\langle\phi_e\rangle$ | $\langle\sigma_x\rangle$ | $\langle\sigma_f\rangle$ | $\langle\gamma_\Pi^2, \phi_e\rangle$ | $V, F, P$ |
| 3 | $\langle\iota\rangle$ | $\langle\sigma_x^2, \theta_{xf}\rangle$ | $\langle\sigma_f^2, \theta_{xf}\rangle$ | $\langle\gamma_\Pi^2, \theta_{xf}\rangle$ | $2\vert\rho_0,\ 2\vert\rho_1,\ 2\vert\rho_0^*,\ 2\vert\rho_1^*$ |
| 4 | $\langle\iota\rangle$ | $\langle\sigma_x\rangle, \langle\sigma_x^2, \theta_{xf}\rangle$ | $\langle\sigma_f^4, \theta_{xf}\rangle$ | $\langle\gamma_\Pi^4, \theta_{xf}\rangle$ | $F, P;\ 2\vert\rho_0$ *or* $2\vert\rho_1;\ 4\vert\rho^*$ |
| 5 | $\langle\iota\rangle$ | $\langle\sigma_x\rangle$ | $\langle\sigma_f^2\rangle$ | $\langle\gamma_\Pi^2\rangle$ | $F, P;\ 2\vert\rho^*$ |
| $4^*$ | $\langle\iota\rangle$ | $\langle\sigma_x^4, \theta_{xf}\rangle$ | $\langle\sigma_f\rangle, \langle\sigma_f^2, \theta_{xf}\rangle$ | $\langle\gamma_\Pi^4, \theta_{xf}\rangle$ | $V, P;\ 4\vert\rho,\ 2\vert\rho_0^*$ *or* $2\vert\rho_1^*$ |
| $5^*$ | $\langle\iota\rangle$ | $\langle\sigma_x^2\rangle$ | $\langle\sigma_f\rangle$ | $\langle\gamma_\Pi^2\rangle$ | $V, P;\ 2\vert\rho$ |
| $4^{\mathrm{P}}$ | $\langle\iota\rangle$ | $\langle\sigma_x^4, \theta_{xf}\rangle$ | $\langle\sigma_f^4, \theta_{xf}\rangle$ | $\langle\gamma_\Pi\rangle, \langle\gamma_\Pi^2, \theta_{xf}\rangle$ | $V, F;\ 4\vert\rho;\ 4\vert\rho^*$ |
| $5^{\mathrm{P}}$ | $\langle\iota\rangle$ | $\langle\sigma_x^2\rangle$ | $\langle\sigma_f^2\rangle$ | $\langle\gamma_\Pi\rangle$ | $V, F;\ 2\vert\rho,\ 2\vert\rho^*$ |

$V$: vertex-transitive ($\rho := \rho_0 = \rho_1$); $F$: face-transitive ($\rho^* := \rho_0^* = \rho_1^*$); $P$: Petrie-transitive.

Table 6.1: Algebraically realizable combinations of stabilizers

5. Graphs of types 1, 2, $2^*$, $2^{\mathrm{P}}$, and 3 are ordinary. All other graphs in $\mathcal{G}$ are extraordinary, but only types 4, $4^*$, and $4^{\mathrm{P}}$ survive elimination as graph-theoretically not realizable. All eight of these surviving types can indeed be realized.

At this point, even though we know all there is to know about the local behavior of the automorphism groups of the graphs in $\mathcal{G}$, Table 6.1 doesn't offer a clue (beyond some parity properties) as to which graphs have automorphism groups of which type. In order to correlate the number of ends with the automorphism group types, we need to correlate them both with the third and last criterion: the "Petrie type". In [C] Coxeter observed that the Petrie walks in the nine finite members of $\mathcal{G}$ are (elementary) circuits of even length. Coxeter called them "Petrie circuits", from which we derived the more general term "Petrie walk". At the other extreme we have already seen (Corollary 5.4) that all Petrie walks are double rays in the one-ended graphs in $\mathcal{G}$ whose edge-symbol majorizes $\langle 4,4;6,6\rangle$. As remarked in Section 5, a particularly useful feature of Petrie walks is the fact that every Petrie walk $\Pi^*$ in the planar dual $\Gamma^*$ of a graph $\Gamma$ consists of the dual edges of a Petrie walk $\Pi$ in $\Gamma$. The following result requires a long technical proof which draws upon information in Table 6.1.

**Theorem 6.4** *In a graph in $\mathcal{G}$, every Petrie walk is either a circuit of even length or a double ray.*

A Petrie walk that is a circuit (respectively, a double ray) is called a *Petrie circuit* (respectively, *Petrie line*). Freed from the possibility that the same edge may be encountered more than once on a given Petrie walk, we now see that each edge lies on exactly two Petrie walks. In the light of edge-transitivity, it makes sense to say that a graph $\Gamma \in \mathcal{G}$ is of *circuit* (resp., *line, mixed*) *type* if the two Petrie walks through any given edge are both circuits (resp., both double rays, one circuit and one double ray). Thus a graph in $\mathcal{G}$ and its planar dual are always of the same Petrie type. If all the Petrie walks in $\Gamma$ belong to the same orbit of $\mathrm{Aut}(\Gamma)$, then we say that $\Gamma$ is *Petrie-transitive.* If $\Gamma$ is not Petrie-transitive, then clearly it admits exactly two orbits of Petrie walks, and the two Petrie walks through each edge $e$ belong to different orbits. Because all valences are $\geq 3$, the Petrie walks through $e$ must cross each other at $e$, and the "crossing" is only one edge long. It follows that if both Petrie walks through $e$ are circuits, then these circuits must "cross" a positive even number of times. It can also be shown that a "crossing" cannot consist just of a single vertex.

**Lemma 6.5** *Let $\Gamma \in \mathcal{G}$. If $\Gamma$ contains either three distinct Petrie walks, every two of which have a common edge, or two distinct Petrie walks that have more than one common edge, then $\Gamma$ is Petrie-transitive.*

The sufficiency of the first condition in Lemma 6.5 for Petrie-transitivity is obvious; the sufficiency of the second condition is less so. Of interest, however, is that the second condition does not hold vacuously. In the graphs characterized by the next theorem, the two Petrie circuits through each edge share exactly one other common edge.

**Theorem 6.6** *An infinite graph $\Gamma \in \mathcal{G}$ is of circuit type if and only if $\Gamma$ or its planar dual has edge-symbol $\langle 3, h; 4, 4\rangle$ for some $h \geq 6$.*

Clearly graphs of mixed type are not Petrie-transitive.

**Corollary 6.7** *Let $\Gamma \in \mathcal{G}$. If $\Gamma$ is of mixed type, then $\epsilon(\Gamma) \geq 2$.*

**Proof** Let $e \in E(\Gamma)$, and let $\Theta$ be the Petrie circuit and $\Delta$ be the Petrie line through $e$. By Lemma 6.5, $\Theta$ and $\Delta$ have no edge other than $e$ in common. By the above discussion (and the Jordan Curve Theorem!), we conclude that one subray of $\Delta$ lies in the interior of $\Theta$ and a disjoint subray of $\Delta$ lies in the exterior of $\Theta$. Thus $\epsilon(\Gamma) \geq 2$. □

The two-ended members of $\mathcal{G}$ are easily seen to be of mixed type. Consider, for example, the strip $\Sigma$ in Figure 3.4 and in particular the Petrie circuit $\Theta$ where $V(\Theta) = \{(0,0), (1,0), (1,1), (2,1), (2,2), (3,2)\}$ is of length $6 = 2\kappa_\infty$. The reader can easily check that graphs that are not Petrie-transitive admit no automorphism of any of the forms $\tau_e, \lambda_e, \sigma_x$ or $\sigma_f$. Clearly $\Sigma$ is vertex-transitive, and being self-dual, is also face-transitive. One easily verifies that $\sigma_x^2, \sigma_f^2 \in \mathrm{Aut}(\Sigma)$ for all $x \in V(\Sigma)$ and $f \in F(\Sigma)$. From Table 6.1, we conclude that $\mathrm{Aut}(\Sigma)$ is of type $2^{\mathrm{P}}$. Note that each of the two 3-subsets of alternate vertices on $\Theta$ is a $\kappa_\infty$-separator. This last observation is borne out by the following theorem, which is the strongest possible converse to Corollary 6.7.

**Theorem 6.8** *If $\epsilon(\Gamma) \geq 2$ and $\Gamma$ is ordinary, then $\Gamma$ contains a Petrie circuit and*

(a) *Every Petrie circuit is a shortest ends-separating circuit;*

(b) *The edge set of every Petrie circuit is a smallest ends-separating edge-cutset.*

The following table gives the correlation among the autmorphism group type, the Petrie type, and the number of ends. (Graphs with zero ends are finite, of course.)

| | 0-ended | 1-ended | 2-ended | $\infty$-ended |
|---|---|---|---|---|
| circuit type | $1, 2, 2^*$ | $2, 2^*$ | – | – |
| line type | – | $1, 2, 2^*, 3$ | – | $4, 4^*, 4^{\mathrm{P}}$ |
| mixed type | – | – | $2^{\mathrm{P}}$ | $2^{\mathrm{P}}, 3, 4^{\mathrm{P}}$ |

Table 6.2: Types for $\mathrm{Aut}(\Gamma)$ when $\Gamma \in \mathcal{G}$

We now briefly sketch a procedure called "interleaving" that yields (1) all ordinary infinitely-ended graphs in $\mathcal{G}$ and (2) all extraordinary infinitely-ended graphs in $\mathcal{G}$ that are of mixed type. The procedure is rather complex as is the proof that it yields all the graphs that we claim it does. Hopefully the following example will give some intuitive idea underlying the general method of construction.

Let $\Lambda_0$ and $\Lambda_1$ be the graphs with edge-symbols $\langle 2, 4; 6, 6\rangle$ and $\langle 2, 4; 4, 6\rangle$ shown in Figure 6.2(a) and (b), respectively. It is significant that these graphs are obtained by subdividing (once) the edges of edge-transitive planar multigraphs with even valences (in this case, the Platonic octahedron graph and the 3-circuit with doubled edges, respectively). Also of significance is that $\Lambda_0$ and $\Lambda_1$ share a common covalence, namely 6. $\mathrm{Aut}(\Lambda_0)$ acts imprimitively on $F(\Lambda_0)$; the faces in one block of imprimitivity are arbitrarily designated as *Z-faces* as are the 6-covalent faces of $\Lambda_1$. (These are the shaded faces in Figure 6.2. Note that the exterior face of each graph is also a $Z$-face.)

The multigraphs $\Lambda_0$ and $\Lambda_1$ together with their sets of $Z$-faces are examples of objects called "modules". Since $\mathrm{Aut}(\Lambda_0)$ is of type 2 and $\mathrm{Aut}(\Lambda_1)$ is of type 3, $\Lambda_0$ and $\Lambda_1$ are *ordinary modules.*

Let $\Gamma_0 = \Lambda_0$, and form $\Gamma_1$ by embedding a copy of $\Lambda_1$ into each $Z$-face of $\Lambda_0$, as shown in Figure 6.3, where the shaded regions indicate the $Z$-faces of $\Gamma_1$. Note that the boundary of the exterior face of $\Lambda_1$ is identified with the boundary of the face of $\Gamma_0$ into which it is embedded. We next form $\Gamma_2$ by imbedding in the same manner a copy of $\Lambda_0$ in each $Z$-face of $\Gamma_1$. Copies of $\Lambda_1$ are then embedded into the $Z$-faces of $\Gamma_2$, and so on, *ad infinitum.* Define $\Gamma_\infty$ to be the union of the nested sequence of graphs:

$$\Gamma_0 \subset \Gamma_1 \subset \Gamma_2 \subset \cdots.$$

We say that $\Gamma_\infty$ is obtained by "interleaving" $\Lambda_0$ with $\Lambda_1$.

The graph $\Gamma_\infty$ is an infinitely-ended member of $\mathcal{G}$ with edge-symbol $\langle 4, 4; 6, 6\rangle$. It is of mixed (Petrie) type, and $\mathrm{Aut}(\Gamma_\infty)$ is of type – wait! Let's figure this out. From Table 6.2, we have three choices. It is not hard to see by inspection that all automorphisms of the form $\theta_{xf}$ are present, and so $\Gamma_\infty$ is ordinary; by Table 6.1 we may eliminate type $4^{\mathrm{P}}$. The reader is left to verify whether some automorphism of the form $\phi_e$ or $\gamma_\Pi$ is present, and thus to determine whether $\mathrm{Aut}(\Gamma_\infty)$ is of type $2^{\mathrm{P}}$ or type 3. It is perhaps counterintuitive but nonetheless true that, had the roles of $\Lambda_0$ and $\Lambda_1$ been reversed in the foregoing interleaving procedure, the resulting graph would be isormorphic to $\Gamma_\infty$.

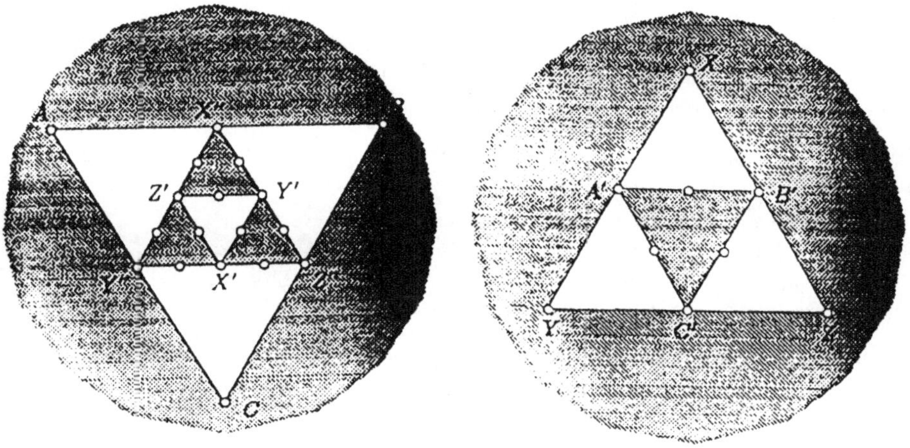


(a) (b)

Figure 6.2: Ordinary modules: (a) $\Lambda_0$; (b) $\Lambda_1$

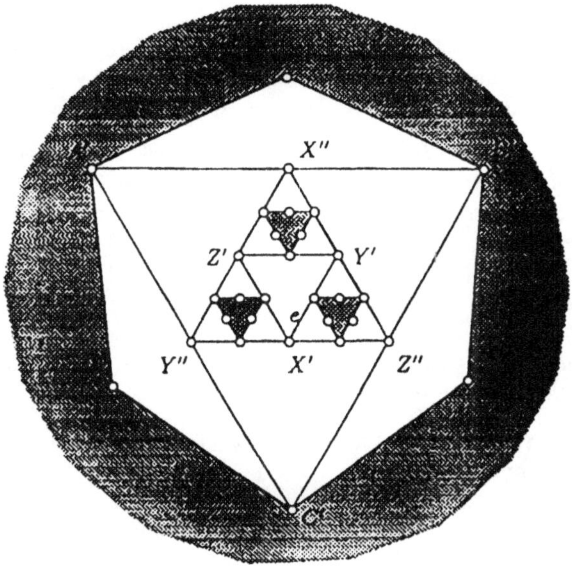


Figure 6.3: The first stage of the interleaving procedure: $\Gamma_1$

We conclude this exposition of the material in [GW] by extending Grünbaum and Shephard's theorem (Theorem 6.1) to all graphs in $\Gamma \in \mathcal{G}$. In order to do this we introduce the *extended edge-symbol* $\langle \rho_0, \rho_1; \rho_0^*, \rho_1^*; \zeta \rangle$, where the new parameter $\zeta$ is defined as follows: $\zeta = 0$ if $\Gamma$ is finite; $\zeta = \infty$ if $\epsilon(\Gamma) = 1$; otherwise $\zeta$ denotes the length of a shortest ends-separating circuit. In this last case, it can be shown that $\zeta$ is always an even integer $\geq 4$.

The 5-tuple $\langle \rho_0, \rho_1; \rho_0^*, \rho_1^*; \zeta \rangle$ is *proper* if all of the following conditions hold:

1. $\rho_0, \rho_1, \rho_0^*, \rho_1^*$ are integers $\geq 3$ and $\zeta$ is a nonnegative even integer or $\infty$.
2. $\zeta = 0$ if $\frac{1}{\rho_0} + \frac{1}{\rho_1} + \frac{1}{\rho_0^*} + \frac{1}{\rho_1^*} > 1$; otherwise $\zeta \geq 4$.
3. Exactly one of the following holds:
   (a) all of $\rho_0, \rho_1, \rho_0^*, \rho_1^*$ are even;
   (b) $\rho_0^* = \rho_1^*$ is even, at least one of $\rho_0, \rho_1$ is odd, and $\zeta = 0$ or $\infty$;
   (c) $\rho_0 = \rho_1$ is even, at least one of $\rho_0^*, \rho_1^*$ is odd, and $\zeta = 0$ or $\infty$;
   (d) $\rho_0 = \rho_1$, $\rho_0^* = \rho_1^*$, they all are odd, and $\zeta = 0$ or $\infty$.

**Theorem 6.9** *The extended edge-symbol of every ordinary member of $\mathcal{G}$ is proper.*

*Conversely, every proper 5-tuple $\langle \rho_0, \rho_1; \rho_0^*, \rho_1^*; \zeta \rangle$ is the extended edge-symbol of a unique ordinary member of $\mathcal{G}$ with the following exception: when $0 < \zeta < \infty, \rho_0 \neq \rho_1$, and $\rho_0^* \neq \rho_1^*$, then it is the edge-symbol of exactly two (nonisomorphic) ordinary members of $\mathcal{G}$.*

**Theorem 6.10** *The graph $\Gamma$ is an extraordinary member of $\mathcal{G}$ if and only if exactly one of the following holds:*

(a) *$\Gamma$ is of mixed type,* $\mathrm{Aut}(\Gamma)$ *is of type* $4^P$, *and $\Gamma$ can be constructed by interleaving two copies of the same extraordinary module.*

(b) *$\Gamma$ is of line type and* $\mathrm{Aut}(\Gamma)$ *is of type* $4, 4^*$, *or* $4^P$.

Examples of each of the three possibilities in part (b) exist. The edges of such graphs may be subdivided (once) to produce the extraordinary modules that may be "interleaved" to produce the graphs described in part (a). We close by posing the following conjecture.

**Conjecture** *Let $\Gamma$ be a locally finite, planar, 2-connected, edge-transitive graph. Then $\Gamma$ is edge-transitive as a plane graph for every one of its planar embeddings. In other words, "2-connected" could have replaced "3-connected" in the definition of $\mathcal{G}$.*

## References


[B] S. Bilinski, Homogene Netze der Ebene, *Bull. Internat. Acad. Yougoslave Cl. Sci. Math. Phys. Tech.* **2** (1949), 63–111.

[BIS] C. P. Bonnington, W. Imrich, and N. Seifter, Geodesics in transitive graphs, *J. Combin. Theory Ser. B* **67** (1996), 12–33.

[BIW] C. P. Bonnington, W. Imrich, and M. E. Watkins, Separating double rays in locally finite planar graphs, *Discrete Math.* **145** (1995), 61–72.

[Bo] I. Z. Bouwer, Vertex and edge transitive but not 1-transitive graphs, *Canad. Math. Bull.* **13** (1970), 231–347.

[C] H. S. M. Coxeter, *Regular Polytopes,* 2nd ed., Macmillan, New York, 1963.

[F] J. Folkman, Regular line-symmetric graphs, *J. Combin. Theory* **3** (1967), 215–232.

[GW] J. E. Graver and M. E. Watkins, *Locally Finite, Planar, Edge-Transitive Graphs,* Memoirs Amer. Math. Soc. **601**, 1997.

[Gr] M. Gromov, Groups of polynomial growth and expanding maps, *Inst. Hautes Études Sci. Publ. Math.* **53** (1981), 53–78.

[GS1] B. Grünbaum and G. C. Shephard, *Tilings and Patterns,* W. H. Freeman and Co., 1987.

[GS2] B. Grünbaum and G. C. Shephard, Edge-transitive planar graphs, *J. Graph Theory* **11** (1987), 141–156.

[Ha1] R. Halin, Über unendliche Wege in Graphen, *Math. Ann.* **157** (1964), 125–137.

[Ha2] R. Halin, Zur häufungspunktfreien Darstellung abzählbarer Graphen in der Ebene, *Arch. Math.* **17** (1966), 239–243.

[Ha3] R. Halin, Automorphisms and endomorphisms of infinite locally finite graphs, *Abh. Math. Sem. Univ. Hamburg* **39** (1973), 251–283.

[Ho] D. F. Holt, A graph which is edge transitive but not arc transitive, *J. Graph Theory* **5** (1981), 201–204.

[I] W. Imrich, On Whitney's theorem on the unique embeddability of 3-connected planar graphs, in: *Recent Advances in Graph Theory* (M. Fiedler, ed.), Academia Praha, Prague, 1975, 303–306.

[IS] W. Imrich and N. Seifter, A note on the growth of transitive graphs, *Discrete Math.* **73** (1988/89), 111-117.

[J] H. A. Jung, A note on fragments of infinite graphs, *Combinatorica* **1** (1981), 285–288.

[JN] H. A. Jung and P. Niemeyer, Decomposing ends of graphs, *Math. Nachr.* **174** (1995), 185–202.

[JW1] H. A. Jung and M. E. Watkins, On the connectivities of finite and infinite graphs, *Monatsh. Math.* **83** (1977), 121–131.

[JW2] H. A. Jung and M. E. Watkins, Fragments and automorphisms of infinite graphs, *European J. Combin.* **5** (1984), 149–162.

[JW3] H. A. Jung and M. E. Watkins, Finite separating sets in locally finite graphs, *J. Combin. Theory Ser. B* **59** (1993), 15–25.

[K1] D. König, Sur les correspondences multivoques des ensembles, *Fund. Math.* **8** (1926), 114–134.

[K2] D. König, *Theorie der endlichen und unendlichen Graphen,* Akad. Verlagsgesellschaft, Leipzig, 1936.

[N1] P. Niemeyer, Verfeinerung des Endenbegriffes bei unendlichen Graphen, Diplom thesis, TU-Berlin, 1992.

[N2] P. Niemeyer, On bounded automorphisms of locally finite, transitive graphs, *J. Graph Theory* **22** (1996), 357–365.

[NW1] P. Niemeyer and M. E. Watkins, Geodetic rays and fibers in one-ended transitive graphs, to appear in *J. Combin. Theory Ser. B.*

[NW2] P. Niemeyer and M. E. Watkins, Geodetic rays and fibers in periodic graphs, in preparation.

[PW] N. Polat and M. E. Watkins, On translations of double rays in graphs, *Period. Math. Hungar.* **30** (1995), 145–154.

[ST] N. Seifter and V.I. Trofimov, Automorphism groups of graphs with quadratic growth, submitted for publication.

[Th1] C. Thomassen, Duality of infinite graphs, *J. Combin. Theory Ser. B* **33** (1982), 137–166.

[Th2] C. Thomassen, The Hadwiger number of infinite vertex-transitive graphs, *Combinatorica* **12** (1992), 481–491.

[TW] C. Thomassen and M. E. Watkins, Infinite vertex-transitive, edge-transitive, non 1-transitive graphs, *Proc. Amer. Math. Soc.* **105** (1989), 258–261.

[Tu] W. T. Tutte, *Connectivity in Graphs,* University of Toronto Press, Toronto, 1966.

[Wa1] M. E. Watkins, Infinite paths that contain only shortest paths, *J. Combin. Theory Ser. B.* **41** (1986), 341–355.

[Wa2] M. E. Watkins, Edge-transitive strips, *J. Combin. Theory Ser. B* **95** (1991), 359–372.

[Wh] H. Whitney, Non-separable and planar graphs, *Trans. Amer. Math. Soc.* **34** (1932), 339–362.S

# Index

GPSR Compliance

*The European Union's (EU) General Product Safety Regulation (GPSR) is a set of rules that requires consumer products to be safe and our obligations to ensure this.*

*If you have any concerns about our products, you can contact us on ProductSafety@springernature.com*

In case Publisher is established outside the EU, the EU authorized representative is:

Springer Nature Customer Service Center GmbH
Europaplatz 3
69115 Heidelberg, Germany

**Batch number: 10809439**

Printed by Printforce, the Netherlands